100 Years of Physical Chemistry

advancing the chemical sciences

ISBN 0-85404-9878 hardback
ISBN 0-85404-9827 softback

A catalogue record for this book is available from the British Library

Published by The Royal Society of Chemistry,
Thomas Graham House, Science Park, Milton Road,
Cambridge CB4 0WF, UK

Registered Charity Number 207890

For further information see our web site at www.rsc.org

Printed by Black Bear Press, Cambridge, UK

Preface

This special volume is published to mark the Centenary of the founding of the Faraday Society in 1903. It consists of 23 papers re-printed from Faraday journals—the *Transactions*, *General Discussions* and *Symposia*—that have been published over the past 100 years. Each article has been selected by an expert in one of the many scientific fields, *bounded by chemistry, physics and biology*, which the Faraday Society and its successor the RSC Faraday Division seek to promote. Each paper is accompanied by a short commentary written by the same expert. They were invited to describe how the paper that they selected influenced the subsequent development of the field, including their own work. As a whole the volume provides a fascinating insight into the wide range of topics that *physical chemists* seek to study and understand, as well as demonstrating the wide range of techniques that they deploy in this quest. In addition, I hope that this volume demonstrates the seminal part that the activities of the Faraday Society/Division have played in the development of so many aspects of physical chemistry.

Of course, the papers chosen are personal choices and the volume makes no claim to being fully comprehensive. No doubt 23 different individuals would have selected 23 different papers, nor, by any means, are the selected papers the only ones published in Faraday journals that have had a lasting impact on physical chemistry. Nevertheless, we hope that they do leave an impression of how important these journals have been in the development of the scientific fields central to the interests of the Faraday Society and its successor.

Not surprisingly, the origins of the Faraday Society are not clearly defined. Our founding fathers were not concerned with giving their successors a particular date on which the Centenary could be celebrated! The idea of such a society seems to have been conceived in 1902 and to have emphasised the study of electrochemistry and electrometallurgy though these interests very soon started to broaden out. It seemed to take about a year (maybe nine months?) for the seed planted in 1902 to gestate and the first meeting of the fledgling society with a scientific content seems to have taken place in June of 1903. This early history is briefly sketched out in an Editorial in *Physical Chemistry Chemical Physics* (the successor to the *Transactions of the Faraday Society*) written by Professor John Simons (President of the RSC Faraday Division 1993–1995)[1] and more details can be found in the splendid history of the Faraday Society written by Leslie Sutton and Mansel Davies and published in 1996.[2] The first volume of the Transactions appeared in 1905, but this publication was apparently preceded by several meetings at which papers were read and discussed.

The idea of larger scientific meetings, at which papers on a particular topic within physical chemistry were read and discussed, and both were subsequently published, was born in 1907. The first General Discussion of the Faraday Society, on *Osmotic Pressure*, was held in London and published in the *Transactions* that same year. For many years, the proceedings of the General Discussions were published as part of the *Transactions*. Only in 1947 was it decided to publish the General Discussions separately and the present numbering of Discussions dates from then. One result is confusion as to how many Faraday Discussions there have now been. By my count, the recent Discussion on *Nanoparticle Assemblies*, held at the University of Liverpool and numbered 125, is actually the 219th Discussion!

The series of General Discussions is perhaps the aspect of its activities in which the Faraday Division continues to take most pride—and not just for their longevity! The meetings are quite unique, on a world-wide basis, in their emphasis on discussion which is recorded and forms part of the published volume. Each General Discussion continues to attract the international leaders in the field under consideration. Each published volume provides a wonderful record of the state of that particular branch of science at the time the meeting was held. For this reason, it is scarcely surprising that about half of those invited to contribute to the present volume have selected to highlight a Discussion paper.

Despite its antiquity the Faraday Society has evolved and will continue to evolve. Of course, a very important change came in the early 1970's when the Faraday Society, which to that point had been run on a shoestring by essentially two dedicated individuals—the Honorary Secretary and the Secretary— amalgamated with the Chemical Society and became part of the much larger Royal Society of Chemistry. The Faraday Society became the Faraday Division of the Royal Society of Chemistry and the *General Discussions of the Faraday Society* became the *Faraday Discussions of the Chemical Society*. The Transactions also underwent some modifications but the most important change came in 1998 when, in a very positive move, the Faraday Division played an important role in joining with its sister societies in Europe to found the journal *Physical Chemistry Chemical Physics*.

This special volume largely looks back—to give a glimpse of how our science has evolved over the past hundred years. The titles of General Discussions give a good impression of how scientific interests have altered over that time. (There is unlikely to be another Discussion on Osmotic Pressure—at least in the foreseeable future!) The topics discussed at General Discussions also give a good idea of the range of scientific interests encompassed by the Faraday Society/Division. The Society/Division has always emphasised interdisciplinarity—even before that word became so fashionable!

At the beginning of its second century, the Faraday Division is in good health and believes that the general areas of its interests remain as scientifically alive and important as ever. As a witness to our faith in the future and as its second major event to celebrate this Centenary, the Faraday Division, in conjunction with the Royal Institution, is to hold a special meeting on October 27th 2003. It will contain two demonstration lectures, by Professors Alex Pines (UC, Berkeley) and Tony Ryan (Sheffield) designed to show post-16 students something of the excitement and relevance of physical chemistry in the 21st century.

Finally, I should like to express my thanks. First, to those who have contributed to this volume, not only for their magnificent contributions but also for co-operating so well that it has been a positive pleasure to bring this volume together. Second, I must thank Dr Susan Appleyard and staff at the Royal Society of Chemistry for their work in preparing the volume in short time and with great skill. Special thanks go to Susan who first had the idea of a volume of this kind and volunteered to do much of the work to make it a reality.

References

1 J. P. Simons, *Phys. Chem. Chem. Phys.,* 2003, **5**(13), i.
2 L. Sutton and M. Davies, *The History of the Faraday Society,* The Royal Society of Chemistry, Cambridge, 1996.

Ian W M Smith
President of the RSC Faraday Division 2001–2003

The RSC Faraday Division is pleased to acknowledge Shell Global Solutions and ICI Group Technology as sponsors of this publication.

Contents

Intermolecular Forces

A. D. Buckingham

Department of Chemistry, University of Cambridge, Lensfield Road, Cambridge, UK CB2 1EW

Commentary on: **The general theory of molecular forces,** F. London, *Trans. Faraday Soc.*, 1937, **33**, 8–26.

The origin of the substantial attractive forces between nonpolar molecules was a serious problem in the early 20th century. While much was known of the strength of these forces from the Van der Waals equation of state for imperfect gases and from thermodynamic properties of liquids and solids, there was little understanding. The difficulty can be illustrated by the fact that the binding energy of solid argon is of the same order of magnitude as that of the highly polar isoelectronic species HCl. Debye[1] suggested in 1921 that argon atoms, while known to be non-dipolar, may be quadrupolar; however, after the advent of quantum mechanics in 1926, it was clear that the charge distribution of an inert-gas atom is spherically symmetric. In 1928 Wang[2] showed that there is a long-range attractive energy between two hydrogen atoms that varies as R^{-6} where R is their separation. Soon afterwards, London presented his 'general theory of molecular forces' and gave us approximate formulae relating the interaction energy to the polarizability of the free molecules and their 'internal zero-point energy'. London showed that these forces arise from the quantum-mechanical fluctuations in the coordinates of the electrons and called them the *dispersion effect*. He demonstrated their additivity and estimated their magnitude for many simple molecules. The paper points out the important role of the Pauli principle in determining the overlap-repulsion force (on p. 21 it associates the Coulomb interaction of overlapping spherical atomic charge clouds with an incomplete screening of the nuclei, causing a repulsion; actually the enhanced electronic charge density in the overlap region between the nuclei would lead to an attraction, so the strong repulsion at short range is due to the Pauli principle).

A feature of London's paper is its emphasis on the zero-point motion of electrons: it is the intermolecular correlation of this zero-point motion that is responsible for dispersion forces. London's Section 9 extends the idea of zero-point fluctuations to the interaction of dipolar molecules. If their moment of inertia is small, as it is for hydrogen halide molecules, then even near the absolute zero of temperature when the molecules are in their non-rotating ground states, there are large fluctuations in the orientation of the molecules and these become correlated in the interacting pair.

London's eqn. (15) for the dipole–dipole dispersion energy is not a simple product of properties of the separate atoms. A partial separation was achieved in 1948 by Casimir and Polder[3] who expressed the R^{-6} dispersion energy as the product of the polarizability of each molecule at the imaginary frequency iu integrated over u from zero to infinity. The polarizability at imaginary frequencies may be a bizarre property but it is a mathematically well behaved function that decreases monotonically from the static polarizability at $u = 0$ to zero as $u \rightarrow \infty$.

Casimir and Polder[3] also showed that retardation effects weaken the dispersion force at separations of the order of the wavelength of the electronic absorption bands of the interacting molecules, which is typically 10^{-7} m. The retarded dispersion energy varies as R^{-7} at large R and is determined by the static polarizabilities of the interacting molecules. At very large separations the forces between molecules are weak but for colloidal particles and macroscopic objects they may add and their effects are measurable.[4] Fluctuations in particle position occur more slowly for nuclei than for electrons, so the intermolecular forces that are due to nuclear motion are effectively unretarded. A general theory of the interaction of macroscopic bodies in terms of the bulk static and dynamic dielectric properties has been presented by Lifshitz.[5] Proton movements in hydrogen-bonded solids and liquids may contribute to the binding energy as well as to the dielectric constant, electrical conductivity and intense continuous infrared absorption.[6]

If one or both of the molecules in an interacting pair lacks a centre of symmetry, e.g. $CH_4 \cdots CH_4$, $Ar \cdots CH_4$, or $Ar \cdots$ cyclopropane, there is, in addition to the dispersion energy terms in R^{-6}, R^{-8}, R^{-10}, ..., an orientation-dependent contribution that varies as R^{-7}. It could be significant for coupling the translation and rotation in gases and liquids and for the lattice energy of solids.[8]

London illuminated the origin of dispersion forces by considering the dipolar coupling of two three-dimensional isotropic harmonic oscillators. He obtained the exact energy and showed that it varies as R^{-6} for large R. Longuet-Higgins discussed the range of validity of London's theory and used a similar harmonic-oscillator model to show that at equilibrium at temperature T there is a lowering of the free energy $A(R)$, though not of the internal energy $E(R)$, through the coupling of two *classical* harmonic oscillators, so their attraction is entropic in nature and vanishes at $T = 0$. It is the quantization of the energy of the oscillators that leads to a lowering of $E(R)$ through the dispersion force. If one of a pair of identical oscillators is in its first excited state the interaction lifts the degeneracy and leads to a first-order dipolar interaction energy proportional to R^{-3}; this is an example of a *resonance* energy which may be considered to arise from the exchange of a photon between identical oscillators.

Since the dispersion energy arises from intermolecular correlation of charge fluctuations, it is not accounted for by the usual computational techniques of density functional theory (DFT) which employ the local density and its spatial derivatives. Special techniques are needed if DFT is to be used for investigating problems where intermolecular forces play a significant role.[9]

References

1 P. Debye, *Phys. Z.*, 1921, **22**, 302.
2 S. C. Wang, *Phys. Z.*, 1928, **28**, 663.
3 H. B. G. Casimir and D. Polder, *Phys. Rev.*, 1948, **73**, 360.
4 J. N. Israelachvili, *Intermolecular and Surface Forces with Applications to Colloidal and Biological Systems*, Academic Press, London, 1985.
5 E. M. Lifshitz, *Sov. Phys. JETP (Engl. Transl.)*, 1956, **2**, 73.
6 G. Zundel, *Adv. Chem. Phys.*, 2000, **111**, 1.
7 H. C. Longuet-Higgins, *Discuss. Faraday Soc.*, 1965, **40**, 7.
8 A. D. Buckingham, *Discuss. Faraday Soc.*, 1965, **40**, 232.
9 W. Kohn, Y. Meir and D. E. Makarov, *Phys. Rev. Lett.*, 1998, **80**, 4153.

8   THE GENERAL THEORY OF MOLECULAR FORCES

THE GENERAL THEORY OF MOLECULAR FORCES.

By F. London (*Paris*).

Received 31st July, 1936.

Following Van der Waals, we have learnt to think of the molecules as centres of forces and to consider these so-called *Molecular Forces* as the common cause for various phenomena : The deviations of the gas equation from that of an ideal gas, which, as one knows, indicate the identity of the molecular forces in the liquid with those in the gaseous state ; the phenomena of capillarity and of adsorption ; the sublimation heat of molecular lattices ; certain effects of broadening of spectral lines, etc.   It has already been possible roughly to determine these forces in a fairly consistent quantitative way, using their measurable effects as basis.

In these semi-empirical calculations, for reasons of simplicity, one imagined the molecular forces simply as rigid, additive central forces, in general cohesion, like gravitation ; this presumption actually implied

F. LONDON   9

a very suggestive and simple explanation of the parallelism observed in the different effects of these forces. When, however, one began to try to explain the molecular forces by the general conceptions of the electric structure of the molecules it seemed hopeless to obtain such a simple result.

§ 1. Orientation Effect.[1]

Since molecules as a whole are usually uncharged the *dipole moment* μ was regarded as the most important constant for the forces between molecules. The interaction between two such dipoles μ_I and μ_{II} depends upon their relative orientation. The interaction energy is well known to be given to a first approximation by

$$U = -\frac{\mu_I \mu_{II}}{R^3}(2\cos\theta_I \cos\theta_{II} - \sin\theta_I \sin\theta_{II} \cos(\phi_I - \phi_{II})) \qquad (1)$$

where θ_I, ϕ_I ; θ_{II}, ϕ_{II} are polar co-ordinates giving the orientation of the dipoles, the polar axis being represented by the line joining the two centres, $R =$ their distance. We obtain attraction as well as repulsion, corresponding to the different orientations. If all orientations were equally often realised the average of μ would be zero.

But according to Boltzmann statistics the orientations of lower energy are statistically preferred, the more preferred the lower the temperature. Keesom, averaging over all positions, found as a result of this preference :

$$\bar{U} = -\frac{2}{3}\frac{\mu_I^2 \mu_{II}^2}{R^6}\frac{1}{kT} \qquad \text{(valid for } \frac{\mu_I \mu_{II}}{R^3} \ll kT\text{).} \qquad . \quad (2)$$

For low temperatures or small distances $\left(kT \lesssim \frac{\mu_I \mu_{II}}{R^3}\right)$ this expression does not hold. It is obvious that the molecules cannot have a more favourable orientation than parallel to each other along the line joining the two molecules, in which case one would obtain as interaction energy (see (1)) :

$$\bar{U} = -\frac{2\mu_I \mu_{II}}{R^3} \qquad \text{(valid for } \frac{\mu_I \mu_{II}}{R^3} \gg kT\text{)} \qquad . \quad . \quad (3)$$

which gives in any case a lower limit for this energy. (2) and (3) represent an attractive force, the so-called *orientation effect*, by which Keesom tried to interpret the Van der Waals attraction.

§ 2. Induction Effect.[2]

Debye remarked that these forces cannot be the only ones. According to (2) they give an attraction which vanishes with increasing temperature. But experience shows that the empirical Van der Waals corrections do not vanish equally rapidly with high temperatures, and Debye therefore concluded that there must be, in addition, an interaction energy independent of temperature. In this respect it would not help to consider the actual charge distribution of the molecules more in detail,

---

[1] W. H. Keesom, *Leiden Comm. Suppl.*, 1912, **24a, 24b, 25, 26**; 1915, **39a, 39b.** *Proc. Amst.*, 1913, **15**, 240, 256, 417, 643 ; 1916, **18**, 636 ; 1922, **24**, 162. *Physik. Z.*, 1921, **22**, 129, 643 ; 1922, **23**, 225.

[2] P. Debye, *Physik. Z.*, 1920, **21**, 178 ; 1921, **22**, 302. H. Falckenhagen, *Physik. Z.*, 1922, **23**, 87.

10   THE GENERAL THEORY OF MOLECULAR FORCES

e.g. by introducing the quadrupole and higher moments.   The average of these interactions also would vanish for high temperatures.

But by its charge distribution alone a molecule is, of course, still very roughly characterised.   Actually, the charge distribution will be changed under the influence of another molecule.   This property of a molecule can very simply be described by introducing a further constant, the *polarisability* α.   In an external electric field of the strength F a molecule of polarisability α shows an *induced* moment

$$M = \alpha . F \qquad . \qquad . \qquad . \qquad . \qquad . \quad (4)$$

(in addition to a possible *permanent* dipole moment) and its energy in the field F is given by

$$U = - \tfrac{1}{2}\alpha . F^2 \qquad . \qquad . \qquad . \qquad . \quad (5)$$

Now the molecule I may produce near the molecule II an electric field of the strength

$$F = \frac{\mu_I}{R^3} \sqrt{1 + 3\cos^2 \theta_I} \qquad . \qquad . \qquad . \quad (6)$$

This field polarises the molecule II and gives rise to an additional interaction energy according to (5)

$$U = - \tfrac{1}{2}\alpha_{II}F^2 = - \frac{\alpha_{II}}{2} \frac{\mu_I{}^2}{R^6}(1 + 3\cos^2 \theta_I) \qquad . \qquad . \quad (7)$$

which is always negative (*attraction*) and therefore its average, even for infinitely high temperatures, is also negative.   Since $\cos^2 \theta = \tfrac{1}{3}$ we obtain :

$$\bar{U}_{I \to II} = - \alpha_{II} \frac{\mu_I{}^2}{R^6}$$

A corresponding amount would result for $\bar{U}_{II \to I}$, *i.e.* for the action of μ_{II} upon α_I.   As total interaction of the two molecules we obtain :

$$\bar{U} = - \frac{1}{R^6}(\alpha_I\mu_{II}{}^2 + \alpha_{II}\mu_I{}^2) \qquad . \qquad . \qquad . \quad (8)$$

If the two molecules are of the same kind ($\mu_I = \mu_{II} = \mu$ and $\alpha_I = \alpha_{II} = \alpha$) we have

$$\bar{U} = - \frac{2\alpha\mu^2}{R^6}. \qquad . \qquad . \qquad . \quad (8')$$

This is the so-called *induction effect*.

In such a way Debye and Falckenhagen believed it possible to explain the Van der Waals equation.   But many molecules have certainly no permanent dipole moment (rare gases, H_2, N_2, CH_4, etc.). There they assumed the existence of quadrupole moments τ, which would of course also give rise to a similar interaction by inducing dipoles in each other.   Instead of (8) this would give :

$$\bar{U} = - \frac{3}{2} \frac{\alpha\tau^2}{R^8}. \qquad . \qquad . \qquad . \quad (9)$$

Since no other method of measuring these quadrupoles was known, the Van der Waals corrections (second Virial coefficient) were used in order to determine backwards τ, which, after μ and α, has been regarded as the most fundamental molecular constant.

§ 3. Criticism of the Static Models for Molecular Forces.

The most obvious objection to all these conceptions is that they do not explain the above mentioned parallelism in the different manifestations of the molecular forces. One cannot understand why, for example, in the liquid and in the solid state between all neighbours simultaneously practically the same forces should act as between the occasional pairs of molecules in the gaseous state. All these models are very far from simply representing a general additive cohesion :

Suppose that two molecules I and II have such orientations of their permanent dipoles that they are attracted by a third one ; then between the two former molecules very different forces are usually operative, mostly repulsive forces. Or, if the forces are due to polarisation, the acting field will usually be greatly lowered, when many molecules from different sides superimpose their polarising fields. One should expect, therefore, that in the liquid and in the solid state the forces caused by induced or permanent dipoles or multipoles should at least be greatly diminished, if not by reasons of symmetry completely cancelled.

The situation seemed to be still worse when wave mechanics showed that the rare gases are exactly spherically symmetrical, that they have neither a permanent dipole nor quadrupole nor any other multipole. They showed none of the mentioned interactions. It is true, that for H_2, N_2, etc., wave mechanics, too, gives at least quadrupoles. But for H_2 we are now able to calculate the value of the quadrupole moment numerically by wave mechanics. One gets only about 1/100 of the Van der Waals forces that were attributed hitherto to suitably chosen quadrupoles.

On the other hand, wave mechanics has provided us with a completely new aspect of the interaction between neutral atomic systems.

§ 4. Dispersion Effect ; a Simplified Model.[3]

Let us take two spherically symmetrical systems, each with a polarisability α, say two three-dimensional isotropic harmonic oscillators with no permanent moment in their rest position. If the charges e of these oscillators are artificially displaced from their rest positions by the displacements

$$\vec{r_I} = (x_I, y_I, z_I) \quad \text{and} \quad \vec{r_{II}} = (x_{II}, y_{II}, z_{II})$$

respectively, we obtain for the potential energy :

$$V = \underbrace{\frac{e^2 r_I^2}{2\alpha} + \frac{e^2 r_{II}^2}{2\alpha}}_{\text{Elastic Energy.}} + \underbrace{\frac{e^2}{R^3}(x_I x_{II} + y_I y_{II} - 2z_I z_{II})}_{\substack{\text{Dipole Interaction Energy} \\ (cf. (1)).}} \quad . \quad . \quad (10)$$

Classically the two systems in their equilibrium position

$$(x_I = x_{II} = \ldots = z_{II} = 0)$$

would not act upon each other and, when brought into finite distance $(R > \sqrt[3]{2\alpha})$, remain in their rest position. They could not influence a momentum in each other.

[3] F. London, *Z. physik. Chem.,* 1930. **B, II,** 222.

12   THE GENERAL THEORY OF MOLECULAR FORCES

However, in *quantum mechanics*, as is well known, a particle cannot lie absolutely at rest on a certain point. That would contradict the uncertainty relation. According to quantum mechanics our isotropic oscillators, even in their lowest states, make a so-called *zero-point motion* which one can only describe statistically, for example, by a probability function which defines the probability with which any configuration occurs; whilst one cannot describe the way in which the different configurations follow each other. For the isotropic oscillators these probability functions give a spherically symmetric distribution of configurations round the rest position. (The rare gases, too, have such a spherically symmetrical distribution for the electrons around the nucleus.)

We need not know much quantum mechanics in order to discuss our simple model. We only need to know that in quantum mechanics the lowest state of a harmonic oscillator of the proper frequency ν has the energy

$$E_0 = \tfrac{1}{2}h\nu \quad . \quad . \quad . \quad . \quad . \quad (11)$$

the so-called *zero-point energy*. If we introduce the following co-ordinates (" normal "-co-ordinates):

$$\vec{r}_+ \equiv \begin{cases} x_+ = \dfrac{1}{\sqrt{2}}(x_{\mathrm{I}} + x_{\mathrm{II}}) \\[2mm] y_+ = \dfrac{1}{\sqrt{2}}(y_{\mathrm{I}} + y_{\mathrm{II}}) \\[2mm] z_+ = \dfrac{1}{\sqrt{2}}(z_{\mathrm{I}} + z_{\mathrm{II}}) \end{cases} \qquad \vec{r}_- \equiv \begin{cases} x_- = \dfrac{1}{\sqrt{2}}(x_{\mathrm{I}} - x_{\mathrm{II}}) \\[2mm] y_- = \dfrac{1}{\sqrt{2}}(y_{\mathrm{I}} - y_{\mathrm{II}}) \\[2mm] z_- = \dfrac{1}{\sqrt{2}}(z_{\mathrm{I}} - z_{\mathrm{II}}) \end{cases}$$

the potential energy (10) can be written as a sum of squares like the potential energy of six independent oscillators (while the kinetic energy would not change its form):

$$V = \frac{e^2}{2\alpha}(r_+{}^2 + r_-{}^2) + \frac{e^2}{2R^3}(x_+{}^2 + y_+{}^2 - 2z_+{}^2 - x_-{}^2 - y_-{}^2 + 2z_-{}^2) \; .$$

$$= \frac{e^2}{2\alpha}\Big[\Big(1 + \frac{\alpha}{R^3}\Big)(x_+{}^2 + y_+{}^2) + \Big(1 - \frac{\alpha}{R^3}\Big)(x_-{}^2 + y_-{}^2) + \\ \Big(1 - 2\frac{\alpha}{R^3}\Big)z_+{}^2 + \Big(1 + 2\frac{\alpha}{R^3}\Big)z_-{}^2\Big] \quad (10')$$

The frequencies of these six oscillators are given by

$$\nu_x{}^\pm = \nu_y{}^\pm = \nu_0\sqrt{1 \pm \alpha/R^3} \approx \nu_0\Big(1 \pm \frac{\alpha}{2R^3} - \frac{\alpha^2}{8R^6} \pm \cdots\Big)$$

$$\nu_z{}^\pm = \nu_0\sqrt{1 \mp 2\alpha/R^3} \approx \nu_0\Big(1 \mp \frac{\alpha}{R^3} - \frac{\alpha_2}{2R^6} \mp \cdots\Big) \quad . \quad (12)$$

Here $\nu_0 = \dfrac{e}{\sqrt{m\alpha}}$ is the *proper frequency* of the two elastic systems, if isolated from each other ($R \to \infty$), and m is their reduced mass. Assuming $\alpha \ll R^3$, we have developed the square roots in (12) into powers of (α/R^3).

The lowest state of this system of six oscillators will therefore be given, according to (11), by:

$$E_0 = \frac{h}{2}(\nu_x{}^+ + \nu_y{}^+ + \nu_z{}^+ + \nu_x{}^- + \nu_y{}^- + \nu_z{}^-)$$

$$= \frac{h\nu_0}{2}\left[6 + (\tfrac{1}{2} + \tfrac{1}{2} - 1 - \tfrac{1}{2} - \tfrac{1}{2} + 1)\frac{\alpha}{R^3} - (\tfrac{4}{8} + \tfrac{2}{2})\frac{\alpha^2}{R^6} + \cdots\right]$$

$$= 3h\nu_0 - \frac{3}{4}\frac{h\nu_0\alpha^2}{R^6} + \cdots.$$

The first term $3h\nu_0$ is, of course, simply the internal zero-point energy of the two isolated elastic systems. The second term, however,

$$U = -\frac{3}{4}\frac{h\nu_0\alpha^2}{R^6} \qquad \cdots \qquad \cdots \qquad (13)$$

depends upon the distance R and is to be considered as an interaction energy which, being negative, characterises an attractive force. We shall presume that this type of force,[4] which is not conditioned by the existence of a permanent dipole or any higher multipole, will be responsible for the Van der Waals attraction of the rare gases and also of the simple molecules H_2, N_2, etc. For reasons which will be explained presently these forces are called the *dispersion effect*.

§ 5. Dispersion Effect ; General Formula.[5]

Though it is of course not possible to describe this interaction mechanism in terms of our customary classical mechanics, we may still illustrate it in a kind of semi-classical language.

If one were to take an instantaneous photograph of a molecule at any time, one would find various configurations of nuclei and electrons, showing in general dipole moments. In a spherically symmetrical rare gas molecule, as well as in our isotropic oscillators, the average over very many of such snapshots would of course give no preference for any direction. These very quickly varying dipoles, represented by the zero-point motion of a molecule, produce an electric field and act upon the polarisability of the other molecule and produce there induced dipoles, which are in phase and in interaction with the instantaneous dipoles producing them. The zero-point motion is, so to speak, accompanied by a synchronised electric alternating field, but not by a radiation field : The energy of the zero-point motion cannot be dissipated by radiation.

This image can be used for interpreting the generalisation of our formula (13) for the case of a general molecule, the exact development of which would of course need some quantum mechanical calculations.

We may imagine a molecule in a state k as represented by an orchestra of periodic dipoles μ_{kl} which correspond with the frequencies

$$\nu_{kl} = \frac{E_l - E_k}{h}$$

of (not forbidden) transitions to the states l. These " oscillator strengths," μ_{kl}, are the same quantities which appear in the " dispersion formula " which gives the polarisability $\alpha_k(\nu)$ of the molecule in the state k when acted on by an alternating field of the frequency ν.

$$\alpha_k(\nu) = \frac{2}{3h}\sum_l \frac{\mu_{kl}{}^2 \nu_{kl}}{\nu_{kl}{}^2 - \nu^2}. \qquad \cdots \qquad \cdots \qquad (14)$$

[4] This type of force first appeared in a calculation of S. C. Wang, *Physik. Z.* 1927, **2B**, 663.
[5] R. Eisenschitz and F. London, *Z. Physik*, 1930, **60**, 491.

14   THE GENERAL THEORY OF MOLECULAR FORCES

If the acting field of the frequency ν_0 has the amplitude F_0, the induced moment M is given by

$$M = \alpha_k(\nu_0) \cdot F_0 = \frac{2}{3h} F_0 \sum_l \frac{\mu_{kl}^2 \nu_{kl}}{\nu_{kl}^2 - \nu_0^2} \qquad . \qquad . \quad (4')$$

and the interaction energy between field and molecule by

$$U = -\tfrac{1}{2}\alpha(\nu_0) \cdot F_0^2 = -\frac{F_0^2}{3h} \sum_l \frac{\mu_{kl}^2 \nu_{kl}}{\nu_{kl}^2 - \nu_0^2}. \qquad . \quad (5')$$

Now this acting field may be produced by another molecule by one of its periodic dipoles $\mu_{\rho\sigma}$ with the frequency $\nu_{\rho\sigma}$ and inclination $\theta_{\rho\sigma}$ to the line joining the two molecules. Near the first molecule (we call it the "Latin" molecule, using Latin indices for its states, and Greek indices to the other one) the dipole $\mu_{\rho\sigma}$ produces an electric field of the strength (compare (6)) :

$$F_{\rho\sigma} = \frac{\mu_{\rho\sigma}}{R^3}\sqrt{1 + 3\cos^2\theta_{\rho\sigma}}. \qquad . \qquad . \quad (6')$$

This field induces in the Latin molecule a periodic dipole of the amount :

$$M_{\rho\sigma}{}^k = \alpha_k(\nu_{\rho\sigma}) \cdot F_{\rho\sigma},$$

and an interaction energy (compare (5')) :

$$-\frac{\alpha_k(\nu_{\rho\sigma})}{2}F_{\rho\sigma}^2 = -\frac{\mu_{\rho\sigma}^2}{3hR^6}(1 + 3\cos^2\theta_{\rho\sigma})\sum_l \frac{\mu_{kl}^2 \nu_{kl}}{\nu_{kl}^2 - \nu_{\rho\sigma}^2}.$$

If we now consider the whole orchestra of the "Greek" molecule in the state ρ we have to sum over all states σ and to average over all directions $\theta_{\rho\sigma}$ ($\cos^2\theta = 1/3$). This would give us the action of the Greek atom upon the polarised Latin atom :

$$U_{\rho\to k} = -\frac{2}{3hR^6} \sum_{l\sigma} \frac{\mu_{kl}^2 \mu_{\rho\sigma}^2 \nu_{kl}}{\nu_{kl}^2 - \nu_{\rho\sigma}^2}.$$

Adding the corresponding expression $U_{k\to\rho}$ for the action of the Latin molecule upon the Greek one, we obtain the total interaction due to the "periodic" dipoles of a molecule in the state k with another in the state ρ :

$$U_{\rho k} = U_{\rho\to k} + U_{k\to\rho} = -\frac{2}{3hR^6} \sum_{l\sigma} \mu_{kl}^2 \mu_{\rho\sigma}^2 \left(\frac{\nu_{kl}}{\nu_{kl}^2 - \nu_{\rho\sigma}^2} + \frac{\nu_{\rho\sigma}}{\nu_{\rho\sigma}^2 - \nu_{kl}^2} \right)$$

$$= -\frac{2}{3hR^6} \sum_{l\sigma} \frac{\mu_{kl}^2 \mu_{\rho\sigma}^2}{\nu_{kl} + \nu_{\rho\sigma}} \qquad . \qquad . \qquad . \quad (15)$$

§6. Additivity of the Dispersion Effect.

Of course this reasoning does not claim to be an exact proof of (15), but it may perhaps illustrate the mechanism of these forces. It can be shown that the formula (15) has the peculiarity of *additivity ;* [3] this means that if three molecules act simultaneously upon each other, the three interaction potentials between the three pairs of the form (15) are simply to be added, and that any influence of a third molecule upon the interaction between the first two is only a small perturbation effect of a smaller order of magnitude than the interaction itself. These attractive forces can therefore simply be superposed according to the parallelogram of forces, and they are consequently able to *represent the fact of a general cohesion.*

If several molecules interact simultaneously with each other, one has to imagine that each molecule induces in each of the others a set of co-ordinated periodic dipoles, which are in constant phase relation with the corresponding inducing original dipoles. Every molecule is thus the seat of very many incoherently superposed sets of induced periodic dipoles caused by the different acting molecules. Each of these induced dipoles has always such an orientation that it is attracted by its corresponding generating dipole, whereas the other dipoles, which are not correlated by any phase relation, give rise to a *periodic* interaction only and, on an average over all possible phases, contribute *nothing* to the interaction energy. So one may imagine that the simultaneous interaction of many molecules can simply be built up as an *additive* superposition of single forces between pairs.

§ 7. Simplified Formula ; Some Numerical Values.

For many simple gas molecules (*e.g.* the rare gases, H_2, N_2, O_2, CH_4), the empirical dispersion curve has been found to be representable, in a large frequency interval, by a dispersion formula of the type (14) consisting of one single term only. That means that for these molecules the oscillator strength μ_{kl} for frequencies of a small interval so far exceed the others that the latter can entirely be neglected. In this case, and for the limiting case $\nu \to 0$ (polarisability in a *static* field) the formula (14) can simply be written :

$$\alpha_k \equiv \alpha_k(0) = \frac{2}{3h} \frac{\mu_k^2}{\nu_k} \quad . \qquad . \qquad . \qquad . \quad (14')$$

(μ_k signifies the dipole-strength of the only main frequency ν_k) and formula (15) for the interaction of the two systems goes over into :

$$U_{\rho k} = -\frac{2}{3hR^6} \frac{\mu_k^2}{\nu_k} \cdot \frac{\mu_\rho^2}{\nu_\rho} \cdot \frac{\nu_k \cdot \nu_\rho}{\nu_k + \nu_\rho}$$

$$= -\frac{3h}{2R^6} \cdot \alpha_k \alpha_\rho \frac{\nu_k \cdot \nu_\rho}{\nu_k + \nu_\rho} \qquad . \qquad . \qquad . \quad (13')$$

This formula is identical with (13) in the case of two molecules of the same kind. It can, of course, only be applied if one already knows that the dispersion formula has the above-mentioned special form. But in any case, if the dispersion formulæ of the molecules involved are empirically known, their data can be used and are sufficient to build up the attractive force (15). No further details of the molecular structure need be known.

We give, in Table I., a list of theoretical values for the attractive constant c (*i.e.* the factor of $-1/R^6$ in the above interaction law) for rare gases and some other simple gases where the refractive index can fairly well be represented by a dispersion formula of one term only. The characteristic frequency ν_D multiplied by h is in all these cases very nearly equal to the ionisation energy $h\nu_I$. This may, to a first approximation, justify using the latter quantity in similar cases where a dispersion formula has not yet been determined. It is seen that the values of

16   THE GENERAL THEORY OF MOLECULAR FORCES

c vary in a ratio from 1 to 1000, and this wide range of the order of magnitude makes even a very crude experimental test of these forces instructive (see § 11).

J. E. Mayer[6] has shown that, for the negative rare-gas-like ions, one is not justified in simplifying the dispersion of the continuum by assuming one single frequency only. He used a simple analytical expression for the empirical continuous absorption and replacing the sums in (15) by integrals over these continua he gets the following list of c values for the 29 possible pairs of ions (Table II.):

Starting from a different method (variation method) and using some simplifying assumptions as to the wave functions of the atoms (products of single electronic wave functions) Slater and Kirkwood[6a] have also calculated these forces. They found the following formula:

$$U = -\frac{1}{R^6}\frac{3eh}{8\pi}\sqrt{\frac{N\alpha^3}{m}} \qquad . \qquad . \qquad . \quad (13'')$$

(N = number of electrons in the outer shell.)

This expression usually gives a somewhat greater value than (13) and may be applied in those cases in which the characteristic frequencies in (13) are not obtainable. But at present it is difficult to say how far one *may rely on formula* (13'').

TABLE I.—DISPERSION EFFECT BETWEEN SIMPLE MOLECULES.

	$h\nu_I$ (e.Volts).	$h\nu_D$ (e.Volts).	$\alpha.10^{24}$ [cm.³].	$c'.10^{48}=\frac{2}{3}\alpha^2h\nu_0.10^{48}$ [e.Volts.cm.⁶.].
He .	24·5	25·5	0·20	0·77
Ne .	21·5	25·7	0·39	2·93
Ar .	15·4	17·5	1·63	34·7
Kr .	13·3	14·7	2·46	69
Xe .	11·5	12·2	4·00	146
H₂ .	16·4		0·81	8·3
N₂ .	17	17·2	1·74	38·6
O₂ .	13	14·7	1·57	27·2
CO .	14·3		1·99	42·4
CH₄	14·5		2·58	73
CO₂		15·45	2·86	94·7
Cl₂ .	18·2		4·60	288
HCl	13·7		2·63	71
HBr	13·3		3·58	128
HI .	12·7		5·4	278
Na .		2·1	29·7	960

TABLE II.—DISPERSION EFFECT BETWEEN IONS.

($c.10^{48}$ in units [e.volts cm.⁶]).

	F⁻.	Cl⁻.	Br⁻.	I⁻.	
Li⁺ .	c_{-+} = 0·13	3·2	4·0	5·4	c_{++} = 0·11
Na⁺ .	7·14	17·8	22·2	30·3	2·68
K⁺ .	31·0	76·3	95·3	130	38·6
Rb⁺ .	49·2	125	157	214	94·3
Cs⁺ .	82·5	205	259	356	247
c_{--} =	23-30	176-206	294-332	600-676	

[6] J. E. Mayer, *J. Chem. Physics*, 1933, **1**, 270.
[6a] J. C. Slater and J. G. Kirkwood, *Physic. Rev.*, 1931, **37**, 682.

§ 8. Systematics of the Long Range Forces.[7]

The formula (15) applies quite generally for freely movable molecules so long as the interaction energy can be considered as small compared with the separation of the energy-levels of the molecules in question ; *i.e.* so long as

$$\frac{\mu_{kl}\mu_{\rho\sigma}}{R^3} < |\,E_k - E_l + E_\rho - E_\sigma\,| \qquad . \qquad . \quad (16)$$

With this restriction, the formula (15) holds for freely movable dipole molecules, as well as for rare gas molecules. There is therefore always a minimum distance for R up to which we can rely on (15).

The difference between a molecule with permanent dipole and a rare gas molecule consists in the following : A rare gas molecule has such a high excitation energy (electronic jump) that for normal temperatures we can assume that all molecules are in the ground state ; therefore we have forces there independent of temperature. For a dipole molecule, on the other hand, we have to consider a Boltzmann distribution over at least the different rotation states, because the energy difference between these states is usually small in comparison with kT.

Let us at first consider an *absolutely rigid dipole* (dumb-bell) molecule (*i.e.* a molecule without electronic or oscillation states). Then the probability $p_{\rho k}$ that the Greek molecule is in the pure rotation state ρ and the Latin one in the pure rotation state k is given by

$$p_{\rho k} = A e^{-\frac{1}{kT}(E_k + E_\rho)}$$

where

$$A^{-1} = \sum_{k\rho} e^{-\frac{1}{kT}(E_k + E_\rho)}.$$

The mean interaction between two such molecules is accordingly

$$\bar{U} = \sum_{\rho k} p_{\rho k} U_{\rho k} = -\frac{2A}{3R^6} \cdot \sum_{\sigma l} \frac{\mu_{kl}^2 \mu_{\rho\sigma}^2}{E_l - E_k + E_\sigma - E_\rho} e^{-\frac{E_k + E_\rho}{kT}}. \quad (17)$$

If in this expression we interchange the notation of the summation indices ρ and k with σ and l, the value of the sum of course remains unchanged. Therefore, taking the average of these two equivalent expressions we can write (since $\mu_{kl} = \mu_{lk}$) :

$$\bar{U} = -\frac{A}{3R^6} \sum_{\substack{\sigma l \\ \rho k}} \mu_{kl}^2 \mu_{\rho\sigma}^2 \frac{e^{-\frac{E_k + E_\rho}{kT}} - e^{-\frac{E_l + E_\sigma}{kT}}}{E_l + E_\sigma - E_k - E_\rho}. \qquad . \quad (17')$$

Developing the exponentials into powers of $1/kT$ we notice that the constant terms cancel each other (no interaction for high temperature as in § 1). The first and the only important term of the development of (17') yields :

$$\bar{U} = -\frac{A}{3R^6}\frac{1}{kT} \sum_{\substack{kl \\ \rho\sigma}} \mu_{kl}^2 \mu_{\rho\sigma}^2 + \ldots = -\frac{2\mu_I^2 \mu_{II}^2}{3kTR^6} + \ldots \quad . \quad (18)$$

Here we designate by μ_I and μ_{II} the permanent moments of the dipole molecule, which for an absolutely rigid molecule are of course independent

---

[7] F. London, *Z. Physik*, 1930, **63**, 245.

18   THE GENERAL THEORY OF MOLECULAR FORCES

of the state.   We therefore obtain exactly the same result as Keesom did from classical mechanics.   One can, by the way, show that whilst the validity of (15) is bounded by the condition (16) the result (18) is only bounded by the weaker condition

$$\frac{\mu_I \mu_{II}}{R^3} < kT$$

which was also the limit for the validity of the classical calculation.

In reality a dipole molecule cannot, of course, be treated as a simple rigid dumb-bell.   It has *electronic* and *oscillation* transitions as well. Let us, for sake of simplicity, assume that kT is big in comparison to the energy differences for pure rotation jumps, but small for all the other jumps.

In this general case we have again formula (17), but here it is sufficient to extend the Boltzmann sum $\underset{\rho k}{\Sigma}$ only over those states which imply pure rotation jumps from the ground state, since the thermo-dynamical probability of the other states being occupied is negligible.   We now divide the sum over σ and l in (17) into four parts

$$\bar{U} = U_{rr} + U_{rg} + U_{gr} + U_{gg}$$

in the following way :

(1) In U_{rr} both, σ and l, shall be restricted to those values which differ from the ground-state only by a pure rotation transition.   For this sum (with certain uninteresting reservations) the above calculation for the rigid dipoles remains valid.   Accordingly we get (18)

$$U_{rr} = -\frac{2}{3R^6}\frac{\mu_I^2 \mu_{II}^2}{kT},$$

i.e. Keesom's orientation effect.

(2) In U_{rg} the summation over σ as before shall be extended only over those terms which differ from the ground state by a pure rotation jump; but l shall designate a great (not a pure rotation-) jump.   Then we may neglect $E_\sigma - E_\rho$ in comparison with $E_l - E_k$ in the denominators of (17) and can write

$$U_{rg} = -\frac{2A}{3R^6}\left(\sum_{kl}\mu_{kl}^2 e^{-\frac{E_k}{kT}}\right)\cdot\left(\sum_{\rho\sigma}\frac{\mu_{\rho\sigma}^2}{E_\sigma - E_\rho}e^{-\frac{E_\rho}{kT}}\right).$$

Comparison with (14) shows that the terms of the second sum on the right-hand side can be represented by the static polarisability $\alpha_\rho = \alpha_\rho(o)$ of the second molecule which will depend very little on the state of rotation ρ of the molecule so that we may signify it simply by α_{II}; whereas the first sum again gives the square of the permanent dipole moment of the first molecule, of which we also may assume that it is approximately independent of the state of rotation   We obtain

$$U_{rg} = -\frac{2}{3R^6}\cdot\mu_I^2\cdot\tfrac{3}{2}\alpha_{II}(o) = -\frac{\mu_I^2\alpha_{II}}{R^6}.$$

(3) Correspondingly

$$U_{gr} = -\frac{\mu_{II}^2\alpha_I}{R^6}$$

(2) and (3) are exactly *Debye's induction effect.*

(4) In U_{gg} finally both, σ and l, shall differ from the ground state by a great (not a pure rotation-) jump.   If we assume that the transition

probabilities of such a jump do not depend noticeably on the state of rotation, we can take simply the ground state for ρ and k and obtain

$$U_{gg} = -\frac{2}{3hR^6} \sum_{l\sigma} \frac{\mu_{ol}^2 \mu_{o\sigma}^2}{\nu_{ol} + \nu_{o\sigma}}.$$

i.e. the *dispersion effect.* If the conditions for (13′) are fulfilled we may join the three effects in the form

$$\bar{U} = -\frac{1}{R^6}\left(\frac{2}{3}\frac{\mu_I^2 \mu_{II}^2}{kT} + \mu_I^2 \alpha_{II} + \mu_{II}^2 \alpha_I + \frac{3h}{2}\alpha_I \alpha_{II} \frac{\nu_I \nu_{II}}{\nu_I + \nu_{II}}\right) \quad (19)$$

We give, in Table III., a short list for the three effects of some dipole molecules :

TABLE III.—THE THREE CONSTITUENTS OF THE VAN DER WAALS' FORCES.

	$\mu . 10^{18}$.	$\alpha . 10^{24}$.	$h\nu_0$ (Volts).	Orientation Effect $\frac{2}{3}\frac{\mu^4}{k293°} . 10^{60}$ [erg cm.⁶].	Induction Effect $2\mu^2\alpha . 10^{60}$ [erg cm.⁶].	Dispersion Effect $\frac{3}{4}\alpha^2 h\nu_0 . 10^{60}$ [erg cm.⁶].
CO .	0·12	1·99	14·3	0·0034	0·057	67·5
HI .	0·38	5·4	12	0·35	1·68	382
HBr .	0·78	3·58	13·3	6·2	4·05	176
HCl .	1·03	2·63	13·7	18·6	5·4	105
NH₃	1·5	2·21	16	84	10	93
H₂O .	1·84	1·48	18	190	10	47

It is seen that the induction effect is in all cases practically negligible, and that even in such a strong dipole molecule as HCl the permanent dipole moments give no noticeable contribution to the Van der Waals' attraction. Not earlier than with NH_3 does the orientation effect become comparable with the dispersion effect, which latter seems in no case to be negligible.

§ 9. Limits of Validity.

We have yet to discuss the physical meaning of the condition (16).

In quantum mechanics, in characteristic contrast to classical mechanics, a freely movable polyatomic molecule has a centrally symmetric and, particularly in its lowest state, a spherically symmetric structure, *i.e.* a spherically symmetric probability function. That means that on the average, even in its lowest state, a free molecule does not prefer any direction, it changes its orientation permanently owing to its zero-point motion. If another molecule tries to orientate the molecule in question a compromise between the zero-point motion and the directing power will be made, but only for

$$\frac{\mu_I \mu_{II}}{R^3} > |E_0 - E_1| . \quad . \quad . \quad . \quad (20a)$$

the directive forces preponderate over the zero-point rotation. Accordingly, in this case, the motion of the dipoles becomes more similar to a vibration near the equilibrium orientation of the dipoles (parallel to each other along the line joining the two molecules) and the interaction will then be of the nature of orientated dipoles, *i.e.* of the order of magnitude of

$$-\frac{2\mu_I \mu_{II}}{R^3}.$$

20 THE GENERAL THEORY OF MOLECULAR FORCES

In quantum mechanics, we learn, in contrast to (3), the condition

$$\frac{\mu_I \mu_{II}}{R^3} > kT \qquad . \qquad . \qquad . \qquad . \qquad (20b)$$

is not sufficient for the molecules being orientated. The orientating forces have not only to overcome the temperature motion but in addition the zero-point motion also. If Θ is the moment of inertia of the molecule the right-hand side of (20a) becomes equal to $\dfrac{h^2}{4\pi\Theta}$; and using this one can easily show that, for example, for HI molecules, at the distances they have in the solid state, the directive forces of the dipoles are still too weak to overcome the zero-point rotation. One has therefore to imagine these molecules always rotating even at the absolute zero in the solid state. But HI is certainly rather an exceptional case.

It is obvious that for larger molecules and for small molecular distances in the solid and liquid state the directive forces are quite insufficiently represented by the dipole action. For these one has simply to replace the left-hand side of (20) by the classical orientation energy in order to obtain a reasonable estimate for the limit of free motion.

As long as we are within the limits of (16) our argument in § 6 as to the additivity holds quite generally for all the three effects collected in formula (19). Only if, in consequence of (20), the free motion of the molecules is hampered does the criticism of § 3 apply, and this concerns the non-additivity of the direction effect as well as of the induction effect.

The internal *electronic* motion of a molecule, however, will not appreciably be influenced when the rotation of the molecule as a whole is stopped. Thus one is justified in applying the formula for the dispersion effect for non-rotating molecules also.

It is obvious, however, that only the highly compact molecules, as listed in Tables I. and II., can reasonably be treated simply as force centres. For the long organic molecules it seems desirable to try to build up the Van der Waals' attraction as a sum of single actions of parts of the molecules. As it is rather arbitrary to attribute the frequencies appearing in (15) or (13) to the single parts of a molecule, it has been attempted [8] to eliminate them by making use of the approximate additivity of the atomic refraction as well as of the diamagnetic susceptibility.

If there is one single " strong " oscillator μ_k only (*cf.* 14') the diamagnetic susceptibility has simply the form :

$$\chi_k = -\frac{\mu_k^2 N_L}{6mc^2} \; (< 0) \qquad (N_L \text{ Loschmidt's number})$$

therefore, because of (14'),

$$\nu_k = \frac{2}{3h}\frac{\mu_k^2}{\alpha_k} = \frac{4mc^2}{hN_L}\frac{\chi_k}{\alpha_k}.$$

We can therefore write, instead of (13'),

$$U_{k\rho} = \frac{3}{2}\frac{h}{R^6}\alpha_k\alpha_\rho \frac{4mc^2}{hN_L}\frac{\dfrac{\chi_k}{\alpha_k}\dfrac{\chi_\rho}{\alpha_\rho}}{\chi_k/\alpha_k + \chi_\rho/\alpha_\rho}$$

$$= \frac{1}{R^6}\frac{6mc^2}{N_L}\frac{\alpha_k\alpha_\rho}{\alpha_k/\chi_k + \alpha_\rho/\chi_\rho} \qquad . \qquad . \qquad . \qquad (13''')$$

[8] J. G. Kirkwood, *Physik. Z.*, 1932, **33,** 57 ; A. Müller, *Proc. Roy. Soc. A*, 1936, **154,** 624.

In this formula the interaction energy is represented by approximately additive atomic constants, and it seems quite plausible to build up in such a way the Van der Waals' attraction of polyatomic molecules from single atomic actions. But the comparison given in Table IV. shows that the exactitude of this method is apparently not very great.

For the *dispersion effect* also, the condition (16) indicates a characteristic limit. The quantity $\dfrac{\mu^2_{kl}}{E_k - E_l}$ is practically identical with the polarisability α, if $E_k \rightarrow E_l$ is the " main " electronic jump (compare 14').

TABLE IV.

	$\dfrac{3mc^2}{N_L}\alpha\chi \cdot 10^{48}$	$c \cdot 10^{48}$ (Table I.).
He .	0.84	0.77
Ne .	4.94	2.93
A .	69.0	34.7
Kr .	180	69
X .	448	146

Accordingly, instead of (16) we may roughly write

$$\alpha < R^3 \qquad . \qquad . \qquad . \qquad . \qquad (16')$$

as condition for the validity of our formulæ for the dispersion effect. What $\chi > R^3$ would mean can easily be inferred from considering our simple model (§ 4) : Some of the proper frequencies (12) would become imaginary, and that indicates that for these short distances the rest positions of the electrons would no longer be positions of stable equilibrium.

Some time ago Herzfeld [9] noticed that if R_0 is the shortest possible atomic distance (atomic diameter) the alternative " $\alpha > R_0^3$ or $\alpha < R_0^3$ " nearly coincides with the alternative " *metal* or *insulator*." Accordingly, for the non-metallic atoms and molecules listed in Table I. one is always within the limits of (16).

§ 10. Higher Approximations. Repulsive Forces.

The formula (15) is very far from completely representing the molecular forces, even of the rare gases, for all distances. It can be considered as a first step of a calculus of successive approximation. The state of a molecule is of course only quite roughly characterised by its orchestra of periodic dipoles ; there are obviously also periodic quadrupoles and higher multipoles, which give rise to similar interactions proportional to R^{-8}, R^{-10}, etc. For big distances these terms are in any case smaller than the R^{-6} forces, and there one may rely on formula (15). For He and H-atoms one [10] could calculate the R^{-8} term and could show that for small distances it can give rise to a contribution comparable with the R^{-6} term. But the R^{-10} term seems always to be negligible.

For these small distances, however, quite another effect has also to be considered. Even if a molecule does not show any permanent multipole but has, on an average, an absolutely spherically symmetrical structure, *e.g.* like the rare gases, quite apart from all effects due to the internal electronic motion, the mean charge distribution itself gives rise to a strong, so to speak " static," interaction, simply owing to the fact that by penetrating each other the electronic clouds of two molecules no longer screen the nuclear charges completely and the nuclei repel each other by the electrostatic Coulomb forces. In addition to this, and simultaneously, a second influence is to be considered. Already the

[9] K. F. Herzfeld, *Physic. Rev.*, 1927, **29**, 701.
[10] H. Margenau, *ibid.*, 1931, **38**, 747.

22   THE GENERAL THEORY OF MOLECULAR FORCES

penetration of the two electronic clouds is hampered by the Pauli Principle : two electrons can only be in the same volume element of space if they have sufficiently *different velocity*. This means that for the reciprocal penetration of the two clouds of electrons the velocity and therefore also the kinetic energy of the internal electronic motion must be augmented : energy must be supplied with the approach of the molecules, *i.e.* repulsion.

This repulsion corresponds to the *homopolar attraction* in the case of unsaturated molecules.   In an unsaturated molecule there are electrons with unsaturated spin and of these, when penetrating the cloud of a corresponding other molecule, the Pauli Principle no longer demands " sufficiently different " velocity but *only different spin orientation*.   In that case, consequently, one has a repulsion only for much smaller distances.

The actual calculation of the repulsive forces needs of course a very exact knowledge of the charge distribution on the surface of the molecules, and therefore presents considerable difficulties ; hitherto, a detailed calculation could only be carried out for the very simplest case [11] of He. The most successful attempts [12] in this direction so far have applied the ingenious Thomas-Fermi method which takes the Pauli Principle directly as a basis and is accordingly able, neglecting many unessential details, to account for just that effect which is characteristic of this penetration mechanism.

It is impossible here to reproduce the results of these numerical methods.   Up to now the repulsive forces have been successfully calculated only for the interaction between the rare gas-like ions, not yet for the rare gases themselves.   This is not because the repulsive forces between the neutral rare gas molecules constitute a very different problem, but because a considerably smaller degree of exactitude of the repulsive forces gives a useful description, when they are balanced by the strong ionic attractive forces instead of the weak molecular forces only.

The chemist at present must be satisfied with the knowledge that the repulsive forces depend on rather subtle details of the charge distribution of the molecules, and that consequently there is no reason to hope that one might connect them with other simple constants of the molecules, as is possible for the far-reaching attractive forces.   Their theoretical determination is in any special case another problem of pure numerical calculations.   But what really will interest the chemist is the fact that it can generally be shown that these homopolar repulsive forces (in characteristic contrast with the above-mentioned homopolar binding forces) have also the property of additivity, in the same approximate sense as the R^{-6} forces are additive, and that, therefore, to a first approximation, it will be quite justified to assume for the repulsive forces also simple analytical expressions, to superpose them simply additively and so to try to determine them from empirical data of the liquid or solid state. Whereas formerly one used to presume a power-law of the form b/R^n for these repulsive forces, quantum mechanics now shows that an exponential law of the form

$$be^{-R/\rho}$$

gives a more appropriate representation of the repulsion.

[11] J. C. Slater, *Physic. Rev.*, 1928, **32**, 349 ; see also W. E. Bleand J. E. Mayer, *Journ. chem. Physics*, 1934, **2**, 252.
[12] H. Jensen, *Z. Physik.*, 1936, **101**, 164 ; P. Gombar, *ibid.*, 1935, **93**, 378.

In this form one now usually presumes the repulsive forces and determines the constants b and ρ empirically.

For the attraction one uses the expression (15) or (13) possibly completed by a term proportional to R^{-8}. The factor R^{-8} supplies a third constant which also has to be empirically determined. For rare-gas like ions of the charges e_1, e_2, finally, one has of course to add the Coulombian term $+\dfrac{e_1 e_2}{R}$.

Thus, altogether one usually now takes, for the simplest molecules, an expression of the form

$$U = be^{-R/\rho} - c/R^6 - d/R^8 + \frac{e_1 e_2}{R} \qquad . \qquad . \qquad . \quad (21)$$

as a reasonable basis, where b, ρ, d are adjustable parameters, whereas c, e_1, e_2 are regarded as theoretically given. Thus, in all applications of the Van der Waals' forces a considerable freedom remains, and this is to be noticed when one wishes to test the theory.

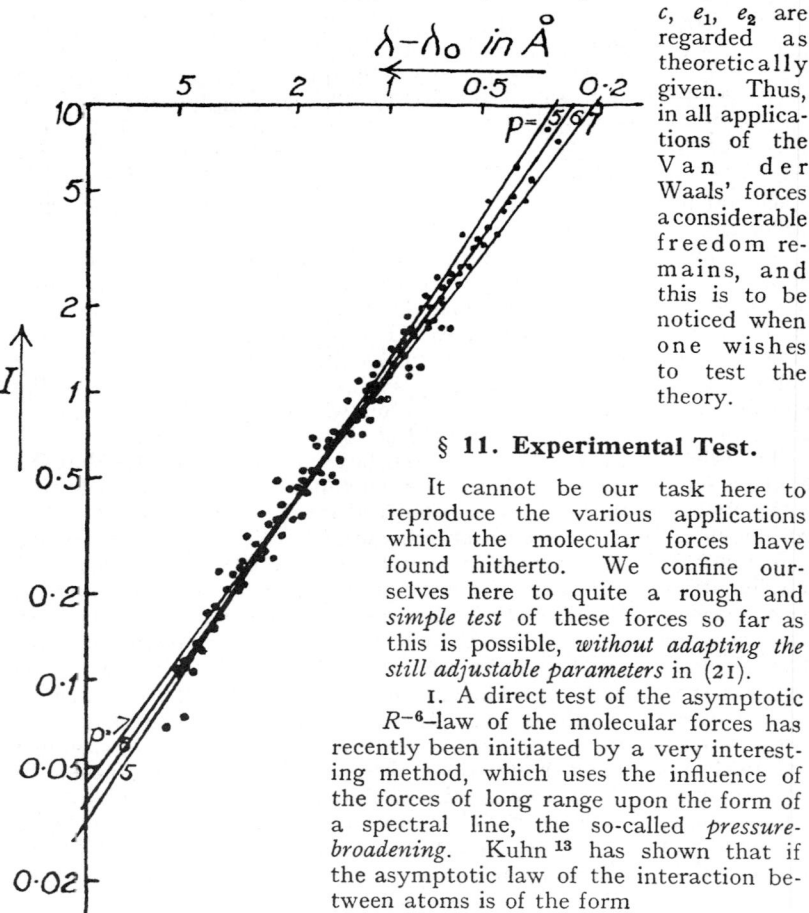

FIG. 1.—Intensity distribution and molecular forces.

§ 11. Experimental Test.

It cannot be our task here to reproduce the various applications which the molecular forces have found hitherto. We confine ourselves here to quite a rough and *simple test* of these forces so far as this is possible, *without adapting the still adjustable parameters* in (21).

1. A direct test of the asymptotic R^{-6}–law of the molecular forces has recently been initiated by a very interesting method, which uses the influence of the forces of long range upon the form of a spectral line, the so-called *pressure-broadening*. Kuhn [13] has shown that if the asymptotic law of the interaction between atoms is of the form

$$U \approx \frac{c}{R^p}$$

[13] H. Kuhn, *Phil. Mag.*, 1934, **18**, 987; *Proc. Roy. Soc. A*, 1936 *in print;* see also H. Kuhn and F. London, *Phil. Mag.*, 1934, **18**, 983.

24   THE GENERAL THEORY OF MOLECULAR FORCES

the intensity in a certain region of the spectral line is given by

$$I(\nu) = \frac{k}{(\nu_0 - \nu)^{\frac{p+3}{p}}}.$$

Thus the inclination of $\log{(I)}$ as a function of $\log{\nu}$ gives immediately the exponent p. Thereupon Minkowski [14] has discussed his measurements of the broadening of the D-lines of Na by Argon. He gives the following figure of his measured values of $\log{(I)}$ (Fig. 1). In addition, we have drawn the lines corresponding to $p = 5$, $p = 6$ and $p = 7$. One sees that the accuracy of the measurements does not yet permit an exact determination of p. But in any case we may say that $p = 6$ fits much better than $p = 5$ or $p = 7$, and that $p = 8$ and $p = 4$ can be excluded with certainty.

2. Testing the theory by the *gas equation* we shall restrict ourselves here to a quite rough check by means of the Van der Waals' a and b only. If this test has a satisfactory result, the exact dependence of the second virial coefficient on temperature may be used for determining *backwards* the still adjustable parameters in (21). But since it is always possible to get a fairly good agreement with the second virial coefficient by adjusting an expression like (21) it seems desirable to simplify the situation in such a way that, if possible, no adjustable parameters would be involved.

Accordingly, we replace (21) by :

$$U = \begin{cases} -c/R^6 \text{ for } R \geq R_0 \\ +\infty \text{ for } R < R_0 \end{cases} \qquad . \qquad . \qquad . \quad (22)$$

That means we idealise the molecules as infinitely impenetrable spheres, and neglect for $R > R_0$ the two adjustable terms $be^{-R/\rho}$ and $-d/R^8$ entirely. For large values of R the term $-c/R^6$ is certainly the only noticeable one. For mean distances $R \gtrsim R_0$ the two neglected terms, having different sign, may to a large extent cancel each other. For $R < R_0$ the very sudden increase of the exponential repulsion is replaced by an infinitely sudden one. By this procedure the order of magnitude of the minimum of U may be affected by a common factor, but will not be completely mutilated. Instead of the three adjustable parameters of (21) we have now only one : the size R_0.

The second virial coefficient B_2 is defined by the development of the gas equation into powers of $\frac{1}{V}$.

$$\frac{pV}{N_LkT} = 1 + \frac{B_2(T)}{V} + \frac{B_3(T)}{V^2} + \cdots,$$

and is given theoretically by

$$B_2 = 2\pi N_L \int_0^\infty (1 - e^{-\frac{U}{kT}})r^2 \, dr \qquad . \qquad . \qquad . \quad (23)$$

In the development of B_2 into powers of $\frac{1}{T}$, the first two terms can be identified with the corresponding terms of Van der Waals' equation :

$$\frac{pV}{N_LkT} = \frac{V}{V-b} - \frac{a}{VN_LkT} \approx 1 + \frac{1}{V}\left(b - \frac{a}{N_LkT}\right) + \frac{1}{V^2}(\ldots) + \cdots$$

[14] R. Minkowski, *Z. Physik*, 1935, **93**, 731.

The comparison gives:

$$B_2 = b - \frac{a}{N_L kT} + \cdots,$$

and if we now substitute (22) into (23) and consider that for high temperatures $U \gg -kT$ for all values of R, we obtain:

$$b = \frac{2\pi N_L R_0^3}{3}$$

$$a = -2\pi N_L^2 \int_{R_0}^{\infty} UR^2 dR = \frac{2\pi N_L^2 c}{3R_0^3}.$$

In these two equations we may eliminate R_0^3 and obtain:

$$ab = \frac{4\pi^2 N_L^3}{9} \cdot c = 1.51 \times 10^{54} \cdot c \qquad . \qquad . \quad (24)$$

Here the numerical factor is so determined that a, as usual, is measured in [atm. cm.⁶ g.⁻²] and c in the units of Table I.

If b is taken from the experimental gas equation the relation (24) can be used for predicting the constant a. These values are listed as $a_{theor.}$ in Table V., where they can be compared with the experimental values a_{exp}.

TABLE V.—Van der Waals a-Constant and Heat of Sublimation.

	$b_{exp.}$ [cm³].	$a_{theor.}$ 10^{-4} [atm. cm.⁶g⁻²].	$a_{exp.}$ 10^{-4} [atm.cm.⁶g⁻²].	ρ.	$L_{theor.}$ 10^{-3} cal./mol.	$L_{exp.}$ 10^{-3} cal./mol.
He	24	4.8	3.5			
Ne	17	26	21	1.46	0.47	0.59
Ar	32.3	163	135	1.70	1.92	2.03
Kr	39.8	253	240	3.2	3.17	2.80
Xe	51.5	430	410			
H₂	26.5	46	24.5			
N₂	39.6	147	135	1.03	1.64	1.86
O₂	31.9	135	136	1.43	1.69	2.06
CO	38.6	166	144	1.05	1.86	2.09
CH₄	42.7	256	224	0.53	2.42	2.70
CO₂	42.8	334	361			
Cl₂	54.8	680	632	2.00	7.18	7.43
HCl	40.1	283	366	1.56	3.94	5.05
HBr	44.2	510	442	2.73	4.45	5.52
HI				3.58	6.65	6.21

It is needless to say how inadequate the use of the critical data is for determining the limiting values for $T \to \infty$ of the second virial coefficient. These inadequacies may produce an uncertainty of perhaps 30 per cent., and our simplified expression (22) may also introduce an error of such an order of magnitude. But these uncertainties will presumably give rise only to a common systematic error for all molecules considered, and though the good absolute agreement found in the list is to be regarded as a lucky chance the relative agreement between theoretical and experimental a-values over such a wide range is certainly not disputable. That may justify trying to improve our knowledge of the Van der Waals

26   THE GENERAL THEORY OF MOLECULAR FORCES

forces by adjusting the expression (21) by means of the empirical second virial coefficient. Hitherto this has only been tried [15] by adding a law of the form b/R^n for the repulsion. But this procedure inevitably gives too small a molecule size, as it must attribute to the R^{-6}-forces what is due to the neglect of the R^{-8}-forces and of the sudden decrease of the exponential repulsion.

3. In Table V. is also listed the *lattice energy* L (sublimation heat extrapolated to absolute zero after subtraction of the zero-point energy) for some molecule lattices, calculated on the basis of the same simplified formula (22). In all cases we have assumed closest packed structure, as this structure is at least approximately realised in the molecular lattices in question. The summation of (22) over the lattice gives

$$L = 8{\cdot}36 \, . \, N_L{}^2 \frac{c}{v^2} 10^4 \left[\frac{\text{cal}}{\text{mol}} \right] = 3{\cdot}04 \times 10^{52} \frac{c \, . \, \rho^2}{M^2} \left[\frac{\text{cal}}{\text{mol}} \right]. \quad . \quad (25)$$

Here c is to be taken from Table I., v is the experimental mol. volume, $\rho = $ density, $M = $ molecular weight.

This test is instructive in so far as it shows plainly the additivity of the forces, and particularly the increase of L from HCl to HI with decreasing dipole moments clearly demonstrates the preponderance of those forces which are not due to the permanent moments.*

When the full expression (21) will be determined, say, from the experimental second virial coefficient it will be possible to calculate all constants (compressibility, elastic constants, etc.) of these molecular lattices.

For the constitution of the *ionic* lattices also, the Van der Waals attraction has been found to be a very decisive factor. We know the forces at present much better for these ions than for the neutral molecules. Using an interaction of the form (21), Born and Mayer [16] have calculated the lattice energy of all alkali halides for the NaCl-type and simultaneously for the CsCl-type and comparing the stability of the two types they could show quantitatively that the relatively great Van der Waals attraction between the heavy ions Cs^+, I^-, Br^-, Cl^- (*cf.* Table II.) accounts for the fact that CsCl, CsBr, CsI, and *these only*, prefer a lattice structure in which the ions of the same kind have smaller distances from each other than in the NaCl-type. The contribution of the Van der Waals' forces to the total lattice energy of an ionic lattice is of course a relatively small one, it varies from 1 per cent. to 5 per cent., but just this little amount is quite sufficient to explain the transition from the NaCl-type to the CsCl-type.

Paris, Institut Henri Poincaré.

[15] K. Wohl, *Z. physik. Chem.* B, 1931, **41**, 36; J. E. Lennard-Jones, *Proc. Physic. Soc.*, 1931, **43**, 461.

* In Table V. the lattice energies of He and H_2 have been omitted, because in these lattices the zero-point energy of the nuclear motion gives such a great contribution that it cannot be neglected. Therefore H_2 and He cannot immediately be compared with the other substances. See F. London, *Proc. Roy. Soc. A*, 1936, **153**, 576.

[16] M. Born and J. E. Mayer, *Z. Physik*, 1932, **75**, J. E. Mayer, *J. Chem. Physics*, 1933, **I**, 270.

Clusters

A. J. Stace

Department of Chemistry, University of Sussex, Falmer, Brighton, UK BN1 9QJ

Commentary on**: Experimental Study of the Transition from van der Waals, over Covalent to Metallic Bonding in Mercury Clusters,** H. Haberland, H. Kornmeier, H. Langosch, M. Oschwald and G. Tanner, *J. Chem. Soc., Faraday Trans.*, 1990, **86**, 2473.

An enduring cliché in cluster science is the continuing quest to 'bridge the gap' between properties characteristic of individual atoms or molecules, and the behaviour of those same substances when in the condensed phase. The type of question a cluster scientist might wish to address would be along the line of: how many water molecules does it take to dissolve sodium chloride? or, how many metal atoms does it take to construct an electrical conductor? In practice it has often proved very difficult to realise the cliché at a molecular level; the length scale over which many physical properties operate is often too large to be investigated using small numbers of atomic or molecular building blocks. For example, the transition from icosahedral to octahedral geometry in solid argon requires ~2000 atoms, and to match the melting temperature of bulk gold, clusters need to contain several million atoms.

Currently, only two bulk physical properties would appear to be accessible using small(!) numbers of atoms or molecules. The first is metal ion solvation,[1] where gas phase experiments show that the essential thermodynamics of this process can be reproduced with approximately six solvent molecules.[2] The paper by Haberland *et al.*[3] addresses a second physical property that is also readily accessible at an atomic level, and that is the transition to metallic bonding in metal clusters. In this case, the basic experiment sought to 'bridge the gap' between the ionisation energy of a single metal mercury atom (IE = 10.4 eV) and the work function (ϕ = 4.49 eV), which represents the energy necessary to remove an electron from the bulk solid. The transition from IE(atom) to ϕ was monitored by measuring the ionisation energies of individual clusters containing between 2 and 100 mercury atoms.

Mercury is an interesting system because the atom is closed-shell (s^2) and the binding energy between pairs of atoms is not too different from that found between pairs of rare gas atoms. What the experiments of Haberland *et al.* revealed is that this rare gas van der Waals bonding persists in mercury clusters containing up to thirteen atoms. Beyond that size the clusters begin to exhibit covalent bonding; but it is not until they contain upwards of ninety atoms that metallic character starts to appear. If clustering did not influence the relative energies of the filled 6s- and vacant 6p-orbitals, then bulk mercury would probably be an insulator. However, the orbitals spread and move towards one another, and for clusters consisting of more than ninety atoms the orbital overlap is sufficient to begin creating a conduction band. Although the measured IE for a cluster of ninety atoms is still ~1 eV away from the work function, a correction allowing for the fact that a positive charge in a cluster is confined to a finite rather than an infinite volume, brings the experimental result in to line with that expected of a classical liquid-drop conductor.

The results of Haberland *et al.* are underpinned by several earlier pieces of work. In particular, Rademann *et al.*[4] measured the ionisation energies of a more limited range of mercury clusters at discrete photon energies. The overall trend in their data is similar to that seen by Haberland *et al.*, but they did not distinguish the van der Waals and covalent contributions to bonding in the smaller mercury clusters. In a slightly different experiment, Bréchignac *et al.*[5] used synchrotron radiation to promote core electrons to valence states in small mercury clusters. The positions of the valence states are considered to be sensitive to changes in the nature of the bonding between atoms in the clusters. This latter experiment is not so strongly influenced by structural changes that may occur as a result of ionisation, and which may create a significant difference between vertical and adiabatic ionisation energy. Bréchignac *et al.* are in agreement with the fact that small mercury clusters are very weakly bound, but the authors also suggest that 20 mercury atoms may be sufficient to initiate electronic band formation. However, it is quite possible that the valence state changes seen by Bréchignac *et al.* actually coincide with the onset of covalent bonding, as identified by Haberland *et al.*

The paper by Haberland *et al.* represents one of the first attempts to use clusters as a means of mapping the development of electronic band structure. The data show clear evidence of a steady progression from the van der Waals behaviour expected of a collection of closed shell atoms, through to the on-set of metallic character. As part of a 'bigger picture' that seeks to understand how collections of atoms or molecules eventually adopt the properties of solids, the work makes an important contribution.

References

1 P. Kebarle, *Annu. Rev. Phys. Chem.*, 1977, **74**, 1466.
2 A. J. Stace, *J. Phys. Chem. A*, 2002, **106**, 7993.
3 H. Haberland, H. Kronmeler, H. Langosch, M. Oschwald and G. Tanner, *J. Chem. Soc., Faraday Trans.*, 1990, **86**, 2473
4 K. Rademann, B. Kaiser, U. Even and F. Hensel, *Phys. Rev. Lett.*, 1987, **59**, 2319.
5 C. Bréchignac, M. Broyer, Ph. Cauzac, G. Delacretaz, P. Labastie, J. P. Wolf and L. Wöste, *Phys. Rev. Lett.*, 1988, **60**, 275.

J. CHEM. SOC. FARADAY TRANS., 1990, **86**(13), 2473–2481 2473

Experimental Study of the Transition from van der Waals, over Covalent to Metallic Bonding in Mercury Clusters

Hellmut Haberland, Hans Kornmeier, Helge Langosch, Michael Oschwald and Gregor Tanner
Fakultät für Physik, Universität Freiburg, Federal Republic of Germany

The following properties have been measured for mercury clusters: (1) ionisation potentials of Hg_n by electron-impact ionisation, (2) dissociation energies of Hg_n^+, and (3) mass spectra for negatively charged mercury cluster ions ($n \geqslant 3$). Cohesive energies for neutral and ionised Hg clusters have been calculated from the data. The transitions in chemical binding are discussed. For small clusters Hg_n is van der Waals bound ($n \leqslant 13$), the binding changes to covalent for $30 \leqslant n \leqslant 70$, and then to metallic ($n \geqslant 100$). A sudden transition from covalent to metallic bonding is observed. It is discussed whether this can be considered as being analogous to a Mott transition for a finite system. The experimentally observed transitions in chemical binding are much more pronounced than those calculated in a tight-binding calculation. This points to strong correlation effects in Hg clusters.

The Hg atom has a $6s^2$ closed electronic shell. It is isoelectronic with helium, and is therefore van der Waals bound in the diatomic molecule and in small clusters. For intermediate sized clusters the bands derived from the atomic 6s and 6p orbitals broaden as indicated in fig. 1, but a finite gap Δ remains until the full 6s band overlaps with the empty 6p band, giving bulk Hg its metallic character. This change in chemical binding has a strong influence, not only on the physical properties of mercury clusters, but also on the properties of expanded Hg,[1] and on Hg layers on solid[2] and liquid[3] surfaces. For a rigid cluster the electronic states are discreet and not continuous as in fig. 1. Also the term 'band' for a bundle of electronic states will be used repeatedly in this paper, although 'incipient band' might be better. As the clusters discussed here are relatively hot, possibly liquid, any discreet structure will be broadened into some form of structured 'band'.

Several groups have studied the transitions in chemical bonding for free Hg_n clusters. Cabauld et al.[4] measured ionisation potentials by electron-impact ionisation for $n < 13$; Rademann et al.[5,6] used a photoionisation and photoelectron coincidence technique to obtain ionisation potentials up to

$n = 78$. We have extended the data up to $n = 100$ using electron-impact ionisation. The excitation of the $5d \rightarrow 6p$ auto-ionising transition has been measured by Bréchignac et al.[7] for $n \leqslant 40$. Measurements of dissociation energies for Hg_n^+ are reported below, and cohesive energies of neutral and ionised clusters are calculated from the data. The mass spectrum of negatively charged mercury clusters is presented for the first time here. The smallest ion observed is Hg_3^-.

From all these data the following general picture arises for the bonding transitions in mercury clusters: Small Hg_n ($n \leqslant 13$) clusters are van der Waals bound, the Δ of fig. 1 is so large that sp hybridisation is energetically unfavourable. After a transition region, the bonding becomes covalent ($30 \leqslant n \leqslant 70$); sp hybridisation leads to an increase of the binding between the atoms. Between $n = 95$ and 100 a rapid decrease of Δ is observed, and $\Delta(n = 100) \approx 0$. The possibility to interpret the rapid decrease of the ionisation potentials as being analogous to a Mott transition in a finite system is discussed.

For small n this picture is consistent with the interpretation given in ref. (7). The overall features are also in agreement with calculations of Pastor et al.,[8,9] although the calculated variations in binding character are smoother than observed experimentally, probably due to neglect of correlation effects. These authors calculate that Δ should go smoothly to zero, and $\Delta(n = 135)$ should lie between 0 and 0.1 eV. All these results contrast with the interpretation given by Rademann,[6] who deduced $\Delta(n = 13) \approx 0$.

Experimental

A continuous Ar-seeded supersonic Hg beam was used to produce the Hg clusters (see fig. 2). Conical and straight nozzles of 40–100 μm diameter were used. Mercury and argon pressures up to 2 and 50 bar could be employed. The electron and photon beams were pulsed. After the ionising pulse had completely terminated, a potential was applied to plate P0, accelerating the ions into the time-of-flight (TOF) mass spectrometer. If the TOF is used in the reflectron mode, the first time focus lies at the slit of the mass selector, the second on detector II. By applying suitable electric pulses to the plates of the mass selector, a single mass can be selected. Ions which decay in the free-flight region have a lower kinetic energy than their parents. They do not penetrate as far into the reflecting field, and can be separated in time at detector II. If the system is operated to function as a linear TOF, detector I is used.

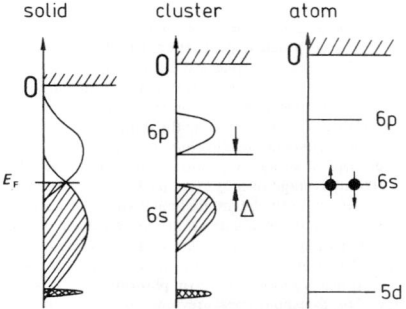

Fig. 1. The Hg atom has a $6s^2$ closed electronic shell. The atomic lines broaden into 'bands' for the cluster. The gap $\Delta(n)$ decreases as a function of the cluster size. The two bands overlap in the solid, giving mercury its metallic character. For large Δ the binding is of the van der Waals type, for intermediate it is covalent, while for vanishing Δ it is metallic. From this experiment it is deduced that the gap closes rather abruptly at *ca.* $n \approx 100$ atoms per cluster. This value is a factor of 2 to 7.7 higher than determined in ref. (1), (6), (28) and (29).

2474   J. CHEM. SOC. FARADAY TRANS., 1990, VOL. 86

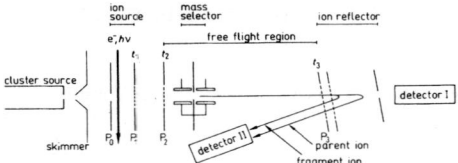

Fig. 2. Schematic diagram of the apparatus. A continuous supersonic Hg/Ar expansion produces the clusters. They can be ionised by electron or photon impact. An electric pulse between the plates P0 and P1 accelerates ions into the time-of-flight mass spectrometer. The first time focus is at the slit of the mass selector, the second is at the detector II. The mass selector can be used to select one mass only, say Hg_{20}^+, which can subsequently decay in the free-flight region into $Hg_{19}^+ + Hg$ due to internal excitation. The reflector can separate parent ($n = 20$) and daughter ($n = 19$) cluster ions.

Ionisation Potentials

For appearance potential measurements of positive ions the cluster beam is crossed by a pulsed electron beam of variable kinetic energy. Fig. 3 shows the yield of Hg^+ and Hg_{10}^+ as a function of the kinetic energy of the electrons. All data show a linear rise, which is interrupted by a 'bend' for clusters with $n \geqslant 5$ and then a second linear rise. Similar linear slopes at the appearance potentials have been observed many times in electron-impact ionisation of clusters.[10] All extrapolations made have been done using a numerical least-squares fitting routine. A fit to the curvature of the data near threshold gives an energy width of ± 180 meV F.W.H.M. of the electron beam, showing that the data presented here have about the same energy resolution as the photoionisation data of ref. (5) and (6). The vertical lines give the location of the $5d \rightarrow 6p$ auto-ionising lines, which completely dominate the photoionisation data.[7] No trace of these two peaks can be seen in fig.

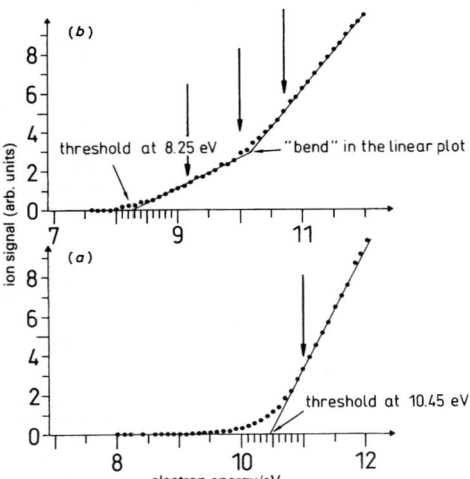

Fig. 3. Intensity on (*a*) Hg^+ and (*b*) Hg_{10}^+ as a function of the kinetic energy of the electrons. All intensities rise linearly above threshold. Above $n = 5$ a 'bend' is observed in the data. The vertical arrows give the energetic locations of the peaks observed in the photoionisation data of ref. (7).

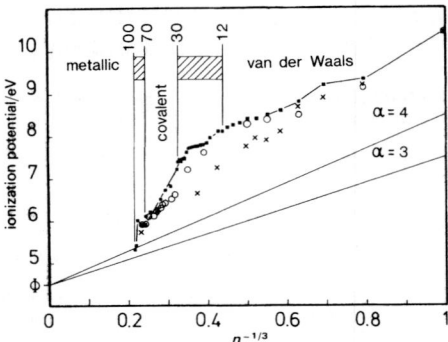

Fig. 4. Ionisation potentials of Hg_n plotted against $n^{-1/3}$. The bulk value ($n^{-1/3} = 0$) is at the left, the atomic value ($n^{-1/3} = 1$) at the right-hand side. The two solid lines starting at Φ (bulk work function) correspond to the two scaling laws proposed for metallic clusters [eqn (7)]. Determined in this experiment; $\bigcirc$, from ref. (4) and (5); $\times$, from the calculation of ref. (8) and (9). From very large clusters down to $n \approx 100$ the binding is metallic. After a transition region (hatched area) one has covalent bonding for $70 \leqslant n \leqslant 30$, after a second transition region van der Waals bonding becomes dominant below $n \leqslant 13$.

3. Obviously a different density of states is measured with electron impact, compared to photoionisation. This is not surprising in view of the different selection rules for the two processes. For photoionisation the dipole selection rule is valid, while for electron-impact ionisation electron-exchange processes may dominate at threshold.

Note that the thresholds can easily be measured by electron impact, while this is not possible for Hg_n, $n > 2$ at the moment by photoionisation owing to the very small oscillator strength at threshold.[11] The electron energy scale was calibrated by the well known ionisation potentials of atomic Hg and Ar. The difference between the two thresholds is reproducible to 0.05 eV. The error in the ionisation potentials (E_i) is difficult to estimate. The overall accuracy is ± 100 meV at $n = 13$ and ± 300 meV at $n = 90$. However, the point-to-point variation of the data, *i.e.* the difference in error between $E_i(n)$ and $E_i(n + 1)$ is much less, and is expected to be $<ca.$ 100 meV.

Fig. 4 shows the ionisation potentials as a function of $n^{-1/3}$, a value which is proportional to R^{-1}, where R is the radius of an assumed spherical cluster. If $V = 4\pi R^3/3$ is the volume of a cluster of radius R one has, neglecting geometric and packing effects, $V = nv$, where v is the volume of an atom. Hence $R^{-1} \propto n^{-1/3}$. For the data points at $n = 90$ and 95 the experimental results had to be averaged over ± 2 cluster sizes in order to obtain accurate threshold data; for $n = 100$ an average over ± 5 cluster sizes was necessary. The cluster density in the beam, the sensitivity of the detector and the electron current available at threshold all decrease for $n > 100$, making a threshold determination impossible.

In order to extend the data to higher n, a pulsed dye laser was frequency-doubled to give photons in the 4.5–6.5 eV range. No threshold measurements were possible as two-photon processes were always present even at the lowest laser fluences employed (< 100 nJ cm^{-2}).[12]

The data of Cabauld *et al.*[4] for $n < 13$ are very similar and partially indistinguishable from ours on the scale of fig. 4, and are therefore omitted. Save at $n = 1$ and in the $n = 50$–79 region the data of Rademann *et al.*[5,6] are generally lower than ours, which could be due to the different density of

J. CHEM. SOC. FARADAY TRANS., 1990, VOL. 86   2475

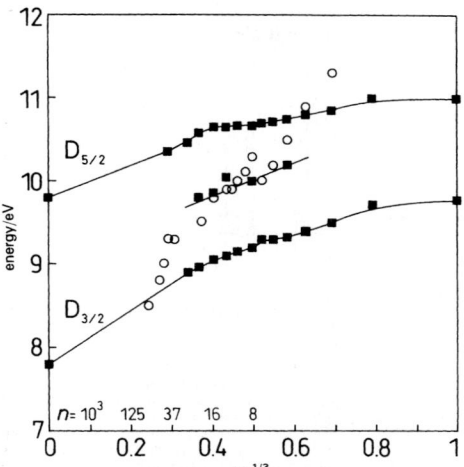

Fig. 5. The upper and lower line of solid squares show the known autoionisation lines of the $5d^{10}6s^2 + h\nu \rightarrow 5d^9 6s^2 6p$ transition.[7] Compared to the ionisation potentials of fig. 4 a more gradual transition from the atom to the bulk is observed. Owing to spin–orbit splitting the $5d^9$ configuration gives rise to two optically allowed lines with $D_{3/2}$ and $D_{5/2}$ cores. A third unidentified line is observed between these two lines. The energetic positions of the 'bends' visible in fig. 3 are given by the open circles. For $6 \leqslant n \leqslant 19$ the 'bends' nearly coincide with the unidentified photon line.

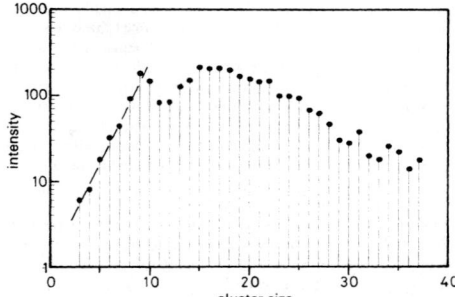

Fig. 6. Mass spectrum of negatively charged mercury clusters. Only above $n = 3$ are the Hg_n^- clusters observed. The intensity rises exponentially at small n, a minimum is always observed at $n = 11$ or 12.

states probed. As mentioned above, photoionisation measures the singlet component and electron impact at threshold the triplet component of the density of states. Although our data agree in general with those of ref. (5) and (6), Rademann[6] arrives at very different conclusions, as discussed below. The large decrease of the ionisation energies for small n is due to the strongly bound dimer ion Hg_2^+. The fragmentation effects due to its strong binding can be expected to be less severe than for the kindred rare-gas case,[13] as the ratio of neutral to ionic dissociation energies is higher. For very small n we expect to have some fragmentation even close to threshold, although we presently do not have a means of quantifying this expectation. For large n some excitation can be tolerated before the cluster breaks up, as discussed in the context of fig. 7 and 8 (later).

Fig. 5 compares the energetic locations of the bends in the linear plots (see fig. 3) to those of the auto-ionising photoionisation lines measured by Bréchignac et al.[7] These lines are due to a $5d \rightarrow 6p$ transition: $5d^{10}6s^2 + h\nu \rightarrow 5d^9(D_{3/2}$ or $D_{5/2})6s^2 6p$. The excited neutral state decays by autoionisation to $5d^{10}6s + e^-$. The 5d hole can either form a $D_{3/2}$ or $D_{5/2}$ state. The authors of ref. (7) conclude that up to $n = 13$ the binding is of the van der Waals type, and that the excitation has consequently an excitonic character. For $n = 6$–19 the energetic positions of the 'bends' coincide with an unidentified peak in the photon data.

Negatively charged Cluster Ions

A continuous, electric-glow discharge is ignited in the region between skimmer and nozzle in order to produce negatively charged cluster ions. This procedure was used earlier to produce negatively charged clusters of other closed-shell atoms and molecules.[14] Fig. 6 shows a mass spectrum. The

minimum cluster size is $n = 3$, the ion intensity rises up to $n \approx 9$ and an intensity minimum is always observed at $n = 11$ Hg atoms per cluster. The relative intensities of the two groups below and above $n = 11$ depend sensitively on expansion conditions, but the minimum is always present. The ion intensities in fig. 6 are given on a logarithmic scale to emphasize the exponential increase at low cluster masses. A similar exponential increase in a negative-ion mass spectrum was recently observed for Xe_n^- clusters, a system which is definitely van der Waals bound.[15] This similarity strengthens the argument that small Hg clusters are purely van der Waals bound, in contrast to the conclusions of ref. (6).

Dissociation Energies

A new method for measuring dissociation energies of cluster ions was recently proposed by Bréchignac et al.[16] The principle of the method is as follows. Neutral clusters are ionised and further excited in the ion source of a TOF spectrometer by an intense laser pulse. It is important that all cluster ions are so highly excited that they evaporate at least one atom in the ion source region. Assuming a statistical model, the internal energies and the decay rates can be calculated. Fig. 7 shows schematically the time evolution of the internal energy of a single cluster. During the 10 ns of the laser pulse a

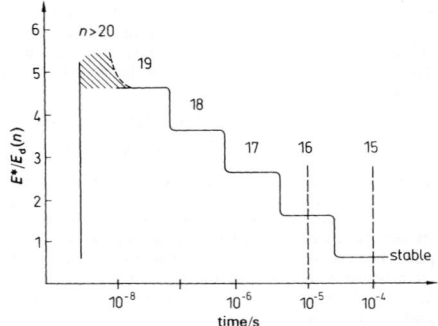

Fig. 7. Energy content E^* in units of the dissociation energy $E_d(n)$ of one single cluster is plotted against time. Hg_n, $n > 20$ is ionised and further excited by a 10 ns laser pulse, at the end of which a Hg_{19}^+ remains. The relative internal energy $E^*/E_d(n)$ decreases each time an atom is evaporated by one unit.

neutral Hg_n cluster, $n > 20$, is ionised and highly excited. If the internal excitation energy E^* is five times the dissociation energy, at most five evaporation processes are possible. Assuming an RRK or RRKM type statistical treatment,[17] one finds that the lifetimes increase by at least a factor of ten between successive evaporations. This allows an approximate separation of the evaporation events. The internal energy, $E^*(n)$, of cluster A_n is determined by its production process. Because of energy conservation, it must have an energy:

$$E^*(n) = E^*(n + 1) - E_d(n + 1) - \varepsilon(n + 1) \equiv E^*$$

Henceforth $E^*(n)$ will be represented as E^*. The dissociation energy $E_d(n + 1)$ is needed to separate one atom from the remaining cluster. The kinetic energy of the recoiling products, $\varepsilon(n + 1)$, is taken to be:[18]

$$\varepsilon(n + 1) = 2E[(n + 1)^* - E_d(n + 1)]/(3n - 4)$$
$$= 2E^*/(3n - 6).$$

In order to derive an analytical equation for the distributions of internal energy in the cluster, the Markov process depicted schematically in fig. 7 has been replaced by a three-step decay: $A_{n+1} \to A_n \to A_{n-1}$. This is valid as all other decays have already finished or have not yet started in the accessible time window, because the decay constant k_n (inverse of the lifetime) depends so strongly on E^*. The first step can be written explicitly as:

$$A_{n+1}[E^*(n + 1)] \xrightarrow{k_{n+1}[E^*(n+1)]} A_n(E^*) + A + \varepsilon(n + 1).$$

If E^* is sufficiently large the cluster A_n will decay further:

$$A_n(E^*) \xrightarrow{k_n(E^*)} A_{n-1}[E^* - E_d(n) - \varepsilon(n - 1)] + A + \varepsilon_n.$$

The probability $P_n(t, E^*)$ that A_n has an energy between E^* and $E^* + dE^*$ at a time t_1 after the ionisation is then given by

$$P_n(t_1, E^*) = \int_0^{t_1} (k_{n+1}[E^*(n + 1)]\exp\{-k_{n+1}[E^*(n + 1)]t\}$$
$$\times \exp[-k_n(E^*)(t_1 - t)]\,dt.$$

The factor in curly brackets is the normalised probability that A_n is produced between t and $t + dt$ from A_{n+1} with energy E^*. The second exponential gives the probability that A_n does not decay in the interval $t_1 - t$. The relevant time t_1 of this experiment is the time the cluster passes through plate P1 in fig. 2. Note that it is not the time to pass through P2, as all ions which decay between P1 and P2 are not focused onto one mass peak, but contribute to a broad background. The probability F_n that A_n^+ fragments into $A_{n-1}^+ + A$ in the field-free drift region (and that A_{n-1}^+ can be focused on one peak by the TOF spectrometer) is

$$F_n(t_3, E^*) = P_n(t_1, E^*)\exp[-k_n(E^*)(t_2 - t_1)]$$
$$\times \{1 - \exp[-k_n(E^*)(t_3 - t_2)]\}.$$

The cluster ion A_n^+ leaves plate P1 at time t_1 with an internal energy distribution given by $P_n(t_1, E^*)$. The second factor gives the probability that it survives the flight between plates P1 and P2, where the fragments are not focused on one mass peak. The last term gives the probability that A_n^+ ejects an atom between plate P2 and the reflector, where it arrives at time t_3. The probability P_n' that A_n^+ is focused on one mass peak is accordingly:

$$P_n'(t_4, E^*) = P_n(t_1, E^*)\exp[-k_n(E^*)(t_4 - t_1)]$$

where t_4 is the flight time to the end of the reflector. Decays between reflector and detector II do not change the mass spectrum.

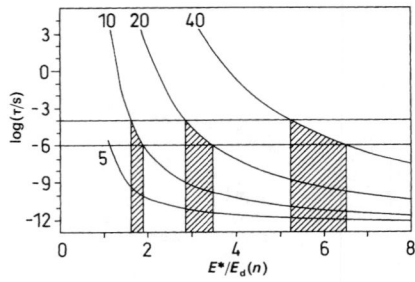

Fig. 8. Lifetimes τ_n of $n = 5, 10, 20$ and 40 atom clusters as a function of the relative excitation energy. For a fixed experimental time window (10^{-6}–10^{-4} s) the clusters have different but relatively narrow internal excitation energies.

To evaluate the integrals defined above, an equation for the decay constants is needed. The simplest choice is the classical RRK expression:[17]

$$k_n(E^*) = vg\left(1 - \frac{E_d(n)}{E^*}\right)^{3n-7}. \qquad (1)$$

Here v is a typical vibrational frequency (*ca.* 10^{12} Hz), g is a degeneracy factor equal to the number of surface atoms, and E^* and $E_d(n)$ are defined above. Fig. 8 shows calculated decay constants using eqn (1). In the actual analysis of the data it was necessary to use the 'quantum-mechanical' form of eqn (1):[17]

$$k_n = vg\prod_{j=1}^{3n-7}\left(1 - \frac{E_d(n)}{E^* + jhv}\right). \qquad (2)$$

In the limit $E^* \gg hv(3n - 7)$ one regains the classical RRK formula. It was tested numerically that for small Hg clusters eqn (2) had to be used. Only if the dissociation energy becomes large, *e.g.* for the larger Hg clusters or the alkali-metal clusters of ref. (16), eqn (2) and eqn (3) give the same result. The treatment due to Engelking[19] was not applicable in our case. The ratios $E_d(n)/E_d(n + 1)$ are lower than those calculated fron eqn (1) and (2). This leads to unrealistically high absolute values of $E_d(n)$ for the larger clusters.

The probability that A_n^+ or A_{n-1}^+ arrives at the detector has to be integrated over all internal energies. The ratio V_n of the integrated probabilities:

$$V_n = \frac{\int F_n(t_3, E^*)\,dE^*}{\int P_n'(t_4, E^*)\,dE^*} \qquad (3)$$

can be calculated from the equations given above, and V_n can be measured in the experiment as the ratio of the parent intensity on mass n and the fragment intensity on mass $n - 1$. This allows determination of the ratios $E_d(n)^+/E_d(n + 1)^+$. The details of the experiment and data reduction will be given elsewhere:[20] only a brief sketch will be given here. The ratios of all cluster ion intensities for $5 \leqslant n \leqslant 30$ were measured for three different kinetic energies of the cluster ions in the drift region. This was used as a check on the consistency of the data, as the times t_i depend on the kinetic energy of the cluster ion. The three sets of data agreed within statistical error and were averaged. The frequency factor was interpolated between the value for the dimer, and that calculated from the bulk Debye temperature. An incorporation of the Marcus improvement using the modified Debye model of Jarrold and Bower[21] is in progress.

Monomer and dimer evaporation is observed experimentally for alkali-metal clusters.[16] This can be used to calibrate

J. CHEM. SOC. FARADAY TRANS., 1990, VOL. 86

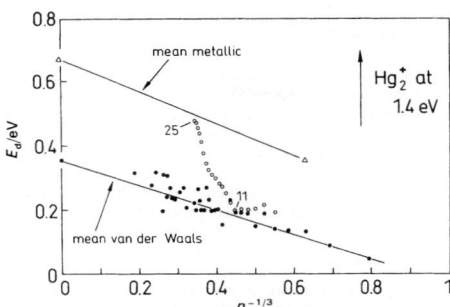

Fig. 9. Experimentally determined dissociation energies of mercury clusters ions (open circles) and calculated dissociation energies of neutral van der Waals clusters, scaled to the Hg_2 dissociation energy.

the absolute energy scale of the dissociation energies, which is necessary since only ratios of dissociation energies are obtained from eqn (3). For Hg clusters no dimer evaporation is observed owing to the low binding energy of Hg_2; however, the physically allowed dissociation energies are bounded from below by the values expected for a pure van der Waals system, and from above by that for a pure metallic system, as indicated in fig. 9 and 11 (later). This puts very stringent limits on the possible dissociation energies. After a variety of simulations, a value of 200 meV was selected for Hg_6^+. A lower value would push the value for $E_d(11)^+$ below the value calculated for a neutral van der Waals cluster. A higher value would put the value for $E_d(25)^+$ above the value expected if Hg would behave like an alkali-metal for all cluster sizes. These conclusions have a somewhat preliminary character, as they are based on a RRK type and not an RRKM type calculation, as discussed above, but large deviations from these conclusions are not expected.

Data Reduction

In this section the original data will be transformed, in order to allow an easier interpretation in the Discussion.

Preliminaries

In this experiment ionisation potentials $E_i(n)$ and dissociation energies for positively charged clusters $E_d(n)^+$ have been measured. Fig. 10 shows the Born–Haber cycle relating these quantities with the electron affinities [$E_{ea}(n)$] and dissociation energies for neutral clusters [$E_d(n)$]. Energy conservation

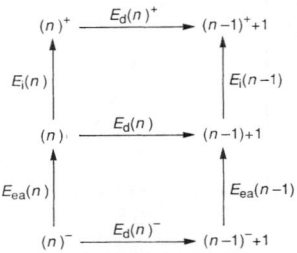

Fig. 10. Ionisation potentials [$E_{ea}(n)$], electron affinities [$E_{ea}(n)$] and dissociation energies for neutral [$E_d(n)$] or positively/negatively charged clusters [$E_d(n)^+/E_d(n)^-$] are interrelated as given by eqn (4)–(6).

gives:

$$E_i(n) - E_i(n-1) = E_d(n) - E_d(n)^+ \quad (4)$$

$$E_{ea}(n) - E_{ea}(n-1) = E_d(n)^- - E_d(n). \quad (5)$$

Adding the two equations one obtains

$$[E_i(n) + E_{ea}(n)] - [E_i(n-1) + E_{ea}(n-1)]$$
$$= E_d(n)^- - E_d(n)^+. \quad (6)$$

Two different scaling laws have been proposed for ionisation potentials and electron affinities:

$$E_i(R) = \Phi_\infty + \frac{\alpha}{8} \frac{e^2}{R} \quad (7)$$

$$E_{ea}(R) = \Phi_\infty - \frac{8-\alpha}{8} \frac{e^2}{R} \quad (8)$$

where Φ_∞ is the bulk work function and R the radius of an assumed spherical cluster. The debate[22] has not finished as to whether $\alpha = 3$ or $\alpha = 4$ is the correct choice in eqn (7) and (8). If $\alpha = 4$, eqn (6) gives $E_d(n)^- = E_d(n)^+$, i.e. the dissociation energies would not depend on the sign of the charge. Experimentally it is observed[23] that $\alpha = 3$ is often a good approximation for cluster diameters larger than 7 Å. With $\alpha = 3$ eqn (6) gives

$$E_d(n)^- - E_d(n)^+ = \frac{e^2}{4} [R(n-1)^{-1} - R(n)^{-1}]$$
$$= \frac{e^2}{4R} \left(1 + \frac{2}{3n} + \dots \right) / 3n > 0.$$

For sufficiently large metallic clusters, one will often have $E_d(n)^- > E_d(n)^+$; the dissociation energy is higher for the negatively than for the positively charged clusters.

Iterating eqn (4) one obtains:

$$E_i(1) - E_i(n) = \sum_{i=2}^{n} [E_d(i)^+ - E_d(i)]$$

The cohesive energies [$E_c(n)$ and $E_c(n)^+$] per atom are defined as

$$E_c(n) = \frac{1}{n} \sum_{i=2}^{n} E_d(i)$$

$$E_c(n)^+ = \frac{1}{n} \sum_{i=2}^{n} E_d(i)^+.$$

Combining the last three equations one obtains[16]

$$E_c(n)^+ - E_c(n) = E_i(1) - E_i(n)/n. \quad (9)$$

Note that the n^{-1} factor enforces a rapid convergence of the cohesive energies of neutral and ionised clusters. For mercury and $n = 100$, the right-hand side of eqn (9) is ca. 50 meV and ca. 5 meV at $n = 1000$. Similarly one obtains by iterating eqn (5):

$$E_{ea}(n) - E_{ea}(j-1) = \sum_{i=j}^{n} [E_d(i)^- - E_d(i)] \quad (10)$$

For Hg and many other closed shell atoms and molecules one has $j > 2$ in eqn (10), as small clusters have a negative electron affinity. For Hg one has $j = 4$.

Dissociation and Cohesive Energies

Fig. 11 compares the cohesive energies $E_c(n)^+$ and $E_c(n)$ determined in this experiment, with expected values for typical van der Waals or metallic bound clusters. The open

2478   J. CHEM. SOC. FARADAY TRANS., 1990, VOL. 86

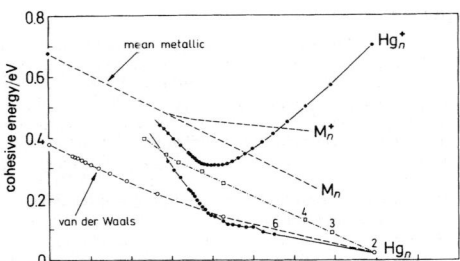

Fig. 11. Cohesive energies for neutral and ionised metal clusters (M_n and M_n^+) and for neutral van der Waals clusters (○) have been scaled to the mercury values as explained in the text. The experimentally obtained cohesive energies for Hg_n and Hg_n^+ are given by the solid circles. For small n the cohesive energy follows the van der Waals line perfectly. The calculated cohesive energies for Hg_n (□, ref. 8) show a smoother behaviour than the experimental values.

circles joined by the dashed line show the calculated cohesive energies for icosahedral Lennard-Jones clusters.[24] The values have been scaled to the newly determined[25] dissociation energy of the dimer: $E_d(2) = 42.7 \pm 2.5$ eV, or $E_c(2) = E_d(2)/2 \approx 21$ meV. As can be expected for a classical calculation the cohesive energies extrapolate well to the classical bulk value[26] of $8.6\, E_d(2) = 0.378$ meV, calculated for an infinite f.c.c. lattice. If Hg were to stay van der Waals bound in the bulk, its cohesive energy would be *ca.* 378 meV. Owing to the non-metal to metal transition the experimental value is a factor of 1.77 higher: $E_c(n \to \infty) = 0.67$ eV.

The dashed line marked 'mean metallic' gives the averaged cohesive energies for the prototype metallic systems of Na and K, also scaled to the bulk cohesive energy of Hg. For Na and K many experimental and theoretical data are available.[16] The individual data points are not given, as they would confuse this figure. The cohesive energies of neutral Na and K clusters agree remarkably well with the classical spherical droplet model

$$E_c(n, \text{metal}) = a_v - a_s n^{-1/3}$$

where a_v is a volume term and a_s gives the surface contribution. The corresponding straight line is marked M_n (M for metal) in fig. 11. The curve for the ionised clusters, M_n^+, tails off from this line and becomes nearly constant for small n. Note that the true Na and K values show structures which have been suppressed in fig. 11. The influence of the charge on the cohesive energy of a metallic system is much smaller than of a van der Waals bound system. Small charged van der Waals clusters have a higher cohesive energy than the bulk. If Hg_n were to remain van der Waals bound all the way from the dimer to the bulk, the curve marked Hg_n^+ in fig. 11 would extrapolate smoothly to the curve marked 'van der Waals'.

No dissociation energies could be obtained in this experiment for $n = 2$–5. The decay constants are too short to be measured by our apparatus. This can be seen from the curve for $n = 5$ in fig. 8. The dimer value $E_d(2)^+ = 1.44$ eV is from ref. (11). The values for $n = 3$, 4 and 5 have been scaled to the kindred Ar_n^+ case.[27] The first fall off of $E_c(n)^+$ is due to the diminishing influence of the charge for increasing cluster size, the increase for small n due to the transition to more strongly bound clusters. The cohesive energies of the neutral clusters, calculated using eqn (9), are given by the curve marked Hg_n.

The curve lies very near the expected van der Waals behaviour up to $n = 16$.

The dissociation energies of alkali-metal clusters show variations due to even–odd effects and shell closures.[16] This is not the case for mercury. The dissociation energies and the mass spectra show no trace of similar rapid variations. The dot-dashed line is from the tight-binding calculation of Pastor *et al.*[8,9] for neutral Hg clusters. It deviates from the experimental result already for $n = 3$ and shows a much smoother transition. The experiment points to a more abrupt transition between the different regions of binding.

Ionisation Potentials

Two transformations are applied to the data of fig. 4. First, the abscissa is transformed to the mean coordination number $\bar{z}$ (= mean number of nearest neighbours). The interpolation formula proposed by Bhatt and Rice,[28]

$$\bar{z} = (n - 1)/[1 + (n - 1)/12] \qquad (11)$$

was used to convert the number n of atoms to $\bar{z}$. This equation agrees well with a direct count in a finite f.c.c. lattice. Bulk Hg has a rhombohedral structure, which is a slightly distorted f.c.c. lattice.[26] These small differences are not expected to play a role in this experiment, as the clusters are relatively hot.[20] The smooth transformation between $n^{-1/3}$ and $\bar{z}$ is shown in fig. 12. Secondly, the difference δ between our data and the expected metallic behaviour

$$\delta = E_i(Hg_n) - \Phi_\infty - \frac{\alpha e^2}{8R}; \quad \alpha = 3 \text{ or } 4 \qquad (12)$$

is used as an ordinate in fig. 13. All the structures on the curves are reproducible. The value of δ vanishes if one of the two scaling laws of eqn (7) is obeyed. In the approximation used by Pastor *et al.*[8,9] one has for sufficiently large clusters: $\delta = -\varepsilon_F(n) - \Phi_\infty$, where $-\varepsilon_F(n) > 0$ is the energy of the highest occupied state of Hg_n. For $n \to \infty$, $\varepsilon_F(n)$ becomes the bulk Fermi energy. We take δ as a measure for the deviation of an electronic property from the metallic behaviour, as the

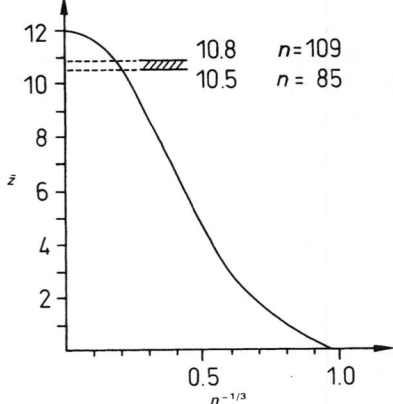

Fig. 12. The mean number of nearest neighbours, $\bar{z}$, as a function of $n^{-1/3}$. In the $\bar{z} = 10.5$–10.8 range the function is smooth, so that the sharp decrease observed in fig. 13 cannot be induced by the transformation of the abscissa.

J. CHEM. SOC. FARADAY TRANS., 1990, VOL. 86

2479

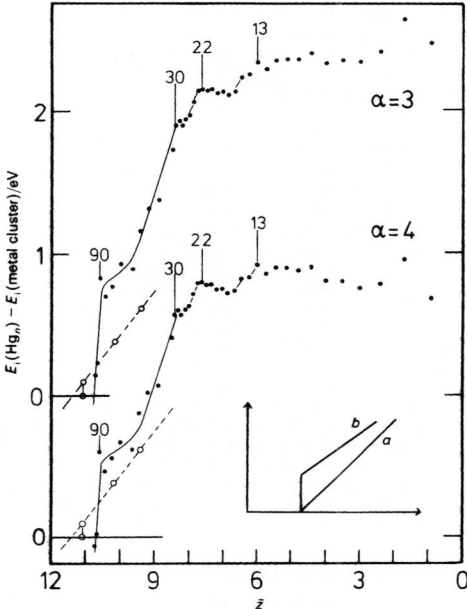

Fig. 13. The difference between the experimental ionisation potentials and the behaviour expected for the two classical scaling laws [see eqn (12)] is plotted as a function of the mean coordination number $\bar{z}$, which increases to the left. The zero of the two curves has been shifted by 1 eV. The open circles joined by the straight, dashed line are from the tight-binding calculation of ref. (8). The experimental results show a much more rapid decrease than calculated in the independent electron theory. This suggests that the transition at $\bar{z} \approx 10.6$ is indeed driven by electronic correlation, as in a Mott transition in bulk material. The insert shows the bandgap ε expected for an independent electron theory (a) and one with enough correlation (b) to induce the discontinuous Mott transition as discussed in the text. The similarity to the theoretical and experimental data near $\bar{z} = 10.6$ is obvious.

usual macroscopic probes of metal to non-metal transition are not yet applicable to free clusters.

Discussion

Bhatt and Rice[28] and Pastor et al.[8,9] have published tight-binding (Hückel) calculations for Hg clusters. In both cases transitions smoother than the experimentally observed transition between the different regions of binding are calculated. Pastor et al. state that an inclusion of the neglected correlation effects would correct this. In their study of expanded, fluid Hg Turkevich and Cohen[29] go beyond an independent-electron approximation. They model the dense vapour by placing Hg atoms on hypothetical lattices, and conclude that a 'single-particle picture even augmented with disorder, cannot explain the metal to non-metal (MNM) transition in liquid mercury'. Including correlation they find an excitonic insulator state near the MNM transition, which goes over to the metallic state via a Mott transition. We cannot presently decide on the excitonic insulator state, but we do find a structure in our data, which could possibly be interpreted similar to a Mott transition, as discussed below.

van der Waals Bonding, $n \leqslant 13$

There is strong evidence now that up to $n \approx 13$ the bonding in mercury clusters is of the van der Waals type. The evidence is: (1) the experimentally determined cohesive energies follow the behaviour calculated for pure van der Waals bonding, as shown in fig. 11, (2) the $5d \rightarrow 6s$ autoionising transition shows an excitonic character[7] incompatible with a band closing at small n, (3) the ionisation potentials behave like those of the rare-gas clusters, and (4) the mass spectrum of the negatively charged mercury clusters behaves similarly to that of Xe_n^-.

The gap Δ between the s and p bands (see fig. 1) is so large that sp hybridisation is not yet energetically favourable. Pastor et al.[8,9] calculate a deviation from van der Waals bonding also for $n < 13$, but they state that inclusion of correlation would correct this. Below it will become apparent that correlation is also important in larger clusters.

Although the ionisation potentials measured in ref. (5) and (6) are not very different from those determined here, Rademann[6] concludes that already at $n = 13$ the two bands overlap, $\Delta(13) = 0$. All experimental evidence is against this conclusion.

Covalent Bonding, $30 \leqslant n \leqslant 70$

In a simple tight-binding calculation[26] the shift and the total width of a band are proportional to $\bar{z}$. We therefore interpret the linear decrease for $30 \leqslant n \leqslant 70$ as being due to the increasing s bandwidth and the concomitant shift of the band to lower energies. At threshold, only the electron with the smallest binding energy can be ejected from the s band, and this energy scales linearly with $\bar{z}$. Around $n = 70$ the sp splitting becomes so small that transitions to metallic bonding must be considered.

Metal to Non-metal Transition, $n \geqslant 90$

Introductory Remarks

The words metal and insulator have well defined meanings for macroscopic samples, where they denote states with finite and vanishing d.c. conductivity at very low temperatures. The study of a transition between these two states of matter, the metal to non-metal (MNM) transition, is an active field of research of theoretical[30,31] and experimental[32] condensed-matter physics today. For macroscopic systems the MNM transition is observed through changes of some electronic property (electrical conductivity, optical absorption, specific heat at low temperature etc.) as a function of pressure, temperature, concentration etc. The MNM transition can be continuous, i.e. the experimental observable depends smoothly on the experimental parameters in the transition region; an independent electron theory gives such a (Wilson) type of transition. If correlation between the electrons is strong, and an independent electron theory is therefore no longer valid, the smooth transition is replaced by an abrupt one, as emphasized repeatedly by Mott[33] and discussed below. Correlation treats that part of the electron–electron repulsion which has not been taken into account by a Hartree–Fock type of calculation.

Interpretation of Data

The sharp decline that occurs between $\bar{z} = 10.57$ ($n = 90$) and $\bar{z} = 10.7$ ($n = 100$) in fig. 13 does not fit into a simple tight-binding picture. The open circles joined by the straight, dashed line in fig. 13 are data from Pastor et al.[8,9] for $n = 55$, 79 and 135. The calculated points have been joined by the expected behaviour linear (in $\bar{z}$). The line goes smoothly through the region where we find the rapid decrease of δ. The tight-binding method is an independent electron theory, which in principle cannot give a Mott transition. The theory of Turkevich and Cohen[29] includes correlation; it does find a Mott transition, but it has been performed on an infinite lattice. No such calculation has appeared for a finite cluster so far.

Two possibilities are suggested as an interpretation of the feature near $\bar{z} \approx 10.6$: (a) a sudden decrease of the nearest-neighbour distance, or (b) the analogue of a Mott transition for a finite system. A combination of both effects seems to be the most plausible explanation. As discussed above, the last three data points for large n are averages, so that the actual transition might be more rapid than can be seen in fig. 13. Note that the sharp transition is independent of which value of α is assumed in the scaling law of eqn (7) or (12).

Decrease of the Nearest-neighbour Distance

The interatomic distance r_{NN} in mercury decreases by *ca.* 10% from the dimer to the bulk.[8,9] If this were to occur rapidly, between $n = 80$ and 100, this could give rise to the effect observed. The decrease in r_{NN} leads to an increase of the hopping integrals (= overlap integrals between nearest neighbours). The bandwidths become larger and the 6s–6p band gap becomes correspondingly smaller. Such a rapid decrease as can be seen in fig. 13 must be driven by an electronic process; thus that a rapid electronic transition seems to be more plausible, accompanied by a sudden shrinkage of the interatomic distance.

Mott Transition

The electronic bands of an infinite crystal can cross as a function of some parameter (pressure, concentration *etc.*). If one treats the e^2/r_{12} term of the electron repulsion correctly, one sees that the crossing transition of the two bands is a first-order phase transition, between the metallic and insulating states. This transition was predicted by Mott[34] in 1946 and has carried his name ever since. In fact, the original Mott criterion does not predict such a transition for Hg, but the criterion was derived for monovalent atoms. For divalent mercury it should not be applicable.[35] Also the semiempirical Herzfeld criterion, which was very successful in predicting the insulator to metal transition in compressed xenon,[36] predicts bulk Hg to be non-metallic. All this seems to imply that Hg is a rather special case.

We expect for the Hg_n clusters either a similar scenario to that calculated by Turkevich and Cohen[29] for bulk Hg, probably augmented by disorder and finite size effects and/or a rapid decrease of r_{NN}, as discussed above. It seems plausible that both effects occur in the same size region. A detailed calculation is needed to clarify this situation.

Mott argued that a small number of free electrons should not be possible at a very low temperature, because a small number of electrons and ions will attract each other and will form neutral pairs. This is not the case if they are strongly screened. The number of free electrons must change discontinuously at a Mott transition. Plotted as in fig. 13 the transition indeed looks 'rapid', making Hg a possible candidate for the study of a phase transition in a finite system. The insert of fig. 13 is taken from an early paper by Mott.[37] It shows the band gap ε as a function of the inverse distance in a crystal. In an independent electron calculation, one obtains the linear rise (a). Including sufficient correlation leads to the discontinuous jump (b), *i.e.* the Mott transition. The similarity of the insert to the experimental result (which includes correlation of course) and the theoretical result (which neglects correlation), respectively, is obvious. The abscissa ($\bar{z}$ or $1/d$) is in both cases a measure of the bandwidth.

Note that what is today known as a Mott–Hubbard transition pertains to a half-filled s band.[33] The situation for Hg clusters is very different. The overlap of the full s band with the empty p band drives the transition. The low-lying d electrons make the problem no easier.

Comparison to the Data for Bulk Liquid Mercury

For the liquid/vapour interface of bulk mercury it is observed that a partial monolayer of non-metallic bound Hg atoms is adsorbed on a metallic Hg surface.[3] Related effects are observed if Hg is adsorbed on solid surfaces.[2] A similar effect is plausible for clusters. The outer Hg atoms have a lower coordination number, and their binding remains covalent, while the inner atoms are metallically bound. This has very recently been calculated by Bennemann *et al.*,[38] and might also explain the difference in the critical coordination number where the metal to non-metal transition occurs in the hot expanded fluid ($\bar{z} = 6$–7, $n = 20$–50) [see ref. (1), (28), (29)] compared to the value observed here for free clusters in vacuum ($\bar{z} \approx 10.6$–10.7, $n = 95$–100). The dielectric constant in the fluid is high, the 'clusters' in the fluid are in constant mass interchange with their surrounding. This effect together with the high temperature and high pressure could induce a metallic bond also on the outer atoms, making it behave like a metal cluster for smaller n. This process is not possible for clusters in vacuum.

Is the Kubo Criterion Applicable?

In any finite-sized metal particle, the energy levels are split. Since we expect the atoms to be mobile within the clusters, the statistical level spacing $\delta \approx 4E_F/3N$ should be applicable,[39] where E_F is the Fermi energy measured from the bottom of the conduction band (7.13 eV) and N is the number of free electrons per cluster. At the metal to non-metal transition $N \leqslant 200$, giving $\delta \geqslant 50$ meV. An estimate[12,20] of the cluster temperature gives $T \leqslant 200$ K and $kT \leqslant 18$ meV. Note that the abrupt decrease near $\bar{z} = 10.6$ in fig. 13 is a factor of 15–20 larger, showing that discreet level effects are probably unimportant for the metal to non-metal transition discussed here. Note that what might be true for electronic binding, can be very different for an electronic transport process, as recently observed.[40]

Conclusion

In summary, ionisation potentials, dissociation and cohesive energies for mercury clusters have been determined. The mass spectrum of negatively charged Hg clusters is reported. The influence of the transition from van der Waals ($n \leqslant 13$), to covalent ($30 \leqslant n \leqslant 70$) to metallic bonding ($n \geqslant 100$) is discussed. A cluster is defined to be 'metallic', if the ionisation potential behaves like that calculated for a metal sphere. The difference between the measured ionisation potential and that expected for a metallic cluster vanishes rather suddenly around $n \approx 100$ Hg atoms per cluster. Two possible interpretations are discussed, a rapid decrease of the nearest-neighbour distance and/or the analogue of a Mott transition in a finite system. Electronic correlation effects are strong; they make the experimentally observed transitions van der Waals/covalent and covalent/metallic more pronounced than calculated in an independent electron theory.

This work was supported by the Deutsche Forschungsgemeinschaft. Stimulating discussions with C. Bréchignac on the method to measure dissociation energies first implemented by her group are acknowledged. We thank O. Ches-

J. CHEM. SOC. FARADAY TRANS., 1990, VOL. 86

novsky, J. Friedel, H. v. Löhneysen and N. F. Mott for discussions or comments on the possibility of a Mott transition in Hg clusters.

References

1 E. V. Frank and F. Hensel, *Phys. Rev.*, 1966, **147**, 109; U. Even and J. Jortner, *Phys. Rev. Lett.*, 1972, **28**, 31; U. El-Hanany and W. W. Warren Jr, *Phys. Rev. Lett.*, 1975, **34**, 276; H. Uchtmann and F. Hensel, *Phys. Lett.*, 1975, **53**, 239; W. Hefner and F. Hensel, *Phys. Rev. Lett.*, 1982, **48**, 1026; H. Ikezeki, K. Schwarzenegger, A. L. Simons, A. L. Passner and S. L. McCall, *Phys. Rev. B*, 1978, **18**, 2494; O. Chesnovsky, U. Even and J. Jortner, *Solid State Commun.*, 1977, **22**, 745.
2 R. G. Jones and A. W-L. Tong, *Surf Sci.*, 1987, **188**, 87; N. K. Singh and R. C. Jones, *Chem. Phys. Lett.*, 1989, **155**, 463.
3 M. P. D'Evelyn and S. A. Rice, *J. Chem. Phys.*, 1983, **78**, 5081.
4 B. Cabauld, A. Hoareau and P. Melinon, *J. Phys. D*, 1980, **13**, 1831.
5 K. Rademann, B. Kaiser, U. Even and F. Hensel, *Phys. Rev. Lett.*, 1987, **59**, 2319; K. Rademann, B. Kaiser, T. Rech and F. Hensel, *Z. Phys. D*, 1989, **12**, 431; Proceedings of the 4. ISSPIC, Aix-en-Provence, July 1988.
6 K. Rademann, *Ber. Bunsenges. Phys. Chem.*, 1989, **93**, 653.
7 C. Bréchignac, M. Broyer, Ph. Cahuzac, G. Delacretaz, P. Labastie, J. P. Wolf and L. Wöste, *Phys. Rev. Lett.*, 1988, **60**, 275.
8 G. M. Pastor, P. Stampfli and K. H. Bennemann, *Phys. Scr.*, 1988, **38**, 623.
9 G. M. Pastor, P. Stampfli and K. H. Bennemann, *Europhys. Lett.*, 1988, **7**, 419.
10 K. Hilpert and K. Ruthardt, *Ber. Bunsenges. Phys. Chem.*, 1987, **91**, 724, and references therein.
11 S. H. Linn, C. L. Liao, C. X. Liao, J. M. Brom Jr and C. Y. Ng, *Chem. Phys. Lett.*, 1984, **105**, 645.
12 G. Tanner, *Diplomarbeit* (University of Freiburg 1989).
13 H. Haberland, *Surf Sci.*, 1985, **156**, 305; *Physics and Chemistry of Small Clusters*, ed. P. Jena, B. K. Rao and S. N. Khann, NATO ASI Series 158 (Plenum Press, New York, 1987), p. 667.
14 H. Haberland, C. Ludewigt, H-G. Schindler and D. R. Worsnop, *Surf Sci.*, 1985, **156**, 157; H. Haberland, in *The Chemical Physics of Clusters*, 107th Enrico Fermi School, ed. G. Scoles (Varenna, 1988) (North Holland, Amsterdam, 1990).
15 H. Haberland, T. Reiners and T. Kolar, *Phys. Rev. Lett.*, 1989, **63**, 1219.
16 C. Bréchignac, Ph. Cahuzac, J. Leygnier and J. Weiner, *J. Chem. Phys.*, 1989, **90**, 1493; C. Bréchignac, Ph. Cahuzac, F. Carlier, M. de Frutos and J. Leygnier, *J. Chem. Phys.*, submitted, and references therein.
17 L. S. Kassel, *J. Phys.*, 1928, **32**, 1065; P. J. Robinson and K. H. Holbrook, *Unimolecular Reactions* (Wiley Interscience, Chichester, 1972).
18 C. E. Klots, *J. Phys. Chem.*, 1988, **92**, 5864.
19 P. Engelking, *J. Chem. Phys.*, 1987, **87**, 936.
20 H. Haberland, M. Oschwald and G. Tanner, to be published.
21 M. F. Jarrold and J. E. Bower, *J. Chem. Phys.*, 1987, **87**, 5728.
22 E. Perdew, *Phys. Rev. B*, 1988, **37**, 6175; G. Makov, A. Nitzan and L. E. Brus, *J. Chem. Phys.*, 1988, **88**, 5076.
23 E. Schumacher, *Chimia*, 1988, **42**, 357.
24 B. W. van de Waal, *J. Chem. Phys.*, 1989, **90**, 3407.
25 A. Zehnacker, M. C. Duval, C. Jouvet, C. Lardeux-Dedonder, D. Solgadi, B. Soep and O. Benoist d'Azy, *J. Chem. Phys.*, 1987, **86**, 6565.
26 N. W. Ashcroft and N. D. Mermin, *Solid State Physics* (Saunders College, Philadelphia, 1976).
27 P. J. Kuntz, J. Valldorf, *Z. Phys. D.*, 1988, **8**, 195; H-U. Böhmer and S. D. Peyerimhoff, *Z. Phys. D*, 1989, **11**, 249.
28 R. N. Bhatt and T. M. Rice, *Phys. Rev. B*, 1979, **20**, 466.
29 L. A. Turkevich and M. H. Cohen, *Phys. Rev. Lett.*, 1964, **53**, 232; *J. Non-cryst. Solids*, 1984, **61&62**, 13; *J. Phys. Chem.*, 1984, **88**, 3751.
30 N. W. Ashcroft, *Science*, 1989, **340**, 345.
31 C. Castellani, G. Kotliar and P. A. Lee, *Phys. Rev. Lett.*, 1987, **59**, 323 and references therein.
32 M. Lakner and H. v. Löhneysen, *Phys. Rev. Lett.*, 1989, **63**, 648.
33 N. F. Mott, *Rev. Mod. Phys.*, 1968, **40**, 667; *Metal-Insulator Transitions* (Taylor & Francis, London, 1974); *The Metallic and Nonmetallic State of Matter* (Taylor & Francis, London, 1985).
34 N. F. Mott, *Proc. Phys. Soc.*, 1949, **62**, 416.
35 N. F. Mott, personal communication, 1989.
36 K. A. Goettel, J. H. Eggert, I. F. Silvera and W. C. Moss, *Phys. Rev. Lett.*, 1989, **62**, 665; R. Reichlin, K. E. Brister, A. K. McMahan, M. Ross, S. Martin, Y. K. Vohra and A. L. Ruoff, *Phys. Rev. Lett.*, 1989, **62**, 669.
37 N. F. Mott, *Rev. Mod. Phys.*, 1968, **40**, 677, fig. 2.
38 K. H. Bennemann, preprint (1989).
39 R. Kubo, *J. Phys. Soc. Jpn*, 1962, **17**, 975; W. P. Halperin, *Rev. Mod. Phys.*, 1986, **58**, 533.
40 P. Marquart, G. Nimtz and B. Mühlschlegel, *Solid State Commun.*, 1988, **65**, 539.

Paper 9/04127C; Received 26th September, 1989

Molecular Spectroscopy

Alan Carrington

Department of Chemistry, University of Southampton, UK SO17 1BJ

Commentary on: **The absorption spectroscopy of substances of short life**,
G. Porter, *Discuss. Faraday Soc.*, 1950, **9**, 60.

The task of choosing the most important development in spectroscopy during the past hundred years is an intriguing one. The task actually set for me was a subset, namely, to identify work originally published in a Faraday Society or Divisional Journal. The more general task is actually the easier one, and I would like to address it before moving to the more restricted exercise.

I find little difficulty in choosing the two most important developments in spectroscopy over the past century; for me they are the development of nuclear magnetic resonance (NMR) in condensed phases, and the invention of the laser. The first papers dealing with condensed phase NMR were published in the *Physical Review* (in 1946) by two groups working independently and headed by F. Bloch[1] and E. M. Purcell[2] respectively. Their work should be regarded as a development of earlier gas phase molecular beam studies by Rabi, Ramsey[3] and others, and their experiments were successful after much earlier attempts, by Gorter[4] in particular, had failed. The techniques of NMR have since undergone an extraordinary process of development, a process which shows no signs of slowing down. The applications of NMR in organic chemistry, biochemistry, and diagnostic medicine affect all of us.

The laser, which is normally operated in the visible and infrared regions of the spectrum, was preceded by its microwave analogue, the maser.[5] The invention rested heavily on our earlier understanding of the fundamental processes involved in the absorption and emission of electromagnetic radiation. Applications of laser technology now permeate much of contemporary scientific research, and life in general. Knowingly or unknowingly, our lives are affected almost every day by the laser. The original work, however, was not published in *Faraday* but in American and Russian journals.

Since I cannot actually choose the above developments, I have, after much deliberation, decided upon the invention of flash photolysis by G. Porter and R. G. W. Norrish. The original work was published in 1950 almost simultaneously in the *Proceedings of the Royal Society*[6] and the *Discussions of the Faraday Society*.[7] Several groups in different parts of the world were attempting to perform successful experiments, but Porter and Norrish, in Cambridge, were the first to succeed. Flash photolysis subsequently became the most important method for obtaining the gas phase electronic spectra of transient molecular species, and removed much of the mystery then associated with free radicals. Apart from the information gained about molecular and electronic structure, it also provided a major route to the study of gas phase kinetics involving free radical species.

I cannot say that the invention of flash photolysis directly influenced my own research; I was still a schoolboy when it was first invented. However the excitement and interest concerning free radicals, generated at the time, might well have affected my first research supervisor, Martyn Symons, and his enthusiasm was certainly transmitted to me. I have since spent my professional life studying the spectroscopy of free radicals in liquids and gases.

References

1 F. Bloch, W. W. Hansen, and M. Packard, *Phys. Rev.*, 1946, **69**, 127.
2 E. M. Purcell, H. C. Torrey, and R. V. Pound, *Phys. Rev.*, 1946, **69**, 37.

3 N. F. Ramsey, *Molecular Beams*, Oxford University Press, Oxford, 1956.
4 C. J. Gorter and L. J. F. Broer, *Physica*, 1942, **9**, 591.
5 J. P. Gordon, H. J. Zeiger, and C. H. Townes, *Phys. Rev.*, 1955, **95,** 282L.
6 G. Porter, *Proc. R. Soc. London, Ser. A*, 1950, **200**, 284.
7 G .Porter, *Discuss. Faraday Soc.*, 1950, **9**, 60.

THE ABSORPTION SPECTROSCOPY OF SUBSTANCES OF SHORT LIFE

By GEORGE PORTER

Received 17th July, 1950

The experimental methods available for the absorption spectroscopy of free radicals are critically discussed with special reference to flash photolysis and spectroscopy. Some free radical spectra which have been obtained in this way are described and values are given for the dissociation energies and vibration frequencies in the upper and lower states of the ClO, SH and SD radicals.

Most of our knowledge of the molecular structure of free radicals and similar short-lived substances is a result of the interpretation of emission spectra obtained from flames and electrical discharges. About 100 di-atomic radicals have now been recognized in this way and many of their molecular constants determined, but there are very many other radicals whose emission spectra have not been observed and about whose structure virtually nothing is known. Chemical methods do not lend themselves readily to high-speed manipulation and other physical methods such as electron diffraction are not at present applicable. The method of absorption spectroscopy is almost alone in its suitability for investigations of this kind and has many advantages ; in particular it can be used to obtain spectra which cannot be observed in emission and it has the great advantage that it makes possible the determination of concentration. Further-more, as more complex radicals are studied, it will be necessary to use

GEORGE PORTER   61

other spectral regions such as the infra-red and far infra-red where absorption techniques become essential.

The difficulties associated with free radical absorption spectroscopy are almost entirely experimental ones and in the first part of this paper a brief account is given of the limitations of the existing methods with particular reference to the flash technique developed by the author. The second part deals with a few of the radical spectra which have been obtained by this method.

Experimental Methods

There are two problems associated with the kinetic absorption spectroscopy of short-lived substances in addition to those encountered with stable molecules. Firstly, the labile molecules must be produced rapidly in high concentration, and secondly, the spectrum must be recorded in a time which is short compared with their average life.

The Preparation of Free Radicals.—The partial pressure of radicals necessary for their observation in a path of 10 cm. varies from 10^{-5} mm. for a radical such as CN with high resolving power to 10 mm. or more under less favourable conditions. In many cases, and especially in the infra-red, a high relative concentration is also necessary if the spectrum is not to be masked by that of the parent molecule.   There are four general methods available for free radical preparation and their uses for our purpose are summarized below.

THERMAL DECOMPOSITION.—This is only available when the dissociation of the radical does not occur much more readily than that of the parent molecule. It has not so far been possible to heat a gas rapidly enough to produce radicals instantaneously at high concentrations for kinetic studies and the method is limited to systems in thermodynamic equilibrium which have, however, one great advantage ; if the thermochemical constants are known the radical concentration and hence its absorption coefficients can be determined.[1, 2, 3]

CHEMICAL REACTION.—Owing to the fact that diffusion is slow compared with the rate of most radical reactions it is difficult to produce mixing rapid enough to give a long absorption path and except in the special case of flames this method has been unsuccessful.[4]   In flames also the reaction zone is narrow and sensitive methods, such as emission line reversal, are usually necessary.[5]

ELECTRICAL DISCHARGE.—By this means accurate synchronization and high percentage decomposition are readily attained and it was, until recently, the only successful method of instantaneous preparation.[6] Its main disadvantages are that it is limited to the low pressures, less than about 1 mm., at which the discharge can be passed, and the fact that the violence of the method results in the production of almost every possible molecular species making the interpretation of the kinetics a matter of great difficulty, except in the very simplest systems.

PHOTOCHEMICAL DISSOCIATION.—This method has many advantages ; the system is relatively simple and well understood, a wide range of conditions may be used and almost all radicals can be produced photochemically. With ordinary light sources, however, the concentration of radicals which can be obtained is so low that it has been impossible to detect them by means of their absorption spectra. The use of high intensity flash sources has completely overcome this difficulty and partial pressures of atoms and radicals of the order of cm. Hg have been produced, making this the most powerful of all methods of preparation.[7]

The Rapid Recording of Absorption Spectra.—If the kinetics of the radical disappearance are to be studied the exposure time must be only a fraction of the radical lifetime which may mean 10^{-4} sec. or less. With continuous sources such as the hydrogen lamp the exposure time required is several seconds at least but three methods of high speed recording are available.

[1] Bonhoeffer and Reichardt, *Z. physic. Chem.*, 1928, **139,** 75.
[2] Oldenberg, *J. Chem. Physics*, 1938, **6,** 439.
[3] White, *ibid.*, 1940, **8,** 439.
[4] Geib and Harteck, *Trans. Faraday Soc.*, 1934, **30,** 139.
[5] Kondratjew and Ziskin, *Acta Physicochim.*, 1937, **6,** 307.
[6] Oldenberg, *J. Chem. Physics*, 1934, **2,** 713.
[7] Porter, *Proc. Roy. Soc. A*, 1950, **200,** 284.

SUBSTANCES OF SHORT LIFE

ELECTRONIC METHODS.—The response time of radiation detectors such as the photocell may be as low as 10^{-8} sec. but for high speed scanning a limit is set by the statistical fluctuations in the photocurrent. The relative fluctuations F_r are given by

$$F_r = 1/(nt)^{\frac{1}{2}},$$

where n = no. of electrons/sec. from the cathode $\simeq$ no. of photons/sec. and t = time interval resolved. For example, if the rate of scanning were 1000 Å/m. sec. and the s/n ratio were 50, the high output current of 100 μA from the 1P. 28 photomultiplier would only give a resolution of 1 Å. Only if resolution or scanning speed can be sacrificed is this method suitable with present sources, and similar considerations apply to infra-red detectors.

PHOTOGRAPHIC INTEGRATION.—If ordinary light sources are used the intensity may be increased by repeating the process many times, 60,000 such exposures being used by Oldenberg for each spectrum.[8] Maeder and Miescher have used a spinning mirror arrangement so that the decay of the radical is recorded at the same time.[9] Such methods are satisfactory if the interval between exposures is short but in flash photolysis the time required for lamp cooling, and for refilling and cleaning the reaction vessel owing to the high decomposition, makes a large number of exposures impracticable.

FLASH SPECTROGRAPHY.—The very high energy which can be dissipated in a single flash, the short duration and easy synchronization coupled with the fact that it gives a very good continuous spectrum makes the high-pressure rare gas filled flash-tube ideal for this purpose. A single flash from a 70 μF condenser at 4000 V across a 15 cm. tube, lasting 5×10^{-5} sec. gives a sufficient exposure on the plate of a Littrow (Hilger E.1) spectrograph down as far as the quartz absorption limit.[7]

Experimental Limitations of the Method of Flash Photolysis.— The construction and properties of flash tubes for photochemical and spectroscopic purposes have been described previously[7] and the present limitations of the method will now be given.

ENERGY DISSIPATION.—Energies of 10,000 J per flash have been dissipated in a 1 m. long tube, corresponding to a useful output of 2×10^{21} quanta in the quartz ultra-violet region. At the higher energies occasional refilling is necessary, for example, about every 100 flashes at 5000 J, and it has been found useful to have a separate pumping system for this purpose. It has also been found that xenon filling is the least troublesome in this respect. In most photochemical systems the pressure of intermediates produced by these energies is of the order of several mm. Hg.

FLASH DURATION.—It has become clear that the flash time of 1 msec. is longer than the lifetime of some of the radicals studied and it will be necessary to use the following methods of reducing its duration.

(*a*) DECREASE OF CAPACITY.—This reduces the duration of the flash without decreasing the intensity but the total output is thereby reduced.

(*b*) INCREASE OF VOLTAGE.—Up to the highest voltage used (8000 V) the energy dissipation of the tube does not appear to vary if CV^2 is kept constant by reducing the capacity, and as there is no consequent reduction in output this is the most useful method.

(*c*) DECREASE OF TUBE RESISTANCE.—This may be accomplished by increasing the diameter or decreasing the length of the tube, the latter being approximately proportional to flash duration. It has been found that the maximum energy dissipation of the lamp is also proportional to its length and again the total output must be reduced, but the output/unit length remains unchanged and, if the length of the reaction vessel can be reduced proportionately, the *percentage* decomposition is also unchanged.

The Determination of Radical Concentrations.—The approximate estimation of the lifetime of a radical by noting the disappearance of its absorption spectrum presents no difficulty and yields much interesting qualitative information about its chemical reactions, but there are many important data, such as rate constants and absorption coefficients, which cannot be obtained without a knowledge of absolute concentrations. At present such information is limited to the few radicals which can be obtained in thermodynamic equilibrium and it is important therefore that the relative simplicity of the photochemical system makes possible, in many cases, the estimation of radical concentration indirectly from the decrease in the absorption of the parent molecule

[8] Oldenberg, *J. Chem. Physics*, 1935, **3**, 266.
[9] Maeder and Miescher, *Helv. physic. Acta*, 1942, **15**, 511 ; 1943, **16**, 503.

GEORGE PORTER   63

and subsequently the appearance of the absorption by the stable products. A complication is introduced by the changing temperature, which is unavoidable in systems containing intermediate products at high concentrations, but the effect of temperature on the absorption spectrum of stable molecules can be determined and suitable corrections applied.

The flash technique was designed primarily for the study of the kinetics of radical reactions but much information is also obtained from the interpretation of the spectra themselves, and two such spectra, of the radicals ClO and SH, will now be considered.

The ClO Radical

A preliminary report and photograph of the spectrum attributed to this radical have been given elsewhere.[7] It appears whenever chlorine is photolyzed in the presence of oxygen and has a half-life of a few milliseconds. The simple vibrational structure strongly suggests a diatomic molecule and, under the circumstances, the only possibility is the ClO radical whose occurrence in chemical reactions has frequently been postulated. Very little is known about the diatomic compounds of Group 6 of the periodic table with Group 7 and the interpretation of this spectrum is therefore of some interest.

VIBRATIONAL STRUCTURE.—The main feature of the spectrum is a series of bands with fairly sharp heads degraded to the red which is clearly a $v'' = 0$ progression and the measurements of these heads with intensities on a scale of 10 are given in Table I. Most of the measurements are accurate to better than 10 cm.$^{-1}$ except for a few bands around $v' = 18$ where the overlapping rotational structure makes the heads difficult to locate.

TABLE I.—$v'' = 0$ PROGRESSION OF ClO

v'	Int.	λ Å	ν cm.$^{-1}$	v'	Int.	λ Å	ν cm.$^{-1}$
4	1	3034·5	32945	14	7	2729·4	36627
5	2	2993·0	33402	15	7	2711·1	36874
6	3	2954·3	33839	16	6	2695·0	37095
7	5	2918·0	34261	17	5	2682·5	37267
8	5	2884·0	34664	18	5	2671·2	37425
9	6	2851·8	35056	19	5	2661·0	37569
10	8	2822·4	35421	20	5	2652·5	37689
11	8	2796·0	35755	21	3	2645·8	37784
12	10	2771·6	36070	22	2	2640·6	37859
13	8	2749·5	36360	23	2	2636·3	37920

An assignment of vibrational quantum numbers in the upper state is not possible on this information alone but a band system attributed to ClO has been obtained in emission from flames containing chlorine or methyl chloride by Pannetier and Gaydon [10] and assuming their interpretation to be correct we may proceed as follows. The average value of the second difference of the wave numbers of the $v' =$ constant progressions in the emission bands is 15 cm.$^{-1}$ and the corresponding value for the upper state, calculated from the absorption spectrum, is 22 cm.$^{-1}$. Starting with the last observed bands of the $v'' = 0$ and $v' = 0$ progressions the positions of further bands heads can be estimated, assuming linear convergence, to give the following values:

Absorption, v = 0 progression   *Emission, v' = 0 progression*
              cm.$^{-1}$ cm.$^{-1}$
Obs.   32945   Obs.   27598
           ⎧ 32466   ⎧ 28406
           ⎪ 31965   ⎪ 29227
Calc.   ⎨ 31442   Calc.   ⎨ 30067
           ⎩ 30897   ⎩ 30920

The value for the last observed emission band has been adjusted slightly and the progression starts in effect from the previous band head which is

[10] Pannetier and Gaydon, *Nature*, 1948, **161**, 242.

64 SUBSTANCES OF SHORT LIFE

probably the more accurate measurement. It will be seen that there is a coincidence within the accuracy of the extrapolations between the last two figures given and a continuation in this way as far as the observed bands of the other system shows no other coincidence within 100 cm.$^{-1}$. The vibrational quantum numbers obtained in this way are given in Table I and the interpretation is supported by intensity considerations and the close agreement between the values for ω_e' obtained from the two spectra. The value of x in the Table of Pannetier and Gaydon is seen to be 4.

These data lead to the following values for the constants of the two states with a probable error in ω_e of 3 %. Account has been taken of a small cubic term in v' in the estimation of ω_e'.

$$\text{Ground state.} \quad \omega_e'' = 868 \text{ cm.}^{-1} \qquad x_e''\omega_e'' = 7 \cdot 5 \text{ cm.}^{-1}.$$
$$\text{Upper state} \quad \omega_e' = 557 \text{ cm.}^{-1} \qquad x_e'\omega_e' = 11 \text{ cm.}^{-1}.$$
$$v_e = 31{,}077 \text{ cm.}^{-1}.$$

FURTHER DETAILS OF BAND STRUCTURE.—Owing to the extended rotational structure the chlorine isotope effect is difficult to observe, the weaker isotope system being obscured by the stronger, and it has not been possible to confirm the above assignment of quantum numbers in this way.

In addition to the bands already discussed a second progression appears at about one-quarter of the intensity, the heads showing a fairly constant separation to higher frequencies from those of the main system. The bands appear to belong to a second multiplet component, the whole forming a doublet system and over the nine measured heads in Table II the doublet splitting decreases from 200 cm.$^{-1}$ to about 185 cm.$^{-1}$.

TABLE II.—SECOND SYSTEM OF ClO

v'	λ Å	ν cm.$^{-1}$	v'	λ Å	ν cm.$^{-1}$
5	2975·0	33603	10	2807·0	35615
6	2937·6	34032	11	2781·4	35943
7	2902·0	34449	12	2758·0	36248
8	2866·7	34874	13	2735·3	36548
9	2835·7	35255	—	—	—

A full analysis of the rotational structure has not yet been carried out, the resolution being rather too low. The P and R branches appear to be the most prominent feature of the bands and the structure suggests a $^2\Pi - {^2\Pi}$ or $^2\Delta - {^2\Delta}$ transition with near case (a) coupling. The former is supported by theoretical considerations which predict a $^2\Pi$ ground state for ClO.[11]

Determination of the Dissociation Energy.—The bands of the main system are observed almost to the convergence limit and a very good extrapolation can be made, the $\Delta v - v$ plot showing slight positive curvature. The value of the dissociation energy to products in the upper state obtained in this way is 37,930 cm.$^{-1}$ or 108·4 kcal./mole, ± about 0·1 %.

To obtain the dissociation energy to unexcited products the only low-lying levels of the atoms to be considered, ignoring multiplet splitting, are the ^2P ground state of chlorine and the ^3P ground and the ^1D and ^1S excited states of oxygen, the latter lying 15,868 and 33,793 cm.$^{-1}$ above the ground state. These lead to three possible values for D_0'' and the lower one is immediately eliminated by the fact that vibrational levels of the ground state are observed which lie above it. The upper one is confirmed by a linear extrapolation of the constants for the ground state which gives a value for D_0'' of 24,680 cm.$^{-1}$. It is clear therefore that the oxygen atom is liberated from the upper level in the ^1D state and the following values for the dissociation energies can now be given unambiguously.

Ground state. Dissociates to Cl ^2P and O ^3P.
$$D_0'' = 22{,}062 \text{ cm.}^{-1} = 63 \cdot 04 \text{ kcal./mole} \pm 0 \cdot 2 \%.$$

Upper state. Dissociates to Cl ^2P and O ^1D.
$$D_0' = 7{,}010 \text{ cm.}^{-1} = 20 \cdot 0 \text{ kcal./mole} \pm 1 \%.$$

[11] Perring (private discussion).

GEORGE PORTER   65

These values assume dissociation to the lowest atomic multiplet levels in each case as the actual levels are unknown. If the atoms are liberated in higher multiplet levels the value of D_0'' would need slight adjustment but the value of D_0' is unchanged. The normal dissociation energy of ClO to chlorine $^2P_{1\frac{1}{2}}$ and oxygen $^3P_2$ depends only on the multiplet level of the Cl atom liberated from the upper state, being the ground state value given above if Cl $^2P_{1\frac{1}{2}}$ is liberated and 881 cm.$^{-1}$ less if the product is Cl $^3P_{\frac{1}{2}}$.

Discussion.—The only existing information about the dissociation energy of ClO is obtained from the spectrum of the ClO_2 molecule.[12] The predissociation at 3750 Å, corresponding to an upper limit of 76 kcal./mole, is not inconsistent with the above value but much lower energies have been suggested from interpretations of the extrapolated convergence limit.

The only existing information about the dissociation energy of ClO is obtained from the spectrum of the ClO_2 molecule.[12] The predissociation at 3750 Å, corresponding to an upper limit of 76 kcal./mole, is not inconsistent with the above value but much lower energies have been suggested from interpretations of the extrapolated convergence limit. The postulated dissociation to normal ClO and $O(^1D)$ now seems quite probable, and although there is some doubt about the heat of formation of ClO_2 it can hardly be high enough to account for this discrepancy. The explanation probably lies in the potential energy surface involved in the dissociation of polyatomic molecules which may lead to products with excess vibrational energy.

It is interesting to compare the bond energy of ClO with the average bond energy in the other chlorine oxides calculated from the heats of dissociation as follows : [13]

Cl_2O	Cl_2O_7	ClO_3	ClO_2	ClO
47·0	50·3	55·6	59·5	63·0 kcal./mole.

The influence of the odd electron in strengthening the bond is clear and the last four molecules form a group of inorganic free radicals showing increasing chemical reactivity as the odd electron becomes localized.

The formation of ClO by the reaction of Cl atoms with O_2 suggests that a radical Cl—O—O· takes part as an intermediate, no trace of other chlorine oxides such as ClO_2 being found in the spectra. Subsequent reactions are then probably

$$Cl—O—O + Cl = (Cl_2O_2 =) \ 2ClO,$$

and at higher temperatures,

$$Cl—O—O + Cl_2 = (Cl_2O_2 + Cl =) \ 2ClO + Cl.$$

Similar experiments with bromine indicate that the bromine atom does not react with O_2.

The SH and SD Radicals

The SH radical is the only diatomic hydride of the first two periods whose vibrational constants are completely unknown and one of the few common hydrides for which not even an approximate value of the dissociation energy is available. Its spectrum does not appear readily and only one band, the o—o band of the $^2\Sigma - {}^2\Pi$ transition, has ever been observed.[14] It is well known that hydrogen sulphide can be decomposed photochemically into its elements, and although other mechanisms have been proposed [15] it is probable that the primary decomposition is to H and SH.[16] For this reason the flash photolysis of H_2S was studied as a probable source of the SH radical.

---

[12] Finkelnberg and Schumacher, *Z. physic. Chem.* (Bod. Fest.), 1931, 704.
[13] Goodeve and Marsh, *J. Chem. Soc.*, 1939, 1332.
[14] Lewis and White, *Physic. Rev.*, 1939, **55**, 894. Gaydon and Whittingham, *Proc. Roy. Soc. A*, 1947, **189**, 313.
[15] Goodeve and Stein, *Trans. Faraday Soc.*, 1931, **27**, 393.
[16] Herzberg, *ibid.*, 1931, **27**, 402.

C

Hydrogen sulphide was prepared from ferric sulphide and sulphuric acid and also by the action of water on an intimate mixture of calcium sulphide and phosphorous pentoxide. It was dried over $CaCl_2$ and P_2O_5 and fractionally distilled *in vacuo*. At pressures of 40 mm. of H_2S one flash produced a partial pressure of 9 mm. H_2, sulphur being deposited on the wall, and there was no overall pressure change. Spectra were taken of the products on a Littrow (Hilger E.I.) spectrograph at increasing time intervals, after the flash. Immediately after the flash the absorption spectrum showed, in addition to the continuum of H_2S and the $^3\Sigma - {}^3\Sigma$ system of S_2, the o—o band of SH at 3236·6 Å identical with that described by Lewis and White,[14] another similar band at 3060 Å, and a diffuse band system between 3168 and 3797 Å.

TABLE III.—SH 1—o BAND

ν (cm.$^{-1}$)	Int.	Branch (J)	
32664·8	8	R_1 head	
32634·4	6	$Q_1(1\frac{1}{2})$	
32618·4	8	$Q_1(2\frac{1}{2})$. $P_1(1\frac{1}{2})$	
32601·0	5	$Q_1(3\frac{1}{2})$	
32587·9	3		$P_1(2\frac{1}{2})$
32580·2	4	$Q_1(4\frac{1}{2})$	
32555·7	6	$Q_1(5\frac{1}{2})$. $P_1(3\frac{1}{2})$	
32528·1	2	$Q_1(6\frac{1}{2})$	
32518·4	2		$P_1(4\frac{1}{2})$
32498·1	2	$Q_1(7\frac{1}{2})$	
32480·0	2		$P_1(5\frac{1}{2})$
32498·1	2	$Q_1(8\frac{1}{2})$	
32469·6	1	$Q_1(9\frac{1}{2})$	
32267·5	3	Q_2 head	

SD o—o BAND | SD 1—o BAND

ν (cm.$^{-1}$)	Int.	Branch (J)	ν (cm.$^{-1}$)	Int.	Branch (J)
30977·4	10	R_1 head	32294·0	10	R_1 head
30959·6	3	$Q_1(1\frac{1}{2})$	32277·0	4	$Q_1(1\frac{1}{2})$
30951·9	3	$Q_1(2\frac{1}{2})$	32269·2	5	$Q_1(2\frac{1}{2})$
30943·6	3	$Q_1(3\frac{1}{2})$	32260·7	3	$Q_1(3\frac{1}{2})$
30934·4	5	$Q_1(4\frac{1}{2})$	32251·1	4	$Q_1(4\frac{1}{2})$
30923·6	2	$Q_1(5\frac{1}{2})$	32238·8	8	$Q_1(5\frac{1}{2})$
30912·1	2	$Q_1(6\frac{1}{2})$	32224·3	5	$Q_1(6\frac{1}{2})$
30899·6	3	$Q_1(7\frac{1}{2})$	32211·4	5	$Q_1(7\frac{1}{2})$
30594·3	5	Q_2 head	32195·4	5	$Q_1(8\frac{1}{2})$
			32174·8	5	$Q_1(9\frac{1}{2})$
			32154·4	4	$Q_1(10\frac{1}{2})$
			31907·1	3	Q_2 head

THE 3060 Å BAND.—The rotational lines of this band appeared with greater intensity than those the o—o band, the relative intensity of the two bands always being the same, and both bands had a half-life of about 1 msec. The resolving power of the spectrograph in this region was 1·5 cm.$^{-1}$ which was not sufficient to separate the satelite $^QP_{21}$ and $^RQ_{21}$ branches from the main Q_1 and R_1 branches nor in some cases the main branch lines from each other. For this reason some of the lines were very broad and the line measurements which are given in Table III are the observed maxima with the probable quantum number assignment for the main branch lines. It is found that, apart from the different spacing owing to the higher value of $B'' - B'$, this band is identical in structure with the

3236·6 Å band and it is fairly certain that it is the 1—0 band of the same system. This was confirmed by comparing the SH spectrum with the spectrum of the SD radical obtained from D_2S when it was found that there was a small isotope shift to shorter wavelengths with SD for the 3236·6 Å band and a greater one to longer wavelengths for the 3060 Å band, the order of magnitude leaving no doubt as to the correctness of the above interpretation. The measurements of the strongest lines of the 0—0 and 1—0 bands of SD are given in Table III.

There are two unusual features about the appearance of the 1—0 band of SH ; firstly, it has a greater intensity than the 0—0 band though a careful search for further bands of the progression shows that they are absent and secondly, Lewis and White observed the weaker 0—0 band only. Both these anomalies, as well as the difficulty experienced in obtaining the bands in emission, would be explained if there were a potential energy curve leading to a lower dissociation limit which crossed the $^2\Sigma$ curve at about the second vibrational level. In this case the 1—0 band might appear stronger at low resolution owing to the broadening of the lines even if the transition were of lower probability, whilst higher transitions might be completely diffuse.

VIBRATIONAL CONSTANTS AND DISSOCIATION ENERGY.—Apart from the upper limit of 93 kcal./mole set by the above-mentioned predissociation these constants cannot be obtained from the spectrum of SH alone as at least three vibrational bands are necessary for their derivation. By using the different zero point energy of the isotopic molecule, however, another relationship is introduced which makes the calculation possible. If we assume the same force constant for the two molecules it can be shown that

$$\Delta^2\nu = (\nu_{1-0} - \nu^i_{1-0}) - (\nu_{0-0} - \nu^i_{0-0}) = \omega_e'(1 - \rho) - 2x_e'\omega_e'(1 - \rho^2)$$

where $\rho = \sqrt{\mu/\mu^i}$, ν_{1-0} is the wave number of the 1—0 band, etc. and the superscript i refers to the SD molecule. We also have

$$\Delta\nu'_{1-0} = \nu_{1-0} - \nu_{0-0} = \omega_e' - 2x_e'\omega_e'$$

and therefore if the separations of the bands are known the values of ω_e' and $x_e'\omega_e'$ can be obtained. The values $\Delta^2\nu = 471$ cm.$^{-1}$ and $\Delta\nu'_{1-0} = 1787$ cm.$^{-1}$ are obtained from the origins of the $^2\Sigma - ^2\Pi_{3/2}$ sub-bands estimated from the Q_1 branches, and substitution in the above equations gives for the $^2\Sigma$ state,

$$\omega_e' = 1950 \text{ cm.}^{-1} \quad \text{and} \quad x_e'\omega_e' = 81 \text{ cm.}^{-1}.$$

A linear extrapolation of these values gives the upper state dissociation energy $D_0' = 10,800$ cm.$^{-1}$ and in deriving a value for the normal dissociation energy two analogies with the $^2\Sigma$ state of OH, which might be expected to be very similar, will be made. The first is that the products in the upper state are H(^2S) and S(^1D) and this is fairly safe as the only alternative of S(^1S) would give a very low value for the normal dissociation energy.

The second analogy is that the linear extrapolation comes about 25 % too high and the final value obtained in this way will be assumed to have a possible error of 20 %. Applying this correction, subtracting the energy of promotion of S from the ^3P to the ^1D state and adding ν_{0-0} we get, for the normal dissociation energy, $D_0'' = 29,700$ cm.$^{-1} = 84·9$ kcal./mole with an error probably less than 5 %.

From the relation

$$\nu_{v-0} - \nu^i_{v-0} = \omega_e'(v' + \tfrac{1}{2})(1 - \rho) - x_e'\omega_e'(v' + \tfrac{1}{2})^2(1 - \rho^2)$$
$$- [\tfrac{1}{2}\omega_e''(1 - \rho) - \tfrac{1}{4}x_e''\omega_e''(1 - \rho^2)]$$

the quantity in square brackets is found to be 369·5 cm.$^{-1}$ and to find the value of ω_e'' a rough estimate of $x_e''\omega_e''$ may be made from the value of $D_0'' + 10$ % (as in OH) the term in $x_e''\omega_e''$ being small. Using the value $x_e''\omega_e'' = 52$ cm.$^{-1}$ obtained in this way we get $\omega_e'' = 2670$ cm.$^{-1}$. The corresponding values of the vibrational constants of the SD radical are $\omega_e' = 1400$ cm.$^{-1}$ and $\omega_e'' = 1910$ cm.$^{-1}$.

THE DIFFUSE BAND SYSTEM.—These bands form a regularly spaced system showing no fine structure, most of them being degraded to the red with fairly sharp heads, the measurements of which are given in Table IV. They appear and disappear with the SH bands and in all the spectra taken, using pressures of H_2S between 1 cm. and 10 cm. Hg their intensity was proportional to that of the SH bands. That they are not bands of sulphur is shown by the fact that they do not appear along with the S_2 bands in the photolysis of CS_2 or S_2Cl_2 and also that they show a shift when D_2S is used in place of H_2S. At first sight they might be another system of SH, their simple vibrational structure suggesting a diatomic molecule, but this is not supported by the isotope shift which occurs to shorter wavelengths with deuterium and is roughly the same, about 50 cm.$^{-1}$ throughout the system.

TABLE IV.—DIFFUSE BAND SYSTEM

λ Å	Int.	ν cm.$^{-1}$	λ Å	Int.	ν cm.$^{-1}$
3168·0	3	31557	3443·9	10	29028
3195·6	4	31284	2479·5	9	28732
3222·1	4	31027	3519·5	8	28405
3249·7	6	30764	3562·5	6	28062
3278·5	8	30493	3604·4	6	27736
3308·0	8	30222	3647·8	6	27406
3340·0	9	29932	3696·5	4	27045
3373·5	10	29634	3745·6	3	26691
3407·0	10	29343	3796·5	1	26332

These facts suggest that the system is that of a molecule, probably polyatomic, containing S and H atoms only. Of these the most likely are the HS_2 radical and the H_2S_2 molecule, the latter compound being quite stable but without a recorded spectrum. One might expect the spectrum of hydrogen persulphide to be entirely continuous by analogy with hydrogen peroxide and it also seems likely that some of it would survive the reaction and be detected in the products but it cannot be entirely ruled out on these grounds. The vibration frequency of something over 350 cm.$^{-1}$ is a reasonable value for the frequency of the —S—S— bond in either molecule, and the only other information which can be obtained from the spectrum is the dissociation energy which a fairly good linear extrapolation gives as 35,600 cm.$^{-1}$ or 102 kcal./mole. The energy of the —S—S— bond in S_8 is 52 kcal.[17] and if it is this bond which is broken dissociation of HS_2 to SH ($^2\Pi$) and S(^1S) gives fair agreement whereas dissociation of H_2S_2 to two SH radicals would not give this value unless there were another state of SH much lower than the $^2\Sigma$ state. Pending an investigation of the spectrum of H_2S_2 it seems more likely that the bands are those of the HS_2 radical which could be formed by the union of H atoms with the S_2 molecule.

Other Diatomic Hydrides of Group 6.—The absorption bands of the $^2\Sigma - ^2\Pi$ systems of OH and OD have been obtained very strongly by the reaction of H or D atoms, prepared by photolysing a small amount of Cl_2 or Br_2, in the presence of H_2 or D_2, with oxygen. The first three bands of the $v'' = 0$ progression appeared at high intensity and this seems to be the first report of the OD bands in absorption. A very complex system of lines, which is the same in both cases and therefore attributable to oxygen

[17] Siskin and Dyatkina, *Structure of Molecules* (Butterworth, 1950), p. 255.

alone, appears from 3000 Å to shorter wavelengths but no diffuse band system similar to that obtained with H_2S is present.

Attempts have been made to obtain the spectrum of the SeH radical in the photochemical decomposition of H_2Se but the only spectrum recorded on the plate was that of Se_2 despite the fact that the amount of decomposition was considerably greater than with H_2S. The absence of the spectrum of SeH may be explained if the predissociation occurs below the first vibrational level in this case, or if the SeH radical is chemically less stable, the latter explanation being probable in view of the decrease in stability from OH to SH, the former being observed for as long as 1/10th sec. after the flash.

The author wishes to thank Prof. R. G. W. Norrish for many helpful discussions and suggestions in connection with this work.

The University,
  Cambridge.

Magnetic Resonance

R. Freeman

Department of Chemistry, University of Cambridge, Lensfield Road, Cambridge, UK CB2 1EW

Fourier Transform Multiple-Quantum Nuclear Magnetic Resonance, Gary Drobny, Alexander Pines, Steven Sinton, Daniel P. Weitekamp and David Wemmer, *Faraday Symp. Chem. Soc.,* 1978, **13**, 49–55.

Selecting a particular paper for re-printing in a centenary edition might be regarded as a hit-or-miss procedure, but at least we can hope that the work in question has already stood the test of time. The full significance of a paper in the general scheme of research is not always obvious at the time of writing. This is partly true of this contribution by Drobny *et al.* to a Faraday Symposium held twenty-five years ago. Their main aim is to study multiple-quantum transitions in nuclear magnetic resonance systems by an indirect method. Normally these "forbidden" transitions are only observed in continuous-wave spectroscopy by applying abnormally intense radiofrequency fields, and the higher-order multiple-quantum transitions are very tricky to detect because of their extreme sensitivity to the strength of the radiofrequency field. Consequently this arcane aspect of magnetic resonance had been neglected for many years.

By allowing multiple-quantum coherence to precess during the evolution period of a two-dimensional experiment, Drobny *et al.* were able to detect its effects indirectly. This idea subsequently blossomed into the new technique of filtration through double-quantum coherence. Multiple-quantum coherence of order n possesses an n-fold sensitivity to radiofrequency phase shifts, which permits separation from the normal single-quantum coherence. This concept inspired the popular new techniques of double-quantum filtered correlation spectroscopy (DQ-COSY)[1] and the carbon–carbon backbone experiment (INADEQUATE),[2] both designed to extract useful connectivity information from undesirable interfering signals.

Occasionally the prosecution of a new experiment necessitates the introduction of new methodology, perhaps regarded as rather trivial at the time. So it was with the selected paper. In order to separate responses from the different orders of coherence, Drobny *et al.* developed "time-proportional phase incrementation" abbreviated "TPPI". The basic idea is simple, even obvious: any radiofrequency that is sampled digitally can be shifted in frequency by linearly incrementing the phase at each successive sampling point. For a digitized signal, discontinuous phase jumps behave in just the same way as a smooth rotation of the phase angle. Due to the n-fold sensitivity of n-quantum coherence to phase shifts, the detected multiple-quantum responses are shifted in frequency and thereby separated according to n, the order of coherence. Drobny *et al.* were thus able to demonstrate the complete separation of responses from all seven orders of coherence (zero through six) of benzene partially aligned in a liquid crystalline solvent. The spectra of the higher-order coherences are progressively easier to interpret, indeed, the method has been likened to selective deuteration of the (aligned) benzene molecule.

This simple phase-incrementation idea, not particularly emphasized by the authors at the time, has more recently had a considerable impact on NMR methodology. First, it was made the basis of one of the standard methods for obtaining "pure-phase" two-dimensional spectra, replacing the undesirable "phase-twist" line shape with a pure absorption-mode signal.[3] Secondly, it has provided a neat way to generate an extensive array of simultaneous soft radiofrequency pulses covering an entire NMR spectrum.   Selective excitation employs a regular sequence of radiofrequency pulses of small flip angle.   By ramping the phase at different rates and combining the results, an entire spectrum of irradiation frequencies can be generated.   Each element of the array can be used to excite an NMR response, and the individual responses can be separated by an encoding scheme based on a Hadamard matrix.[4] An extension of this concept of "Hadamard spectroscopy" can speed up the gathering of multidimensional NMR data by orders of magnitude, rendering these spectra accessible in a reasonable time.[5]

Both aspects of this paper by Drobny *et al.* have had a wide impact, perhaps not fully appreciated at the time, and the seemingly "obvious" trick of linear phase incrementation has influenced a flurry of new experiments.   It certainly deserves to be reprinted.

References

1 U. Piantini, O. W. Sørensen, and R. R. Ernst, *J. Am. Chem. Soc.*, 1982, **104**, 6800.
2 A. Bax, R. Freeman and T. A. Frenkiel, *J. Am. Chem. Soc.*, 1981, **103**, 2102.
3 D. Marion and K. Wüthrich, *Biochem. Biophys. Res. Commun.*, 1983, **113**, 967.
4 E. Kupče and R. Freeman, *J. Magn. Reson.*, 2003, **162**, 158.
5 E. Kupče and R. Freeman, *J. Biomol. NMR*, 2003, **25**, 349.

Fourier Transform Multiple Quantum Nuclear Magnetic Resonance

By Gary Drobny, Alexander Pines, Steven Sinton,
Daniel P. Weitekamp and David Wemmer

Department of Chemistry,
University of California, Berkeley, California, U.S.A.

Received 18th December, 1978

The excitation and detection of multiple quantum transitions in systems of coupled spins offers, among other advantages, an increase in resolution over single quantum n.m.r. since the number of lines decreases as the order of the transition increases. This paper reviews the motivation for detecting multiple quantum transitions by a Fourier transform experiment and describes an experimental approach to high resolution multiple quantum spectra in dipolar systems along with results on some protonated liquid crystal systems. A simple operator formalism for the essential features of the time development is presented and some applications in progress are discussed.

The energy level diagram of a system of coupled spins 1/2 in high field is shown schematically in fig. 1. The eigenstates are grouped according to Zeeman quantum number m_i with smaller differences in energy within a Zeeman manifold due to the couplings between spins and the chemical shifts. For any eigenstate $|i\rangle$ of the spin Hamiltonian H (in frequency units)

$$\begin{aligned} H \,|\, i \rangle &= \omega_i \,|\, i \rangle \\ I_z \,|\, i \rangle &= m_i \,|\, i \rangle. \end{aligned} \tag{1}$$

The single quantum selection rule[1] of the low power c.w. experiment and the one dimensional Fourier transform experiment arises because $\langle i \,|\, I_x \,|\, j \rangle$ vanishes unless

$$q_{ij} = m_i - m_j = \pm 1. \tag{2}$$

Simple combinatorial considerations show that the number of eigenstates decreases as $|m_i|$ increases and the number of transitions decreases as $|q_{ij}|$ increases. The highest order transition possible is the single transition with $|q| = 2I$ where I is the total spin. For a system of N spins 1/2, transitions up to order N are possible.

Detection of multiple quantum transitions in c.w. experiments is well known.[2-4] Extension to high order transitions is not promising, since the transitions observed are a sensitive function of r.f. field strength. This leads to difficult spectral simulations and experimental problems of saturation and sample heating.

The alternative time domain experiment is the determination of multiple quantum transition frequencies by following the time development of multiple quantum coherences point by point.[5-7] This work treats a class of such multidimensional

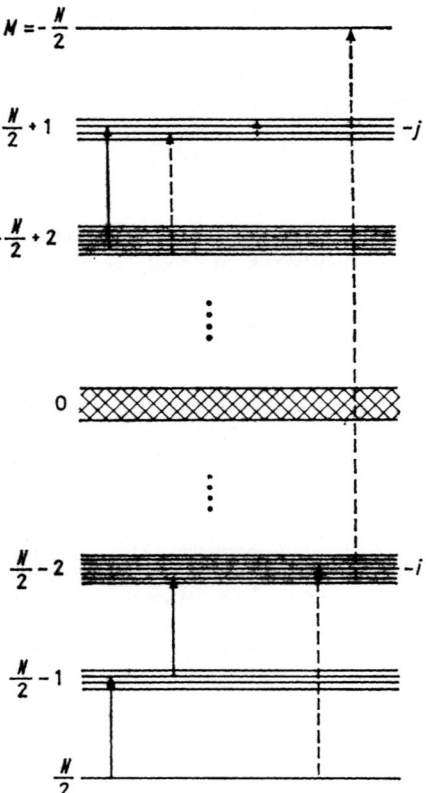

Fig. 1.—Schematic representation of the high field energy level diagram of coupled spins 1/2. Broken arrows indicate the forbidden types of transition observed in Fourier transform multiple quantum experiments.

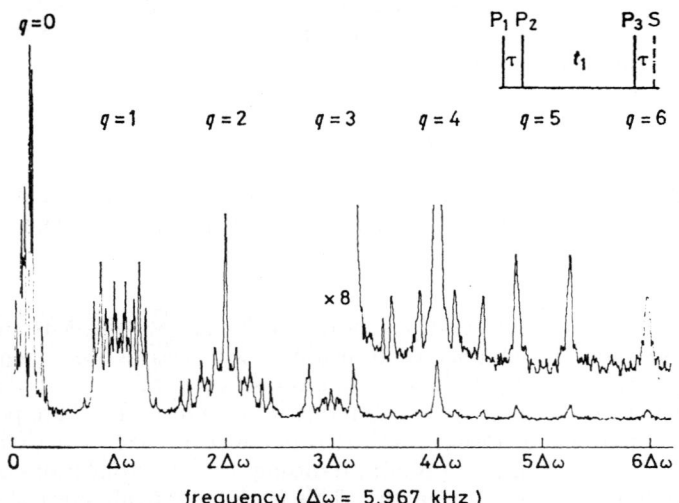

Fig. 2.—Multiple quantum spectrum of benzene (15 mol %) in *p*-ethoxybenzylidene-n-butylaniline (EBBA) at 20 °C. The three pulse sequence was $P1 = \pi/2_x$, $P2 = \pi/2_{\bar{x}}$, $P3 = \pi/2_x$. The magnitude spectra obtained for 11 values of τ spaced at 0.1 ms intervals from 9.6 to 10.7 ms were added. The value of t_1 ranged from 0 to 13.824 ms in 13.5 μs increments for each τ. A single sample point was taken at $t_2 = \tau$ after P_3. One half of a symmetrical spectrum is shown.

DROBNY, PINES, SINTON, WEITEKAMP AND WEMMER 51

experiments in which the irradiation consists of pulses at the Larmor frequency. Time proportional incrementation of the r.f. phase (TPPI) allows separate determination of the spectra of all orders free from effects of magnet inhomogeneity.

EXPERIMENTAL

The spectrometer is of pulsed Fourier transform design with super-conducting magnet (Bruker) operating in persistent mode at a proton frequency of 185 MHz. Phase shifting was performed at 185 MHz by a digitally controlled device (DAICO 100D0898) under control of the pulse programmer.

Samples were approximately 400 mg sealed in 6 mm glass tubing after degassing by repeated freezing and evacuation. All observations are in the nematic phase. Synthesis of 4-cyano-4'-[^{2}H]$_{11}$pentyl-biphenyl was by the procedure of Gray and Mosley.[8]

RESULTS

The spectrum of benzene dissolved in a liquid crystal served as a prototype in the development of the single quantum n.m.r. of complex spin systems in ordered phases.[9] The multiple quantum spectrum of ordered benzene is shown in fig. 2. The resolution is limited by magnetic homogeneity and the inhomogeneous linewidth is proportional to $|q|$.

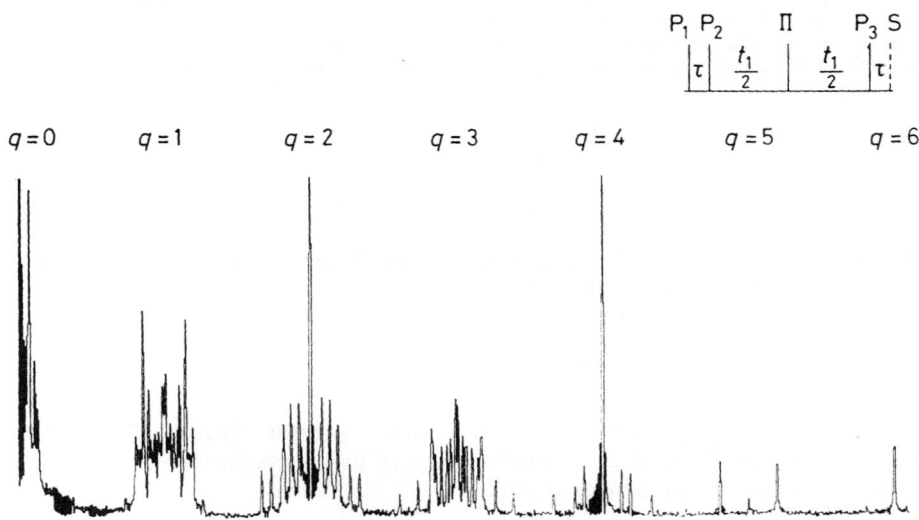

FIG. 3.—Multiple quantum spectrum of benzene at 22 °C and TPPI pulse sequence. The sample is the same as in fig. 2. The pulses are P1 = $\pi/2_\varphi$, P2 = $\pi/2_{\bar\varphi}$ and P3 = $\pi/2_x$, where $\varphi = \Delta\omega t_1$. The increment in φ was 29.5 degrees and the increment in t^1 was 10 μs for each of 1024 points. The magnitude spectra for eight values of τ between 9 and 12.5 μs were added. The magnetization was sampled at $t = \tau$.

The spectrum of fig. 3 demonstrates the use of the spin echo to remove inhomogeneous line broadening and the use of time proportional phase increments (TPPI) to restore the offset. Resolution is limited by truncation of the multiple quantum free induction decay and scale of reproduction. The actual linewidth is less than 2 Hz for all orders and suffices to resolve all allowed transitions of all orders.

An application of the TPPI method to the eight proton system of an alkyl-deuterated cyano biphenyl liquid crystal is shown in fig. 4. All eight orders are observed. Resolution is limited by truncation. Actual linewidths are <100 Hz in a spectral width of ≈ 40 kHz for each order.

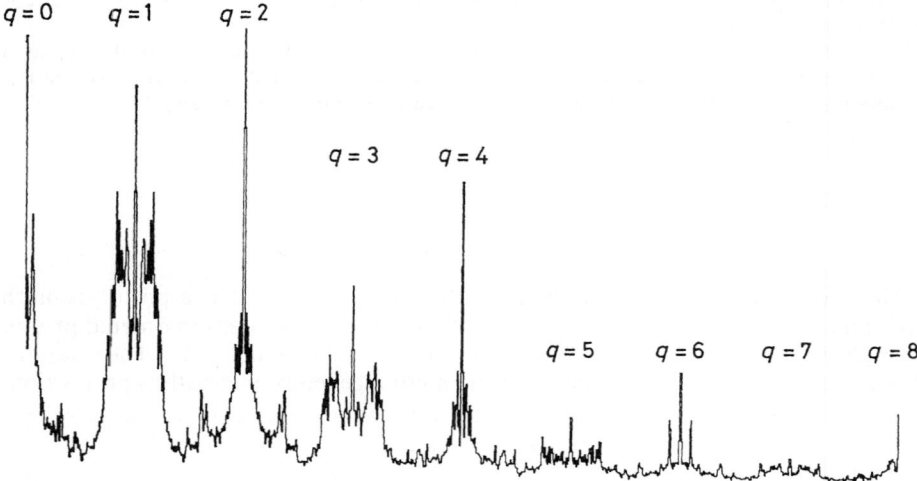

FIG. 4.—Proton multiple quantum spectrum of 4-cyano-4′-[^{2}H$_{11}$]pentyl-biphenyl at 23.3 °C by TPPI. The increments are 22.5 degrees in φ and 1.5 μs in t_1 for each of 1024 points. The time τ took five values between 0.5 and 1.0 ms and the magnetization was sampled at 64 intervals of 5 μs starting at $t_2 = (\tau + 0.1)$ ms.

DISCUSSION

THREE PULSE EXPERIMENT

The time development of the spin system during the pulse sequences of fig. 2 and 3 is conveniently discussed in terms of a spherical tensor operator expansion of the density matrix. For any time,

$$\rho(t) = \sum_{k,\alpha,q} C_q^{k\alpha}(t) T_q^{k\alpha} \tag{3}$$

where T_q^k is the qth component of a spherical tensor operator of rank k.[10] The label α completes the specification of a complete basis of tensor operators.

The initial equilibrium density matrix is

$$\rho(0) = C_0^{11}(0) I_z = C_0^{11}(0) T_0^{11} \tag{4}$$

and immediately after a $\pi/2_y$ pulse ($\omega_1 \gg H$, $\omega_1 t_{P1} = \pi/2$)

$$\rho(t_{P1}) = -C_0^{11}(0) I_x = C_1^{11}(t_{P1}) T_1^{11} + C_{-1}^{11}(t_{P1}) T_{-1}^{11}. \tag{5}$$

The Zeeman quantum number q is conserved during evolution under a high field Hamiltonian. Thus neglecting relaxation

$$\rho(t) = \sum_{k,\alpha,\pm 1} C_q^{k\alpha}(t) T_q^{k\alpha} \qquad (6)$$

at any time t after the initial $\pi/2$ pulse and no multiple quantum coherence has been created. After a period of time on the order of the inverse of the coupling frequencies, terms with $k \leqslant 2I$ will be present. A second strong pulse may then rotate $T_{\pm 1}^{k\alpha}$ into $T_q^{k\alpha}$ with $-k \leqslant q \leqslant k$. Thus at time τ, after the second pulse, all orders with $-2I \leqslant q \leqslant 2I$ have, in general, been created and evolve during the time t_1 at the eigenfrequencies $\omega_{ij} = \omega_i - \omega_j$.

If the only observable measured is the transverse magnetization corresponding in the rotating frame to the $|q| = 1$ operators I_x and I_y, it is not possible to follow the evolution of orders with $|q| \neq 1$ directly. Rather, a third pulse at time t_1 after the second pulse is needed to rotate the various tensor components back to $T_{\pm 1}^{k\alpha}$. These may then evolve into the signal observed at time t_2 after this detection pulse.

The signal may be written then as

$$S(\tau, t_1, t_2) = C_0^{11}(0)\mathrm{Tr}[I_+\rho(\tau + t_1 + t_2)]. \qquad (7)$$

Viewed as a function of t_1 this is the multiple quantum free induction decay. It is collected pointwise by variation of t_1 on successive shots and is an example of a multi-dimensional n.m.r. experiment.

SEPARATION OF ORDERS BY FREQUENCY OFFSET

In order to obtain an increase in resolution over the single quantum experiments it is necessary to separate the spectra of different order. Consider the decomposition of the density matrix at time $\tau(t_1 = 0)$ into Zeeman components

$$\rho(\tau) = \sum_{q=-2I}^{2I} \rho_q(\tau). \qquad (8)$$

By construction, $\rho_q(\tau) = \sum_{k\alpha} C_q^{k\alpha}(\tau) T_q^{k\alpha}$ and it follows from the definition of the spherical tensor operators that

$$\rho_q(\tau, \varphi) = \exp(i\varphi I_z)\rho_q(\tau) \exp(-i\varphi I_z) = \exp(iq\varphi)\rho_q(\tau). \qquad (9)$$

If we let $\varphi = \Delta\omega t_1$, then eqn (9) corresponds to a calculation of the effect of a frequency offset term $-\Delta\omega I_z$ in the rotating frame Hamiltonian. Thus a modulation by $\exp(iq\Delta\omega t_1)$ is caused by the offset, shifting the spectrum to $q\Delta\omega$ as seen in fig. 2.

The difficulty with this approach to separation of orders is that the effect of magnet inhomogeneity is proportional to $|q|$. The sample volume at position r with offset $\Delta\omega(r)$ contributes a term $\rho_q(\tau, t_1, r) = \exp(iq\Delta\omega(r)t_1)\rho_q(\tau + t_1)$ and integration over r leads to damping of the coherence with a time inversely proportional to $|q|$.

HIGH RESOLUTION SPECTRA BY TIME PROPORTIONAL PHASE INCREMENTS (TPPI)

The Hahn spin echo technique may be used to remove the inhomogeneous broadening. A π pulse at $t_1/2$ changes q to $-q$, allowing a refocusing of the coherence at t_1 for all orders. However, this π pulse also removes the separation of the orders which was a result of the frequency offset.

A solution to this dilemma is to view eqn (9) not as the effect of a frequency offset, but as a shift of the r.f. phase. In particular, consider the effect of preparing the system with pulses of phase φ and $\bar{\varphi} = \varphi + \pi$ at time zero and τ, respectively, with $I_\varphi = I_x \cos \varphi - I_y \sin \varphi$. Then

$$
\begin{aligned}
\rho(\tau, \varphi) &= \exp{(i\theta I_\varphi)} \exp{(-iH\tau)} \exp{(-i\theta' I_\varphi)} \times \\
&\quad \rho(0) \exp{(i\theta' I_\varphi)} \exp{(iH\tau)} \exp{(-i\theta I_\varphi)} \\
&= \exp{(i\varphi I_z)} \exp{(i\theta I_x)} \exp{(-iH\tau)} \exp{(-i\theta' I_x)} \times \\
&\quad \rho(0) \exp{(i\theta' I_x)} \exp{(iH\tau)} \exp{(-i\theta I_x)} \exp{(-i\varphi I_z)} \\
&= \exp{(i\varphi I_z)} \rho(\tau) \exp{(-i\varphi I_z)} \\
&= \sum_q \exp{(iq\varphi)} \rho_q(\tau).
\end{aligned}
\tag{10}
$$

The result is that again

$$
\rho_q(\tau, \varphi) = \exp{(iq\varphi)} \rho_q(\tau)
$$

but this modulation is now an artifact of the r.f. phase of the first two pulses and does not depend on the evolution of the system during t_1. However, since each point t_1 of the signal is collected separately, we may set $\varphi = \Delta\omega t_1$ for some parameter $\Delta\omega$, thereby recovering an apparent offset.

The actual evolution during t_1 may include an echo pulse to remove the effect of field inhomogeneity, as in fig. 3, or a train of π pulses to remove the effect of small chemical shifts and heteronuclear couplings from the dipolar spectra.

This TPPI technique may be compared to the method of phase Fourier transformation (PFT) discussed elsewhere.[11,12] The PFT method generates a signal array $S(\tau, t_1, t_2, \varphi)$ by repeating the experiment for each t_1 and φ, where again φ describes the phase of the preparation pulses as in eqn (10). Fourier transformation with respect to phase with q as the conjugate variable separates the orders. A second Fourier transformation with respect to t_1 gives the spectra. In the TPPI experiment the phase and time dimensions are collapsed into one by the relation $\varphi = \Delta\omega t_1$. A single time Fourier transformation gives the spectrum of each order with apparent offset $q\Delta\omega$ as in fig. 3 and fig. 4.

MULTIPLE QUANTUM SPECTROSCOPY OF LIQUID CRYSTALS

Although the spectroscopy discussed is of general applicability, it is worthwhile to note the particular suitability of the m.f.t.n.m.r. method to the study of liquid crystals. The diffusion present in such systems sharply reduces the intermolecular dipolar couplings. This allows one to obtain resolved spectra reflecting only the intramolecular couplings without dilution of the spin system. Combinatorial arguments suggest that it suffices in general to analyse only the $|q| = N - 2$ and $N - 1$ spectra to determine all couplings in a system of N spins 1/2. Since these transitions involve only the relatively small $|m| = \dfrac{N}{2}, \dfrac{N}{2} - 1$ and $\dfrac{N}{2} - 2$ Zeeman manifolds of

the basis of kets, the diagonalizations are simpler than those needed for single quantum spectroscopy. Should the study of mixtures be of interest, partial deuteration of the background species will reduce its contribution to the high order spectra more rapidly than to the single quantum spectrum and thus the requirement for isotopically pure synthesis is reduced.

Applications in progress include the configurational analysis of both ring and chain regions of liquid crystals, the study of relaxation effects in multiple quantum spectra and improvement of signal to noise by removal of the dipolar Hamiltonian during detection and sampling of magnetization at many times t_2.

We would like to acknowledge the help of Mr. Sidney Wolfe in synthesis of the liquid crystals and of Terry Judson in preparation of the manuscript. Support for this work was by the Division of Materials Sciences, Office of Basic Energy Sciences, U.S. Department of Energy.

D.P.W. held a Predoctoral National Science Foundation Fellowship.

[1] I. J. Lowe and R. E. Norberg, *Phys. Rev.*, 1957, **107**, 46.
[2] W. A. Anderson, *Phys. Rev.*, 1956, **104**, 850.
[3] J. I. Kaplan and S. Meiboom, *Phys. Rev.*, 1957, **106**, 499.
[4] K. M. Worvill, *J. Mag. Res.*, 1975, **18**, 217.
[5] (a) H. Hatanaka and T. Hashi, *J. Phys. Soc. Japan*, 1975, **39**, 1139.
  (b) H. Hatanaka, T. Terao, and T. Hashi, *J. Phys. Soc. Japan*, 1975, **39**, 835.
[6] W. P. Aue, E. Bartholdi and R. R. Ernst; *J. Chem. Phys.*, 1976, **64**, 2229.
[7] (a) S. Vega, T. W. Shattuck and A. Pines, *Phys. Rev. Letters*, 1976, **37**, 43.
  (b) S. Vega and A. Pines, *J. Chem. Phys.*, 1977, **66**, 5624.
[8] G. W. Gray and A. Mosley, *Mol. Cryst. Liq. Cryst.*, 1976, **35**, 71.
[9] A. Saupe, *Z. Naturforsch*, 1965, **20a**, 572.
[10] A. R. Edmonds, *Angular Momentum in Quantum Mechanics* (Princeton University Press, Princeton, N.J.; 2nd Edn. 1960), Chap. V, p. 68.
[11] A. Wokaun and R. F. Ernst, *Chem. Phys. Letters*, 1977, **52**, 407.
[12] A. Pines, D. Wemmer, J. Tang and S. Sinton, *Bull. Amer. Phys. Soc.*, 1978, **23**, 21.

Quantum Chemistry

Nicholas C. Handy

Department of Chemistry, University of Cambridge, Lensfield Road, Cambridge UK CB2 1EW

Commentary on: **Independent Assessments of the Accuracy of Correlated Wave Functions for Many-Electron Systems**, S. F. Boys, *Symp. Faraday Soc.*, 1968, **2**

A Symposium on Molecular Wave Functions was held at the Royal Institution on December 12–13 1968, attended by 144 members. Amongst these was Samuel Francis (Frank) Boys, the pioneer of molecular wave function calculations using the computer. Although Boys is exceedingly well-known for his work on (a) Gaussian basis function integrals,[1] (b) configuration interaction,[2] (c) localised orbitals,[3] (d) the transcorrelated wave function[4] and (e) the Counterpoise Correction,[5] he only published 38 papers (he died at the age of 60 in 1972).[6] Here is one of his papers, which demonstrates Boys' foresight in the field.

It is written in his 'nearly impossible' style: he invented his own language for many of the standard items in the subject, such as 'detors', which are 'determinants of orthonormal orbitals'. He introduces Definitions and gives Theorems, an approach which was not attractive to the students who attended his Mathematics Part III course 'The Quantum Theory of Molecules' in Cambridge.

Now let us look at the paper. Eqn. (1) gives the form of the transcorrelated wave function $C\Phi$, where $C = \Pi_{i>j} f(r_i, r_j)$ is a Jastrow factor, and Φ is a determinant. This compact wave function includes the effects of electron correlation through the introduction of r_{ij} in C. The form $C\Phi$ is taken as the trial wave function in quantum Monte Carlo (QMC) molecular computations today. Indeed the explicit form for $f(r_i, r_j)$ is most often used by the QMC community. The transcorrelated wave function was obtained by solving $(C^{-1}HC - W)\Phi = 0$, which Boys called the transcorrelation wave equation. Because $C^{-1}HC$ is a non-Hermitian operator, it was important to devise independent assessments of the accuracy of the wavefunction $C\Phi$.

Frost was probably the first to suggest an examination of the variance through numerical procedures. Boys suggests a numerical investigation of the variance d^2 for the transcorrelation equation, as a means of examining which of many transcorrelated wavefunctions was the more accurate. The essence of the paper was to use the fact that $C^{-1}HC$ contains at most three-electron operators, and he proved that it is possible to compute the variance through a numerical scheme which only increases as N^3, where N is the number of electrons.

There are many forward-looking aspects. In the QMC approach the variance is routinely calculated (although not using the scheme of this paper). In the later parts of his career Boys was convinced of the importance of numerical techniques: here we see mention of numerical integration in multi-dimensional space. Boys was impressed by the fact that he had devised a scheme which increased in cost as N^3, against the expected N^4 (much effort today is spent today on reducing the order of cost). At the end of the paper Boys discusses whether it will be possible to transfer the form of transcorrelated wave functions from one molecule to another: using localized orbitals in the determinant, and directly transferring the Jastrow factor. Computational chemists remain intrigued by this idea.

I was extremely fortunate to be a research student of S. F. Boys; this paper is an excellent demonstration of the considerable thought which lay behind all of his innovative ideas in the early days of computational quantum chemistry.

References

1 S. F. Boys, *Proc. R. Soc. London, Ser. A*, 1950, **200.** 542.
2 J. M. Foster and S. F. Boys, *Rev. Mod. Phys.*, 1960, **32**, 305.
3 J. M. Foster and S. F. Boys, *Rev. Mod. Phys.*, 1960, **32**, 300.
4 S. F. Boys and N. C. Handy, *Proc. R. Soc. London, Ser. A,* 1969, **310**, 63.
5 S. F. Boys and F. Bernardi, *Mol. Phys.*, 1970, **19**, 553.
6 C. A. Coulson, *Biogr. Mem. Fellows. R. Soc.*, 1973, **19**, 95.

Independent Assessments of the Accuracy of Correlated Wave Functions for Many-Electron Systems

By S. F. Boys

Theoretical Chemistry Department, University Chemical Laboratory, Cambridge

Received 11th September, 1968

The assessment of the accuracy of fully correlated wavefunctions by means of variance methods requires computation which only varies as N^3, a lower power of N, the number of electrons, than had been expected. This seems to be dependent on an indirect approach first constructed for a transcorrelated method. This means that various different variance tests could be used for the assessment of the accuracy of wavefunctions calculated by the transcorrelated method developed by Handy and Boys. These would require much less equipment in programmes and computer facilities, than the original calculations of such wavefunctions. Supplementary investigations on correlated wavefunctions at this level might make possible a whole variety of informative experiments on very exact wavefunctions and energies.

1. INTRODUCTION

This investigation was made to try to find the simplest method of testing the accuracy of a correlated wavefunction of the form

$$C\Phi \equiv \prod_{i>j} f(\mathbf{r}_i, \mathbf{r}_j)\Phi \equiv C\mathscr{A}\phi_1\phi_2\ldots\ldots\phi_N, \tag{1}$$

where Φ is detor of space spins orbitals ϕ. This would be valuable because Handy and Boys [1] have developed a method applicable to all molecules which has already given an energy value for LiH of $-8\cdot063$ a.u. compared with the experimental $-8\cdot070$ a.u. and the SCF value $-7\cdot987$. It is likely that many such wavefunctions will be obtained but there are some special uncertainties because the convergence is not monotonic. The result of examining the briefest test which can be applied to these wavefunctions, by other workers with much less facilities than necessary for the original calculation, has turned out to be the variance type of criterion. However, in the examination of this from a different approach, in which this result was not foreseen, a method of evaluation has been developed which appears to increase only as N^3, whereas the shortest direct method which had been found for the variance increased as N^4. N here denotes the number of electrons. Hence this method of calculation will be reported although it can be regarded as a special approach to the methods which have already been used by others (Conroy,[2] Frost [3]) for systems of few electrons. It is hoped that by these techniques, it may be used for much larger systems.

The guiding principle was to obtain the variance for the transcorrelation wave equation

$$(C^{-1}HC - W)\Phi = 0, \tag{2}$$

solved by Handy and Boys instead of that for the simple Schrödinger equation,

$$(H - W)\psi = 0. \tag{3}$$

Actually at the final stage of the calculation these can all be changed into each other and into other types of variance.

For an equation

$$(\bar{H} - W)\psi = 0 \tag{4}$$

the variance of an approximate solution ψ' is d^2, obtained by

$$\bar{W} = \langle \psi' \mid \bar{H} \mid \psi' \rangle / \langle \psi' \mid \psi' \rangle.$$
$$d^2 = \langle (\bar{H} - \bar{W})\psi' \mid (\bar{H} - \bar{W})\psi' \rangle / \langle \psi' \mid \psi' \rangle. \tag{5}$$

If $d = 0$, the ψ' must be a true solution of (4) and ψ' with small d can be regarded as a high accuracy approximation to the true solution.

For $\bar{H}$ we can write either the Schrödinger H or the transcorrelated operator $C^{-1}HC$ and the convergence conditions are the same. It was the use of the latter which was originally found to give a shorter computation, and then it was found that this could be changed into the ordinary variance. If the integration is now performed by taking the values of the functions at some many-electron space-spin points R_I and by the multiplication of these values by suitable weights h_I, then approximate values of all the above integrals can be obtained. This will be referred to as the method of numerical integration. If the values at these points are represented by E_I, F_I with

$$E_I = \bar{H}\psi'(\mathbf{R}_I) ; \quad F_I = \psi'(\mathbf{R}_I), \tag{6}$$

then the variance may be written as

$$\bar{W} = \sum_I h_I E_I F_I / \sum_I h_I F_I^2,$$
$$d^2 = \sum_I h_I E_I^2 / \sum_I h_I F_I^2 - \bar{W}^2. \tag{7}$$

In the next section the values of E_I and F_I will be obtained for the transcorrelated operator by a method which gives the N^3 dependence. After that stage it is not necessary to restrict the analysis to the transcorrelated case because the value of E corresponding to the ordinary variance is obtained by

$$E_I^S = C(\mathbf{R}_I)E_I^T; \; F_I^S = C(\mathbf{R}_I)F_I^T, \tag{8}$$

where E_I^S denotes the E_I for the Schrödinger equation and E_I^T the corresponding transcorrelated quantity. These inserted in eqn. (7) give the corresponding variance.

If any weighting function p^2 is used to form another variance integral such as

$$d^2 = \langle p(\bar{H} - \bar{W})\psi' \mid p(\bar{H} - \bar{W})\psi' \rangle / \langle \psi' \mid \psi' \rangle, \tag{9}$$

this can be evaluated by using

$$E_I^p = p_I E_I; \quad F_I^p = p_I F_I,$$

in eqn. (5)

The essential property of all these variances is that they can only be zero if the wave equation is satisfied at all $(\mathbf{R}_I)$. In practice, the valuable use for these is to assess two approximate wavefunctions such as $C_A\Phi_A$ and $C_B\Phi_B$. These could be results obtained either by different workers or in one calculation at different stages of the iteration. The comparison of two such results by an independent method would be very informative. If

$$d_A^2 > 2d_B^2,$$

then it would generally be reasonable to assume $C_B\Phi_B$ and W_B would be the more accurate solution.

2. THE DETERMINATION OF THE TRANSCORRELATED AND DIRECT VARIANCES FOR MANY ELECTRONS

There are three possible definitions of the quantity which it is convenient to call the transcorrelated variance. The first given in definition (A) defines it in effect as the variance of the operator $C^{-1}HC$ in what would be considered the orthodox way for this operator. This is the definition which is simplest in concept and corresponds to the direct approach of sampling at many points. The second definition is formulated to give exactly the same numerical value but in a form which is the result of the calculation of the energy by a particular numerical integration grid. The third definition is a transform of the latter which has the valuable property that it can be reduced to three electron sums by the well-known detor integral theorem. It is considered that this latter is important because only a multiple of N^3 basic operations are required in the computation whereas the direct approach appears to lead to N^4.

It is necessary to define the integration points in electron space-spin variables. If the spin variable is not included at this stage a useful aspect is omitted. These four definitions will be given as compactly as possible and, after each, the relevant identity will be established.

DEFINITION

A WEIGHTED POINT SYSTEMS FOR N ELECTRONS. A set of values of $(h_I, X_{Ii}, Y_{Ii}, Z_{Ii} V_{Ii})$ for $i = 1, 2, \ldots$. N and $I = 1$ to M where M can be finite, or infinite, and where the spin variables V_{Ii} can only take the values $+\frac{1}{2}$ or $-\frac{1}{2}$ will be called an N-electron point system. Such a set may frequently be obtained by selecting for each value of I a sub-set of N points out of a set (x_i, y_i, z_i, v_i) for a single electron but this does not affect the principle of the way in which the N-space points enter the theory. An example is given in eqn. (27) and (28).

It will be assumed that there are no values of $i \neq j$ for any value of I where

$$X_{Ii} = X_{Ij}, \; Y_{Ii} = Y_{Ij}, \; Z_{Ii} = Z_{Ij}, \tag{10}$$

When $V_{Ii} = V_{Ij}$ this is not a deliberate exclusion but merely a matter that such a point would give a zero value in any antisymmetric functions and hence zero in all the integrals which occur here. For a case $V_{Ii} \neq V_{Ij}$, this is a deliberate exclusion because, if this is not made, the points with $r_{12} = 0$ can easily occur with disproportionate probability. Since these are the points at which cusps occur this increases the errors due to the use of finite sets of points.

For the set of points associated with a given value of I we shall define the repeated delta operation which selects the value of a function at these points. Thus if

$$Q_I = \delta(x_1, X_{I1})\delta(y_1, Y_{I1})\delta(z_1, Z_{I1})\delta(v_1, V_{I1})\delta(x_2, X_{I2}) \ldots .$$
$$= \prod_i \delta(x_i, X_{Ii})\delta(y_i, Y_{Ii})\delta(z_i, Z_{Ii})\delta(v_i, V_{Ii}). \tag{11}$$

Hence for any function F

$$F(X_{Ii}, Y_{Ii}, Z_{Ii}, V_{Ii}) = \langle Q_I \mid F \rangle, \tag{12}$$

and from this it follows for any G that

$$\langle Q_I \mid FG \rangle = \langle Q_I \mid F \rangle \langle Q_I \mid G \rangle. \tag{13}$$

Also if the whole I set is sufficient and has correct weights to perform a numerical integration with an error of order $O(\mu_Q)$

$$\sum_I h_I \langle Q_I \mid F \mid G \rangle = \langle F \mid G \rangle + O(\mu_Q). \tag{14}$$

DEFINITION (A)

Let E_I and F_I be defined to be

$$E_I = \langle Q_I \mid C^{-1}HC \mid \Phi \rangle,$$
$$F_I = \langle Q_I \mid \Phi \rangle. \tag{15}$$

By relations (7),

$$\overline{W} = \sum_I h_I E_I F_I / \sum_I F_I^2 + O(\mu_Q),$$

$$d^2 = \sum_I h_I E_I^2 / \sum_I F_I^2 - \overline{W}^2 + O(\mu_Q). \tag{16}$$

Hence when a method of calculating E_I and F_I are given the integral transcorrelated variance can be directly estimated to within $O(\mu_Q)$. As explained before the ordinary variance can be obtained by using $C_I E_I$ and $C_I F_I$ in place of E_I and F_I where $C_I = \langle Q_I \mid C \rangle$:

DEFINITION (B)

If ϕ_i' for which $\phi_i'(R_{Ij}) = \delta_{ij}$ let

$$F_I' = \langle \overline{Q}_I \mathscr{A} \phi_1' \phi_2' \dots \phi_N' \mid \Phi \rangle,$$
$$E_I' = \langle \overline{Q}_I \mathscr{A} \phi_1' \phi_2' \dots \mid C^{-1}HC \mid \Phi \rangle, \tag{17}$$

where

$$\overline{Q}_I = \prod_i \sum_j \delta(x_i, X_{Ij})\delta(y_i, Y_{Ij})\delta(z_i, Z_{Ij})\delta(v_i, V_{Ij}). \tag{18}$$

Here the ϕ' would generally be dependent on I but they are only a device to multiply some terms in an expansion by 1 and others by zero and it is simpler to write as ϕ_i'. The $\overline{Q}_I$ is the operation which would be obtained if it were decided to do an extremely crude numerical integration over only N points in all dimensions just those of X_{Ii}, ..., for $i = 1, \dots N$.

THEOREM 1. If q_t is written for the same constant quantity, then

$$E_I = q_t E_I'; \qquad F_I = q_t F_i'. \tag{19}$$

Hence the variance formula with E_I, F_I gives the same value as with E_I', F_I'.

PROOF. E_I' can be written as

$$E_I' = \langle \mathscr{A} \phi_1' \phi_2' \dots \overline{Q}_I \mid A \rangle,$$

where A denotes the relevant antisymmetric quantity and hence $\sigma_u P \overline{Q}_{uI} A = \overline{Q}_I A$, where $\sigma_u P_u = \pm P_u$ is any permutation of the electronic variables with the sign according to its parity. It then follows

$$E_I' = (N!)^{-\frac{1}{2}} \sum_u \langle \sigma_u P_u \phi_1' \phi_2' \dots \mid \overline{Q}_I A \rangle$$

$$= (N!)^{-\frac{1}{2}} \sum_u \langle P_u^{-1} P_u \phi_1' \phi_2' \dots \mid \sigma_u P_u^{-1} \overline{Q}_I A \rangle$$

$$= (N!)^{-\frac{1}{2}} \sum_u \langle \phi_1' \phi_2' \dots \overline{Q}_I \mid A \rangle$$

$$= (N!)^{\frac{1}{2}} \langle Q_I \mid A \rangle. \tag{20}$$

In the second equality $P_{\bar{U}}{}^1$ permutes all variables in the complete integral and hence does not alter its value. For the last equality, ϕ_1' cancels all but the term $\delta(x_1, x_{11})$ ..., out of the first factor of $\bar{Q}_I$, and so with the other ϕ_i' obtains the result which from the definition (A) is the first statement of the theorem.

DEFINITION (C)

Let some intermediary orbitals $\phi_i'' = \Sigma_r U_r^i \phi_r$ be defined, where the coefficients are the solutions of $\Sigma_r U_r^i \phi_r(\mathbf{R}_{Ij}) = \delta_{ij}$ so that

$$\langle \bar{Q}_I \phi_i' \mid \phi_j'' \rangle = \delta_{ij}, \tag{21}$$

Let $\Delta_I = \mid \phi_r(\mathbf{R}_{Ii}) \mid = \mid U_r^i \mid^{-1}$, where this determinant and the coefficients can all be evaluated from the original i, j coefficients $\phi_r(\mathbf{R}_{Ii})$ with a multiple of N^3 arithmetic operations. Let the final definitions be

$$E_I'' = \langle \bar{Q}_I \Phi' \mid C^{-1}HC \mid \Phi'' \rangle \Delta_I,$$
$$F_I'' = \langle \bar{Q}_I \Phi' \mid \Phi'' \rangle \Delta_I, \tag{22}$$

where

$$\Phi'' = \mathscr{A} \phi_1'' \phi_2'' \ldots \tag{23}$$

THEOREM 2 $$E_I'' = E_I', \qquad F_I'' = F_I'. \tag{25}$$

PROOF. E_I'' becomes E_I' by the insertion of

$$\Phi'' = (N!)^{-\frac{1}{2}} \mid \sum_r U_r^i \phi_r(\mathbf{R}_{Ij}) \mid$$

$$= \mid U_r^i \mid \Phi = \Delta_I^{-1} \Phi$$

COMMENT. The value of the definition (C) is that the matrix elements occuring as expressions for the N-space integrals can be immediately reduced by the detor integral theorem. In fact, because the ϕ^i orbitals only choose the values each at a single point, another level of simplicity is introduced. In the most laborious integral for a three-electron interaction which occurs in $C^{-1}HC$ there are only a multiple of N^3 terms. This simplification is dependent on the fact that the detor integral theorem holds exactly for approximate numerical integrations which are obtained by using the same points of integration for every electron. The proof follows just the same steps as in the theorem for exact integration and the theorem will be stated here without proof for the case when $T(\mathbf{r}_1, \mathbf{r}_2, \mathbf{r}_3)$ is an operator dependent on three electrons.

THEOREM 3. If $\langle \bar{Q}_I \phi_i' \mid \phi_j'' \rangle = \delta_{ij}$,

$$\langle \bar{Q}_I \Phi' \mid \sum_{ijk} T(\mathbf{r}_i, \mathbf{r}_j, \mathbf{r}_k) \mid \Phi_j'' \rangle = \sum_{ijk} \langle \bar{Q}_I \phi_i' \phi_j' \phi_k' \mid T(\mathbf{r}_1, \mathbf{r}_2, \mathbf{r}_3) \mid \sum_u^6 \sigma_u P_u \phi_i'' \phi_j'' \phi_k'' \rangle, \tag{26}$$

where P_u are the permutations of the three sets of electron variables which follow this operator.

COMMENT. The above analysis has shown that the transcorrelated variance can be evaluated by use of the E_I'' by computation which only increases as N^3 for N electron problems. If the direct variance is obtained by using $C_I E_I''$ in place of E_I'', since C_I only involves computation varying as N^2, the computation again increases as N^3.

Without this analysis the lowest rate of increase of the computation for the direct variance which the author had been able to obtain had been N^4. This is due to the evaluation of N determinants in each of which one column had been altered from the original wavefunction by a ∇ operator. So although this analysis had been commenced with a mistaken intention of demonstrating a shorter evaluation for the transcorrelated wave equation, it appears to have provided the shortest method for either of these two variances.

Any variance with another weighting say p^2, in the form

$$\langle p^2 \mid ((H - W)\psi)^2 \rangle$$

could also be obtained by this analysis since it is merely necessary to evaluate $p(\mathbf{R}_{Ii})$ and use $p(\mathbf{R}_{Ii})E_I = E_I^p$ in the original variance formula. It appears, that there is still interesting exploration to be made on the possibility of some p providing a better practical method. The author thinks that the use of $p = \Phi$ has some promising features.

There are many sets of the N-space-spin points previously denoted by $\mathbf{R}_{ij}$, and explained in detail in definition (1). But to illustrate the essential properties of these, one example will be given. It may have practical value but it appears to be the simplest for a schematic demonstration. Let T_m for $m = 1, 2, ..$, be the set of single electron points whose components are

$$q_{m3}/(1 - q_{m3}^2), \ q_{m5}/(1 - q_{m5}^2), \ q_{m7}/(1 - q_{m7}^2), \ \tfrac{1}{2}(-1)^m, \tag{27}$$

where $q_{m3} = (m\sqrt{3}) \bmod(1)$ which is the fractional part of $(m\sqrt{3})$; and q_{m5} similarly for $(m\sqrt{5})$ etc. Let $\mathbf{R}_{Ik}$ be all selections of N of these points arranged in the order that if the Ith set consists of

$$T(m_{I1}), \ T(m_{I2}), \ T(m_{I3}), \ ..., \tag{28}$$

with $m_{I1} > m_{I2} > m_{I3}$,

and so that if

$$(m_{I1} - m_{J1})(m_{I1} + m_{J1})^N + (m_{I2} - m_{J2})(m_{I1} + m_{J1})^{N-1} + (m_{I3} - m_{J3})(m_{I1} + m_{J1})^{N-2} ...$$
$$\tag{29}$$

is positive then $\mathbf{R}_{Ik}$ occurs after $\mathbf{R}_{Jk}$. This is just a formal way of defining an order of increasing " seniority ". The weights of these are complicated to write explicitly in algebraic form but are easily obtained from the Jacobian of the transformations in (27) when the initial weights in the q space are all taken to be equal. The dependence on the surds ensures that these tend to a uniform distribution over all q space.

DISCUSSION

The number of point sets denoted by M for $I = 1, 2, ... M$ has not been discussed because this will generally be completely a matter of expediency. If by some M value the variance of $C_A\Phi_A$ is persistently greater than for $C_B\Phi_B$, for the same progression of points, then it will be sufficient to regard the test as complete and W_B as the better prediction of the energy, etc.

There are several circumstances in which these assessments will be interesting and there is one which is forward-looking but which is particularly inspiring. It appears that it will be possible to transfer localized orbitals from one molecule to another and also transfer the correlation functions. To a first approximation these latter are $\prod_{i>j} \exp(\tfrac{1}{2}r_{ij})$. In this case, many estimated wavefunctions may become available.

S. F. BOYS 101

If these can be tested directly, even inspired empirical estimates of correlated wave-functions may become a valuable subject. At a more realistic level, however, the resolution of instabilities in predictions of excited states; the assessment of the claims of different systematic calculations; and the assessment of rates of convergence with different systems of expansion functions will probably form the major use of the direct assessment of the accuracy of correlated wavefunctions.

[1] S. F. Boys and N. C. Handy, in course of publication.
[2] H. Conroy, *J. Chem. Physics*, 1967, **47**, 5307.
[3] A. A. Frost, R. E. Kellogg and E. C. Curtis, *Rev. Mod. Physics*, 1960, **32**, 313.

Photochemical Dynamics

Richard N. Dixon

School of Chemistry, University of Bristol, Bristol, UK BS8 1TS

Commentary on: **Excited Fragments from Excited Molecules: Energy Partitioning in the Photodissociation of Alkyl Iodides**, S. J. Riley and K. R. Wilson, *Discuss. Faraday Soc.*, 1972, **53**, 132.

A major object of modern chemical physics has been to obtain a detailed understanding of the factors that control chemical reactions. This is a field in which there has been a particularly strong interplay between ongoing experimentation and parallel theoretical developments. Photodissociation is inherently one of the simpler types of reaction in that the products are already in close contact in the reactant: so called "half collisions".[1] Technical developments in molecular beams and lasers have been particularly instrumental in many recent advances in this field.

The above paper[2] is one of the earliest of a series of seminal papers by Kent Wilson and his colleagues which laid the foundations not only for modern photochemical experiments, but also for their interpretation. Their photofragment spectrometer introduced several new features that greatly influenced subsequent studies. The primary interaction was between an effusive molecular beam and the 266 nm fourth harmonic of a Nd laser, at which wavelength the alkyl halides all absorb into dissociative continua. This was, by contemporary standards, an heroic experiment, since the laser only operated at one shot per minute, so that each run lasted six hours. But the novel experimental features were the use of field-free photofragment time-of-flight measurements with a post-flight mass spectrometer for detection; and a geometry which exploited the polarisation properties of the laser beam, thereby conferring spatial anisotropy to the molecular excitation and thence to the fragmentation.

At the time of this paper I was studying the spectroscopy of photofragments generated by broad band flash photolysis. Wilson's papers were an inspiration when I later embarked on the study of the dissociation itself. By then tuneable lasers conferred greater versatility in the choice of molecular systems and in detection methods, thus facilitating state to state photochemical experiments. Our first experiments, on the photodissociation dynamics of nitrous acid, were carried out while I was on sabbatical leave in Richard Zare's group at Stanford University, supported by a NATO Grant for International Collaboration. These experiments exploited the polarisation properties of not only the photolysis laser, but also a probe laser which was used to detect OH photofragments by laser induced fluorescence, and which was capable of resolving recoil Doppler broadened line shifts. This led to the determination of the anisotropic nature of both the translational and rotational motions of the ground-state hydroxyl radical,[3] and later the NO fragment.[4] An alternative development from the basic Wilson photofragment spectrometer has been the use of multiphoton ionisation of the fragments within the source region. This was first combined with an acceleration field to give efficient product yield spectra; but can also be used in the field free Wilson mode to give time of flight spectra, and thus recoil velocity analysis.

These types of study have been extended to a large number of molecular systems in numerous laboratories around the world.[5] One consequence was to stimulate the need for more detailed theories of vector correlations in the motions of photofragments.[6,7]

The more recent development of "Rydberg tagging" came about following collaboration between my colleague Michael Ashfold and Karl Welge in Bielefeld. One of the photofragments, H-atoms in the first instance, was promoted to a very high Rydberg state *via* a two step laser excitation within a field free source region. Those atoms that flew towards the detector subsequently became ionised post flight by the strong electric fields around the detector. The great advantage of this approach over multiphoton ionisation at source is that it avoids the unwanted scrambling of recoil velocities that can be brought about through ion–ion Coulomb repulsion in the source region. The light mass of H-atoms confers the added advantage that their recoil velocities are often very much higher than the velocity spread of the parent molecule and, when combined with a jet-cooled seeded beam to minimise the internal energy of the parent, may result in very high resolution of the time-of-flight spectra. This technique has been applied to many hydride molecules, of which water vapour and ammonia are examples which possess very rich fragmentation dynamics.[8]

A further feature of the chosen paper[2] is the parallel development of theoretical models of possible fragmentation dynamics for comparison with the experimental data; in this case statistical, soft impulsive, and rigid impulsive models for energy partitioning within the alkyl radical products. Many later authors have adopted this approach, which has greatly enriched our theoretical understanding of reaction dynamics generally. It is now appreciated that the non-adiabatic effects, recognised by Wilson and others in the alkyl halides, play a role in the photodissociation of very many molecules.

References

1 K. R. Wilson, in *Excited state chemistry*, ed. J. N. Pitts, Gordon and Breach, New York, 1970.
2 S. J. Riley and K. R. Wilson, *Discuss. Faraday Soc.*, 1972, **53**, 132.
3 R. Vasudev, R. N. Zare and R.N. Dixon, *Chem. Phys. Lett.*, 1983, **96**, 399; R. Vasudev, R. N. Zare and R. N. Dixon, *J. Chem. Phys.* 1984, **80**, 4863.
4 R. N. Dixon and H. Rieley, *J. Chem. Phys.*, 1989, **91**, 2308.
5 See for example: the papers presented on "Dynamics of Molecular Photofragmentation" at *Faraday Discuss.*, 1986, **82**; R. Schinke, *Photodissociation dynamics*, Cambridge University Press, 1993.
6 C. H. Greene and R. N. Zare, *Annu. Rev. Phys. Chem.*, 1982, **33**, 119; C. H. Greene and R. N. Zare, *J. Chem. Phys.*, 1983, **78**, 6741.
7 R. N. Dixon, *J. Chem. Phys.*, 1986, **85**, 1866.
8 D. H. Mordaunt, M. N. R. Ashfold and R. N. Dixon, *J. Chem. Phys.*, 1994, **100**, 7360; D. H. Mordaunt, M. N. R. Ashfold and R. N. Dixon, *J. Chem. Phys.*, 1996, **104**, 6460; D. H. Mordaunt, M. N. R. Ashfold and R. N. Dixon, *J. Chem. Phys.*, 1996, **104**, 6472.

Excited Fragments from Excited Molecules:
Energy Partitioning in the Photodissociation of Alkyl Iodides

By Stephen J. Riley and Kent R. Wilson

Dept. of Chemistry, University of California, San Diego, La Jolla, California 92037
U.S.A.

Received 9th March, 1972

Photofragment spectroscopy has been applied to the photodissociation of methyl, ethyl, n-propyl and isopropyl iodide at 266.2 nm, both to study the process of unimolecular break-up of electronically excited molecules and to determine the energy distributions of the resulting excited fragments. By crossing a molecular beam with a powerful pulsed laser beam in high vacuum and monitoring the arrival times of the recoiling photofragments with a mass spectrometer detector, the translational energy distribution of the photodissociation products is measured. From energy balance, the distribution of fragment internal energy is then calculated. Two peaks are seen in these distributions for both methyl and ethyl iodide, corresponding to formation of ground- and excited-state iodine atoms. The fraction of the energy available after exciting the I atom which goes into internal excitation of the alkyl fragments increases from ~12 % for methyl iodide to ~50 % for the propyl iodides. Thus, the " hot " methyl radicals formed in methyl iodide photolysis are predominantly translationally, rather than internally, excited. The experimental results are disscused in relation to the processes involved in alkyl iodide photodissociation lasers. Dynamic models for energy partitioning in molecular photodissociation are compared with the observed data. The simplest statistical models predict much too high excitation in the alkyl radical. The measured results fall in between the predictions of direct impulsive models based on " rigid " and " soft " radicals, and thus a possible picture for the photodissociation of the alkyl iodides is quasi-diatomic excitation of the C—I bond, followed by recoil of the I atom and a partially deformable alkyl radical. An illustrative example is given of how one might use these unimolecular results in modelling the bimolecular reactions of alkali metal atoms with alkyl iodides, following the analogy drawn by Herschbach and co-workers between photodissociation and " harpooning ", i.e., electron transfer, reactions.

Excited species of two types are involved in molecular photodissociation, parent molecules and photofragments. Following photon absorption to an unstable state, the parent molecule undergoes a unimolecular " reaction " or decomposition. The resulting fragments carry off excess energy in the form of translational and internal (electronic, vibrational, and rotational) excitation. Their reactivities are usually enhanced by this extra energy, and many subsequent reactions become possible that do not proceed at thermal energies.

Traditional photochemical studies have concentrated on identifying these initial product fragments and following their subsequent reactions. In spite of the richness of this literature, the process of photodissociation itself has received comparatively little study. We believe that it deserves more, that the molecular dynamics of photodissociation offer a fertile and largely untilled testing ground for theories of unimolecular processes, and that much of the knowledge gained will also be applicable to bimolecular processes such as energy transfer and chemical reaction.

Photodissociation offers a " clean " system to investigate theories of molecular decomposition. The amount of energy in the excited state can be well defined, yet varied at will over a considerable range, by using narrow-band, tunable, light sources.

The initial geometry of the " half collision " can be well characterized, in contrast to bimolecular scattering. The variety of molecules convenient for study is very broad. One would expect to find a correspondingly broad variety of appropriate dissociation models, covering, for example, long-lived excited states with internal randomization calling for statistical treatment, as well as rapid (before energy randomization is possible) break-up on steeply repulsive excited-state potential-energy surfaces which are best treated by direct, perhaps even impulsive, models.

Unfortunately, photodissociation has not been easy to study in detail. Spectro-scopically, dissociative transitions ordinarily appear as broad, almost structureless, continua, and in photochemical systems the dissociation fragments (or photofrag-ments) often lose their initial energy or react before they can be directly observed. A fuller characterization of the primary photodissociation process is needed for com-parison with theory, for example, the nature and lifetime of the excited state leading to dissociation and the partitioning of available energy among fragment translational and internal modes. We have used photofragment spectroscopy to study these parameters in the ultra-violet photodissociation of methyl, ethyl, n-propyl and iso-propyl iodides,[1]

$$RI + h\nu \ (266.2 \ nm) \rightarrow R + I, \tag{1}$$

as will be described in this and a subsequent paper.[2]

We have chosen to study the alkyl iodides for three reasons : their importance in classical photochemistry, their importance in photodissociation laser systems, and the " chemical virtuosity " of the alkyl group which can be widely varied to test models of unimolecular decomposition. Alkyl iodides have been the subject of photolysis studies for many years.[3] The overlap of their moderately strong u.-v. dissociative continua by the 253.7 nm Hg line has facilitated investigation of the reactions of " hot " alkyl radicals, in particular H atom abstraction from organic molecules by " hot " methyl radicals.[4] These studies have been unable to elucidate the nature of the excess energy contained in the alkyl fragments, and the validity of proposed reaction mechanisms has remained in question. Attention was focused on the other dissociation fragment, atomic iodine, in 1964 when Kasper and Pimentel[5] reported the first photodissociation lasers, operating on the $I(^2P_{\frac{1}{2}} \rightarrow {}^2P_{\frac{3}{2}})$ transition following methyl iodide and perfluoromethyl iodide photolysis. This confirmed an early spectral analysis by Porret and Goodeve[6] which concluded that excited I atoms predominate in the photodissociation. Subsequent studies[7, 8] of other alkyl iodides and fluorinated alkyl iodides have also shown efficient population inversion and laser action, yet absolute initial ratios of excited to ground state I atom production have not been determined. Such chemically intriguing questions as why n-propyl iodide leads to lasing, but isopropyl iodide does not,[7a] have also remained unanswered.

The experiments presented in this paper directly measure the energetics of the molecular photodissociation. From the distribution of fragment recoil translational energies, the extent of internal fragment excitation can be determined by energy balance. It is found that the " hot " methyl radicals are produced largely transla-tionally rather than internally excited, but that the internal excitation of the alkyl fragment increases sharply with the size of the radical. These results are compared with both statistical and impulsive models of unimolecular decomposition. In addition, the relation of these experiments to the understanding of alkyl iodide photo-dissociation lasers and to the bimolecular reactions of alkali metal atoms with alkyl iodides is discussed. A subsequent paper[2] will discuss the symmetry, configuration, and lifetime of the dissociative states of the excited parent molecules as elucidated by the fragment recoil angular distributions.

134   ALKYL IODIDE PHOTODISSOCIATION

EXPERIMENTAL

The apparatus, a photofragment spectrometer, is described in detail elsewhere.[9]   A schematic drawing is given in fig. 1.   The molecules to be studied are formed into a molecular beam and injected into a high vacuum chamber.   This molecular beam is perpendicularly crossed by short pulses of polarized laser light which are focused into the vacuum chamber through a quartz window.   Photodissociation fragments recoiling upward from the ~ 0.3 cm diam. intersection volume are detected by a mass spectrometer detector mounted a known distance away in a separately pumped chamber.   The fragments are monitored as a function of (i) mass, (ii) flight time, and (iii) recoil angle, measured with respect to the polarization direction (electric vector) of the light.   From these measurements the identity and translational and internal energies of the photofragments can be determined.   The use of high vacuum assures collisionless recoil of the photofragments, so the true primary photodissociation process can be characterized.

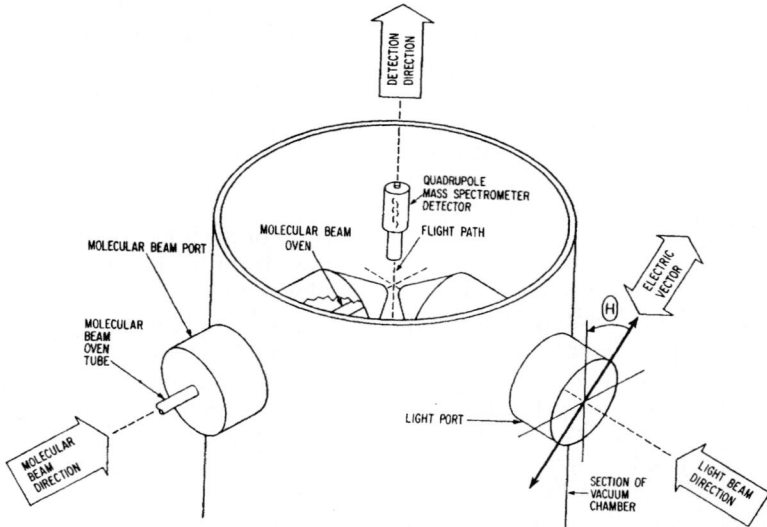

FIG. 1.—Schematic view of the photofragment spectrometer.   In a high-vacuum chamber a molecular beam is perpendicularly crossed with pulses of polarized light from a laser.   Photofragments recoiling upward from the intersection region are monitored by an electron bombardment quadrupole mass spectrometer detector as a function of fragment mass, recoil time (after the laser pulse) over a known flight path, and angle Θ between the electric vector of the laser light and the detection direction.

A photon wavelength of 266.2 nm is used in the experiments reported here, the fourth harmonic of a special neodymium laser system which can generate powerful ultra-violet pulses of a joule or more.[1a]   Experimental conditions are (see also table 1) : molecular beam density at the interaction volume, $\sim 6 \times 10^{13}$ molecules cm^{-3}; laser energy, 0.3 J per pulse (4×10^{17} photons per 10 ns pulse, i.e., 30 MW) ; laser repetition rate, one pulse per minute ; and flight path length, 5.63 cm.

The experiment is controlled [9] by an on-line time-shared computer which collects, processes, stores and displays the raw data : the number of detected fragments per microsecond for the first 400 μs after the laser pulse at a particular recoil angle.   The running average of many consecutive laser pulses at the same recoil angle is calculated by the computer as the experiment progresses, so that a real-time indication of the signal-to-noise ratio can be obtained.   Typically 350 laser firings were used to provide sufficient signal-to-noise ratios for the energy distribution results presented here.

The alkyl iodides, from Eastman Kodak Co., were used without further purification. The principal impurity, molecular iodine, has a much lower vapour pressure and absorbs only weakly at 266.2 nm.

STEPHEN J. RILEY AND KENT R. WILSON   135

RESULTS

The energetics of photodissociation are governed by the energy conservation expression

$$E_{avl} = h\nu + E_{int}^P - D_0^0 = E_t + E_{int}. \tag{2}$$

E_{avl}, the available energy, is the photon energy $h\nu$ plus the initial parent molecule internal energy, E_{int}^P, minus D_0^0, the energy of dissociation from the ground state of the parent molecule to the ground states of the observed products. After photodissociation, E_{avl} partitions between E_t, the total centre-of-mass (c.m.) translational energy of both products, and E_{int}, the total internal energy of both fragments. Since, for the alkyl iodides, photodissociation in the lowest continuum can result in only two discrete values of I atom internal excitation, it is also convenient to define an available energy E_{avl}^* for dissociation to excited $I^*(^2P_{\frac{1}{2}})$ atoms. E_{avl}^* is less than E_{avl} defined in eqn (2) by the 21.7 kcal mol^{-1} splitting between $I^*(^2P_{\frac{1}{2}})$ and $I(^2P_{\frac{3}{2}})$. Once the I atom electronic state is known (or assumed) for a particular photodissociation process, use of the corresponding available energy (E_{avl} or E_{avl}^*) in eqn (2) permits direct identification of E_{int} as E_{int}^R, the internal energy of the alkyl fragment. In the discussion that follows we will often refer to the *effective available energy*, denoted by E_{avl}', to mean either E_{avl} or E_{avl}^*, depending on the photodissociation process under consideration.

The distributions of E_t from the photodissociation of the alkyl iodides at 266.2 nm are shown in fig. 2. These distributions are derived from the laboratory (lab) flight time distributions for recoiling I atoms and alkyl fragments by an approximate procedure, described elsewhere.[10] The principal approximations are the assumption of a single parent molecular beam velocity (derived from the molecular beam temperature[11]) and the assumption of point interaction and detection regions. The first approximation is best when the fragment laboratory recoil velocity is much larger than the beam velocity,[10] as is evident from the enhanced resolution seen in fig. 2 for the curves derived from measurements on the lighter, higher-recoil velocity alkyl fragments. Each symbol in fig. 2 (circles for detected I fragment, triangles for detected alkyl fragment) represents the transformation of the lab data in a single 1 μs channel in time space, except at low translational energies where the density of data points becomes too high to plot, and thus averages over several adjacent time channels are shown. While the approximate nature of the lab→c.m. transformation casts some uncertainty on the widths of the distributions, their shapes are, where resolved, roughly symmetrical, so peak values will be considered in this paper as representative of the average energies involved in the dissociations.

Analysis of these results in terms of fragment internal excitation requires evaluation of the available energy E_{avl} in eqn (2). The photon energy $h\nu$ at 266.2 nm is 107.3 kcal mol^{-1} (37 550 cm^{-1} or 4.65 eV). The distribution of parent internal energy, E_{int}^P, can in principle be calculated from the molecular beam oven temperature and the moments of inertia and vibrational modes of the parent molecules.[11] In practice, a classical formulation for the distribution in total rotational energy E_r,

$$P(E_r) \propto (E_r)^{\frac{1}{2}} \exp(-E_r/kT) \tag{3}$$

can be used with considerable accuracy.[10] For ethyl and n-propyl iodides, which are assumed to have in addition one free internal rotation, eqn (3) is modified to

$$P(E_r) \propto E_r \exp(-E_r/kT) \tag{3'}$$

Summing such distributions, weighted by the Boltzmann factor for vibrational excitation, over all vibrational states yields the distribution of E_{int}^P. The average

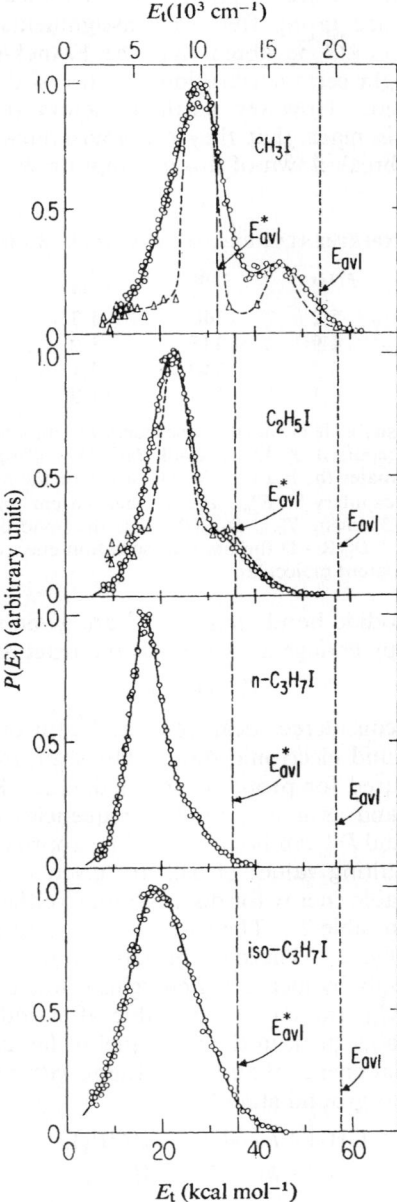

FIG. 2.—Flux density distributions of total (both fragments) centre-of-mass recoil translational energy E_t for alkyl iodide photodissociation at 266.2 nm. Circles ($\bigcirc$) and solid curves are derived from the photofragment spectra of detected iodine atoms at a laboratory angle Θ (see fig. 1) of 15°, the maxima in the angular distributions. The energy distribution is very similar at other angles. Triangles ($\triangle$) and dashed curves are for detected alkyl fragments, also at $\Theta = 15°$. Vertical lines show the available energy values from table 2, short-dashed lines indicating E_{avl} for dissociation to ground-state I atoms and long-dashed lines indicating E_{avl}^*, which is E_{avl} minus the I atom excitation energy. The distributions of total fragment internal energy E_{int} are just the E_t distribution curves with energies instead measured from right to left, beginning at E_{avl}.

values of E_{int}^{P} shown in table 1 are calculated (at the molecular beam oven temperatures given) from published group frequency assignments,[12] and are used in the calculation of E_{avl}. It should be noted that the Franck-Condon factors may, in reality, preferentially weight certain vibrational states of the parent molecule at the particular wavelength used. However, as these factors are unknown for the alkyl iodides, the assumption is made that they are approximately equal. The energies involved are such that a break-down of this assumption will not materially affect the results.

TABLE 1.—EXPERIMENTAL PARAMETERS (ENERGIES IN kcal mol^{-1})

	T_rK	P/Torr	T_c/K	E_{int}^{P}	D_{298}(R—I)	D_{0}^{0}(R—I)
CH_3I	218	5.2	340	1.3	56.3±1	55.0±1
C_2H_5I	239	4.9	335	1.9	53.0±1	51.7±1
n-C_3H_7I	253	3.4	340	2.1	54.0±2	52.6±2
iso-C_3H_7I	253	6.1	335	1.9	53.0±2	51.7±2

The molecular beam intensity is determined by the reservoir temperature T_r resulting in a pressure P (interpolated from values reported by D. R. Stull, *Ind. Eng. Chem.*, 1947, **39**, 517) behind the multicapillary array that collimates the beam. For effusive flow the molecular beam temperature is the temperature T_c of the capillary.[11] E_{int}^{P}, the average parent molecule internal energy, was calculated from eqn (3) and (3′) using T_c. D_{298}(R—I) is the bond dissociation enthalpy at 298 K to form ground-state iodine,[13] D_{0}^{0}(R—I) the bond dissociation energy at 0 K to form ground-state fragments from ground-state parent molecules.

Values for the alkyl iodide bond strengths [13] are given in table 1 as D_{298}(R—I). These refer to the enthalpy change at 298 K for the reaction

$$RI \rightarrow R + I \tag{4}$$

in which all species are considered ideal gases at 1 atm pressure and the I atom is assumed to be in its ground electronic state. However, in eqn (2), D_{0}^{0}, the energy change at 0 K, must be used for proper energy balance. Knowing the alkyl iodide vibrational frequencies, and assuming similar frequencies in the alkyl radicals, the difference between D_{298} and D_{0}^{0} can be calculated by approximate statistical-mechanical methods.[14] The resulting values of D_{0}^{0} are given in table 1. Average values of E_{avl}', the effective available energy for dissociation to either ground- or excited-state iodine atoms, are given in table 2. The values of E_{avl} and E_{avl}^{*} are also indicated by vertical dashed lines in fig. 2. The distributions of E_{int} can be read from the E_t distributions of fig. 2, simply by measuring energies toward the left from E_{avl}.

For methyl iodide, E_{avl}^{*}, the energy available after iodine atom excitation, lies well to the left of the smaller peak in the top panel of fig. 2. Thus this peak, representing *lower* total internal energy than is consistent with process (5) corresponds to process (6), dissociation to ground state I:

$$CH_3I + h\nu \rightarrow CH_3 + I^*(^2P_{\frac{1}{2}}), \tag{5}$$
$$CH_3I + h\nu \rightarrow CH_3 + I(^2P_{\frac{3}{2}}). \tag{6}$$

This is consistent with kinetic spectroscopy measurements [15] which indicate the presence of some ground-state iodine atoms following flash photolysis of methyl iodide. The main peak in the methyl iodide energy distribution presumably corresponds largely to I* production, although one cannot rule out *a priori* some I contribution. If one assumes (as we do for simplicity) that the main peak is due entirely to I* production, then the ratio of excited- to ground-state iodine atom production is approximately 3.5 to 1. A more refined lab→c.m. transformation might somewhat alter this estimate, by altering relative peak shapes. In addition to reactions (5) and

(6) there are other possible minor dissociation pathways [16, 17] producing $CH_2 + HI$ and $CHI + H_2$. The first minor pathway has too small a quantum yield to be detected in our measurements, while the products of the second would have been rejected by the c.m.→lab transformation and by the mass spectrometer.

TABLE 2.—ENERGETICS OF DISSOCIATION (ENERGIES IN kcal mol^{-1})

	Iodine fragment state	E'_{avl}	E_t	E^R_{int}	E^R_{int}/E'_{avl}	E^R_{int}/E^R_{tot}
CH_3I	I*	31.9	28.0	3.9	0.12	0.135
	I	53.6	45.0	8.6	0.16	0.18
C_2H_5I	I*	35.8	22.6	13.2	0.37	0.42
	I	57.5	34.3	23.2	0.40	0.45
$n\text{-}C_3H_7I$	I*	35.1	16.6	18.5	0.53	0.60
	I	56.8	—	—	—	—
$iso\text{-}C_3H_7I$	I*	35.8	[18.9]	[16.9]	[0.47]	[0.545]
	I	57.5	[18.9]	[38.6]	[0.67]	[0.73]

I* denotes $I(^2P_{\frac{1}{2}})$, I denotes $I(^2P_{\frac{3}{2}})$. E'_{avl} is the effective available energy (see discussion following eqn (2)). E_t is the total translational energy of both fragments (peak values from fig. 2). E^R_{int} is the internal excitation energy of the alkyl radical, and E^R_{tot} is its total energy (translational plus internal). These assignments assume that for CH_3I and C_2H_5I the main and smaller peaks are I* and I respectively and that the $n\text{-}C_3H_7I$ peak is I*. For $iso\text{-}C_3H_7I$, where the probable iodine excitation is less clear, the peak is analyzed twice: first, assuming it is I* and secondly assuming it is I.

For the higher alkyl iodides, branching ratios between I* and I production are less clear. Spectroscopic observation of ground state atoms in the flash photolysis of these iodides has been hampered by strong parent molecule absorption at the vacuum u.-v. atomic iodine transitions which were monitored.[18] In ethyl iodide photodissociation (panel 2, fig. 2), two overlapping peaks appear in the energy distribution. While the energetic arguments are no longer as compelling as for methyl iodide, it seems reasonable to assume that again the smaller peak corresponds to I production and the larger peak to I*. For n-propyl iodide (panel 3), there is perhaps a slight shoulder in the distribution which might again correspond to a smaller I peak overlapping a main I* peak. Such an interpretation is consistent with the fact that methyl, ethyl, and n-propyl iodide all form photodissociation laser systems operating on the I*→I transition,[5, 7] and this interpretation is used here. The distinctly broader curve for isopropyl iodide (which does not show laser action [7a]) is considered in more detail in the next section. For the present it will be assumed that dissociation to both I and I* is represented in the distribution of panel 4.

The energetics of the dissociation processes are summarized in table 2. The next-to-last column shows the fraction of the effective available energy E'_{avl}, which appears as E^R_{int}, the alkyl fragment internal excitation. The final column shows the alkyl internal energy as a fraction of the total (translational plus internal) alkyl energy. Two conclusions are clear: (i) the fraction of the effective available energy which appears as alkyl internal excitation does not depend drastically on whether I* or I is produced, and (ii) this fraction rises dramatically from a very low value for methyl to approximately one half for n-propyl.

The angular distributions of recoiling iodine atoms were also measured for all four alkyl iodides studied. As is discussed elsewhere,[19] such distributions provide information about the symmetry, configuration, and lifetime of the parent dissociative excited state. The details of these results will be presented in a future paper.[2] Briefly, the angular distributions show that the transition dipole moment lies along the C—I bond and that the excited state breaks up on a time-scale short compared to a rotational period.

DISCUSSION

IODINE ATOM EXCITATION

Observation [5, 7] of stimulated $I(^2P_{\frac{1}{2}} \rightarrow {}^2P_{\frac{3}{2}})$ emission following flash photolysis of the smaller alkyl iodides has led to increased interest in the mechanisms of production and reactions of electronically excited iodine atoms. These studies have encompassed spectroscopic and photochemical techniques, recently reviewed by Husain and Donovan,[8] as well as our present photofragment spectroscopy experiments.[20]

The basic theoretical and experimental question is the ratio of I* to I production as a function of alkyl group. Stimulated emission indicates population inversion, but any I*/I branching ratio greater than $\frac{1}{2}$ provides, statistically, such an inversion. Threshold concentrations of excited atoms necessary for laser action are difficult to determine precisely, and quenching collisions affect lasing efficiency, so absolute branching ratios cannot easily be found from stimulated emission studies. Spectroscopic observation of photodissociation products, especially via vacuum u.-v. and near-vacuum u.-v. absorptive transitions in I* and I, have confirmed excited iodine atom production for several alkyl iodides,[8, 15, 18] but unfortunately, except for CH_3I and CF_3I, parent molecule absorption has precluded direct observation of ground-state iodine atoms.

The technique of photofragment spectroscopy has, potentially, the ability to determine branching ratios. The present results for methyl iodide, if we assign the major peak to I*, indicate an I*/I ratio of three or four to one, and, with parallel assignments, a similar figure for ethyl iodide. These values certainly correlate with the observations of strong stimulated emission in these systems. For n-propyl iodide, the resolution in the iodine fragment translational energy distribution is insufficient for such an estimate. We have also detected the recoiling propyl fragments, and, while the signal-to-noise levels are considerably reduced (due probably to increased cracking of the radical in the ionizer and subsequent reduced intensity at any one ion mass), there is evidence for a similar I*/I ratio, again in agreement with the laser studies.

It is emphasized that the present results apply only for photodissociation at 266.2 nm, near the absorption maximum. The ratio of I*/I production may well be a function of photolyzing wavelength. According to the spectral analysis of Porret and Goodeve for methyl iodide,[6] the ratio decreases drastically at longer wavelengths. Future photofragment spectroscopy studies with tunable ultra-violet light sources might provide a complete map of I* against I production throughout the absorption band, especially if a detector were used which could distinguish I* and I.

The lack of laser action in the photolysis of isopropyl iodide [7a] raises intriguing questions. As Husain and Donovan [8] point out, this does not *necessarily* indicate the absence of population inversion, since under the laser experimental conditions there could instead be an insufficient absolute concentration of I* atoms. Spectroscopic studies show that excited iodine atoms are produced from isopropyl iodide photodissociation, but at lower relative concentrations than for n-propyl iodide under similar conditions.[18] Since the two propyl iodides show similar I* quenching rates,[7a, 18] it would appear most likely that a decreased I*/I ratio is the reason stimulated emission is not seen. The present experiments, unfortunately, cannot provide a more quantitative explanation. The distinct broadness of the isopropyl iodide distribution in fig. 2 indicates a departure from the methyl→ethyl→n-propyl trend, and might represent comparable amounts of I* and I atom production, with overlapping translational energy distributions, at least when viewed with our present resolution. Such a lowering of the branching ratio, even if not drastic enough to prevent population inversion, might still preclude lasing when coupled with other factors such as collisional quenching and the weakness of the I-atom electronic transition. The difference in the photodissociation of the n- and isopropyl iodides thus offers a fascinating unsolved chemical puzzle, and a good opportunity to investigate theoretically and experimentally the effect of molecular structure on molecular fragmentation.

INTERNAL EXCITATION OF ALKYL FRAGMENT

The present results indicate (with the I* and I peak assignments given above, which we assume in this section) that translational excitation of the alkyl fragment strongly predominates over internal excitation in methyl iodide photolysis at 266.2 nm, but that the radical internal excitation rapidly increases for ethyl and n-propyl. The results for methyl iodide are particularly surprising, as they contradict previous energy partitioning estimates. If the CH_3 maintains its tetrahedral configuration in the excited parent molecule, it has been estimated that 10 kcal mol^{-1} of vibrational excitation should result as it transforms to a planar structure on dissociation.[21] If, as has been reasoned,[22] the HCH bond angle is *less* than tetrahedral in the excited methyl iodide state, even more internal excitation of the CH_3 fragment could result. As we observe only ~ 4 kcal mol^{-1} of internal methyl excitation in the major peak, this picture clearly does not adequately describe methyl iodide dissociation. Either the estimates of the bond distortion energy are incorrect or the dynamics do not correspond to simple release of a bent radical. For instance, the potential minimum of the methyl radical in excited CH_3I when the I or I* is close might still be pyramidal, with the planar potential surface only being assumed gradually as the iodine moves away. If so, one might not expect to see the radical assume such a high vibrational excitation.

The existence of " hot " methyl radicals in methyl iodide photochemical systems has been inferred from the production of methane via such reactions as

$$CH_3 + CH_3I \rightarrow CH_4 + CH_2I, \tag{7}$$

even in the presence of efficient thermal methyl radical scavengers.[4] Reaction (7) is believed [3a, 23] to have an activation energy of about 9 kcal mol^{-1}, an energy easily provided by the excess energy to be partitioned in the photodissociation. The details of this partitioning must be known before the mechanism of methane formation can be elucidated. The reduction of methane formation upon addition of inert foreign gases shows an " abnormal " mass effect,[24] suggesting the excess methyl energy is not translational. However, it has also been argued [4] that the decreased reactivity of deuterated methyl radicals from photolysis of CD_3I might indicate that the excess energy is predominantly translational. The present results clearly support this latter conclusion and indicate that alternate proposed reaction mechanisms [3a] not dependent on internally "hot " photodissociation methyl fragments must be considered more likely.

The trend toward increased internal excitation with size of the alkyl radical is an excellent example of the effect of structure on the dynamics of molecular processes. In general, two limiting cases of molecular dissociation mechanisms can be distinguished, one involving a long-lived activated parent molecule in which the excess energy is partitioned statistically among the available degrees of freedom prior to dissociation, and the other involving a break-up that is direct, occurring on a time scale comparable to that of internal molecular motions. We now consider specific models based on these pictures that lead to specific predictions of energy partitioning in alkyl iodide photodissociation.

STEPHEN J. RILEY AND KENT R. WILSON   141

STATISTICAL MODEL

As the size of the alkyl iodide increases so does its number of vibrational modes, and a statistical partitioning of the available energy should result in greater internal excitation of the alkyl fragment, such as we observe. This model can be made quantitative [25] by considering the activated parent molecule to be a collection of oscillators whose total energy (above their zero point levels) is E'_{avl}. The probability that the system will have a total vibrational energy between E_v and $E_v + dE_v$ is

$$P(E_v)dE_v/W(E'_{avl}) \qquad (8)$$

in which $W(E'_{avl})$ is the total number of states with energy $\leqslant E'_{avl}$ and

$$P(E_v) = dW(E_v)/dE_v \qquad (9)$$

is the density of states at E_v. The average vibrational energy of the collection of N oscillators is then

$$\langle E_v \rangle = \int_0^{E'_{avl}} E_v P(E_v)dE_v. \qquad (10)$$

The modified form [26] of the " semiclassical " expression [27] for the total number of states with energy $\leqslant E'_{avl}$ in a collection of N oscillators of frequencies v_i is

$$W(E'_{avl}) = (E'_{avl} + aE_z)^N/\Gamma(N+1)\prod_i hv_i, \qquad (11)$$

in which E_z is the total zero-point energy ($E_z = \sum_i hv_i/2$) and a is an empirical correction factor. If the parent molecule has N vibrational degrees of freedom, the activated molecule has $N-1$, one oscillation having been converted to a translation of the separating fragments. Thus, applying eqn (8-11) to the activated molecule and assuming the factor a is independent of E_v, yields

$$\langle E_v \rangle = \frac{(N-1)E'_{avl} - aE_z}{N} + \frac{(aE_z)^N}{N(E'_{avl} + aE_z)^{N-1}}, \qquad (12)$$

in which calculation shows the second term usually to be negligible for $N \geqslant 8$. For alkyl iodide dissociation, the iodine atom carries off no vibrational energy and eqn (12) gives the average vibrational excitation of the alkyl radical.

This statistical model can be evaluated as follows. The value of $a = 1$ will normally (except [28] at very low values of E'_{avl}) give a lower limit to the true value of $\langle E_v \rangle$, and $a = 0$ will give an upper limit. Other approximations exist but their results will usually fall between these two limits. The total zero-point energy is calculated from the spectral frequency assignments,[12] minus the contribution from C—I stretch. For ethyl and n-propyl iodide, in which one degree of free internal rotation is assumed to exist,[29] N is replaced [30] by $N+\frac{1}{2}$ in eqn (12). These results are graphed in fig. 3, together with the experimental energy partitioning values for the major peaks, given as the ratio of the alkyl radical internal energy to the effective available energy. As can be seen, this statistical model consistently predicts, even in its lower limit, too large a fraction of the available energy appearing in internal excitation of the alkyl fragment. It should be recognized that we have assumed no repulsion between the recoiling fragments, and that such repulsion, taking place after the statistical partitioning, would decrease the radical internal excitation.[25a]

The failure of a simple statistical model is not too surprising, since most evidence indicates alkyl iodide dissociation is a rapid, direct process. The recoil angular distributions which we have measured show that photodissociation occurs prior to significant molecular rotation,[2] and the lack of any vibrational structure in the ultraviolet absorption spectra is also consistent with dissociation before significant vibrational motion.

IMPULSIVE MODELS

Impulsive models [10, 31] can readily be applied to the photodissociation dynamics of alkyl iodides.   Photon absorption is assumed to be quasi-diatomic,[31, 32] affecting initially only the breaking C—I bond.   This assumption is made plausible by the close similarity in all the saturated alkyl mono-iodide absorption spectra of the transition we are exciting.[3, 6, 33]   Thus, in this picture, the available energy appears originally entirely in C—I repulsion.   In the pure impulsive, " soft " radical limit, the α-carbon is assumed to be so weakly attached to the rest of the alkyl fragment in relation to the sharp C—I repulsion that it alone initially absorbs the energy as the fragments repel one another.   A certain momentum (consistent with energy conservation for recoiling I and C only) is thus imparted to the α-carbon, and this momentum must be conserved as the carbon interacts with the rest of the fragment and imparts vibrational and rotational excitation to it.   Final translational energy is determined purely by conservation of energy and momentum, and the result for the extent of internal excitation is [10, 31]

$$E_{int}^{R}/E_{avl}' = 1 - (\mu_a/\mu_f), \tag{13}$$

in which E_{int}^{R} is the alkyl radical vibrational plus rotational energy, E_{avl}' is, as before, the effective available energy, μ_a is the reduced mass of the atoms C and I at either end of the breaking bond, and μ_f is the reduced mass of the R and I fragments.   Values of E_{int}^{R}/E_{avl}' from eqn (13) are plotted in fig. 3.   Although they follow a trend similar to the experimental values, they are consistently higher, indicating that this picture must be somewhat too simplified.

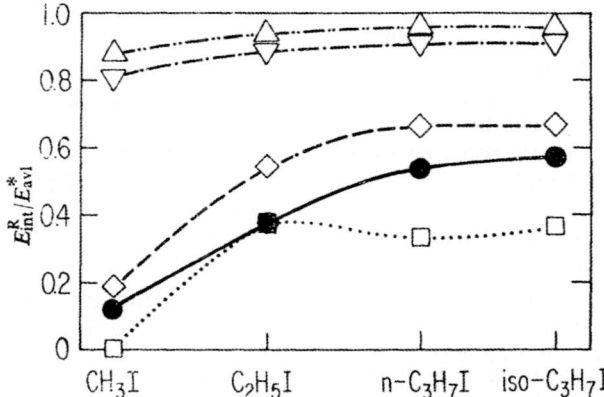

FIG. 3.—Fraction of E_{avl}^{*}, the energy available after iodine atom excitation, which appears in internal excitation of the alkyl fragment, plotted against identity of the parent molecule (arbitrary scale). The points are : △, statistical model, eqn (12), $a = 0$; ▽, statistical model, eqn (12), $a = 1$; ◇, " soft " radical impulsive model, eqn (13) ; ●, experimental ; □, " rigid " radical impulsive model, eqn (14).   The experimental points are derived from the major peaks in fig. 2, assumed to represent I* production for CH_3I, C_2H_5I and n-C_3H_7I.   For iso-C_3H_7I the peak may be a mixture of I and I*, and the experimental point is the energy disposal averaged between the values for I and I* production. The curves linking the points are intended only as a visual guide.

In another limiting form of the impulsive model, the " rigid " radical limit, the alkyl radical is assumed to recoil as a rigid body, so that only rotational excitation can occur.   The extent of excitation is determined by conservation of energy as well as linear and angular momentum.   The result is

$$E_{int}^{R}/E_{avl}' = E_{rot}^{R}/E_{avl}' = \left[1 + (\mathcal{I}/\mu_f r_{cm}^2 \sin^2 \chi)\right]^{-1}, \tag{14}$$

in which again the effective available energy is used, E_{rot}^R is alkyl rotational energy, r_{cm} is the distance from the α-carbon to the centre-of-mass of the alkyl fragment, $\mathcal{I}$ is the moment of inertia of the alkyl fragment about an axis through its centre-of-mass and perpendicular to the plane defined by the centre-of-mass, α-carbon, and the iodine atom, and χ is the angle between the α-carbon to I bond and the line from the α-carbon to the alkyl centre-of-mass. If we assume, in harmony with the Franck-Condon principle, that the initial excited state geometry is similar to the ground state, then eqn (14) can be readily evaluated from the ground-state structure of the alkyl iodides.[34] The results are graphed in fig. 3. The experimental results, which were somewhat below the " soft " radical model, are generally somewhat above the " rigid " radical model, indicating that *a possible model for alkyl iodide photodissociation would involve quasi-diatomic C—I excitation followed by recoil of a " semi-rigid " radical.* Interestingly, eqn (14) comes quite close to matching the experimental results for ethyl iodide, where the absence of any low frequency " soft " bending modes in the radical (in contrast to propyl) would support the picture of rotational excitation being the principal mechanism for supplying the alkyl fragment with internal energy.

As shown in table 2, the fraction of effective available energy which goes into alkyl internal excitation is similar for both the I* and I peaks, where they can be distinguished. One is thus tempted in the future to try to find models, like the statistical and impulsive ones discussed above, in which this fraction is independent of the magnitude of the effective available energy.

COMPARISON WITH ELECTRON TRANSFER REACTIONS

The reactions

$$M + RI \rightarrow MI + R, \tag{15}$$

in which M is Na, K, Rb or Cs and RI is an alkyl iodide, have been extensively studied in crossed molecular beam experiments.[35] Considerable evidence supports a predominantly direct mechanism for these reactions. Product angular distributions do not show the symmetry to be expected from break-up of a long-lived collision complex,[35g, 35h, 36] and recoil energy distributions are inconsistent with statistical energy partitioning.[37] These reactions have been modelled as an electron transfer from the alkali atom to the iodide as the reactants approach, the dynamics then being largely determined by the forces within the RI⁻ ion.[36]

The similarity of this electron transfer process to charge transfer spectra was pointed out several years ago by Herschbach.[36] He has subsequently refined this picture,[38] emphasizing the close analogy linking such electron transfer reactions (and other reactions [39]) with photodissociation. For the alkyl iodides, the Herschbach picture might be stated in the following way. RI photodissociation involves the promotion of an electron from an essentially non-bonding orbital to a low-lying anti-bonding C—I orbital,[40] presumably the same orbital the transferred electron enters in the alkali atom reactions. Thus, the RI* excited molecule and the RI⁻ molecular ion might be expected to have similar potential energy surfaces, and one can argue that reaction and photodissociation dynamics should be essentially equivalent. As Herschbach and co-workers point out,[38, 39, 41] one might thus relate the continuous absorption spectra of these molecules, via an essentially impulsive model, to the observed reaction product recoil energy distributions.

The present photodissociation results can be used to help interpret such a picture. For illustration, we choose a definite model. The general energy balance for the reactions is

$$\mathcal{E}_{int}^R + \mathcal{E}_{int}^{MI} + \mathcal{E}_t = \mathcal{E}_{avl} \tag{16}$$

144 ALKYL IODIDE PHOTODISSOCIATION

in which $\mathcal{E}_{int}^{R}$ and $\mathcal{E}_{int}^{MI}$ are internal excitation energies of the alkyl radical and alkali iodide products respectively, $\mathcal{E}_t$ the total c.m. recoil translational energy, and $\mathcal{E}_{avl}$ the energy available in the reaction (equal to the difference of bond strengths plus initial internal and collisional energies). We model the partitioning of the available energy in the following manner. The initial velocity of the M atom is assumed to be negligible at the moment of electron transfer. The nascent M⁺—I⁻ bond, when the electron is transferred, will contain an initial energy, $\mathcal{E}_0^{MI}$, representing the potential energy (above ground state) due to the M—I displacement from its equilibrium value. The "remaining energy", $\mathcal{E}_{avl} - \mathcal{E}_0^{MI}$, is assumed, upon electron transfer, to be "switched on" in the RI⁻ in much the same manner as the available energy is "switched on" in RI photon absorption. Applying the results of the present photodissociation experiments, RI⁻ will quickly dissociate and a fraction κ of this "remaining energy" will go into alkyl internal excitation. Thus,

$$\mathcal{E}_{int}^{R} = \kappa(\mathcal{E}_{avl} - \mathcal{E}_0^{MI}). \tag{17}$$

We take this fraction κ as given by the equivalent E_{int}^{R}/E_{avl}' photodissociation values in table 2. The I⁻ ion then collides (impulsively) with the M⁺ ion, and conservation of energy and momentum require [10, 31] the final total c.m. recoil energy to be

$$\mathcal{E}_t = (\mu_a/\mu_f)(1-\kappa)(\mathcal{E}_{avl} - \mathcal{E}_0^{MI}), \tag{18}$$

and the final total MI internal energy to be

$$\mathcal{E}_{int}^{MI} = \mathcal{E}_0^{MI} + (1 - \mu_a/\mu_f)(1-\kappa)(\mathcal{E}_{avl} - \mathcal{E}_0^{MI}), \tag{19}$$

in which μ_a is now the reduced mass of R and I, and μ_f the reduced mass of R and MI.

Evaluation of eqn (17)-(19) requires some estimate of $\mathcal{E}_0^{MI}$. If there is negligible interaction between M and RI prior to the electron transfer, the transfer radius r_t is approximately given by the energy of the M⁺, RI⁻ ion pair,[36]

$$e^2/r_t = \text{IP(M)} - \text{EA(RI)}, \tag{20}$$

in which IP(M) is the ionization potential of M and EA(RI) the vertical electron affinity of RI. Alternatively, if the total reaction cross-section σ is known the transfer radius may be approximately given by

$$\sigma = \pi r_t^2. \tag{21}$$

The energy $\mathcal{E}_0^{MI}$ can then be estimated from known or assumed (e.g., partially coulombic) MI potential curves. No great accuracy can be expected from either of the above schemes. The assumption of $\mathcal{E}_0^{MI} = 0$, however, should provide an instructive test of this model, as it will yield, from eqn (18), an upper limit to the predicted translational energy. Coupled with $\kappa = E_{int}^{R}/E_{avl}^{*}$ values from table 2, this assumption allows us to calculate $\mathcal{E}_t/\mathcal{E}_{avl}$, the fraction of available energy appearing in translation, which we find to be almost independent of M identity: approximately 0.84, 0.58, and 0.42 for methyl, ethyl, and n-propyl iodide, respectively. These figures fall above the preliminary results of crossed molecular beam experiments,[37, 42] consistent with their interpretation as upper limits, and they agree with the experimentally observed trend toward lower recoil energy for the larger alkyl iodides.

It must be emphasized that this model is only one of a large family inspired by the picture drawn by Herschbach and co-workers, and thus they should not be blamed for its particular faults. We do not wish to imply that it is a necessary consequence of the original picture, but rather we have taken it as a simple illustration of one possible way of integrating unimolecular photofragment spectroscopy results into bimolecular dynamic models of electron transfer reactions. More sophisticated

STEPHEN J. RILEY AND KENT R. WILSON   145

treatments appear promising, and we hope that fruitful and more realistic quantitative comparisons linking photodissociation and bimolecular reaction dynamics will soon follow.

Support of this work by the National Science Foundation and the Office of Naval Research (Contract #N00014–69A–0200–6020), and use of computer facilities supported by the National Institutes of Health and the National Science Foundation are gratefully acknowledged. We also thank R. Clear for help in the "rigid" radical impulsive model calculations.

[1] Preliminary reports of these results have been presented at two meetings. See (a) G. Hancock and K. R. Wilson, *Fundamental and Applied Laser Physics—Proc. Esfahan Symp.* (*Esfahan, Iran*, 1971), ed. M. Feld, N. Curnit and A. Javan (John Wiley and Sons, New York, 1972), in press; (b) S. J. Riley and K. R. Wilson, *Bull. Amer. Phys. Soc.*, 1972, **17**, 110.

[2] S. J. Riley and K. R. Wilson, to be submitted to *J. Chem. Phys.*,

[3] (a) J. R. Majer and J. P. Simons, *Adv. Photochem.*, 1964, **2**, 137, and references cited therein; (b) R. C. Mitchell and J. P. Simons, *Disc. Faraday Soc.*, 1967, **44**, 208.

[4] R. D. Doepker and P. Ausloos, *J. Chem. Phys.*, 1964, **41**, 1865, and references cited therein.

[5] J. V. V. Kasper and G. C. Pimentel, *Appl. Phys. Letters*, 1964, **5**, 231.

[6] D. Porret and C. F. Goodeve, *Proc. Roy. Soc. A*, 1938, **165**, 31.

[7] (a) J. V. V. Kasper, J. H. Parker and G. C. Pimentel, *J. Chem. Phys.*, 1965, **43**, 1827; (b) M. A. Pollack, *Appl. Phys. Letters*, 1966, **8**, 36.

[8] D. Husain and R. J. Donovan, *Adv. Photochem.*, 1971, **8**, 1.

[9] G. E. Busch, J. F. Cornelius, R. T. Mahoney, R. I. Morse, D. W. Schlosser and K. R. Wilson, *Rev. Sci. Instr.*, 1970, **41**, 1066.

[10] G. E. Busch and K. R. Wilson, *J. Chem. Phys.*, 1972, **56**, 3626.

[11] R. B. Bernstein and A. Rulis of the Theoretical Chemistry Institute, University of Wisconsin, have pointed out to us in private communication that there is an upward shift in the velocity distribution of parent RI due to non-effusive flow through the multicapillary array of the molecular beam oven. The magnitude of this shift, estimated from their data, is small enough not to affect greatly our approximate transformation, as can also be seen from the agreement of our separate R and I fragment measurements. It is, however, sufficiently large that it should be taken into account in more accurate transformations. The non-effusive flow should also have a slight effect on the parent molecule internal energy, but this too has been ignored here.

[12] N. Sheppard, *J. Chem. Phys.*, 1949, **17**, 79; N. Sheppard, *Trans. Faraday Soc.*, 1950, **46**, 533; J. K. Brown and N. Sheppard, *Trans. Faraday Soc.*, 1954, **50**, 1164; N. T. McDevit, A. L. Rozek, F. F. Bentley and A. D. Davidson, *J. Chem. Phys.*, 1965, **42**, 1173; G. Herzberg, *Molecular Spectra and Molecular Structure III. Electronic Spectra and Electronic Structure of Polyatomic Molecules* (Van Nostrand Reinhold Co., New York, 1966), p. 621.

[13] J. A. Kerr, *Chem. Rev.*, 1966, **66**, 465.

[14] T. L. Hill, *An Introduction to Statistical Thermodynamics* (Addison-Wesley Publishing Co. Inc., Reading, Massachusetts, 1960), p. 166; S. W. Benson, *J. Chem. Ed.*, 1965, **42**, 502.

[15] R. J. Donovan and D. Husain, *Nature*, 1966, **209**, 609.

[16] C. C. Chou, P. Angelberger and F. S. Rowland, *J. Phys. Chem.*, 1971, **75**, 2536.

[17] C. Tsao and J. W. Root, *J. Phys. Chem.*, 1962, **76**, 308.

[18] R. J. Donovan, F. G. M. Hathorn and D. Husain, *Trans. Faraday Soc.*, 1968, **64**, 3192.

[19] G. E. Busch and K. R. Wilson, *J. Chem. Phys.*, 1972, **56**, 3638.

[20] For a discussion of the observation of I and I* in the photofragment spectroscopy of other molecules (I_2, IBr and ICN) see G. E. Busch, R. T. Mahoney, R. I. Morse and K. R. Wilson, *J. Chem. Phys.*, 1969, **51**, 837; R. J. Oldman, R. K. Saunder and K. R. Wilson, *J. Chem. Phys.*, 1971, **54**, 4127; and ref. (1a).

[21] C. D. Bass and G. C. Pimentel, *J. Amer. Chem. Soc.*, 1961, **83**, 3754.

[22] A. D. Walsh, *J. Chem. Soc.*, 1953, 2321.

[23] J. G. Calvert and J. N. Pitts, Jr., *Photochemistry* (John Wiley and Sons, Inc., New York, 1966), p. 647.

[24] R. D. Souffie, R. R. Williams and W. H. Hamill, *J. Amer. Chem. Soc.*, 1956, **78**, 917.

[25] (a) C. E. Klots, *J. Chem. Phys.*, 1964, **41**, 117; (b) M. A. Haney and J. L. Franklin, *J. Chem. Phys.*, 1968, **48**, 4093.

[26] B. S. Rabinovitch and R. W. Diesen, *J. Chem. Phys.*, 1959, **30**, 735.

146   ALKYL IODIDE PHOTODISSOCIATION

[27] R. A. Marcus and O. K. Rice, *J. Phys. Chem.*, 1951, **55**, 894; R. A. Marcus, *J. Chem. Phys.*, 1952, **20**, 359.

[28] M. Vestal, A. L. Wahrhaftig and W. H. Johnston, *J. Chem. Phys.*, 1962, **37**, 1276.

[29] The other internal degrees of rotation in the propyls are assumed not free and are assigned in the spectra.

[30] G. Z. Whitten and B. S. Rabinovitch, *J. Chem. Phys.*, 1964, **41**, 1883.

[31] K. E. Holdy, L. C. Klotz and K. R. Wilson, *J. Chem. Phys.*, 1970, **52**, 4588.

[32] F. E. Heidrich, K. R. Wilson and D. Rapp, *J. Chem. Phys.*, 1971, **54**, 3885.

[33] D. Porret and C. F. Goodeve, *Trans. Faraday Soc.*, 1937, **33**, 690; M. Ito, P. C. Huang and E. M. Kosower, *Trans. Faraday Soc.*, 1961, **57**, 1662; K. Kimura and S. Nagakura, *Spectrochim. Acta.*, 1961, **17**, 166.

[34] *Tables of Interatomic Distances and Configuration in Molecules and Ions*, ed., L. E. Sutton (The Chemical Society, London, 1958), p. M134.

[35] (a) P. R. Brooks and E. M. Jones, *J. Chem. Phys.*, 1966, **45**, 3449; (b) R. J. Beuhler, R. B. Bernstein and K. H. Kramer, *J. Amer. Chem. Soc.*, 1966, **88**, 5331; (c) J. R. Airey, E. F. Greene, G. P. Reck and J. Ross, *J. Chem. Phys.*, 1967, **46**, 3295; (d) C. Maltz and D. R. Herschbach, *Disc. Faraday Soc.*, 1967, **44**, 176; (e) R. J. Behuler and R. B. Bernstein, *Chem. Phys. Letters*, 1968, **2**, 166; (f) Ch. Ottinger, *J. Chem. Phys.*, 1969, **51**, 1170; (g) J. H. Birely, E. A. Entemann, R. R. Herm and K. R. Wilson, *J. Chem. Phys.*, 1969, **51**, 5461; (h) G. H. Kwei, J. A. Norris and D. R. Herschbach, *J. Chem. Phys.*, 1970, **52**, 1317; (i) R. M. Harris and J. F. Wilson, *J. Chem. Phys.*, 1971, **54**, 2088 : (j) M. E. Gersh and R. B. Bernstein, *J. Chem. Phys.*, 1971, **55**, 4661. For a review of work prior to 1966, see ref. (36).

[36] D. R. Herschbach, *Adv. Chem. Phys.*, 1966, **10**, 319.

[37] unpublished work of Ch. Ottinger, P. M. Strudler and D. R. Herschbach, preliminary results summarized in ref. (35f).

[38] D. R. Herschbach in *Potential Energy Surfaces in Chemistry*, ed. W. A. Lester, Jr. (IBM Research Laboratory, San Jose, Calif., 1971), pp. 44-57.

[39] J. D. McDonald, P. R. LeBreton, Y. T. Lee and D. R. Herschbach, *J. Chem. Phys.*, 1972, **56**, 769.

[40] R. S. Mulliken, *Phys. Rev.*, 1935, **47**, 413.

[41] D. R. Herschbach, private communication.

[42] R. B. Bernstein, private communication.

Gas-Phase Kinetics

Robin Walsh

Department of Chemistry, University of Reading, Whiteknights, Reading, UK RG6 6AD

Commentary on: **Rates of pyrolysis and bond energies of substituted organic iodides (Part 1),** E. T Butler and M. Polanyi, *Trans Faraday Soc.*, 1943, **39**, 19.

This paper, published under some pressure during World War 2, was considered incomplete by its authors at the time (and indeed a Part 2 was published two years later). What it achieved despite its authors' reservations, was to present some rate measurements which brought together almost all of the contemporary ideas which linked reaction kinetics to the measurement and understanding of bond dissociation energies (BDEs). It generated values for more than 36 BDEs in alkanes, alcohols and alkyl halides, some of which are remarkably close to current best values. This work provided the springboard in postwar years for a vast expansion and refinement of the methods of using kinetic measurements of activation energies to provide an increasingly reliable basis for the strengths of chemical bonds.

Butler and Polanyi investigated the rate of pyrolysis of a series of organic iodides by a flow technique. They calculated the fraction of iodide decomposed by measuring the amount of free iodine formed in the reaction. In their opinion the use of a flow technique was advantageous, since the accumulation of the products resulting from an extended period of flow made it possible to work with very small partial pressures of organic iodide and to limit the total decomposition to a very small percentage. Thus by maintaining low concentrations of both reactants and products, the chances of secondary reactions were considerably reduced. This was thought to be further minimised by the brief reaction time of less than a second during which gases passed through the reaction vessel.

Since the C-I bond is the weakest bond in organic iodides, it was thought obvious that the first step in the pyrolysis of these compounds should involve its rupture:

$$RI \rightarrow R + I$$

The formation of I_2 was attributed to this reaction (via the subsequent rapid combination process: $I + I \rightarrow I_2$). First order rate constants for RI decomposition were calculated and shown (in most cases) to be relatively independent of conditions (although the authors were aware of the complicating HI elimination reaction for some iodides).

Measurements over a temperature range allowed activation energies to be calculated in some cases but to cover all the molecules under investigation the authors chose to assume a pre-exponential (*i.e.* A) factor of 10^{13} s^{-1} which allowed the calculation of E_a values for all the organic iodides. The E_a values were assumed to be identical with the bond dissociation energies. After listing the values the paper goes on to discuss the differences between bond energies and bond energy terms, the general relationship between activation energies and bond energies, the resonance stabilisation of radicals (such as allyl and benzyl) and the connection between bond energies and substitution heats used to generate the full table of BDEs referred to earlier.

This publication stimulated much discussion, particularly relating to the assumptions about the mechanism of decomposition of iodides. Especially contentious was the assumption of no secondary reactions. In the 1950s Szwarc[1] developed the toluene carrier technique to try to ensure the complete scavenging of the initially formed radicals from an organic bond fission reaction. A further refinement of this method by Trotman-Dickenson[2] was the aniline carrier technique.

In 1961 Benson and O'Neal[3] reinterpreted the kinetics of organic iodide decomposition and thereby laid the basis for what became known as the iodination technique, *viz.* measurement of E_a for the reaction:

$$I + RH \underset{r}{\overset{f}{\rightleftharpoons}} R + HI$$

This important dissociation energy $D(R-H)$ was obtained from the thermochemical relationship: $D(R-H) = D(H-I) + E_f - E_r$. This method was exploited by Benson and co-workers[4] to great effect to obtain large numbers of BDE values. Parallel but less extensive work was done on bromination kinetic studies (by Kistiakowsky,[5] Whittle[6] and others).

Work in my group[7] extended the use of the iodination method into the area of organosilane and organogermane BDEs. Direct pyrolytic dissociation measurements of BDEs, *i.e.* via simple bond fission, analogous to the original Butler and Polanyi mechanism, has been further exploited using the shock tube studies (Tsang[8]) and the VLPP technique (Golden and Benson[9]). Further refinements of the iodination and bromination methods, *viz.* the direct measurement of E_a for R + HX (the so called 'back reaction') were developed, particularly by Gutman.[10]

I began my research career in the 1960s when I first became familiar with this field. With hindsight working through the past 40 years it has often seemed that progress has been frustratingly slow and fragmented. Yet there have been many contributions too numerous to record in such a brief appreciation. However it is abundantly clear that much of the inspiration and many of the arguments about bond dissociation energies and their values were captured in this landmark paper. As an indication of how close Butler and Polanyi came to currently accepted values their estimated alkane C–H bond strengths are compared with current best values in the table below.

Bond	Butler and Polanyi	Current best value[a]
CH$_3$–H	102.5	104.8
C$_2$H$_5$–H	97.5	101.1
i-C$_3$H$_7$–H	89.0	98.6
t-C$_4$H$_9$–H	86.0	96.5

[a] Ref 10.

References

1 M. Szwarc, *Chem. Rev.*, 1950, **47**, 75.
2 G. L. Esteban, J. A. Kerr and A. F. Trotman-Dickenson, *J. Chem. Soc.*, 1963, 3873.
3 S. W. Benson and H. E. O'Neal, *J. Chem. Phys.*, 1961, **34**, 514.
4 D. M. Golden and S. W. Benson, *Chem. Rev.*, 1969, **69**, 125.
5 G. B. Kistiakowsky and E. R. Van Artsdalen, *J. Chem. Phys.*, 1944, **12**, 469.
6 J. C. Amphlett, J. W. Coomber, and E. Whittle, *J. Phys. Chem.*, 1966, **70**, 593.
7 R. Walsh, *Acc. Chem. Res.*, 1981, **14**, 246.
8 W. Tsang, In *Shock Waves in Chemistry*, ed. A. Lifshitz, Dekker, New York, 1981 p.59.
9 G. N. Spokes, D.M. Golden and S. W. Benson, *Angew. Chem., Int. Edn. Engl.*, 1973, **12**, 534.
10 J. Berkowitz, G.B.Ellison and D. Gutman, *J Phys. Chem.* 1994, **98**, 2744.

RATES OF PYROLYSIS AND BOND ENERGIES OF SUBSTITUTED ORGANIC IODIDES (PART I).

By E. T. Butler and M. Polanyi.

Received 11th September, 1942.

There is much evidence of an indirect nature, in the wide variation in reactivity of related organic compounds, which suggests that the nature of a chemical bond is greatly influenced by substitution. This would seem to conflict with the additivity rule of Bond Energies based on thermo chemical evidence, which assumes that the energy of a given kind of bond is a constant. The present work was begun with the object of clarifying this situation by observing the effect of substitution on organic bond strength as manifested in the rates of pyrolysis of a series of organic iodides.*

R. A. Ogg and **Ogg and Polanyi** have studied the pyrolysis of a number of aliphatic iodides by static methods.[1] Under their conditions the kinetics of the decomposition are rather complex, and we have come to the conclusion from our own work reported below, that the rate of the primary bond breaking process $RI = R + I$ cannot be deduced correctly from their results. Pyrolysis as a means of establishing bond energies was used by Rice and Johnston [2] who measured the temperature coefficients of the rates of decomposition of several organic compounds and calculated the corresponding activation energies. They used a flow method in which the free radicals formed during the process were allowed to remove metallic mirrors. We used a similar arrangement for pyrolysing organic iodides, while measuring the rate of reaction by the amount of iodine (and in certain cases the hydrogen iodide) formed. The advantages of the flow method are twofold : by the accumulation of the product resulting from an extended period of flow it is possible to use

---

* A preliminary communication of our results was made in *Nature*, 1940, **146**, 121.

How far we have, here, succeeded in establishing the correct values of bond-energies may be open to doubt. However, the confirmation which our results for (CH_3—I) and C_2H_5—I) have recently gained by quite independent observations (see p. 29 and p. 35 below) ; the close correspondence of the variations of estimated bond-strength with predicted forms of resonance ; the theory which could be built up on this basis for the variations of formation heats and the accompanying changes in dipole strength for a series of hydrocarbons and their derivatives ; [21] the correspondence between estimated bond energies and activation energies of the Na-reaction—all these features together have convinced us that we may be permitted to disregard for the time certain gaps in the kinetic evidence of the pyrolysis experiments. Nevertheless, we would have preferred to extend this evidence before publication—especially as this would have saved much tiresome discussion of details—if the extension were at all possible. However, the experiments had to be discontinued in consequence of the war for an indefinite time and postponement of their publication would involve a serious risk of their complete loss. In view of the variety of subjects on which they seem to throw light, we have felt their publication in the present form to be desirable.

[1] Ogg, *J.A.C.S.*, 1934, **56**, 526 ; Ogg and Polanyi, *Trans. Faraday Soc.*, 1935, **31**, 604 ; see also Jones and Ogg, *J.A.C.S.*, 1937, **59**, 1931, 1939, 1942 ; Jones, *ibid.*, 1938, **60**, 1877 ; 1939, **61**, 3284.

[2] Rice and Johnston, *ibid.*, 1934, **56**, 214.

very large volumes of gas in the reaction and thereby (*a*) maintain very small partial pressures of the organic substance, and (*b*) limit the total decomposition to a very small percentage. The low concentrations of the initial and final products thus achieved reduce considerably the chances of secondary reactions. These are suppressed further by the brief duration of the reaction which is over in a second or less as the gas passes through the reaction chamber.

Experimental.

The complete apparatus is illustrated in Fig. 1. The reaction vessel R_1R_2 was of Pyrex, R_1 being 6 cm. in diameter and R_2 2 cm., each being 15 cm. long. It was possible to heat either the whole vessel R_1R_2 (volume 440 c.c.) or the narrow part R_2 (40 c.c.) alone, and in this way a tenfold variation in contact time could be effected without changing the rate of flow; adjustment of the latter gave further variation of contact time.

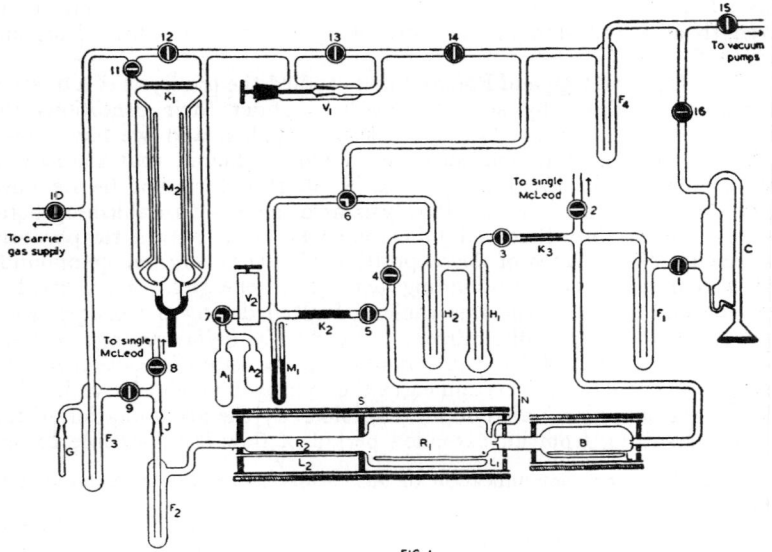

FIG. 1.

The temperature of the reaction vessel was measured by means of copper-advance thermocouples placed in the narrow inlet tubes L_1 and L_2. When steady conditions had been attained the maximum difference in temperature along the whole length of the reaction vessel was 10° at 500°, and less at lower temperatures. The average temperature inside the reaction vessel was found by moving the thermocouples along the tubes and reading the temperatures at 1 cm. intervals. By careful control the temperature could be kept constant within ± 1·5° during a run of several hours duration.

The carrier-gas stream was circulated by the three-stage mercury vapour pump C. After removal of mercury vapour by passage through the liquid air trap F, the stream divided into two parts. One part went first through the vessel B (where mercury was introduced in those experiments in which its effect was investigated) and entered the reaction vessel through a short capillary, while the other part after picking up the vapour of the organic iodide entered the reaction vessel at N. The decomposition products and undecomposed iodide were condensed in F_1 and F_2 immersed in a mixture of acetone and solid CO_2 and liquid air respectively. The

E. T. BUTLER AND M. POLANYI 21

rate of flow of the carrier-gas was measured by the pressure gradient across the calibrated capillary K_1 recorded on the double McLeod gauge M_2, and could be varied by adjustment of the Hauschild valve V_1.

Two methods were employed for controlling and measuring the rate of entry of the iodide into the reaction vessel. Iodides of high vapour pressure (> 20 mm. at room temperature) were introduced into the storage vessel A_1 and outgassed by freezing and evacuating several times. The iodide vapour flowed through the needle-valve V_2 the fine capillary K_2 and the tap 5 into the carrier-gas stream. The flow of the iodide was controlled by adjusting V_2 and measured by the fall in pressure across K_2 which was calibrated for each iodide used. Iodides of lower vapour pressure were introduced by a method devised by Warhurst.[3] The stream of carrier-gas passed first through the trap H_1 containing the iodide and then through a second trap H_2 maintained at a temperature about 20° below that of H_1. This arrangement ensured saturation of the carrier-gas with the iodide at the temperature of H_2. The rate of flow of the iodide was again found by preliminary calibration.

Before beginning a run the reaction vessel was baked out for two hours at a pressure less than 10^{-5} mm. When the compound undergoing pyrolysis gave hydrogen iodide as well as free iodine, the latter was condensed in F_2 at $-78°$, while the former passed through F_2 and collected in F_3 at $-195°$. At the end of the experiment F_2 was cut off, the contents washed out with absolute alcohol and the iodine titrated against standard sodium thiosulphate. The hydrogen iodide was titrated against standard alkali. When mercury was present, the mercury halides collected in F_2 were (after removal of the volatile contents of the trap) reduced by boiling with zinc dust and water and the iodide estimated with standard silver nitrate.

The ethyl, *n*-propyl, *n*-butyl and allyl iodides were commercial products, carefully purified before use, while the other iodides were prepared and purified by standard methods. Cylinder nitrogen was freed from oxygen by passage over sodium at 350°, and cylinder hydrogen was purified by passage through a palladium thimble. Pure nitric oxide was made by the method of Giauque and Johnston.

Results.

In Table I are presented the results for ethyl iodide in hydrogen as carrier-gas. We will discuss—to begin with—only the experiments carried out at 492°-494°. The rates of reaction are expressed in terms of " first order constants " k_1; which must not prejudice the question of the actual kinetic mechanism involved. Comparison of the pairs grouped together shows that except in one case reproducibility was better than ± 5 %. In view of the whole investigation we may say that this degree of reproducibility within one run is not uncommon where the quantities of iodine produced are sufficient. However in some cases, to which attention will be called in due course, reproducibility was considerably less.

The introduction of Hg-vapour which was used from Exp. 3 onwards had no noticeable effect. This admixture was continued throughout the experiments with ethyliodide and then left away. A more than twofold increase in H_2 pressure in Exps. 3-4 and 9-10 produced a rise of k_I by about 40 %, while a more than sixfold increase in Exps. 13-14 and 11-12 caused a rise of k_I by 115 %. In the latter comparison we disregard the change in ethyliodide pressure since in subsequent experiments its variation proved entirely ineffective. Comparing (with the same proviso) Exps. 9-10 and 14-13 we note a considerable fall of k_I with increasing time of contact.

The tendency for k_I to rise with increasing pressure and to fall with increasing contact time was much less noticeable—in fact, not definitely

[3] Warhurst, *Trans. Faraday Soc.*, 1939, **35**, 674.

TABLE I.

Exp. No.	Temp. °C.	Total Press. mm.	Iodide Press. mm. × 10^{-3}.	Contact Time. sec.	k_I. sec.$^{-1}$	k_{HI}. ×10^{-2}.
1. Ethyl Iodide in Hydrogen.						
NO ⎰ 1	492	5·95	1·40	0·275	4·62	
Hg. ⎱ 2	492	6·46	1·21	0·258	4·67	
3	493	5·74	1·22	0·284	5·60	
4	492	5·92	1·24	0·285	4·04	
9	493	13·4	1·07	0·252	6·60	
10	493	14·1	1·02	0·256	6·77	
13	493	13·3	8·11	0·586	3·35	
Hg. ⎰ 14	492	13·2	9·33	0·660	3·70	
11	493	2·06	1·30	0·641	1·52	
12	493	2·02	1·21	0·659	1·52	
6	462	6·90	1·04	0·260	1·37	
7	462	5·56	1·18	0·292	0·82	
8	461	5·52	1·21	0·304	0·90	
2. Ethyl Iodide in Nitrogen.						
Varying ⎰ 27	495	12·8	1·47	1·39	1·10	1·02
iodide ⎱ 26	493	13·3	1·64	1·04	1·22	—
pressure ⎰ 30	493	9·32	30·4	1·69	1·00	1·07
29	494	9·58	33·5	1·76	1·06	0·98
Varying ⎰ 15	493	1·79	1·75	1·61	0·87	—
total ⎱ 17	492	4·92	1·56	1·56	0·95	—
pressure ⎰ 18	493	5·01	1·52	1·52	0·93	—
19	493	9·00	2·11	1·92	0·99	—
28	494	13·3	1·54	2·33	1·01	0·97
20	492	12·8	1·43	1·72	1·08	—
Varying ⎰ 25	492	12·5	1·59	0·728	1·69	—
contact ⎱ 22	494	12·7	1·38	0·641	1·67	—
time ⎰ 24	493	12·4	1·57	0·140	1·32	—
31	494	10·7	26·4	0·139	0·50	0·63
23	492	11·8	1·52	0·060	1·04	—
NO ⎰ 32	493	9·38	31·7	1·74	1·90	0·89
present ⎱ 33	493	9·45	30·9	1·80	2·66	0·85
3(a). n-Propyl Iodide in Nitrogen.						
112	492	6·50	23·7	1·08	4·94	4·22
109	492	6·28	11·0	0·516	5·20	3·69
108	491	7·32	8·21	0·473	5·19	3·47
113	492	6·12	11·2	0·0412	3·69	6·91
115*	493	6·06	11·0	0·528	5·35	4·14
114†	492	5·71	15·7	0·710	5·38	5·01
110	440	6·96	10·0	0·521	0·466	0·408
111	442	6·39	10·2	0·550	0·472	0·444

* NO present.   † Carrier-gas : hydrogen.

3(b). n-Butyl Iodide in Nitrogen.

Exp. No.	Temp. °C.	Total Press. mm.	Iodide Press. mm. × 10^{-3}.	Contact Time. sec.	k_I. sec.$^{-1}$	k_{HI}. ×10^{-2}.
121	492	6·49	6·32	1·40	8·08	11·2
125	491	5·39	7·78	0·565	8·38	—
118	492	6·63	2·57	0·502	10·05	17·4
122	492	6·36	7·38	0·0435	7·30	14·0
119	440	7·01	1·59	0·506	0·85	2·30
120	440	6·53	2·85	0·568	0·59	1·44
124	441	6·54	8·28	0·646	0·65	0·99

E. T. BUTLER AND M. POLANYI 23

TABLE I.—*Continued.*

Exp. No.	Temp. °C.	Total Press. mm.	Iodide Press. mm. × 10⁻³.	Contact Time. sec.	k_I.	k_{HI}. sec.⁻¹ × 10⁻².

4(a). Allyl Iodide in Nitrogen.

76	494	6·14	56·2	0·520	169·0	
77	425	5·88	57·9	0·588	52·1	
87	425	61·0	7·53	0·570	52·8	
80	427	6·20	64·0	0·0487	223·0	
84	356	6·23	183	1·86	5·61	
78	354	6·39	50·8	0·625	6·21	
82	357	6·35	67·1	0·612	8·83	
86	358	6·52	7·08	0·588	6·37	
89	358	6·52	6·60	0·0460	8·9	
81	355	6·61	63·0	0·0475	20·5	
83	296	6·13	65·2	0·681	1·08	
79	298	6·49	47·9	0·663	0·92	
85	297	5·99	66·9	0·562	1·05	
88	300	6·27	7·18	0·6600	0·53	

4(b). Vinyl Iodide in Nitrogen.

91	506	5·20	50·8	1·27	0·073	
92	508	5·44	35·2	0·545	0·21	

5(a) Benzyl Iodide (Carrier-Gas : Hydrogen).

61	492	6·28	1·22	0·242	253	
67	493	6·01	1·24	0·245	249	
60	492	5·80	1·22	0·254	234	
64	492	6·29	1·50	0·023	465	
66	433	5·49	3·82	2·11	10·1	
62	429	6·13	1·18	0·266	37·8	
63	429	6·57	1·389	0·249	39·3	
65	430	6·22	1·32	0·0224	46·9	

5(b). Benzyl Iodide (Carrier-Gas : Nitrogen).

68	495	5·68.	1·73	0·591	70·4	
69	494	6·73	1·38	0·0439	322·0	
70	430	5·61	1·70	0·640	18·5	
71	431	5·94	1·56	0·0512	26·8	
73†	431	5·85	1·56	0·635	19·6	

5(c). Phenyl Iodide in Hydrogen.

49*	511	7·51	30·8	0·224	0·89	
50†	509	6·53	30·2	0·577	1·76	

* Hg vapour present.   † Carrier-gas : NO.

6(a). Acetyl Iodide in Nitrogen.

73	492	6·18	84·8	0·517	3·54	
74	392	5·91	61·7	0·563	1·84	
75	364	6·28	47·5	0·597	1·41	

6(b). Benzoyl Iodide in Nitrogen.

98	430	5·37	3·40	0·619	9·87	
103	430	6·76	5·00	0·600	10·4	
102	430	5·59	4·04	0·0468	36·4	
104	431	5·75	5·36	0·0468	36·4	
99	390	6·91	3·75	0·630	3·63	
100	390	5·99	3·64	1·26	2·35	

24   RATES OF PYROLYSIS AND BOND ENERGIES

TABLE I.—*Continued.*

Exp. No.	Temp.	Total Press.	Iodide Press.	Contact Time.	k_I.	k_{HI}.
	°C.	mm.	mm. $\times$ 10^{-3}.	sec.		sec.$^{-1}$ $\times$ 10^{-2}.

7(a). Acetonyl Iodide in Nitrogen.

34*	493	7·13	87·2	2·08	5·98	
35*	494	7·70	42·6	0·500	21·3	
36*	494	7·23	33·5	0·0430	97·0	
37	451	6·71	25·4	0·530	33·8	
93	429	6·48	37·3	0·584	2·52	
97	382	6·37	33·3	0·596	0·90	
96	383	5·16	23·8	0·73	0·45	

* Hg vapour present.

7(b). Methyl Iodide in Nitrogen.

130	494	6·32	3·32	0·615	2·33	
131	495	4·97	3·40	0·585	5·17	
132	494	5·50	2·76	0·592	2·70	
133	494	7·00	3·59	1·51	7·80	
134	493	6·88	12·3	1·63	3·05	

observable at all—for ethyl iodide in nitrogen, to which Table I (2) refers. We would not venture to ascribe any of the variations of k_I here shown (apart from Exps. 32-33) to the corresponding changes in the reaction variables. A distinct rise to about twofold values of k_I is caused by the admixture of NO in Exps. 32-33.

In Table I (2) there also appear values for k_{HI} expressing the rate of HI-formation which was not tested previously. This reaction proceeds at a rate perceptibly equal to that of the formation of iodine, except in Exps. 32-33 where it shows itself unaffected by the admixture of NO.

Taken by themselves the experiments in nitrogen would form a sufficient proof for the presence of a pair of truly monomolecular reactions

$$\text{(1) } C_2H_5I = C_2H_5 + I$$
and
$$\text{(2) } C_2H_5I = C_2H_4 + HI,$$

the first of which leads to the formation of I_2. The process $2I = I_2$ may be assumed to occur with less than 100 % yield owing to losses likely to be caused by the back reaction $C_2H_5 + I = C_2H_5I$. It may be thought that the undiminished primary decomposition (1) is observed in Exps. 32-33 where the NO present removes the free radicals.

This picture is consistent with all the later experience on pyrolysis of iodides both in the presence of nitrogen and hydrogen, but it is not easy to reconcile with some of the results of Table I (1). Even admitting the possibility of a reaction sequence on the lines $C_2H_5 + H_2 = C_2H_6 + H$; $H + C_2H_5I = C_2H_6 + I$, the yield in Exps. 9-10 still appears to be inexplicably high, especially taking into account that further reduction in contact time and increase in pressure seems to increase k_I even more. Such difficulties do not recur throughout the whole of the investigation recorded in this and the following paper. In view of this fact (and considering also that the experiments in Table I (1) were the very first results obtained by the new technique utilised throughout the rest of the investigation) we did not think it right to suspend indefinitely the detailed evaluation of all the material hitherto collected, but preferred rather to disregard for the time being the difficulties which the interpretation of Table I (1) presents and to lay all the stress on the mass of later experiments from Table I (2) onwards.

Measurements on *n*-propyl iodide and *n*-butyl iodide, are recorded in Table I (3). Exps. 112-109-108-113 show that neither the variation of the iodide pressure nor the very large changes in contact time have any definite effect on k_I. The formation of HI continues to occur as in ethyl iodide at about the same rate as that of I_2. The admixture of NO in Exp. 115 had no effect ; nor did the replacement of N_2 by H_2 in 114 produce any change. We will return to the measurements at lower temperature (110 and 111) later.

For *n*-butyl iodide there is again a wide variation of contact times in Exps. 121-125-118-112 without any definite change in k_I. HI-formation is somewhat larger in proportion to I_2-formation and is less reproducible.

Comparison of ethyl iodide in nitrogen with *n*-propyl and *n*-butyl iodide shows a definite increase in the rate of pyrolysis both in the production of I_2 and of HI. For ethyl : *n*-propyl : *n*-butyl k_I at 492° = 1 : 4·3 : 7·4 k_{HI} at 492° = 1 : 5·0 : 11·6. Extrapolating the data of EtI to 440° by use of temperature coefficient evaluated from Table I (1) we obtain for the ratio of k_I at 440° 1 : 4·7 : 7·0.

Table I (4) (*a*) shows that allyl iodide decomposes much more readily than any of those previously described. At 494 decomposition was almost 60 %, so that in order to obtain more moderate decomposition lower temperatures had to be chosen. Exps. 77-78 show the absence of any effect of iodide-pressure on the rate. The comparison of these results with Exp. 80 reveals a fourfold increase of k_I caused by a twelvefold reduction of the contact time. Exps. 84, 78, 82 and 86, carried out at the lower temperature of about 356° again show k_I as independent of the iodide pressure. Comparing Exp. 84 with Exps. 78, 82 and 86, and the latter with Exps. 89 and 81, we see also that the rise of k_I with diminishing contact time has become less marked—if not altogether negligible. With a further lowering of the temperature to about 298° (Exps. 83, 79, 85, 88) the dependence of k_I on contact time becomes quite imperceptible. Thus with decreasing temperature the kinetics of the reaction appear to conform increasingly well to the monomolecular scheme. Extrapolating the rate of pyrolysis of *n*-propyl iodide down to 298°, the ratio of k_I for *n*-propyl : allyl can be calculated at about 1 : 14000.

The striking increase in the rate of pyrolysis observed in passing from the saturated iodides to allyl iodide is contrasted by a considerable variation in the opposite direction in the case of vinyl iodide—Table I (4) (*b*). At the standard temperature of 493° no decomposition could be observed. The highest accessible temperatures (506°-508°) had to be applied in order to obtain a measurable result. While no great accuracy can be claimed in these circumstances, Exps. 91-92 leave no doubt that k_I is considerably reduced here as compared *e.g.* with ethyl iodide—the rates for both substances being extrapolated to the same temperature, 507°. For the ratio ethyl : vinyl we calculate k_I about 16 : 1.

Next we come—in Table I (5)—to benzyl iodide and phenyl iodide. Benzyl iodide was measured both in hydrogen and in nitrogen at two temperatures ; in Exp. 73 NO was used as carrier gas. At 493° there is both for H_2 and N_2 carrier gas a deviation from the monomolecular form in the sense of higher k_I values being obtained at shorter contact times : but again this behaviour is much reduced by lowering the temperature, as the experiments at 430°-431° indicate. The presence of hydrogen increased the production of I_2 almost twofold both at 493° and 431° ; while NO shows no effect at all. Comparison of the rate of pyrolysis with that of ethyliodide at 431° (taking both iodides in nitrogen) yields for ethyl : benzyl k_I about 1 : 4800.

In the experiments with phenyl iodide there was again no observable decomposition at the standard temperature of 493° ; but we obtained pyrolysis at 510°. We may estimate for phenyl : ethyl k_I about 1 : 3.

Acid iodides were tested by the examples of acetyl iodide and benzoyl iodide. The sensitivity to light of the former compound caused traces of iodine to be formed even without pyrolysis. The observations in Table I (6) can justify nothing more than the claim that at the temperature of 493° k_I for acetyl iodide is around 2 to 3 × 10^{-2}. Benzoyl iodide is seen to be far more readily decomposed than acetyl iodide. However, at 430

26   RATES OF PYROLYSIS AND BOND ENERGIES

there occurs an almost fourfold increase of k_I when contact time is reduced twelvefold, and a similar feature is present at 390° ; which makes k_I as a monomolecular reaction constant very uncertain. Yet we may note that while the decomposition of acetyl iodide is about twice as fast as that of ethyl iodide, there is about a 2000-fold increase of the rate from ethyl to benzoyl iodide—comparing these two substances at 390° and taking the shortest measured time of contact from Table I (6) for benzoyl iodide.

Finally we add some observations on acetonyl iodide and methyl iodide. We had considerable difficulty with the former substance on account of the decomposition which it undergoes by light and even on standing. Yet Table I (7) may reasonably indicate the position of this compound among other iodides. It lies between benzoyl and *n*-butyl, so that *n*-butyl < acetonyl < benzoyl ; roughly in the ratio 1 : 4 : 70. While methyl iodide gave no reproducible results, its pyrolysis proved very distinctly slower than that of ethyl iodide. The mean rate constant $4·2 \times 10^{-3}$ sec.$^{-1}$ is about 3 times less than that of ethyl iodide.

We may use the observations listed above to derive in some cases the temperature coefficients of the rate of pyrolysis and hence the activation energy of the reaction. For *n*-propyl iodide and *n*-butyl iodide we have sufficiently reliable data for two temperatures from which we calculate : for *n*-propyl $Q = 52$ kcal. and for *n*-butyl $Q = 53$ kcal. This lends support to the value of similar magnitude, $Q = 55$ kcal., obtained from the otherwise less reliable Exps. 3, 4, and 6, 7, 8 for ethyl iodide.

The measurements for allyl-, benzyl-, and benzoyl iodide and acetonyl iodide all show at higher temperatures a variation of k_I with contact time which makes the calculation of an activation energy uncertain. Since in view of the conditions of our experiments we can exclude errors which would increase the rate of decomposition, we consider the higher values which are observed at the shorter contact times as nearer to the truth and have selected the data for calculation in accordance with this view. When we combine for allyl iodide the average of the experiments made at 298° with (*a*) Exp. 81 (*b*) Exp. 89 (*c*) Exp. 80 we obtain for Q (*a*) 39 (*b*) 27·5 (*c*) 34·0. For benzoyl iodide we may rationally choose (*a*) Exps. 64 and 65, and (*b*) Exps. 69 and 71 as representing the highest values of k_I for both temperatures and carrier gases. We obtain for Q (*a*) 47·5, and (*b*) 48·5. For benzoyl iodide the most rational seems to be, on similar grounds, to choose Exp. 99 combined with the average of 102 and 104 which leads to $Q = 53$; alternatively we obtain from Exp. 99 and the mean of 98 and 103 $Q = 24$. For acetonyl iodide Exps. 36 and 37 yield $Q = 27$.

Before discussing these values, we may try to derive activation energies by the alternative method of using the equation for the rate of monomolecular decomposition $k = \nu e^{-Q^*/RT}$ where ν is usually about 10^{13}. The difficulty is here that we do not know for certain what fraction of the primarily formed iodine is lost by recombination either while the gas is at the reaction temperature or during the subsequent phase of cooling. Tentatively assuming that recombination is negligible, at least at the lowest temperatures and pressures when conformity to the monomolecular form was closer—and limiting ourselves for the sake of uniformity to the results in nitrogen—we obtain from $k_I = 10^{13}e^{-Q^*/RT}$ the series of Q^* values listed in Table II.

Now if the yields of iodine were much less than the assumed value of unity, the energies Q^* would be considerably higher than the true activation energies ; a yield of 0·1 for example would cause Q^* to be too large by 3 kcal. For a number of compounds such an assumption seems unacceptable. For *n*-propyl iodide and *n*-butyl iodide the activation energies calculated from $k_I = 10^{13}e^{Q^*/RT}$ at the temperature of 493° are 49·9 and 48·9 kcal. respectively, which values scarcely differ from Q^* calculated for the lower temperature and listed in Table II. This fact is reflected once more in the satisfactory correspondence between the energies Q^* and the activation energies $Q = 52$ and $Q = 53$ calculated from the

E. T. BUTLER AND M. POLANYI **27**

TABLE II

Iodide.	Q^* kcal.	$k_I(430°) \times 10^5$ sec.$^{-1}$,	$T(1\%$ per sec.) in °C.
Methyl	(54)	14	516
Ethyl	52·2	50	489
n-Propyl	50·0	250	457
n-Butyl	49·0	560	442
iso-Propyl	46·1	4000	398
tert Butyl	45·1	9000	385
Allyl	39·0	630000	297
Vinyl	55·0	6·3	529
Benzyl	43·7	23000	366
Phenyl	54·0	14	516
Acetyl	(50·7)	160	445
Benzoyl	43·9	18000	369
Acetonyl	45·0	9000	385

temperature coefficient derived from the same data. It leaves no room for the assumption that the true activation energy is noticeably less than Q^*.

We could try to get nearer to the truth about the reduction of the yield due to the back reaction by using our observations in the presence of NO. Its ineffectiveness in case of *n*-propyl iodide might be taken to confirm the assumption that the yield of iodine is unreduced in this case ; while from Exps. 29, 30 and 32, 33 on ethyl iodide it might be suggested that for this substance the yield is reduced to about 0·5. This would mean that Q^* for ethyl iodide exceeds the activation energy by 1 kcal. These considerations, however, would strain the present evidence too far— particularly in view of the uncertainty of the value of the factor v and of its possible variations from one substance to another. Comparing the value $Q^* = 44$ for benzyl iodide with the Q-values from the temperature coefficient ((*a*) 47·5, and (*b*) 48·5) there again seems little possibility for the activation energy to be markedly less than Q^*.

In the cases of allyl and benzoyl iodide the Q values are too scattered to justify comparisons with Q^* ; but there is nothing to suggest that the behaviour of these compounds differs much from that of the others already discussed. We have assumed this behaviour to hold also for vinyl and phenyl iodide for the purpose of constructing Table II for the extrapolation of the k_I values for vinyl and phenyl iodide to 430° by use of a temperature coefficient derived from Q^*.

The mean value of 54 kcal. given from Q^* of methyl iodide results from the experiments listed in Table I (7) from which values ranging from 53 to 55 kcal. were obtained.

Comparisons between the values of Q^* and Q can also be used to examine the possibility of the observed rate, as postulated by the theory of monomolecular reactions, falling short of the limiting rate for high pressures. The kind of argument we have just used applies here with even greater force, since if the rate were limited by energy transfer we should have to postulate that the correct activation energy is less than Q^* and higher than Q. We see that, at least for the two compounds *n*-propyl and *n*-butyl iodide there is no room for such an assumption, and that it is also contrary to the less reliable data for ethyl, allyl, benzyl and benzoyl iodides.

We conclude that though some inaccuracy of our results on account of the recombination of R + I and of the possible insufficiency of energy transfer cannot be excluded, these influences cannot be of sufficient magnitude to account to any considerable extent for the range of variations in the rate of pyrolysis which we observed. We attribute therefore this range —which is represented in Table II by the k_I (430°) values calculated for the temperature of 430° (at which the largest number of common observations were made) as well as by the calculated temperatures T (1 % per sec.) at which k_I would have the common value of 1 % per second—to the variation in the primary rates of decomposition of the C—I bond in the different compounds.

Bond Energies and "Bond Energy Terms".

Our use of the equation $k_1 = \nu e^{-Q^*/RT}$ implies that Q^* is the energy required for breaking the C—I bond. We will consider, therefore, that the Q^* in Table III represent the Bond Energies for the various iodides in question ; *the C—I Bond Energy being defined as the dissociation energy of the molecule into a free radical and an iodine atom.*

In this identification of the activation energy Q^* with the Bond Energy (D) we apply the usual conception of a chemical bond derived from the theory of diatomic molecules which postulates an energy curve monotonously rising from the normal state to complete dissociation. We note, however, that for hexaphenylethane Q^* has been observed to be 8 kcal. in excess of D for the ethane linkage. But this case may be considered as exceptional in various respects. For one thing there is likely to be present (as a Fisher model of hexaphenylethane clearly demonstrates) a very considerable steric hindrance opposing the formation of the ethane linkage, and this may cause considerable repulsion to occur before the valence force begins to become effective. In any case we do not feel that the deviation from $Q^* = D$ for hexaphenylethane necessitates any serious reservation in the identification of these two magnitudes for bonds of a more usual character.

Turning now to the theorem first postulated by Fajans [4] and elaborated much further by Sidgwick [5] and Pauling [6] according to which the heat of formation of organic compounds from free atoms can be represented as the sum of constant contributions characteristic of each chemical link, we note first that the validity of this theorem constitutes no evidence for the existence of a constant heat of formation of the bonds covered by its scope. Take two kinds of bonds formed, say, by carbon with the two different atoms X and Y. Whatever the variation may be in the energies of the C—X and C—Y bonds with the position of the C-atom, no deviation from the additivity rule would result, so long as the variations are equal for both kinds of bonds ; the additivity rule merely expresses the constancy of the substitution heat $- \Delta H_s$ in the reaction

$$C—X + Y = C—Y + X - \Delta H_s$$

independently of the position of C.

It seems logical to express this state of affairs by designing the constant contribution attributed to one chemical link the *Bond Energy Term* of that link as distinct from the *Bond Energy* (D) as defined, *e.g.*, above for the C—I bond.

The evaluation of Bond Energy Terms starts from a compound containing only one kind of bond (*e.g.* CH_4, CCl_4) and proceeds by dividing up the heat of formation of the molecule by the number of bonds present. The *Bond Energies* of C—H, C—Cl, etc., in CH_4, CCl_4, etc., on the other hand, are defined for the breaking of *one* bond only, and may thus differ considerably from the corresponding Bond Energy Term.

The investigations of Rossini [7] on the heats of combustion of the aliphatic hydrocarbons and of primary alcohols derived from these have proved the existence of some remarkable deviations from the additivity rule. They can be expressed by a steady decrease of the substitution heats

$$(C_nH)_{2n+1} - H + CH_3 = C_nH_{2n+1} - CH_3 + H$$
and
$$(C_nH)_{2n+1} - H + OH = C_nH_{2n+1} - OH + H$$

[4] Fajans, *Ber.*, 1920, **53**, 643 ; 1922, **55**, 2826.
[5] Sidgwick, *The Covalent Link in Chemistry*, Cornell, 1933.
[6] Pauling, *The Nature of the Chemical Bond*, Cornell, 1939.
[7] Rossini, *Bull. Bur. Standards J. Research*, 1934, **13**, 29, 189 ; Knowlton and Rossini, *ibid.*, 1939, **22**, 115.

E. T. BUTLER AND M. POLANYI 29

with n increasing from 1 onwards, the main change being between CH_4 and C_2H_6 and approximate constancy being reached after $n = 5$. Other deviations from the additivity rule were discovered by Kistiakowsky [8] in the case of the heats (1) of hydrogenation, and (2) of bromination of olefines. These can be expressed again, as Baughan and Polanyi [9] have suggested, as a variation of the substitution heats of a C—H link with varying position of the carbon. The observed changes in the heat of hydrogenation are expressed in the reaction

$$\underset{\alpha}{\overset{\beta}{\underset{C_nH_{2n+1}-C \, . \, R-H}{H_2C-}}} = \underset{C_nH_{2n+1}-C \, . \, R}{H_2C} + H$$

for the same α carbon position as explored by hydrogenation.

We thus have evidence that the differences between bond energies vary in simple hydro-carbons and in substituted hydrocarbons : but so far this leaves us in the dark as to the variations of the individual values. Definite indications—if only of an approximate nature—of wide variations in bond energies were first postulated by Ogg and Polanyi [10] from the variations in the rates of reaction between organic halides and sodium vapour observed by Hartel and Polanyi.[11] More recently H. S. Taylor and Smith [12] derived similar conclusions from the marked variations in the rate of reaction of methyl radicals with hydrocarbons. A fall in the bond energy of C—H was quite recently confirmed and quantitatively fixed by more direct methods. D. P. Stevenson [23] has given the values as $D(CH_3—H) = 101$ and $D(C_2H_5—H) = 96$ while independently and by different methods Anderson, Kistiakowsky and van Arstdalen [13] obtain $D(CH_3—H) = 102$ and $D(C_2H_5—H) = 98$ kcal.

Bond Energies and Activation Energies.

It was one of the main purposes of this investigation to test the theo-retical conclusion that the ob-served gradation in the rate of the reaction between Na and organic halides is due to a cor-responding gradation in the energy of the halide bond. This postulate is illustrated by Fig. 2, taken from the above cited paper of Evans and Polanyi,[10] which reflects the simplification achieved by neglect of repulsion between the Na and Cl atoms in (the initial half of) the transi-tion state. The variation ΔQ in the activation energies is connected here with the varia-tion ΔH of the heat of reaction by the proportionality relation

$$\Delta Q = -\alpha \Delta H$$
$$0 < \alpha < 1$$

FIG 2. $Q - Q' = \Delta Q = -\alpha \Delta H$.

[8] Kistiakowsky et alii, *J.A.C.S.*, 1936, **58**, 137 ; 1937, **59**, 831 ; 1938, **60**, 440 2764. [9] Baughan and Polanyi, *Nature*, 1940, **146**, 685.
[10] Ogg and Polanyi, *Trans. Faraday Soc.*, 1935, **31**, 1375. Further developed by Evans and Polanyi, *ibid.*, 1938, **34**, 11. Calculations based on this theory were carried out for the Na + Methyl halide reaction by Evans and Warhurst. *ibid.*, 1939, **35**, 593.
[11] Hartel and Polanyi, *Z. physik. Chem. B*, 1930, **11**, 97 ; 1932, **19**, 139.
[12] H. S. Taylor and Smith, *J. Chem. Physics*, 1939, **7**, 390 ; 1940, **8**, 543.
[13] Anderson, Kistiakowsky and Artsdalen, *ibid.*, 1942, **10**, 305.

where the value of α depends on the relative inclinations of the attraction curve R—Cl and the repulsion curve R . . . Cl$^-$ at the point of mutual crossing. Since the latter curve is bound to be much steeper than the former, it follows that α is always smaller than $\frac{1}{2}$.

This derivation implies that changes in the bond energy will be reflected, in the region of the "crossings" and below that, by vertical parallel displacements of the potential curve. This would be strictly true if the variations of bond energy were due entirely to variations in the resonance energy of the free radical, which would leave the potential curve unaffected in regions near the hollow of the curve. Actually there is always a certain amount of resonance present in the undissociated bond and in certain cases, like the vinyl halides, this resonance—and its absence in the freeradical—may be entirely responsible for the variation in bond strength. In this case, and in general whenever substitution affects resonance in the undissociated state, this influence must vary to some extent with nuclear separation and a change in the shape of the energy cruve in the region of the equilibrium distance will follow.

It seems, however, scarcely worth while to attempt the very uncertain evaluation of these deviations from the simple scheme of Fig. 2, since they are not likely to affect the first approximation attempted here. We have therefore proceeded to evaluate the scheme indicated by Fig. 2, for which purpose we calculated the curves shown in Fig. 3. These relate to the reaction Na + ClCH$_3$ = NaCl + CH$_3$. The chlorine compounds are taken here instead of the iodine compounds because the gradation of the rates of reaction with Na-vapour is much more widely spaced for the chlorides than for the iodides, and hence it is desirable to transfer the argument to the former. This can be done by using the gradation of the C—I bond strength as representing also the gradation of the C—Cl bond strength. The implied assumption of a constant substitution heat of C—I by Cl is sufficiently well supported to justify this move—which certainly can cause no serious error in our conclusions.

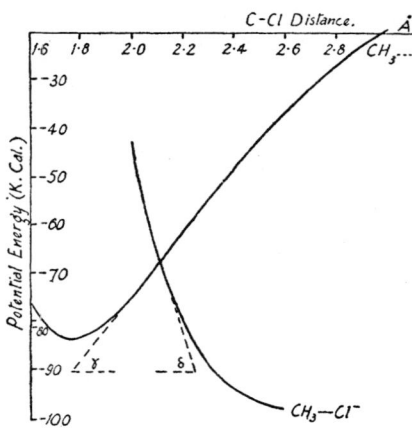

FIG 3.

The curve in Fig. 3 represents the C—Cl bond, according to a Morse function with $D = 83.5$, $r_0 = 1.76A$, $\omega_0 = 710$ cm.$^{-1}$ and hence $a = 1.66 \times 10^8$. The value of D corresponds in view of the known thermochemical data to $D = 54$ for CH$_3$I. The repulsion curve between Cl$^-$ and CH$_3$ was calculated in accordance with M. G. Evans and E. Warhurst [10] by the equation

$$E = -\frac{\alpha e^2}{r^4} + br^{-9},$$

using $\alpha = 2.2 \times 10^{-24}$ c.c., and (for the determination of b) $r_0 = 3.01A$.* The activation energy as indicated by the crossing point in Fig. 3 would

---

* This value is the sum of the CH$_3$ radius 1.40A as used by Evans and Warhurst [10] and the Cl$^-$ radius of 1.6A estimated from electron-diffraction data of gaseous alkyl-halides by Maxwell, Hendricks and Mosley, *Physic. Rev.*, 1937, **52**, 968 (see Baugham and Polanyi, *Trans. Faraday Soc.*, 1941, **37**, 648).

E. T. BUTLER AND M. POLANYI 31

be 16·0 kcal., which is somewhat higher than the experimental value of about 10 kcal. The difference of 6 kcal. may be accounted for by transition state resonance, which in the theory represented by Fig. 2 is supposed not to vary with the nature of R and to have therefore no influence on changes of E with varying R. Accordingly the coefficient α in $\Delta E = - \alpha \Delta H$ remains represented by the ratio of the inclinations of the two curves at the "crossing point," and can be estimated from Fig. 3 at the value 0·27.†

An attempt to plot the observed activation energies of the Na + ClR reaction as functions of the corresponding C—I bond energies (as representing the gradation of the C—Cl bond energies) leads to Fig. 4. Here are included the two compounds *iso*-propyl iodide and *t*-butyl iodide for which the bond energies are derived in the following paper and the four compounds acetyl, acetonyl, benzoyl and benzyl iodides [14] —for which the activation energies are 5·0, 2·0, ∼ 0 and 0·9 respectively—are left away because the values fall into an altogether different region.

The compounds shown in the figure are allyl, *t*-butyl, *iso*-propyl, *n*-butyl, *n*-propyl, ethyl, phenyl, vinyl iodides. They comprise *all* the pure hydrocarbons measured in this paper, and the next one for which the rate of the Na reaction is known. For this group the evidence in the figure is consistent with a linear relationship of the form $\Delta E = - \alpha \Delta H$ where $\alpha = 0.28 \pm 0.015$ corresponds to the minimum of squared deviations. This value of α seems remarkably close to the value $\alpha = 0.27$ derived theoretically from Fig. 3, but even without relying in any way on this close correspondence, it would seem that strong evidence has been found here for the mechanism of the reactions of alkali metals with or-

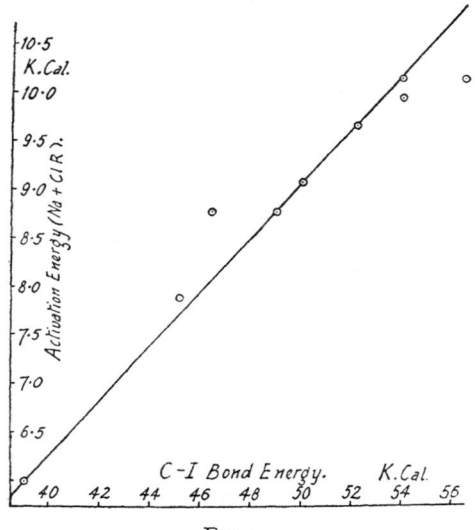

FIG 4.

ganic halides as postulated by Ogg and Polanyi. The evidence will also strengthen the case for considering the kind of linear relationship of which Fig. 4 is a particular instance as fundamentally determined by the relationship between reaction heat and activation energy in the sense suggested by Evans and Polanyi.[15]

It may be objected that the compounds in Fig. 4 represent a selection which is not justified by the theory. This is true of the original form of the theory as represented by Fig. 2 ; but a note of Evans and Polanyi [16] has recently pointed out theoretical and experimental reasons for extending this theory by taking into account changes in transition state resonance due to variations of R. In a reaction Na + XRX = NaX + RX there are three structures, instead of the usual two, resonating in the transition state :

† If tan γ and tan δ be the inclinations at the crossing point $\dfrac{1}{\alpha} = \left|\dfrac{\tan \delta}{\tan \gamma}\right| + 1.$

[14] From unpublished work in these laboratories.
[15] Evans and Polanyi, *Trans. Faraday Soc.*, 1936, **32**, 1333.
[16] Evans and Polanyi, *Nature*, 1941, **148**, 436.

32   RATES OF PYROLYSIS AND BOND ENERGIES

$$(1) \ \text{Na} \quad \text{XRX}_1,$$
$$(2) \ \text{Na}^+\text{X}^- \quad \text{RX}_1,$$
as well as $$(3) \ \text{Na}^+ \quad \text{XR} \quad \text{X}_1{}^-$$

The effect of the negative substituent X^- is thus to cause an extra depression of the activation energy below the " crossing point," and thus to accelerate the reaction beyond the rate which it would have in view of the bond-energy if the relationship $\Delta E = - \alpha \Delta H$ held. This effect of negative substituents may well explain all the deviations—from the linear relationship established in Fig. 4. We note in particular that all deviations represent *faster* rates than given by the straight line ; that all the deviating compounds (except benzyl iodide) contain substituents, known to possess negative character, and that conversely all compounds containing negative substituents show marked deviations from the straight line relationship in the expected direction. This statement includes the experimental material both of this and the next paper. The fact that benzyl falls into the negatively substituted group of radicals is surprising, and will require explanation ; but we do not feel that this fact goes far to invalidate the scheme from which it forms an apparent exception.

Bond Energy and Resonance.

We follow here the lead of the theory first proposed as an explanation of the existence of free radicals by Hückel [17] and by Pauling,[18] according to which variations in bond energy are to be attributed to changes in the difference between the resonance energies of a molecule held together by the bond in question and the two halves of the molecule resulting from the splitting of the bond. With few exceptions the changes in bond energies which we have observed can be accounted for in this sense by some previously known form of resonance.

A fall in Bond Energy from CH_3—I to CH_3CH_2—I was first postulated by Wheland [19] on account of an additional resonance due to the substitution of one H atom by a CH_3 group. This can be written :

$$(1) \ \alpha \ CH_3\text{—}CH_2. \quad \beta \ H \cdot CH_2 = CH_2 \ (3 \ \text{times}).$$

We take this to be the explanation for the observed fall of D from methyl to ethyl iodide, and we would extend the scheme to include the further fall of D observed for *n*-propyl and *n*-butyl iodide, as follows :

$$\begin{aligned}
&(2) \ \alpha \ CH_3\text{—}CH_2\text{—}CH_2 \quad \beta \ CH_3 \cdot \ CH_2 = CH_2 \quad\quad \overset{\displaystyle H \cdot}{\gamma \ CH_3\text{—}CH} = CH_2 \ (2 \ \text{times}). \\
&(3) \ \alpha \ CH_3\text{—}CH_2\text{—}CH_2\text{—}CH_2 \quad\quad\quad\quad\quad\quad\quad \beta \ CH_3\text{—}CH_2 \cdot \ CH_2 = CH_2
\end{aligned}$$

$$\overset{\displaystyle H \cdot}{\gamma \ CH_3\text{—}CH_2\text{—}CH} = CH_2 \ (2 \ \text{times}) \quad\quad \delta \ H \cdot CH_2 = CH_2 \ CH_2 = CH_2 \ (3 \ \text{times}).$$

It seems likely that the β form postulated for the *n*-propyl radical adds more to the resonance energy than the corresponding β form of ethyl, since the energy of the former is lower by the difference between a C—H and a C—C bond. In case of resonance (3) we have the number of resonating forms increasing from 4 to 7 which may well increase the resonance of *n*-butyl over that of *n*-propyl. It seems reasonable to assume that the magnitude of this kind of resonance increases further on these lines with further increasing chain length. This seems indicated by the fact that in the case of *n*-aliphatic hydrocarbons and alcohols Rossini found variations in the substitution energy with increasing normal chain length even as far as the fifth carbon atom.

[17] E. Hückel, *Z. Physik*, 1933, **83,** 632.
[18] Pauling and Wheland, *J. Chem. Physics*, 1933, **1,** 362.
[19] Wheland, *ibid.*, 1934, **2,** 474.

E. T. BUTLER AND M. POLANYI 33

To avoid repeating the argument on this point later we include here a reference to the *iso*-propyl and *t*-butyl radicals which we have already briefly discussed in connection with Fig. 4. It has been pointed out by Wheland that the number of resonating forms increases in the series ethyl, *iso*-propyl, *t*-butyl from three to six and to nine. This should cause a steady increase in resonance energy and may explain the observed steady fall in the bond energy.

Our results for allyl and benzyl iodides can be compared with the predictions of Pauling [18] and of Hückel, [17] concerning the resonance between a free valence and a double bond in β-position to it. The resonance energies were calculated by Coulson [20] and by Pauling and Wheland [18] at 15·4 kcal. for allyl and 15 kcal. for benzyl radicals ; while we found a reduction in the corresponding C—I bond energy of 15 kcal. and 11 kcal. respectively. To complete the comparison we should, however, yet consider the ionic resonance in the undissociated bond, to which we refer in the last chapter. Our values for vinyl and phenyl iodides confirm the existence of the following kind of resonance assumed by Pauling [21] to account for the shortening of the C-Halogen bond-length and the reduction in the dipole moments of the vinyl and phenyl halides when compared with the methyl halides :

$$CH_2 = CH—X \qquad \bar{C}H_2—CH = \overset{+}{X} \qquad \text{and}$$

This degeneracy should cause the C—I bond to be strengthened as compared with the CH_3—I bond, as we, in fact, find it to be for vinyl iodide. In the case of phenyl iodide there is a possible offsetting effect due to the following degeneracy of the phenyl radical :

M. G. Evans and E. Warhurst [10] suggest that the magnitude of this resonance energy is of the order of 10 kcal., but our experiments indicate a much lower value.

In spite of the inaccuracy of our value for acetyl iodide we think that we have established that the C—I bond energy in this compound is not greater but rather less than that in ethyl iodide, which raises some important points. In view of its α double bond, acetyl iodide should show the same resonance as vinyl iodide,

and hence its C—I bond should be much stronger than that of ethyl iodide. The fact that this is not observed suggests that the effect is offset by some other factor. Observations on dichloro- and dibromoiodo methane to be described in the following paper indicate that the presence of halogen substituents lowers the C—I bond energy in these compounds. If this were due to the electron-attractive nature of the halogen atoms, a similar effect could be postulated for the O-atom in acetyl iodide, and this would offer the required explanation.

[20] Coulson, *Proc. Roy. Soc. A*, 1938, **164**, 383.
[21] Brockway, Beach and Pauling, *J.A.C.S.*, 1935, **57**, 2693 ; Brockway and Palmer, *ibid.*, 1937, **59**, 2181.

2

34   RATES OF PYROLYSIS AND BOND ENERGIES

If X is a negative substituent, the following resonance may arise in the free radical and contribute to its stability as follows

$$
\begin{array}{ccc}
\text{H} & \text{H} & \text{H} \\
\ddot{}\, & \ddot{}\, & \ddot{}\, \\
\text{X} : \ddot{\text{C}} \cdot & {}^{-}\text{X} : \;\ddot{\text{C}} \cdot {}^{+} & \text{X} \cdot \;\ddot{\text{C}} : \\
\ddot{}\, & \ddot{}\, & \ddot{}\, \\
\text{H} & \text{H} & \text{H} \\
\alpha & \beta & \gamma
\end{array}
$$

which can also be written by the use of a three-electron bond (using the symbol of three dots suggested by Pauling for this kind of link) :

$$
\begin{array}{cc}
\text{XCH}_2 . & \text{X} \cdots \text{CH}_2 \\
\alpha & \beta
\end{array}
$$

The same kind of resonance will be much less marked in the absence of a lone electron, *i.e.*, when the radical forms part of a molecule. The difference reduces the energy of the bond. We may assume that the effect will depend on the stability of the form β in the above resonance formula, *i.e.*, on the negativity manifested by the substituent X, and we may expect therefore that another strongly negative substituent like oxygen will act in the same sense and weaken the strength of an associated bond C—I. This would explain that the bond energy in acetyliodide is weakened—instead of strengthened—in comparison to CH_3I.

In benzoyl iodide the situation is even more complex. In addition to the two opposing resonance effects manifested in CH_3COI, we may expect to find here also a weakening of the bond due to resonance with a β unsaturation. If the effects corresponding to the acetyl-structure and to β conjugation were additive, the bond energy should fall somewhat short of that observed in benzyl iodide ; our results indicate that, in fact, it about coincides with the latter.

In acetonyl iodide we might expect to see the appearance of resonance due to a β conjugated double bond. To this will be added any effects that β substitution by a negative particle may exercise. Evidence that substitution by a halogen atom in β position does weaken a bond will be given in the next paper. We may explain this by the resonance :

$$
\begin{array}{cc}
\text{X—C—C} & \overset{+\frac{1}{2}}{\text{X}} \cdots \overset{-\frac{1}{2}}{\text{C}}\text{—C} \\
\alpha & \beta
\end{array}
$$

The oxygen atom in β-position—as in acetonyl iodide—may be expected to have an influence similar to that of X. The reduction of bond energy by about 10 kcal. observed for acetonyl iodide is thus seen to represent the sum of two types of resonance.

Bond Energies and Substitution Heats.

Substitution heats are differences of bond energies. Following Baughan and Polanyi, we have set out in Table III a system of substitution heats, all of which were related to the C—H link as the initial state. Once the absolute values for any one kind of bonds become available, all the bond energies represented in it can forthwith be calculated. While the C—I bonds which we investigated here do not form part of the system as it stands, yet the results obtained for these bonds can be used to throw light on the gradation of the C—Br bonds which do form part of it. This was done by Baughan and Polanyi by using the assumption (for which they adduce evidence) of a constant substitution heat of the reaction

$$
\text{C—I} + \text{Br} = \text{C—Br} + \text{I} + 14.7 \text{ kcal.}
$$

We have thought it useful to reproduce the resulting table of bond energies in the present context, with such revisions as more recent observations and

E. T. BUTLER AND M. POLANYI 35

TABLE III.—SUBSTITUTION HEATS (FROM BAUGHAN AND POLANYI).

RH.	X = CH₃.	X = Br.	X = OH.	X = β.
CH₃H .	. 15·4	(33·9)	16·0	—
C₂H₅H .	. 13·0	30·5	10·2	68·0
n-C₃H₇H .	. 12·7	30·1	8·8	67·5
n-C₄H₉H*	. 12·7	30·0	8·2	—
iso-C₃H₇H	. 10·8	27·9	4·5	66·0
tert-C₄H₉H	. 10·0	25·9	0·6	64·6

TABLE IV.—BOND ENERGIES OF BAUGHAN AND POLANYI (REVISED).

(To the nearest half calorie.)

R.	C—I.	C—Br.	C—H.	C—CH₃.	C—OH.	α—β.
CH₃ .	. 54·0	68;·5	102·5	87·0	86·5	—
C₂H₅	. 52·0	67·0	97·5	85·0	87·0	57·5
n-C₃H₇	. 50·0	64·5	95·0	82·0	86·0	—
n-C₄H₉*	. 49·0	63·5	94·0	81·0	85·5	—
iso-C₃H₇	. 46·5	61·0	89·0	78·0	84·5	52·5
tert-C₄H₉	. 45·0	60·0	86·0	75·5	85·0	52·0

α-β signifies the " second half" of a double bond.

* *n*-butyl was not included in the original table of Baughan and Polanyi The figures given here for substitution heats are calculated from the same sources the more detailed analysis of the experiments allowed us now to make. The main difference between our table and the previously published one comes from the fact that we refrained from smoothing the observed values because we felt doubtful whether this rather considerable correction is sufficiently supported on theoretical grounds.

The table brings out the fact that our estimate of the bond energies is consistent with the previously suggested values for the C—H bond energy in methane, which various authors have put at 100-108 kcal., and is well in excess of the average bond energy of CH₄ calculated at 87 kcal. by Sidgwick and Pauling.* This point has gained considerable precision by the more recent publications already mentioned above, by D. P. Stevenson [23] and Andersen, Kistiakowsky and Artsdalen,[13] who—by entirely different methods—arrive at values for the CH₃—H bond energy of 101 and 102 ± 1 kcal. respectively, which come quite close to the figure suggested by us. The force of this argument seems to decide that the older determinations of the C—I bond energy from kinetic experiments [1] which gave a value of 44 kcal. for CH₃—I cannot be maintained.

The unequal gradation of bond energies in the alkyl bonds was explained by Baughan, Evans and Polanyi,[22] by the presence of ionic resonance in the undissociated bond which tends to offset the resonance of the free radicals. Ionic resonance is small in the C—H bond, and we therefore observe here almost the whole resonance of the free radical as a reduction of the bond energy. Our previous comparison of the resonance of the free radical with the reduction in the strength of the corresponding bond energies—for example in the case of allyl and benzyl—is now seen to be incomplete ; owing to ionic resonance we should expect the reduction of the C—I bond energy to be generally somewhat smaller than the resonance energy of the free radical.

[22] Baughan, Evans and Polanyi, *Trans. Faraday Soc.*, 1941, **37**, 377.
[23] Stevenson, *J. Chem. Physics*, 1942, **10**, 291.
[24] Baughan, *Nature*, 1941, **147**, 542.

* Baughan [24] (*Nature*, 1941, **147**, 542) has suggested that this value for the average bond energy is in error on account of the use of too low a value for the heat of sublimation of carbon. Herzberg [25] supports the earlier view in *J. Chem. Physics*, 1942, **10**, 306.
[25] Herzberg, *J. Chem. Physics*, 1942, **10**, 306.

2 *

36   ELASTICITY OF LONG–CHAIN MOLECULES

Summary.

An attempt has been made to determine the C—I bond energy and its variations under the influence of various substituents by measuring the rates of pyrolysis of the compounds in question. Variations in bond energy are found to be very marked. For a number of simple hydrocarbons the previously known activation energy of the reaction RCl + Na appears to be proportional to the bond energy R—I, the proportionality factor being 0·28 in fair agreement with theory. Negative substituents depress the activation energy below the value corresponding to the bond energy.

The above experiments were carried out during the sessions 1938/9 and 1939/40. The authors wish to thank Dr. E. Warhurst and D. J. N. Haresnape for their help in constructing the apparatus. One of us (E. T. Butler) is indebted to the University of Wales for a Fellowship.

The University, Manchester.

Ultrafast Processes

David Phillips

Imperial College London, South Kensington, London, UK SW7 2AZ

Commentary on: **Picosecond jet spectroscopy and photochemistry: Energy redistribution and its impact on coherence, isomerisation, dissociation and solvation,** Ahmed H. Zewail, *Faraday Discuss. Chem. Soc.*, 1983, **75**, 315-330

Ultrafast techniques have really come into their own since the development over the past decade of reliable, relatively inexpensive femtosecond lasers, but in the 1980's, ultrafast still meant picoseconds, and the chosen paper typifies work in this time-domain. The whole subject of course owes its inspiration to George Porter, who with Norrish pioneered flash photolysis,[1] first in the milli-, then micro-, and ultimately nano- and picosecond time-domains.

In the 1960's and 70's, the basic theories of non-radiative decay of photo-excited molecules was put on a firm footing by the pioneering work of Hochstrasser, Jortner, Robinson and many others,[2] and this permitted a broad understanding of decay processes observed experimentally in simple di- and tri-atomic molecules through to large polyatomic systems. However almost all data on large molecules up to this time had been obtained in the condensed phase, or in "bulb" experiments on gases at low pressures in the so-called isolated molecule limit, *i.e.* pressures such that collision with other molecules would not occur during the lifetime of the excited state. These experiments were bedevilled by the fact that at room temperature or the often higher temperatures necessary to maintain a sufficient concentration of molecules in the gas-phase, even a very narrow-line laser excitation source resulted in simultaneous population of a large number of excited -state ro-vibrational levels, since the Boltzmann distribution of ground-state levels led to severe sequence congestion in the spectra.

"Supersonic jet" technology developed by Levy and others[3] was extremely effective in simplifying the electronic *spectroscopy* of complex polyatomic molecules because the narrowing of the Boltzmann distribution by nozzle expansion, reaching effective temperatures well below 1 K allowed single vibrational and even single rotational level excitation for the first time. The technique was rapidly used in conjunction with narrow line-width cw laser excitation to produce high-resolution excited state spectra of many complex molecules, and in some cases their van der Waals complexes and clusters. Ahmed Zewail was one of the first to realise that the use of short-pulse lasers in conjunction with jet cooled molecules could provide similarly revealing information about the *dynamics* of decay of excited states, and the early paper selected here reviews some of the startling successes his group had in this field. The paper acted as an inspiration to many others in the field, and ultimately to work on the femtosecond timescale,[4] and of course, to the award of the Nobel Prize to the author.

Zewail acknowledged early on that he was inspired to work in the dynamics area by amongst others, George Porter's development of "fast" reaction techniques, *viz.* "Flash Photolysis" which is reported elsewhere in this volume. In the early experiments outlined in the present paper, three detection techniques were employed; time-correlated single photon counting, with 30–50 ps time resolution; streak camera detection of fluorescence, with 10 ps resolution, and multiphoton ionisation with resolution determined by the pulse width of the laser, 1 or 15 ps.

Using the former technique, the most significant result was the observation of "quantum beats" in the fluorescence decay of jet-cooled anthracene At low excess energies, the fluorescence and fluorescence excitation spectra of anthracene are very sharp, and the fluorescence decay of single vibronic levels is exponential. At an excess energy of 1400 cm^{-1} however, clear quantum beats were seen, arising from the interference between the initially populated vibronic state, and a state produced by vibrational redistribution into the small number of weakly coupled background vibrational levels.[5] At higher excess energies, the large number of states interfering renders the quantum interference effects invisible. This first clear observation of these quantum beats led to studies on rotational coherence by Felker and Zewail[6] which in favourable cases permitted estimation of rotational constants, and hence structural information, in large polyatomic molecules where conventional spectroscopy was impossible. After the development of shorter-pulse lasers the way was opened for elegant studies on wave-packet dynamics in simple dissociating systems, such as NaI,[4] and gradually more complex systems in jets, and ultimately to wave-packet dynamics in the condensed phases.

Another major goal of many groups in the late 1970's had been the understanding of the influence of solvent upon the dynamics of excited states in the liquid state. Real progress has of course come with the arrival of femtosecond techniques, and direct studies in solution, but around 1980, it was felt that advances could be made by investigating the spectroscopy and decay characteristics of jet-cooled molecules solvated sequentially by one, two, three, *etc.*, solvent molecules, and higher clusters, which might ultimately mimic condensed-phase behaviour. Large numbers of investigators, including ourselves, entered this field in the 1980's,[7,8] earning from Ed Schlag the irreverent collective title "The Solvation Army", but Zewail was a pioneer, and the work reported in this early paper on dynamics of isoquinoline molecules solvated by water, methanol, and acetone motivated many to follow this path.

Ultrafast dynamics moved on dramatically from the picosecond regime over the 1990's and into this century, but the early work of Zewail outlined in this 1983 review paper of his provided a first glimpse of some of the coherence effects to be studied with ever faster lasers.

References

1 G. Porter, *Proc. R. Soc London, Ser. A*, 1950, **200**, 284
2 J. Jortner, S. A. Rice and R. M. Hochstrasser, *Adv. Photochem.*, 1969, **7**, 149, and references therein.
3 R. E. Smalley, L. Wharton and D. H. Levy, *J. Chem. Phys.*, 1975, **63**, 4977.
4 A. H. Zewail, in *Femtosecond Chemistry*, ed. J. Manz and L.Woste, VCH, Weinheim, 1995, pp. 15–128, and references therein.
5 P. M. Felker and A. H. Zewail, *Adv. Chem. Phys.*, 1988, **70**, 265.
6 P. M. Felker and A. H. Zewail, in *Jet Spectroscopy and Molecular Dynamics*, ed. J. M. Hollas and D. Phillips, Blackie Academic, 1995, pp. 181–221.
7 A. G. Taylor, T. Burgi and S. Leutwyler, in *Jet Spectroscopy and Molecular Dynamics*, ed. J. M. Hollas and D. Phillips, Blackie Academic, 1995, pp. 151–180.
8 E. M. Gibson, A. C. Jones, A. G. Taylor, W. G. Bouwmann, D. Phillips and J.Sandell, *J. Phys. Chem.*, 1988, **92**, 5449.

Faraday Discuss. Chem. Soc., 1983, **75**, 315–330

Picosecond-jet Spectroscopy and Photochemistry

Energy Redistribution and its Impact on Coherence, Isomerization, Dissociation and Solvation

By Ahmed H. Zewail *

Arthur Amos Noyes Laboratory of Chemical Physics,
California Institute of Technology, Pasadena, California 91125, U.S.A.

Received 22nd December, 1982

The development of the picosecond-jet technique is presented. The applications of the technique to the studies of coherence (quantum beats), photodissociation, isomerization and partial solvation of molecules in supersonic-jet beams are detailed with emphasis on the role of intramolecular energy redistribution. Experimental evidence for intramolecular *threshold effect* for rates as a function of excess molecular energy is given and explained using simple theory for the redistribution of energy among certain modes. Comparison with R.R.K.M. calculation is also made to assess the nature of the statistical behaviour of the energy redistribution.

INTRODUCTORY REMARKS

The dynamics of vibrational energy flow in *large* and *isolated* molecules following selective laser excitation is very interesting and challenging for many reasons. From a theoretical point of view one would like to know how the coupling between bond vibrations influences the energy flow, and at what energy threshold does this flow or redistribution of energy from one mode to another occur. The energy region in which randomization or "chaotic behaviour" dominates is relevant to mode-selective chemistry.[1]

Experimentally, a number of classical studies have been designed to examine the dynamics of vibrational energy in the S_1 state [2] of molecules in gas "bulbs" at low pressures, *i.e.* isolated in the sense that time between molecular collisions is longer than the experimental timescale. However, these large molecules at the bulb temperature have very large thermal energy, and coherent (or incoherent) excitation by a light source will suffer from the problem of sequence congestion. With the more recent (revisited) beam or jet techniques, one can vibrationally and rotationally cool these molecules and obtain the dispersed fluorescence or the excitation spectra. As shown from a number of important studies,[3-6] excitation and fluorescence spectra at all times (more precisely using tens of nanoseconds resolution or using c.w. laser excitation) provide information on the high-resolution excited-state spectra, the geometry of van der Waals complexes and clusters, and in some cases the gross features of interstate coupling (line broadening etc.).

Our own interest has focused on the development [7-13] of the picosecond-jet technique to study the time-resolved dynamics following the picosecond excitation of isolated molecules in supersonic jet beams. This technique was applied to a number

---

* Camille and Henry Dreyfus Foundation Teacher–Scholar.

of problems dealing with isomerization, dissociation, stepwise-solvation, hydrogen bonding and coherence.   Here we shall highlight the new findings we obtained from these different studies, with particular emphasis on the relevance of these observations to vibrational energy redistribution (statistical R.R.K.M.-type as against non-statistical) and dephasing in isolated molecules.

THE PICOSECOND-LASER–MOLECULAR-BEAM APPARATUS

The arrangement we used for interfacing the picosecond laser to the molecular beam (or free jet) is shown schematically in fig. 1.   The laser is a synchronously pumped dye-laser system whose coherence width, time and pulse duration were characterized [14] by the SHG autocorrelation technique.   The pulse widths of these lasers are typically 1–2 ps, or 15 ps when a cavity dumper is used.   For detection one of three techniques

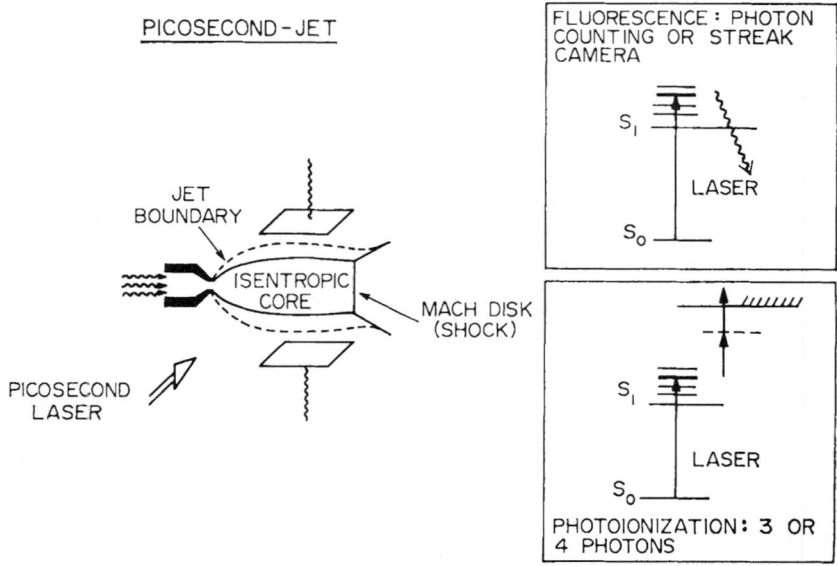

Fig. 1. An outline of the picosecond-jet technique and detection methods.

can be used: (a) Time-correlated single-photon counting (dispersed fluorescence) using a fast photomultiplier (or microchannel plate)—this way we achieved a resolution of ca. 100 ps (h.w.h.m.) *without* deconvolution.   With deconvolution, one obtains, if required, the 30–50 ps time constants of single exponential decays.   (*b*) Streak camera—a synchroscan streak camera gives a resolution of better than 10 ps with a repetition rate of up to 200 MHz.   (*c*) Multiphoton ionization—we are using this technique (see fig. 1) to obtain the short-time behaviour, being limited by the pulse width of the laser (1 or 15 ps).   This was applied to the stilbene isomerization problem.

Finally, the actual laser bandwidth was varied by using a combination of intra-cavity filters and etalons.   In all these experiments the vibrational and rotational temperatures were typically <20–40 K and <10 K, respectively, depending on the molecular-beam conditions.

APPLICATIONS

(A) QUANTUM BEATS AND COHERENCE: ANTHRACENE, STILBENE AND PYRAZINE

The spectra of anthracene in a gas bulb at ca. 480 K reveal very little intensity of quasi-sharp lines (I_s) and a large background intensity that is very broad (I_b) or diffuse in nature. When anthracene is jet-cooled, I_s/I_b varies dramatically depending on the excess vibrational energy in S_1. As shown in fig. 2, this ratio varies from 100 or

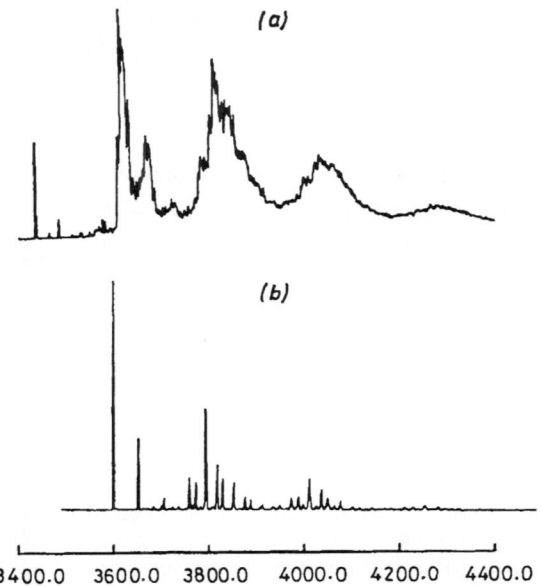

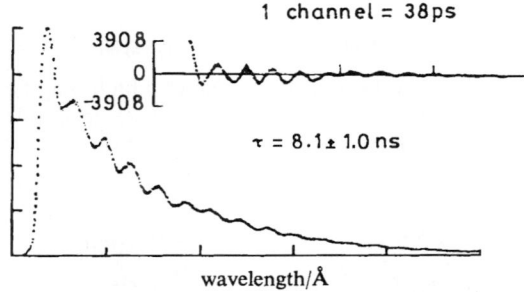

Fig. 2. Jet-cooled fluorescence spectra of anthracene at two excitation energies and quantum beats: (*a*) $E_x \approx 1400$ cm^{-1}; (*b*) $E_x = 0$.

more at zero excess energy to 10^{-2} or less at 5600 cm^{-1}; the spectra in the high-excess-energy (E_x) region are very similar to the bulb spectrum.

At $E_x \approx 1400$ cm^{-1} of excess energy in anthracene, we observed quantum beats in the dispersed fluorescence with a large modulation depth. The beats' modulation depth is sensitive to the excess energy and to the fluorescence detection wavelength.

(This work is now completed and will be published elswhere.) For stilbene, the quantum beats appeared at a number of different E_x. Clearly, the observation of beats in these large molecules is related to the coherence of the vibrational–rotational states excited, a point that we shall discuss later. The beats, due to rovibronic states, in anthracene are not expected to be sensitive to magnetic fields. In a recent work we reported[8] on the Zeeman effect on quantum beats in pyrazine, and on the importance of beats in unravelling the coupling of singlet and triplet rotational levels. In a more recent study by Kommandeur's group[15] the high-resolution frequency spectra of pyrazine were observed and compared with our Fourier-transform beat spectra; excellent agreement was found. Finally, it should be emphasized that the origin of the beats in stilbene or anthracene is different from pyrazine; the former being directly related to the rotational–vibrational energy redistribution problem to be discussed later.

(B) ISOMERIZATION OF STILBENE

The isomerization process (I) involves some torsional vibrational modes for the twisting to finalize. This is a classical problem in chemistry, and much work has been done in solution[16] and more recently under gas phase isolated molecule conditions.[17] We have shown recently that the jet-cooled spectra[10] of *trans*-stilbene displays the isolated molecule low-frequency modes which are optically active.

(I)

(trans)   *(cis)*

Furthermore, the observed[10] rates measured as a function of excess energy (fig. 3) exhibit a *threshold* at ca. 1200 cm^{-1}, which is in very good agreement with the estimated barrier height for isomerization from S_1 in solution. The jet studies clearly indicate the absence of thermal energy influence on rates, and the involvement of the redistribution in the isomerization. We shall discuss this point later.

C. HYDROGEN-BONDED SYSTEMS

With the same technique we examined molecules which may be exhibiting proton transfer in the excited state. In other words in these molecules there exists the possibility of " tunnelling " between different configurations. The results for methyl

(II)

(blue form)   (u.v. form)

salicylate (blue form) in the jet show a drastic change in the equilibrium configuration upon excitation involving the OH, CO and low-frequency mode of 180 cm^{-1}. Furthermore, the observed fluorescence rate exhibits again a threshold at certain E_x; from $(12 \text{ ns})^{-1}$ at zero E_x to $(160 \text{ ps})^{-1}$ at $E_x \approx 2000$ cm^{-1} (see fig. 4). We concluded from these studies that for the $E_x = 0$ excitation the distorted excited state is created simultaneously with the absorption of a photon in the *isolated* molecule. In the high

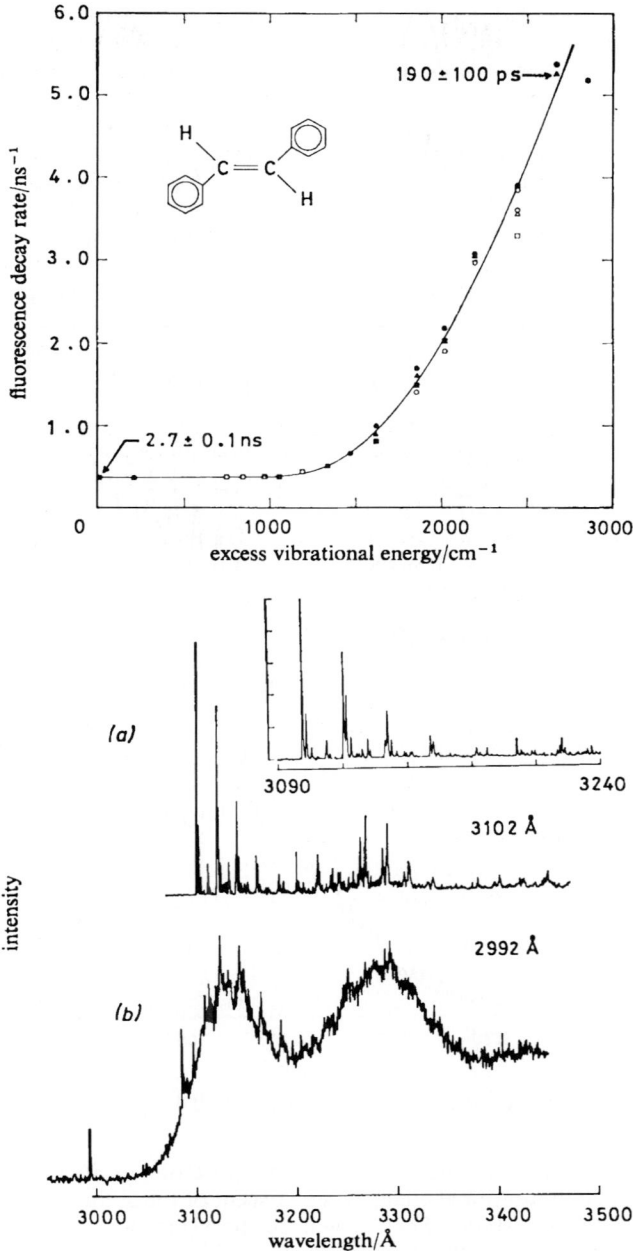

Fig. 3. Jet-cooled fluorescence spectra of stilbene [bottom (*a*) and (*b*)] and the threshold effect (top). Spectrum (*a*) is for 0,0 excitation.

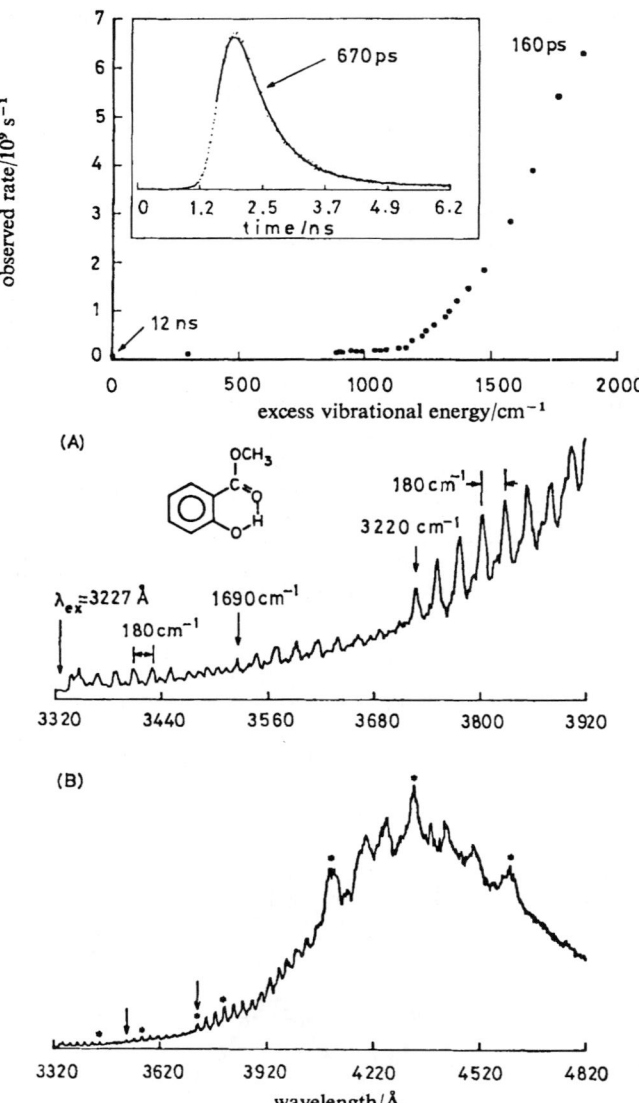

Fig. 4. Jet-cooled fluorescence spectra [bottom, (A) and (B)] and threshold effect (top) in MS.[9]

excess-energy limit, the low-frequency motions promote excited state electronic states coupling or configurational changes reminiscent of the stilbene problem. Thus the prevailing of the threshold effect as will be discussed later. We are currently studying the isotope effect on the dynamics.

(D) CHROMOPHORE-SELECTIVE PICOSECOND EXCITATION

The idea of this experiment is to excite a single chromophore in a molecule (jet-cooled) and to observe the real energy flow from the optically pumped vibrations of this chromophore to other modes of the rest of the molecule. We chose the following molecule:

$$A^* \!-\! CH_2 \!-\! CH_2 \!-\! CH_2 \!-\! \varphi$$

where A* stands for excited anthracene and φ for an aniline group. This molecule was chosen because we already know a great deal about the redistribution problem in the " optically active " chromophore, anthracene, and also because this type of molecule has been extensively studied by the Columbia group and by others in solutions.[18] In the jet we excite the different modes of the anthracene species with picosecond pulses and follow the time-resolved dispersed fluorescence of anthracene-like emission and product emission (red shifted emission due to the folding of the aniline species towards anthracene, charge-transfer or exciplex emission; henceforth referred to as " product"). At a given excess energy we observed a *decay* of the an-thracene-like emission and a buildup in the product emission (fig. 5). These rates were found to be very sensitive to the excess vibrational energy in the anthracene chromophore! Thus, for the first time one can obtain the real-time measurement of product formation due to energy flow in an isolated large molecule.

Since the conformation change needed to give the red emission requires that the vibrational modes of the propyl linkage become populated, our results indicate that the excess vibrational energy in the anthracene species redistributes to vibrations (and " rotations ") in the side chain, thus triggering the temporal behaviour of the molecule. The threshold for the excess-energy dependence of the product formation is ca. 1000 cm^{-1} (2.9 kcal mol^{-1}), consistent with a reaction barrier involving C—C type bonds (see fig. 5). Finally, as expected, the decay and buildup time constants at ca. 3000 cm^{-1} of excess energy (ca. 400 ps) are much different from the decay time constant of bare anthracene at similar excess energies (5.7 ns). The question then is: what does this rate in the isolated molecule mean and what is the role of the solvent?

Using an assumed set of vibrational frequencies for the molecule along with our jet lifetimes at different E_x, we have calculated the decay of the anthracene-like emission for a thermal distribution at 298 K. The calculated lifetime (ca. 600 ps) is appreciably shorter than actually observed in cyclohexane solution (1.4 ns).[18] Thus it appears that the geometrical changes needed for product formation are hindered by the solvent.

Marcus and Noyes [19] have discussed the effect of diffusion on observed reaction rates. Near steady state, $\tau_{obs} \approx \tau_{activ} + \tau_{diff}$. Consequently, the real τ_{act} can be shorter than that observed in solution. Since the diffusion coefficient is on the order of a few times 10^{-5} cm^2 s^{-1}, it is perhaps not surprising for a diffusion distance of order of 10 Å that $\tau_{diff} \approx 1$ ns, and τ (isolated molecule) ≈ 500 ps. More tests of these ideas are in progress. The important point, however, is that energy redistri-bution to the propyl linkage occurs and product formation has a threshold at ca. 1000 cm^{-1}. As shown in fig. 5 the rate of product buildup is the same as that of the anthracene chomophore decay.

(E) STEPWISE SOLVATION AND PHOTODISSOCIATION

Last year we extended the application of the picosecond-jet technique to the study of the dynamics of isolated molecules in various stages of solvation with various solvents (water, alcohol, etc).[13,20] The idea was to study this " controlled " solvation and its dependence on the energy redistribution. Also we wanted to examine the photodissociation of these different solvated species or complexes following selective pumping by the picosecond laser. The systems we studied in some detail are azine–solvent complexes made in the jet with He or Ar as the carrier gas.

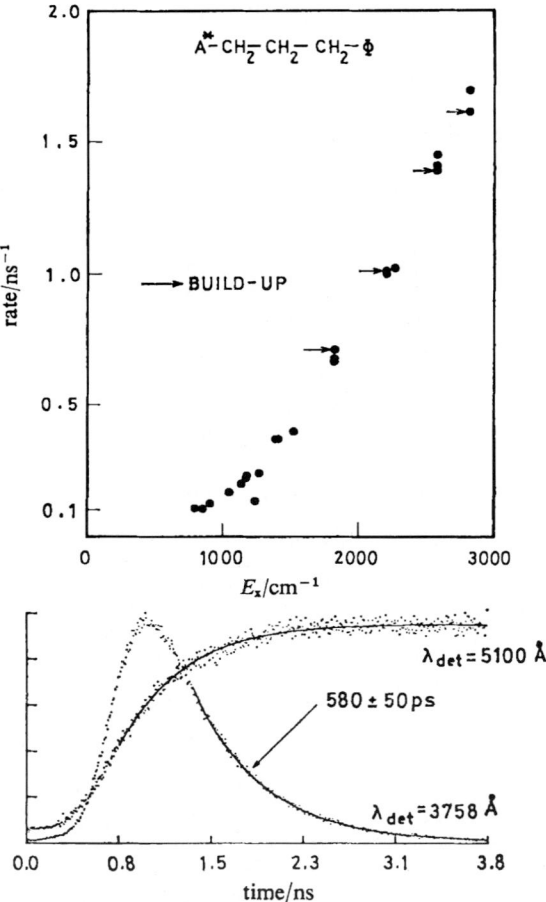

Fig. 5. For caption see opposite.

From the dispersed fluorescence, excitation spectra and their dependence on solvent concentration we identified the different solvated species and obtained the frequency of the new vibrational modes that result from complexation (typically 170 cm^{-1} and below). For isoquinoline (IQ) three solvents were used (water, methanol and acetone) to deduce some particular effects regarding the nature of hydrogen bonding in the species:

$$\text{>N:} \cdots (\text{HOR})\,(\text{HO—R})_n \quad .$$

Fig. 6 and 7 display typical results for the effect of solvent on the excitation spectra, fluorescence and lifetimes. We used these observations to measure bond dissociation energy of the different species by varying the excess vibrational energy of the " parent " jet-cooled molecule type modes. The threshold effect found for the dissociation of the 1 : 1 IQ (methanol) is evident and occurs at ca. 3 kcal mol^{-1}, in agreement with solution-phase enthalpies. (The effect of higher degree of solvation on the dynamics will be discussed elsewhere.)

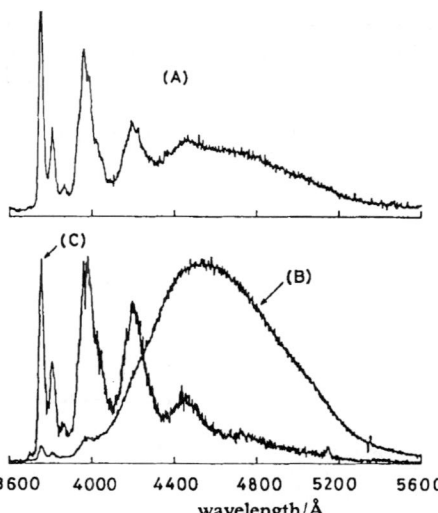

Fig. 5. Threshold effect (facing page, top) and fluorescence spectra at different E_x.   Facing page, bottom: decay of anthracene-like emission and build-up of product emission.   This page: fluorescence spectra at early and long gating times after the pulse.   Note the evolution of the product emission with time.[11]   (A) All times, $E_{exs} \approx 1400$ cm^{-1}; (B) all times, $E_{exs} \approx 2800$ cm^{-1}; (C) early time $E_{exs} \approx 2800$ cm^{-1}.

SOME THEORETICAL IDEAS

(A) HOMOGENEOUS BROADENINGS IN ISOLATED MOLECULES

As is well known, one cannot observe quantum beats unless the separation between energy states is larger than or comparable to the width of these states (or bunch of states).   In anthracene the results of Lambert et al.[7] (see previous section) must therefore indicate that the rovibronic states prepared by the laser pulse (at the measured excess vibrational energy) are not severely overlapping.   In fact, a limit on the homogeneous width is now available: <500 MHz.   The apparent width of the excitation spectra is, of course, much larger (implying picosecond relaxation) than this width, but, in principle, by using single-mode laser excitation we should be able to resolve bands that correlate with the beat frequency, especially in the resonance fluorescence region.   In this region there is no complexity due to the presence of many levels in the final state of the emission.

The interesting findings in anthracene are: (*a*) the uniqueness of the observation of beats at certain excess energy and the sensitivity of the observation to detection wavelengths; (*b*) the shortening of lifetimes at higher E_x.   These lifetimes do not vary drastically in magnitude even though the dispersed fluorescence broadens in a dramatic way at higher E_x values.   Actually, the " broad " feature of the dispersed fluorescence is *not* a real continuum.   This suggests that the final state of the emission could be highly structured.   To clarify the involvement of the final state (or states) in the observed spectra (hence the nature of vibrational redistribution) Keelan[21] is examining the effect of density-of-states in the initial and final manifolds, and couplings on the overall redistribution.

The reason that these beats are sensitive to excitation and detection energies is clear.   For the excitation process we must form linear combinations of these rovibronic states in order to see the interference (coherent) effects.   The laser source, which was varied with respect to coherence time and width, has pulse substructure with each of the " noise " spikes of the substructure having large enough bandwidth to span the excited states to form the superposition.   If the damping is faster than the period of the beats, as it will be at high excess energies, then no beats will be observed.   In the

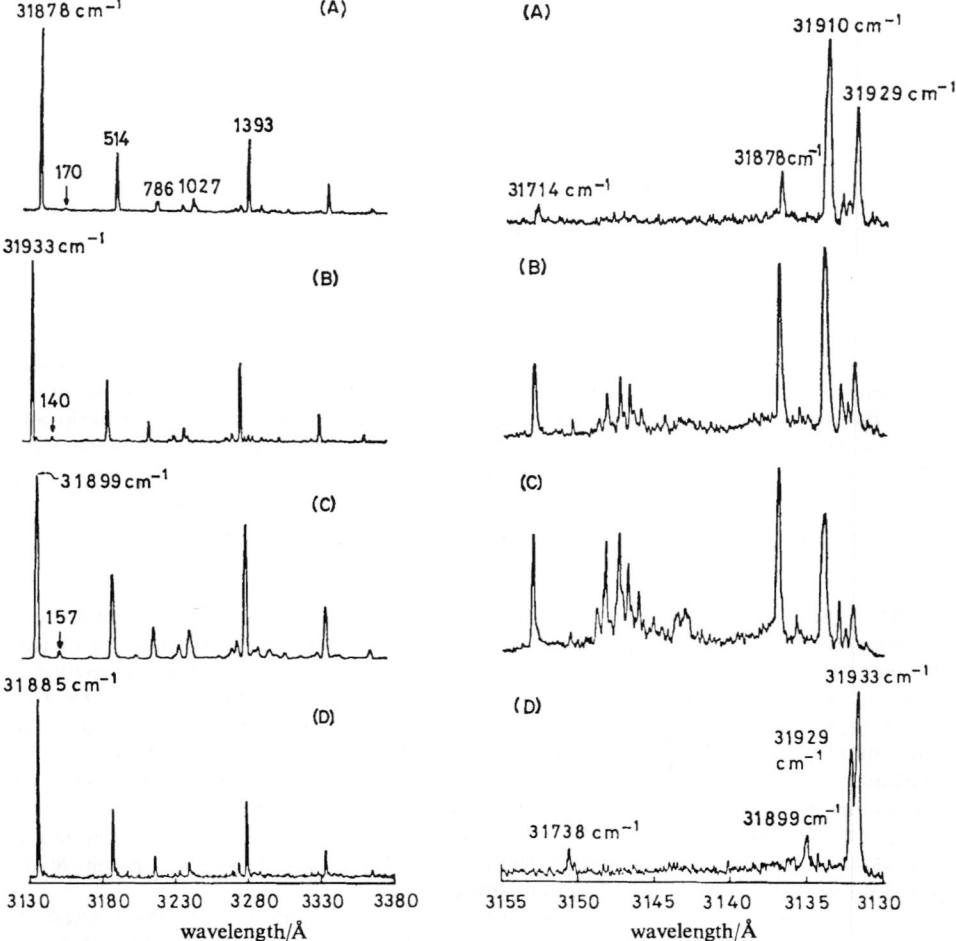

Fig. 6. Jet-cooled fluorescence spectra of IQ–solvent (left) and the excitation spectra (right) at different solvent concentrations.   (A)–(C) on the right give excitation spectra at 4, 10 and 25 Torr methanol, respectively.   (D) is the spectrum at 10 Torr water.   On the left, (A) is for methanol-2, (B) and (C) for water-1 and water-2, and (D) for acetone.   Details are given in ref. (13) and (20).

decay process, on the other hand, one must have branching transitions, i.e. a common level to the emitting states.   If the oscillator strength of one of the beating levels to the final state is very different, the modulation depth will be altered.   The modulation depth varies in our experiment, and one plausible explanation of the *nature* of the beating states is that there is a mode-specific Coriolis coupling at this excess energy.

This coupling will alter the relative oscillator strength of the emitting states, and will be specific depending on the K rotational structure. This should be evident in the very high-resolution Doppler-free spectra of anthracene, reminiscent of the benzene work of Schlag's group.[22] It is also possible that rotationally averaged anharmonic coupling is present. The points will be addressed in a forthcoming paper (by

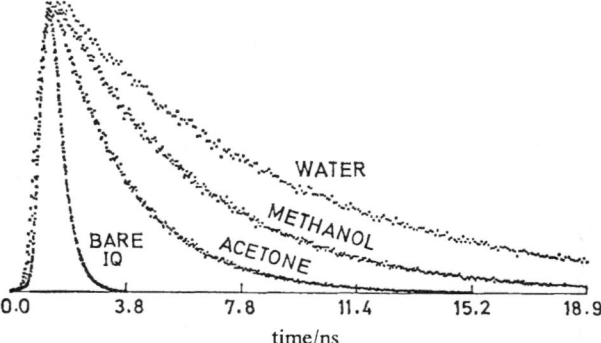

Fig. 7. Lifetimes of IQ–solvent complexes in the jet: water, 7.9 ns; methanol, 4.6 ns; acetone, 2.5 ns; bare IQ, 380 ps.

Lambert, Felker and Zewail) from this labaratory. The important point here is that whatever the nature of the states involved, it is clear that some strongly coupled vibrational states in the molecule at certain excess vibrational energies are coherent on subnanosecond time-scales, as evident from the data on anthracene and stilbene. Recently similar findings have been found for molecules in their ground electronic state but also with excess vibrational energy.[23]

(B) THE " THRESHOLD " EFFECT AND REDISTRIBUTION BY LOW-FREQUENCY MODES

In our studies of anthracene, stilbene, methyl salicylate, $A\!-\!(CH_2)_3\!-\!\varphi$ and photo-dissociation in the jet, the following general features were found: (*a*) A threshold for an order-of-magnitude change in the rate at E_x between 1000 and 2000 cm^{-1}, depending on the molecule. In anthracene there is a levelling off of this rate at $E_x \approx 1500$ cm^{-1} and the change in the rate is only a factor of 4. (We have also studied the effect of deuterium substitution.) (*b*) Quantum beats in anthracene and stilbene are absent above the energy region for this abrupt change in the rate. (*c*) The dispersed fluorescence of anthracene around this threshold area displays two features—a sharp resonance emission and a " diffuse " red-shifted [from a (0,0) excitation] emission. The spectra become completely diffuse at high excess energies and resemble the trend found in other large molecules.

An interesting question arises from these studies: What is the origin of this threshold effect? In what follows some ideas are presented.

The initial temperature of the molecule is very low and certainly is much lower than $\hbar\omega/k$ of totally symmetric modes. We may divide the modes of the molecule into those which are optically active (predominantly totally symmetric or " relevant " R) and those which are not excited directly by the laser (bath modes B). This division of the system–bath interactions accounts for dephasing and energy relaxation by T_2 and T_1 time constants as discussed elsewhere.[1,24]

The laser excites the totally symmetric R modes of anthracene but the molecule has a number of low-frequency (including non-totally symmetric) B modes (about ten) which range in frequency from 100 to 500 cm^{-1}. The redistribution of excitation of R

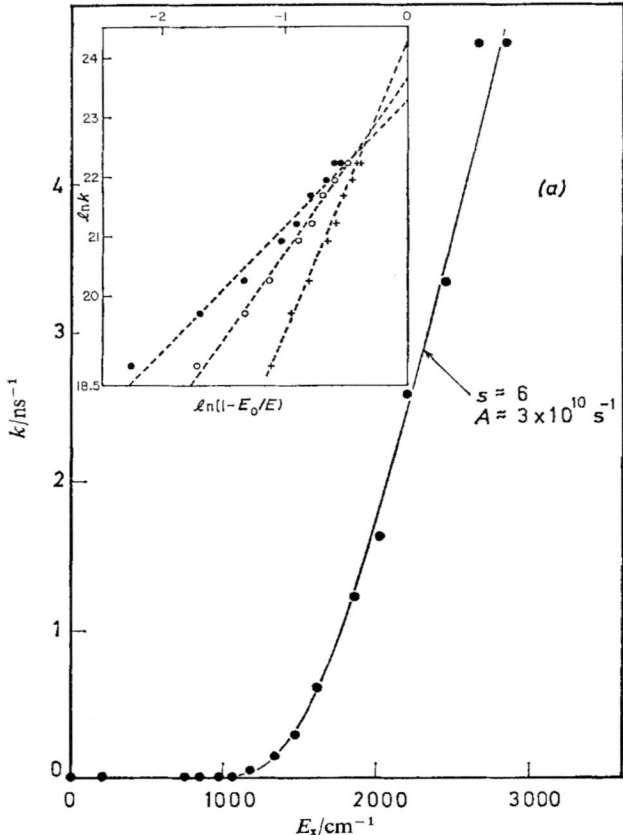

Fig. 8. For caption see opposite.

to B will have a threshold depending on the number of quanta that can be populated in B (density of states) and the degree of coupling. Invoking these low-frequency modes in the redistribution suggests the use of restricted density of states since high frequency modes are not efficiently populated. It is interesting that in all molecules we studied such modes do exist, and it is perhaps a universal character of these large molecules to involve these modes in the redistribution. The change in anthracene rates around the threshold is not as dramatic as in the other molecules whicn enjoy more of these low-frequency modes. The " suppression " of the activity of these modes by the Shpolskii effect in matrices may be the reason for obtaining sharp spectra, or in our language, in enhancing the I_s/I_b ratio. The question is now: Knowing the mode structure, can we predict theoretically the energy threshold effect?

(C) STATISTICAL AND SIMPLE THEORIES FOR THE REDISTRIBUTION

Here we shall outline an approach that is simple, but perhaps too simple. For the redistribution, we will define a redistribution parameter (" temperature ") which describes the rate according to the following equation:

$$k = Ae^{-E_0/kT_{eff}} = Ae^{-\bar{n}E_0/E_x}$$

where $\bar{n}$ is the average number of modes (restricted number of modes) that the molecule " heats up " in the redistribution. If this is a real heating process, the analogy with

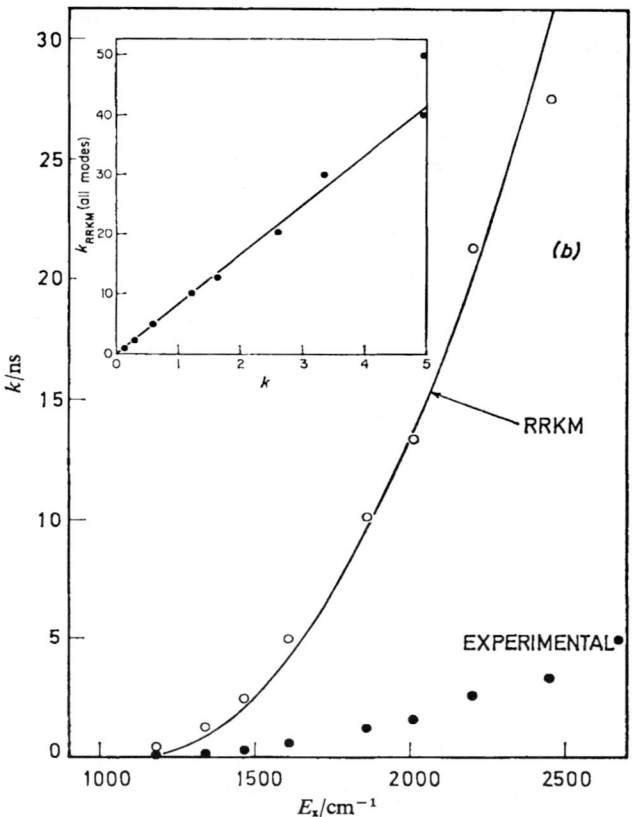

Fig. 8. (*a*) The fit of the stilbene data to R.R.K. theory.   The insert shows the sensitivity of the fit to the E_0 value.   +, $E_0 = 900$ cm^{-1}; O, $E_0 = 1100$ cm^{-1}; ●, $E_0 = 1200$ cm^{-1}. (*b*) Comparison of the R.R.K.M. calculation for stilbene with our data.   When the experimental data are scaled by α, the agreement with R.R.K.M. theory is quite good.   We ascribe α as due to reversible isomerization process at the E_x of interest. ●, Observed rates; O, scaled rates (αK_{obs}; $\alpha = 8.3$). The plot of ln k against $1/E_x$, not shown, gives a straight line.

solids for the use of specific-heat will give an effective number of modes.   When we plotted experimental ln$k$ against $1/E_x$ for stilbene (see data of fig. 3) we obtained a very good straight-line fit with a " pre-exponential " of ca. 1.5×10^{11} s^{-1}, and an apparent activation energy of ca. 9000 cm^{-1}.   Since E_0 is known from our work to be ca. 1000 cm^{-1}, which is in good agreement with solution-phase studies of the isomerization rate, $n \approx 9$.   This number is in surprisingly good agreement with the above " naive " prediction and with the number of low-frequency modes (<500 cm^{-1}) which are known for stilbene.   It should be mentioned that when we used the classical R.R.K. expression * for k, we obtained from the fit of the data $S \approx 6$ and $E_0 \approx 1000$ cm^{-1}

* The R. R. K. expression is simply

$$k = v \left(\frac{E - E_0}{E} \right)^{S-1}$$

where S is the number of vibrational oscillators.

(fig. 8).   (One can show that the R.R.K. expression can be written in an exponential form.)   Applications of this idea to other systems are underway.

In a collaborative effort with Marcus, we used quantum calculations of the density of states to obtain the rates that are in accord with the R.R.K.M. theory. We have found the following: First it can be shown that the E_0 obtained from our experiments (k against E_x) can be correlated with that obtained from solution-phase studies ($k \approx A e^{-E_0/kT}$).   Secondly, deviations from R.R.K.M. can be handled by

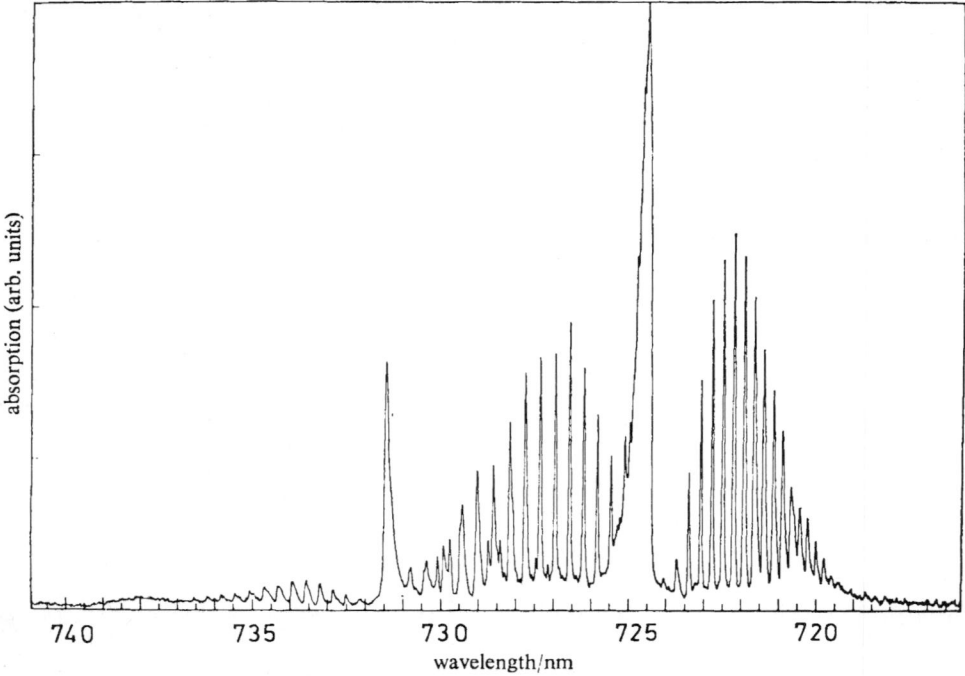

Fig. 9. The overtone spectrum of CHD_3 in the gas phase at $\Delta v_{CH} = 5$.   Note the well resolved P. Q, R branches.   The presence of two Q branches is due to Fermi resonance interaction.

using the appropriate modification in rates in order to suit these new observations. These modifications include reversible rate processes for the isomerization at these excess energies and trapping of energy in certain coordinates in the transition-state configuration.

For the dissociation of IQ–water and IQ–methanol good agreement with R.R.K.M. theory was found.   In these systems there are numerous low-frequency modes due to complexation, and the density of B-type modes (states) is very large.   This work will be published elsewhere.

In conclusion, the measurements of rates as a function of E_x in the beam provide a way of obtaining the threshold energy (" solvent free ") and the redistribution among modes.   It is possible that B-type modes are the main constituents of the phase space for the redistribution, and that this effect is almost universal.   It is interesting that the effective-temperature idea for k as a function of E_x will predict that at ca. 2000 cm^{-1} of excess energy the effective temperature is approaching room temperature, consistent with redistribution and congestion of spectra.   These studies raise interesting new questions regarding these possibilities and more experiments are in progress to test these ideas. Finally, the studies of stepwise solvation in beams promise interesting new avenues in that we may learn about linkages between gas-phase and solution chemistry from a microscopic point of view.

(D) WHAT ABOUT THE HIGH-ENERGY REGION?

Until now we have not obtained real-time measurements in the high-energy region (2 eV or so). However, in a recent experiment by Perry et al.[25] we have examined the spectra (fig. 9) of the molecule CHD_3 where there is only *one* CH local-mode oscillator that can couple to all other modes. Several interesting points emerge from the studies. First, the spectra $\Delta v_{CH} = 6,7 \ldots$ etc. are incredibly sharp even in the very high-energy region (ca. 16 000 cm^{-1}). Secondly, the rotational pattern basically prevails except for Fermi resonance type interactions. Thirdly, the transition linewidth of the isolated molecule (low-pressure) gives a relaxation time *longer* than ca. 1 ps. Using the same theoretical ideas discussed above for the threshold effect ρ_t at this excess ground-state vibrational energy (16 000 cm^{-1}) is only ca. 40 cm, which is the range at which we expect the redistribution rate to be different in magnitude from that at lower energies (the number of states at 5000 cm^{-1} is few). This means that the threshold should be at much higher energies than those described for the other large molecules in S_1. (In contrast, benzene at 16 000 cm^{-1} shows $\rho_t \approx 2 \times 10^8$ cm.) Consistent with this idea is the fact that the rotational constant as a function of excess energy undergoes a marked change about this threshold energy. More discussion of the energy distribution problem at high energies is given in ref. (26).

This work was supported by grants from the National Science Foundation (CHE8112833 and DMR8105034). I thank Prof. R. Marcus and Prof. J. Hopfield for very enlightening discussions. I would also like to aknowledge the dedicated efforts of William Lambert, Peter Felker, Joe Perry and Jack Synge.

[1] See, e.g., A. H. Zewail, *Physics Today*, 1980, **33**, 27.

[2] For a recent review see: C. S. Parmenter, *J. Phys. Chem.*, 1982, **86**, 1735.

[3] P. S. H. Fitch, L. Wharton, and D. Levy, *J. Chem. Phys.*, 1979, **70**, 2018.

[4] J. B. Hopkins, D. E. Powers and R. E. Smalley, *J. Chem. Phys.*, 1979, **71**, 388; 1980, **72**, 2950.

[5] A. Amirav, U. Even and J. Jortner, *J. Chem. Phys.*, 1981, **74**, 3745.

[6] D. H. Levy, in *Photoselective Chemistry*, ed. J. Jortner, R. Levine and S. A. Rice (Wiley-Interscience, New York, 1981), p. 323.

[7] Wm. R. Lambert, P. M. Felker and A. H. Zewail, *J. Chem. Phys.*, 1981, **75**, 5958.

[8] P. M. Felker, Wm. R. Lambert and A. H. Zewail, *Chem. Phys. Lett.*, 1982, **89**, 309.

[9] P. M. Felker, Wm. R. Lambert and A. H. Zewail, *J. Chem. Phys.*, 1982, **77**, 1603.

[10] J. A. Synge, Wm. R. Lambert, P. M. Felker, A. H. Zewail and R. M. Hochstrasser, *Chem. Phys. Lett.*, 1982, **88**, 266.

[11] P. M. Felker, J. A. Synge, Wm. R. Lambert and A. H. Zewail, *Chem. Phys. Lett.*, 1982, **92**, 1.

[12] A. H. Zewail, in *Picosecond Phenomena III*, ed. K. Eisenthal, R. Hochstrasser, W. Kaiser and A. Lauberau (Springer-Verlag, New York, 1982), p. 184.

[13] P. M. Felker and A. H. Zewail, *Chem. Phys. Lett.*, 1983, **94**, 448; 454.

[14] D. Millar and A. H. Zewail, *Chem. Phys.*, 1982, **72**, 381.

[15] J. Kommandeur, to be published.

[16] See, e.g., J. Saltiel and J. Charlton, in *Organic Chemistry* (Academic Press, New York, 1980), vol. 42-3, p. 25.

[17] R. M. Hochstrasser, *Pure Appl. Chem.*, 1980, **52**, 2683.

[18] Y. Wang, M. Crawford and K. B. Eisenthal, *J. Phys. Chem.*, 1980, **84**, 2696; *Chem. Phys. Lett.*, 1981, **79**, 529; M. Migita et al., *Chem. Phys. Lett.*, 1980, **72**, 229.

[19] R. A. Marcus, *Discuss. Faraday Soc.*, 1960, **29**, 129; R. M. Noyes, *Prog. React. Kinet.*, 1961, **1**, 129.

[20] P. M. Felker and A. H. Zewail, *J. Chem. Phys.*, 1983, in press.

330   PICOSECOND-JET SPECTROSCOPY AND PHOTOCHEMISTRY

[21] B. Keelan and A. H. Zewail, work in progress.
[22] E. Riedle, H. Neusser and E. Schlag, *J. Phys. Chem.*, to be published.
[23] R. Sharp, E. Yablonovitch and N. Bloembergen, *J. Chem. Phys.*, 1982, **76**, 2147.
[24] N. Bloembergen and E. Yablonovitch, *Physics Today*, 1978, **31**, 23.
[25] J. W. Perry, D. Moll, A. Kuppermann and A. H. Zewail, to be published.
[26] J. W. Perry and A. H. Zewail, *J. Phys. Chem.*, **86**, 5197.

Molecular Reaction Dynamics

G. Hancock

Physical and Theoretical Chemistry Laboratory, University of Oxford, UK OX1 3QZ.

Commentary on: **Crossed-Beam Reactions of Barium with Hydrogen Halides: Measurement of Internal State Distributions by Laser-induced Fluorescence,** H. W. Cruse, P. J. Dagdigian and R. N. Zare, *Faraday Discuss. Chem. Soc.*, 1973, **55,** 277.

By the end of the 1960s, the study of molecular reaction dynamics was well underway through the development of techniques such as molecular beam scattering and infrared chemiluminescence, both described in Faraday Discussions of 1962[1] and 1967.[2] Although such methods are still in vigorous use, it is rare nowadays to find an experimental article on reaction dynamics which does not describe the use of a laser to prepare reagents or to probe the products. Lasers in the late 60s and early 70s were certainly available, and although they were generally limited to fixed frequencies, they were not feeble devices. Kent Wilson's group were using a monster Nd:YAG oscillator with a 1 m glass amplifier to produce 1 J per pulse of frequency quadrupled light at the photochemically useful wavelength of 266 nm, (admittedly at the experimentally tedious repetition rate of 1 pulse per minute), and the first results were described at a Faraday Discussion in 1972 on the photofragment spectroscopy of alkyl iodides.[3] The paper that I have chosen however shows one of the first demonstrations of the use of tunable lasers to probe nascent reaction products, by the then new technique of laser induced fluorescence (LIF), and it was presented by Cruse, Dagdigian and Zare at the Faraday Discussion on Molecular Beam Scattering in 1973.[4] A brief foretaste of the method had appeared in *J. Chem. Phys.* the previous year,[5] but the data presented at the Faraday Discussion were far more comprehensive. The reactions studied by Richard Zare's group in Columbia were those of Ba with the four hydrogen halides, and the results were truly outstanding. Nascent vibrational populations, envelopes of the rotational distributions, and relative cross sections were all measured, and the spectroscopy had to be sorted out as well: no look up tables of spectroscopic constants and Franck–Condon factors here.

The method combined molecular beams and lasers. A beam of Ba atoms was crossed with a beam of the hydrogen halide. A pulsed nitrogen laser at 337 nm pumped a tunable dye laser at a repetition rate of 10 Hz and an output energy of 10 µJ pulse^{-1}, and the laser radiation was tuned over the barium monohalide absorption spectrum in the range 495–500 nm, with resultant fluorescence detected. Although the pulse energy was low in comparison with that delivered by modern tunable lasers, it is salutary to note that the authors recognised that the resultant laser power could saturate these strong transitions, and needed attenuation if relative populations were to be extracted with confidence. The resultant fluorescence was essentially all at the excitation wavelength (the Franck–Condon factors are diagonal), and "carefully aligned baffles" were used to cut the scattered light to the equivalent of one photoelectron per pulse as detected by the photomultiplier. Many future investigators would have done well to heed the advice given in this early paper.

Although laser detection is now so ubiquitous that we tend to forget its origins, we should not forget that the questions that the modern versions of the technique are able to answer were being posed in those early days. The idea of vector correlations involving velocity and rotational angular momentum was being mooted in the molecular beam scattering experiments described by Herschbach in the 1962 Discussion,[6] and this blossomed into the study of quantum state resolved stereodynamics by LIF, reviewed by John Simons in his 1999 Spiers Memorial Lecture.[7] But the 1973 paper was foreseeing this, with the suggestion made that velocity measurements on nascent products would be possible through Doppler line profile determination.[4] Laser excitation through quantised electronically excited states, first shown through LIF studies of this kind, led to the ideas of REMPI and its extraordinary success in ion imaging, particularly in photodissociation and more recently in reaction dynamics. Equipment development has influenced this field more than most. A nitrogen laser of thirty years ago used to probe reaction dynamics now seems extraordinarily dated in comparison for example with femtosecond lasers probing reactions as they happen, and with the use of lasers with the largest commercial market, cw diode lasers, in sensitive absorption techniques exploiting their exceptionally narrow bandwidths.

Another aspect of the Cruse *et al.* paper[4] deserves mention. The Ba + HI reaction works its way nowadays into University lecture courses on reaction dynamics as an example of how conservation of angular momentum can be used to extract further detailed information about the process. For this combination of atomic masses—a light particle eliminated from a reaction which also involves two heavy atoms—orbital angular momentum dominates in the reagents, whilst for the products the angular momentum is all in rotation. This one-to-one mapping allows determination of the reaction *opacity function*, the way in which reaction probability depends upon the impact parameter. This aspect of the dynamics did not feature in the early experiments of Cruse *et al.*,[4] requiring far more sophisticated molecular beam and narrow band laser techniques for its evaluation, but it must have dwelt heavily in the mind of at least one of the authors, as the latest (final?) paper on this reaction ("Energy and angular momentum control of the specific opacity functions in the Ba + HI → BaI + H reaction.") emerged from his laboratory as recently as 1996.[8]

So what personal influence did this paper have? I first met one of its authors (RNZ) and learnt about the power of LIF methods for studying reaction dynamics a couple of years before its publication, when I was a post-doc in California. In 1973 I embarked on using LIF systems with Karl Welge in Bielefeld, getting hold of the early commercially available tunable dye lasers pumped by a pulsed nitrogen or a cw Ar ion laser. The 1970s saw many other research groups starting to probe reaction products through LIF. A sabbatical in the Zarelab in 1982 led me to experiments on measuring, by LIF, the energy distributions in the products of the Ca + HF($v = 1, J$) reaction, and the way in which they changed with the rotational state J of the vibrationally excited HF. I shared an office with Richard Dixon, also on sabbatical at the time, and he was then developing methods for using LIF to measure and understand vector correlations in the products of photodissociation.[9] He bounced his ideas off me. Most of the collisions were totally elastic, but a certain amount of energy transfer occurred. The trail from the original Cruse *et al.* paper[4] leading to more sophisticated technique development for the applications of tunable lasers to study quantum state resolved species has certainly had a major influence on my involvement with the area of molecular reaction dynamics.

References

1 "Inelastic Collisions of Atoms and Simple Molecules". *Faraday Discuss. Chem. Soc.,* 1962, **33**.
2 "Molecular Dynamics of the Chemical Reactions of Gases". *Faraday Discuss. Chem. Soc.,* 1967, **44**.
3 S. J. Riley and K.R. Wilson, *Faraday Discuss. Chem. Soc.,* 1967, **72**, 132.
4 H. W. Cruse, P.J. Dagdigian and R.N. Zare, *Faraday Discuss. Chem. Soc.,* 1973, **55**, 277.
5 A. Schultz, H. W. Cruse and R. N. Zare, *J. Chem. Phys.,* 1972, **57**, 1354.
6 D. R. Herschbach, *Faraday Discuss. Chem. Soc.,* 1962, **33**, 281.
7 J. P. Simons, *Faraday Discuss.,* 1999, **113**, 1.
8 K. S. Kalogerakis and R. N. Zare, *J. Chem. Phys.,* 1996, **104**, 7947.
9 R. Vasudev, R. N. Zare and R. N. Dixon, *J. Chem. Phys.,* 1984, **80**, 4863.

Crossed-Beam Reactions of Barium with Hydrogen Halides

Measurement of Internal State Distributions by Laser-Induced Fluorescence

By H. W. Cruse, P. J. Dagdigian and R. N. Zare

Department of Chemistry, Columbia University, New York,
New York 10027, U.S.A.

Received 22nd January, 1973

The use of laser-induced fluorescence as a molecular beam detector for the measurement of internal state distributions of reaction products is presented and applied to the reactions of barium with the hydrogen halides. It is found that most of the reaction exoergicity appears as translational energy of the products and that the total reactive cross section is positively correlated with the average fraction of the exoergicity appearing as vibrational excitation.

Conventional molecular beam reactive scattering studies have excelled in the determination of the angular and velocity distributions of reaction products, but direct information on the internal state distributions has been sparse.[1] One of the most important of the non-beam methods for learning about the partitioning of reaction energy into the internal degrees of freedom of the products has been infra-red chemiluminescence studies.[2] Unfortunately, this technique has hitherto been limited to hydride compounds, principally hydrogen halides. We present an alternative technique based on electronic fluorescence spectroscopy.

In this method, a tunable narrow-band light source is swept in wavelength. When it coincides with a molecular absorption line, the molecule makes a transition to an excited electronic state and is detected by observing the subsequent light emission of the molecule. One of the principal advantages of this technique is its specificity, for it is possible to identify unequivocally the absorbing molecule and to assign internal quantum numbers. Moreover, the sensitivity of electronic fluorescence detection using presently available tunable laser sources appears to exceed that of a universal ionizer.

In this paper we describe the use of laser-induced fluorescence as a molecular beam detector and apply it to the reactions of barium atoms with hydrogen halides,

$$Ba + HX \rightarrow BaX(X^2\Sigma^+) + H ; \quad X = F, Cl, Br, I. \qquad (1-4)$$

Reactions (1)-(3) have not been studied using beam techniques; however, an angular distribution has been measured for reaction (4).[3] We report the vibrational population distributions of the BaX products, approximate rotational distributions in reactions (1) and (2), and relative total reactive cross sections for the four reactions.

EXPERIMENTAL

The experimental arrangement consists of a simple molecular beam apparatus, a pulsed tunable dye laser for exciting fluorescence, and a gated optical detection system.

MOLECULAR BEAM SYSTEM

The beam apparatus is housed inside a cylindrical stainless steel chamber (70 cm long and 25 cm diam.) that is evacuated by two 4-inch diffusion pumps (effective pumping speed

700 l. s^{-1}).   A metal atom beam and a gas beam intersect to form the reaction zone through which the laser beam is directed (see fig. 1).   The laser-induced fluorescence is viewed at right angles to the laser beam by a photomultiplier.   In some experiments, the hydrogen halide beam source is replaced by a gas jet so that the vacuum chamber serves essentially as a scattering box, as in earlier preliminary experiments.[4]

The barium atom beam effuses from a molybdenum oven (orifice diam. 0.3 cm) that is resistively heated to between 1000 and 1100 K using tantalum wire windings.   The oven is

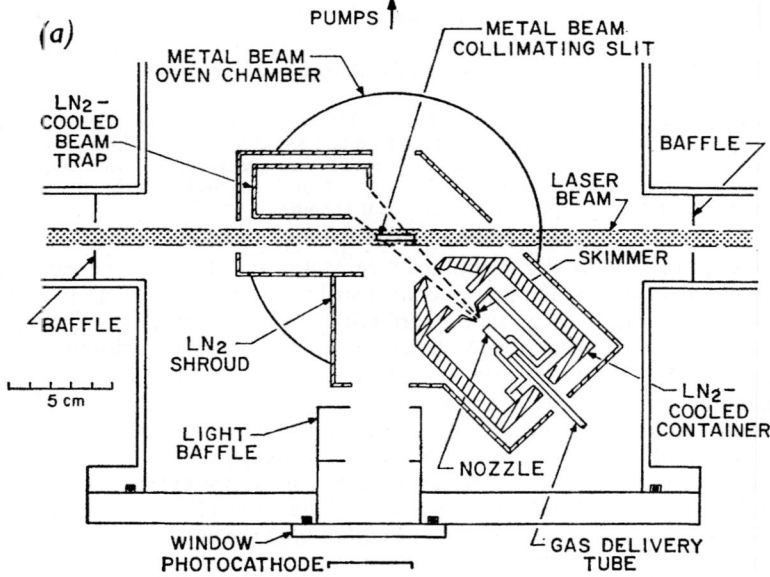

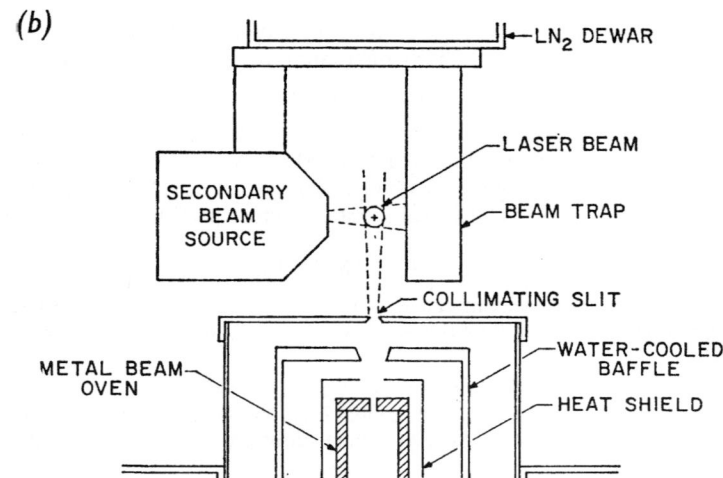

FIG. 1.—The molecular beam apparatus: (*a*) top view, (*b*) side view. In (*b*) the laser beam is perpendicular to the plane of the figure and the secondary beam source has been rotated by 45° about the Ba beam axis for clarity.

situated beneath the main vacuum chamber in a separate chamber evacuated by another 4-inch diffusion pump. The barium beam passes into the main chamber through a rectangular slit, 0.2×1.9 cm in size, located 4 cm from the oven orifice. When the gas beam source is used, the collimating slit is 5 cm below the reaction zone through which the laser beam passes. With the scattering box arrangement, the slit is 2.5 cm below the laser beam. The metal beam flux is measured using a film thickness monitor (Granville–Phillips) and is typically 2×10^{15} atom cm^{-2} s^{-1}, corresponding to a density in the reaction zone of 5×10^{10} atom cm^{-3}.

The hydrogen halide beam source consists of a stainless steel nozzle (orifice diam. 0.025 cm) with a conical-shaped stainless steel skimmer, 0.13 cm diam. and 0.5 cm from the nozzle. The nozzle and skimmer assembly is surrounded by a liquid-nitrogen-cooled copper shroud. A Nichrome heater is used to maintain the nozzle and skimmer assembly near room temperature. Final collimation of the beam is provided by a 0.60 cm diam. hole in the copper shroud. The beam angular divergence is about $10°$ FWHM. After traversing the reaction zone, the hydrogen halide beam enters a liquid-nitrogen-cooled " catcher ". Both the beam source and beam catcher are suspended from a liquid nitrogen dewar situated directly above the barium oven. Additional liquid-nitrogen-cooled copper shields partially surround the beam source and catcher (see fig. 1) to reduce heating by room-temperature black-body radiation. In addition, the shielding serves to reduce extraneous scattered light from the oven and from the laser beam. The hydrogen halide source is normally operated at 20 Torr and the calculated beam density in the reaction zone is 6×10^{11} molecule cm^{-3}. Note that the laser beam passes through the long axis of the reaction zone in order to maximize the absorption path length.

With both molecular beams in operation, the pressure in the main chamber is 1×10^{-6} Torr. The maximum rise in pressure due to the secondary beam source is 3×10^{-7} Torr for HCl, while for the other gases it is 1×10^{-7} Torr or less. For the scattering box arrangement with a pressure of 2×10^{-5} Torr of reactant gas in the main chamber, the pressure in the oven chamber is 3×10^{-6} Torr of which 1×10^{-6} Torr is attributable to the reactant gas. The quoted pressures are uncorrected ionization gauge measurements.

All hydrogen halide samples are purified before use: HCl, HBr, HI (Matheson Gas Products), and DCl (Merck) are condensed at 77 K, pumped on, and then distilled to remove parent halogen impurities; HF (Matheson), which is handled in a copper vacuum line fitted with Monel valves, is condensed at 196 K and pumped on to remove SiF$_4$.[5]

FLUORESCENCE EXCITATION AND DETECTION

The internal state distributions of the barium halide reaction products are determined by monitoring the fluorescence intensity as a function of excitation wavelength. Fig. 2 shows a block diagram of the tunable dye laser and fluorescence detection equipment. A pulsed nitrogen laser (Avco–Everett Research Laboratories) is used to pump a noncirculating dye solution contained in a closed cell. The dye laser cavity is formed by a 50 % reflecting (output) mirror and a 600 groove mm^{-1} diffraction grating (Bausch and Lomb) used in 5th order to select the desired laser wavelength. An intra-cavity 20 power beam expander narrows the bandwidth of the laser beam.[6] The dyes 3-aminophthalimide and brilliant sulphaflavine (saturated solutions in ethanol) cover the required wavelength range in two steps: 485-530 and 515-565 nm. The energy output of the dye laser is typically 10^{-5} J per pulse, corresponding to 3×10^{13} photons per pulse. The duration of a pulse is about 5 ns and the repetition rate is 10 pps.

The laser beam passes through a double polarizer that both selects the polarization of the light source (usually perpendicular to the barium beam) and serves as a convenient and continuous method for attenuating the laser intensity. The beam then passes into the main vacuum chamber through a 40 cm focal length lens sealed to the end of a 60 cm long side arm. This lens collimates the slightly divergent laser beam, which at the reaction zone is about 1 cm diam. The laser beam exits from the vacuum chamber through a window also sealed to the end of a sidearm.

During an experimental run, the laser wavelength is scanned by a variable-speed clock motor (Erwin Halstrup, Kirchzarten, West Germany) connected to the grating tilt control, and the laser intensity is monitored continuously using a photodiode combined with a gated amplifier or is monitored at approximately 1 nm intervals by a 1 m spectrometer (Interactive Technology) that provides laser wavelength and bandwidth data. The laser bandwidth is typically 0.05 nm and the variation in the laser wavelength at a fixed grating setting is not detectable ($<$0.005 nm), while under the same conditions the variation in the laser intensity is less than 5 % over a 10 min period (detector output time constant of 3 s, corresponding to 30 pulses).

An RCA 7265 photomultiplier (S-20 photocathode) detects the reaction product fluorescence. The photomultiplier is located 15 cm from the reaction zone behind a quartz window in the main chamber. Only 0.5 % of the fluorescence impinges on the photocathode which has a quantum efficiency of about 20 %. Thus, at best only one laser-excited molecule in a thousand is detected. Following the short excitation pulse, the intensity of the molecular fluorescence decays exponentially with a lifetime which is characteristic of the excited state.

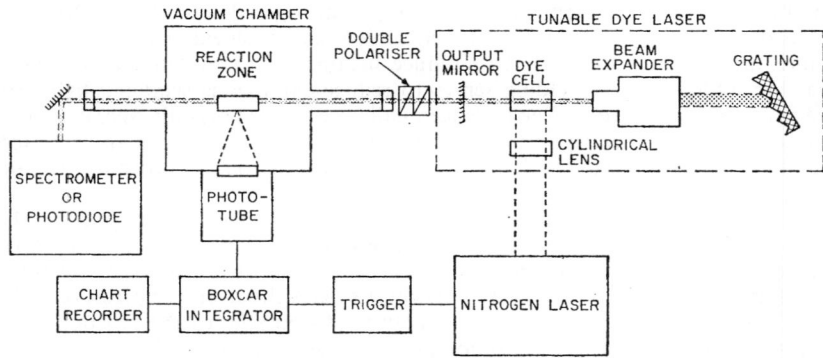

FIG. 2.—Schematic of the tunable dye laser and fluorescence detection equipment.

The photomultiplier output is detected, therefore, either by a fast oscilloscope (Tektronix model 454 or 7904) or by a boxcar integrator (Keithley Instruments models 881 and 882). The latter device (1) operates as a gated amplifier, detecting that signal appearing during and immediately after the laser pulse, and (2) averages the signals associated with many laser pulses. In these experiments, where the product molecule excited-state lifetimes are all approximately 15 ns,[7] a gate width of 100 ns is used.

In previous experiments, unwanted scattered light, which can be many times more intense than the desired fluorescence signal, was removed either by delaying the opening of the boxcar gate or by using a cutoff filter at the laser wavelength.[4] However, these remedies are not well suited to the present experiment because (1) the product molecule radiative lifetimes are comparable to the laser pulse duration and the discrimination time of the detection electronics and (2) most fluorescence occurs at wavelengths close to the laser line. Instead, carefully aligned baffles (shown in part in fig. 1) are placed in the long sidearms and in front of the photomultiplier and the interior of the main chamber is blackened with Nextel paint (3M Company). In this manner the scattered light level is reduced to the equivalent of about one photoelectron per pulse.

RESULTS

The BaX reaction products are detected by observing the laser-induced fluorescence of the $C^2\Pi$–$X^2\Sigma^+$ band systems that lie near 500 nm. Fig. 3-6 present typical scans showing the variation of fluorescence intensity with excitation wavelength. The scan rates are between 0.1 and 0.5 nm min^{-1}, the detector time constant being 1 s. The variation of laser intensity with wavelength is small for all these scans ($<$10 %). Within experimental error identical results are obtained using the secondary beam source and scattering chamber arrangements, the latter operating at HX gas pressures of $1\text{-}2 \times 10^{-5}$ Torr. Since no discernible alteration of the fluorescence spectra occurs upon increasing (or decreasing) the HX pressure by a factor of four, we conclude that rotational or vibrational collisional relaxation is negligible under the conditions used. The data are also unaffected by vibrational radiative decay since we estimate that the vibrational lifetimes, for all levels observed in these experiments, are larger by more than two orders of magnitude than the time delay between product formation and detection ($< 10^{-4}$ s).

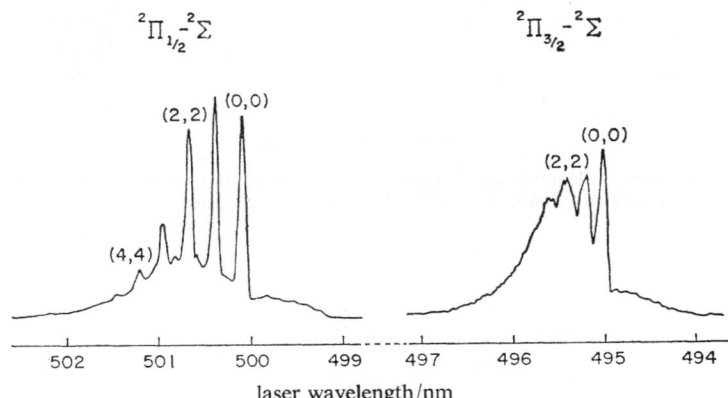

FIG. 3.—Variation of BaF fluorescence intensity with laser wavelength for the reaction Ba+HF→ BaF+H. The band heads of the BaF $C^2\Pi\text{-}X^2\Sigma^+$ system are designated by (v', v'').

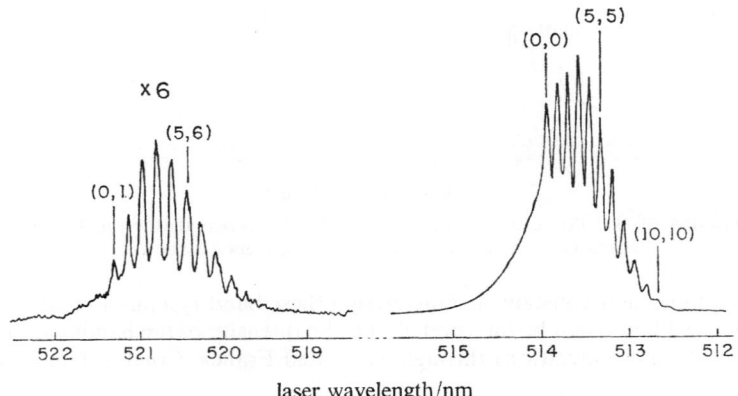

FIG. 4.—Variation of BaCl fluorescence intensity with laser wavelength for the Ba+HCl→BaCl+H reaction. The $C^2\Pi_{\frac{3}{2}}\text{-}X^2\Sigma^+$ $\Delta v = 0$ and -1 sequences are shown, the latter with a 6 times more sensitive ordinate scale.

The outstanding features of these band systems are the small spacing between the bands of a vibrational sequence and the dominant intensity of the $\Delta v = 0$ (diagonal) sequence. These facts indicate that the potential curves for the C and X states are nearly identical and help to explain why no rotational analysis has been carried out in the spectroscopic studies of these BaX band systems.[8-12] In order to extract meaningful internal state populations from the data shown in fig. 3-6, it is necessary to

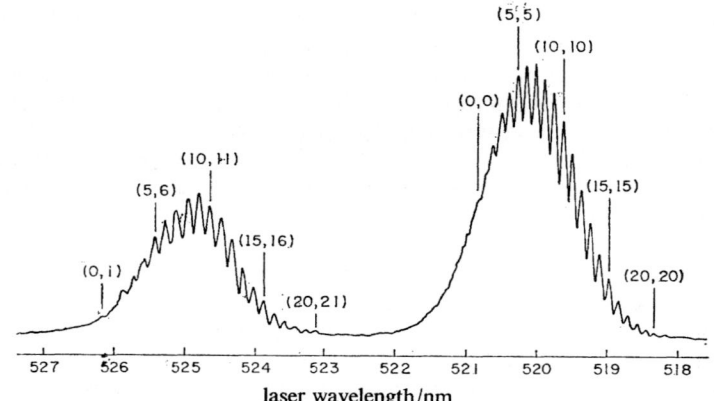

FIG. 5.—Variation of BaBr fluorescence intensity with laser wavelength for the Ba + HBr→BaBr + H reaction. The $C^2\Pi_{\frac{3}{2}}-X^2\Sigma^+ \Delta v = 0$ and -1 sequences are shown.

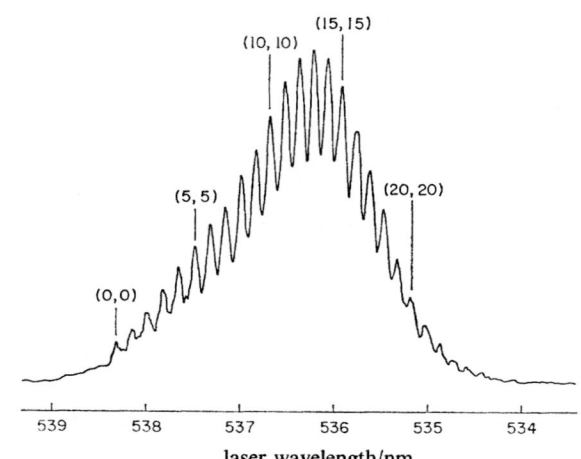

FIG. 6.—Variation of BaI fluorescence intensity with laser wavelength for the Ba + HI→BaI + H reaction. The $C^2\Pi_{\frac{3}{2}}-X^2\Sigma^+ \Delta v = 0$ sequence is shown.

estimate the molecular constants characterizing these band systems so that the formation of the band heads can be interpreted and the intensity of the bands can be related to the internal state populations through calculated Franck–Condon factor arrays.

ESTIMATES OF MOLECULAR CONSTANTS FOR THE
BaX $C^2\Pi-X^2\Sigma^+$ BAND SYSTEMS

For a $^2\Pi-^2\Sigma^+$ transition there are a total of 12 branches whose members have line positions given by

$$F_i(J') - F_j(N'') = v_0^{(i)} + B_{eff}^{(i)}J'(J'+1) - B''N''(N''+1) \tag{1}$$

where $i = 1$ or 2 denotes the $^2\Pi_{\frac{1}{2}}$ or $^2\Pi_{\frac{3}{2}}$ fine structure components of the $C^2\Pi$ state and $j = 1$ or 2 denotes the $J'' = N'' + \frac{1}{2}$ and $J'' = N'' - \frac{1}{2}$ spin components of the

$X^2\Sigma^+$ state. In eqn (1) centrifugal distortion corrections, Λ doubling in the upper state and spin splitting in the lower state have been disregarded. Here $v_0^{(i)}$ is the band origin of the $^2\Pi_i$–$^2\Sigma^+$ subband and $B_{\mathrm{eff}}^{(i)}$ is an effective rotational constant that is related to the mechanical rotational constant B_v by $B_{\mathrm{eff}}^{(i)} = B_v + (-1)^i B_v^2/A_v$, for $A_v \gg B_v$, where A_v is the spin-orbit constant (assumed to be positive [13]). Because of the spin-orbit splitting in the C state, each (v', v'') band shows four band heads. Since the bands are degraded to the red, the band heads are formed by the $Q_1 + R_{12}$ and R_1 branches in the $^2\Pi_{\frac{1}{2}}$–$^2\Sigma^+$ subband and by the $R_2 + Q_{21}$ and R_{21} branches in the $^2\Pi_{\frac{3}{2}}$–$^2\Sigma^+$ subband.[14]

The molecular constants for the C–X system are obtained by assuming that the C and X states are described by Morse oscillators for the low-lying vibrational levels. Then the vibrational spacings are exactly represented by $G(v) = \omega_e(v+\frac{1}{2}) - \omega_e x_e(v+\frac{1}{2})^2$ and the rotational constant α_e may be found from a knowledge of B_e, ω_e and $\omega_e x_e$.[15]

For BaF, the rotational constant B_0'' is known from an analysis of other band systems,[16] which together with approximate values of ω_e'' and $\omega_e x_e''$ permit the determination of α_e''. Values of $B_{\mathrm{eff}}^{(i)}$ are then obtained from eqn (1) by using the separation between the two heads of each (v', v'') subband measured by Jenkins and Harvey.[8] The errors introduced into $B_{\mathrm{eff}}^{(i)}$ by using band head separations should be small since the separation of the band heads is large compared to their experimental uncertainties. The $B_{\mathrm{eff}}^{(i)}$ values are found to differ from the B'' values by only 1 %. The band origins, $v_0^{(i)}$, are then obtained from the measured band head positions using the B'' and $B_{\mathrm{eff}}^{(i)}$ values. The rotational constant B_e' is obtained by averaging the $B_{\mathrm{eff}}^{(i)}$ values. Finally, a linear least squares fit is made to these derived band origins to obtain vibrational constants for both states. In this last step, ω_e' and $\omega_e x_e'$ are forced to be the same for both subbands by allowing A_v to vary linearly with $(v+\frac{1}{2})$, i.e., $A_v = A_e + a_e(v+\frac{1}{2})$. The spectroscopic constants found in this manner for BaF are listed in table 1.

TABLE 1.—ESTIMATED SPECTROSCOPIC CONSTANTS (cm^{-1}) OF THE BARIUM MONOHALIDE $C^2\Pi$–$X^2\Sigma^+$ BAND SYSTEMS

		BaF	BaCl	BaBr	BaI
$X^2\Sigma^+$	ω_e''	469.26(9)	279.92(10)	189.3(20)	163.0(2)
	$\omega_e x_e''$	1.822(1)	0.785(3)	0.22(5)	0.42(1)
	B_e''	0.216 44	0.080 78		
	α_e''	0.001 17	0.000 33		
$C^2\Pi$	ω_e'	459.32(9)	283.08(10)	193.5(20)	168.7(2)
	$\omega_e x_e'$	1.679(1)	0.772(3)	0.19(5)	0.43(1)
	B_e'	0.214 28	0.079 77		
	α_e'	0.001 22	0.000 34		
	$T_{00}(^2\Pi_{\frac{1}{2}})$	19 987.7	19 062.5	18 651	17 816.2
	$B_{\mathrm{eff}}^{(1)}(v=-\frac{1}{2})$	0.213 79	0.079 57		
	$T_{00}(^2\Pi_{\frac{3}{2}})$	20 185.7	19 451.6	19 193	18 571.5
	$B_{\mathrm{eff}}^{(2)}(v=-\frac{1}{2})$	0.214 77	0.079 97		
	A_e	195.6(2)	387.6(3)	542.1(20)	755.3(25)
	a_e	4.74(3)	2.87(3)	1.50(6)	0.1(6)

The numbers in parenthesis are 3σ of the fit in units of the last quoted digit.

A similar analysis is carried out for BaCl, whose band head separations have been measured by Parker.[9] The B_e'' value is estimated from electron diffraction data on $BaCl_2$ using the empirical observation [18] that the bond length of an alkaline earth monohalide is approximately 3 % smaller than that of the dihalide. The spectroscopic constants obtained for BaCl are also listed in table 1.

The spectroscopic data presently available for BaBr [10, 11] and BaI [12] preclude a similar treatment. Only single band heads have been reported in the case of BaI, while for BaBr the errors in the band head separations reported by Harrington [11] are deemed too large for the calculation of reliable Franck–Condon factors. The vibrational constants describing the band heads are also included in table 1. The single heads reported by Hedfeld [10] are used for BaBr. In the case of BaI, the ground state parameters are obtained from a fit to the $D^2\Sigma^+\text{--}X^2\Sigma^+$ band heads reported by Reddy and Rao,[19] while the upper state parameters are derived from a fit to the reported band heads of the C–X $\Delta v = 0$ sequence.[12]

Morse Franck–Condon factors are numerically calculated for BaF and BaCl by standard means. The results for BaF and BaCl are listed in table 2. For BaF the $\Delta v = 0$ sequence is by far the strongest and the three sequences $\Delta v = 0, \pm 1$ account for a minimum of 99 % of the $q_{v'v''}$ sum for v'' values up to 9. For BaCl the $\Delta v = \pm 1$ sequences become more important as v'' increases. It can be seen that even for the highest v'' value listed ($v'' = 9$), the sum of the $q_{v'v''}$ for the three sequences $\Delta v = 0, \pm 1$ is about 94 %. It is to be emphasized that the Franck–Condon factors depend much more sensitively on the difference $B'_v - B''_v$ than on the absolute values of B'_v and B''_v. Thus the calculational procedure described above is expected to give fairly reliable Franck–Condon factors for the $\Delta v = 0$ sequence even if the absolute values of B'_v and B''_v values are somewhat uncertain.

TABLE 2.—FRANCK–CONDON FACTORS CALCULATED FROM THE SPECTROSCOPIC CONSTANTS OF TABLE 1

BaF $C^2\Pi$–$X^2\Sigma^+$

v''	$v' = v'' - 1$	$v' = v''$	$v' = v'' + 1$
0		0.9868	0.0130
1	0.0131	0.9614	0.0251
2	0.0255	0.9375	0.0362
3	0.0371	0.9151	0.0464
4	0.0479	0.8943	0.0557
5	0.0581	0.8749	0.0642
6	0.0675	0.8568	0.0718
7	0.0764	0.8401	0.0788
8	0.0846	0.8247	0.0849
9	0.0923	0.8105	0.0904

BaCl $C^2\Pi$–$X^2\Sigma^+$

v''	$v' = v'' - 1$	$v' = v''$	$v' = v'' + 1$
0		0.9667	0.0317
1	0.0333	0.9034	0.0586
2	0.0648	0.8450	0.0813
3	0.0945	0.7910	0.1002
4	0.1230	0.7411	0.1158
5	0.1498	0.6950	0.1285
6	0.1753	0.6524	0.1385
7	0.1994	0.6131	0.1463
8	0.2224	0.5768	0.1520
9	0.2442	0.5432	0.1559

DETERMINATION OF INTERNAL STATE DISTRIBUTIONS

The detection method employed in these experiments measures the number density of a reaction product rather than the flux. Since product populations are proportional to flux, we must consider the laboratory velocity v_{LAB} of the recoiling product, which factor relates flux to number density. In particular, we must consider the possible variation of average lab. velocities for products formed with different internal energies. In these experiments where no angular distribution measurements have been attempted, we are interested in obtaining the total flux associated with a specified set of internal states, e.g., one vibrational level. Thus populations refer to total cross sections (detailed rate constants) for formation of this specified set.

Fig. 7 shows Newton diagrams for reactions (1)-(4). The concentric circles centered on the tip of the centre of mass velocity vector v_{CM} indicate the magnitudes of the BaX recoil speeds for various internal energies. Reaction exoergicities are calculated using BaX dissociation energies estimated by Gole, Jonah and Zare [20] and HX dissociation energies from Darwent.[21] It is clear that for all possible internal energies v_{LAB} lies close to v_{CM}. Indeed, in the worst case, Ba + HI, v_{LAB} differs from v_{CM} by a maximum of $\pm 15 \%$, and the typical difference is expected to be much less. This situation is well-known to be unfavourable to conventional angular distribution studies [3]; however, it facilitates the analysis of this experiment since we are justified in assuming that the magnitude and angular distribution of v_{LAB} is the same for each internal state, and hence no correction is necessary to relate measured number density to population.

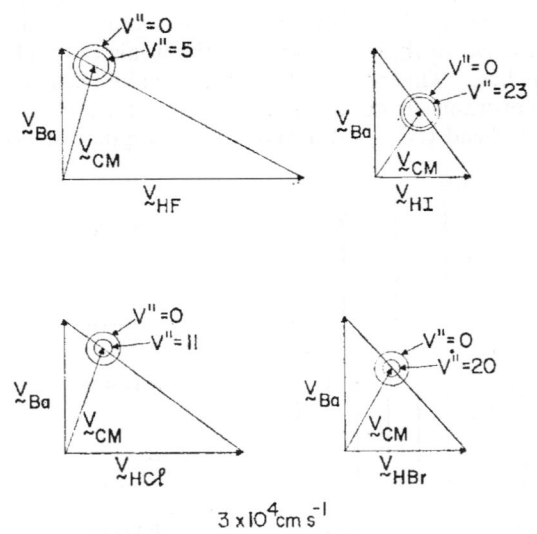

FIG. 7.—Newton diagrams for the Ba + HX reactions. Most probable velocities are used. The Ba beam is assumed to be effusive, while for the supersonic HX beam it is assumed that $\frac{1}{2} m v_{HX}^2 = \frac{5}{2} k T_0$, where T_0 is the upstream temperature. The concentric circles show the range of BaX speeds corresponding to the range of observed vibrational states v''. (In the text, $\underset{\sim}{V}$ is printed as v.)

The principal information determined from the fluorescence spectra is the values of vibrational band intensities $I_{v'v''}$ (energy s^{-1}). These are related to vibrational populations $N_{v''}$ by

$$I_{v'v''} = k[N_{v''}\rho(\lambda_{v'v''})q_{v'v''}] \sum_v v_{v'v}^4 q_{v'v} S(\lambda_{v'v})$$ (2)

where $\rho(\lambda_{v'v''})$ is the laser power density (energy cm^{-3}) at the wavelength of the (v', v'') band, q is a Franck–Condon factor for the transition designated by the subscripts,

$S(\lambda_{v'v})$ is the phototube sensitivity at the wavelength of the (v', v) band, $v_{v'v}$ is the frequency of the same band and k is a proportionality constant that includes parameters such as geometrical factors and the electronic transition moment (assumed to be constant). The term in brackets in eqn (2) represents the absorption process, while the summation represents the subsequent fluorescence pathways (assumed to occur only to the ground state) terminating in the various vibrational levels v.

The application of eqn (2) assumes that $I_{v'v''}$ is linearly proportional to $\rho(\lambda_{v'v''})$. This is valid in the limit of no radiation trapping and no optical pumping [22] of the molecular sample. Under our experimental conditions, the former poses no problem, whereas the latter requires consideration because of the high laser intensity used. Indeed, we observe a departure from linearity, which decreases as the laser intensity is reduced. Data, such as that shown in fig. 3-6, are taken with the laser intensity attenuated by typically fourfold, at which condition the deviation from linearity is $\sim 15\,\%$. To simplify data analysis, the same laser intensity (to within 20 %) is used for all scans, and only linear corrections to $I_{v'v''}$ are made for variation in laser intensity.

(a) $Ba + HF \rightarrow BaF + H$

The $\Delta v = 0$ sequence of the $C^2\Pi_{\frac{1}{2}}-X^2\Sigma^+$ band system is used to obtain the vibrational state populations of BaF since the band heads (formed by the $Q_1 + R_{12}$ branches) are well separated (see fig. 3). The vibrational band intensities are taken to be proportional to the areas of these heads above the weaker " background " formed by other headless branches. This procedure should be valid because (1) it is known from the spectroscopic constants in table 1 that $Q_1 + R_{12}$ lines with N'' values $\leqslant 150$ fall within 0.1 nm of the head ($N''_{head} \approx 40$) and (2) the populations of high N'' levels are

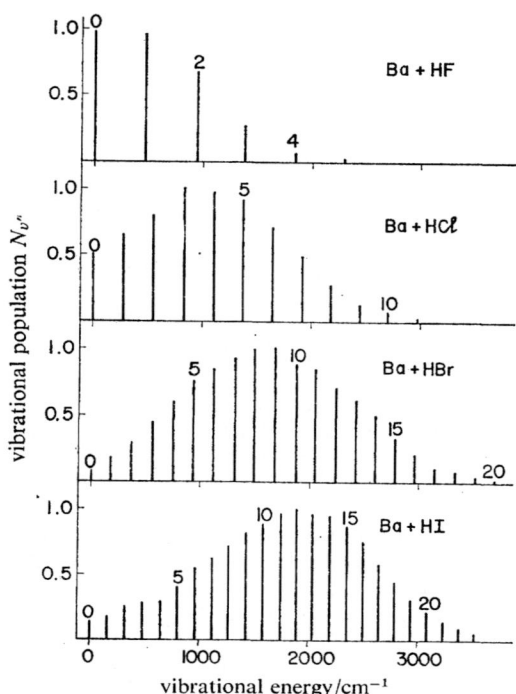

FIG. 8.—Relative vibrational populations for the reactions $Ba + HX$.

small since the R_1 heads, which would occur at 499.3 nm for the (0, 0) band, 499.6 nm for the (1, 1) band, etc., are not formed ($N''_{head} \approx 119$). Hence, fluorescence from all significantly populated rotational levels is included in the measured band head intensities. The vibrational populations are listed in table 3 and displayed in fig. 8.

TABLE 3.—VIBRATIONAL POPULATIONS $N_{v''}$ FOR THE Ba + HX REACTIONS [a]

v''	$N_{v''}$(BaF)	$N_{v''}$(BaCl)	$N_{v''}$(BaBr) [b]	$N_{v''}$(BaI)
0	(1.00)	0.51	0.09	0.16
1	0.97	0.66	0.19	0.19
2	0.68	0.80	0.30	0.26
3	0.28	(1.00)	0.46	0.29
4	0.06	0.98	0.59	0.29
5	0.01	0.91	0.76	0.41
6		0.72	0.85	0.55
7		0.49	0.92	0.63
8		0.26	1.00	0.71
9		0.12	(1.00)	0.81
10		0.05	0.90	0.89
11		0.02	0.84	0.97
12			0.70	(1.00)
13			0.60	0.97
14			0.49	0.96
15			0.32	0.87
16			0.21	0.76
17			0.12	0.59
18			0.07	0.45
19			0.04	0.32
20			0.02	0.23
21				0.16
22				0.09
23				0.06

[a] The populations are normalized to the most probable level, indicated by parentheses. The estimated errors in $N_{v''}$ for BaF and BaCl are ±5 % for the most highly populated levels; for BaI and BaBr the errors are larger, perhaps ±10 % (excluding the unknown systematic error in the $q_{v'v''}$ for BaBr).
[b] Data for the $\Delta v = 0$ sequence, see text.

While individual rotational lines are not observed, it is still possible to estimate the $v'' = 0$ rotational distribution from the envelope of the (0, 0) band R_1 branch in the region 499.3 to 499.6 nm, which is not overlapped by other branches. The intensity $I(\lambda)$ at a wavelength λ is given by

$$I(\lambda) = k' N_{N''} S(N'') / [|dv/dN''|(2J'' + 1)] \tag{3}$$

where $N_{N''}$ is the population and $S(N'')$ is the line strength [23] of the R_1 branch of the rotational level $N'' = J'' - \frac{1}{2}$, $[dv/dN'']^{-1}$ expresses the variation of the number of rotational lines per unit frequency and k' is a proportionality constant. Since no R_1 head is observed, the N'' values greater than N''_{head} make a negligible contribution to $I(\lambda)$. For the N'' levels in the isolated region, $50 \leqslant N'' \leqslant 110$, a fit to a rotational temperature $T_{rot} = 800 \pm 100$ K is obtained.

We have attempted to simulate the observed fluorescence spectrum using the determined vibrational populations and various rotational distributions. A calculated spectrum is convoluted with the measured laser bandwidth profile and compared with the experimental scans to determine the " best " rotational distribution. We find that if all vibrational states are assumed to have the same T_{rot}, the value of about 800 K provides the best fit. Moreover, it is found that several other distribution functions cannot reproduce the experimental spectrum; these functions include (1) $N_{N''} = $ constant for $N'' \leqslant N_{cutoff}$ (step function), (2) $N_{N''} = (2N'' + 1)$ for $N'' \leqslant N_{cutoff}$ (linear ramp), (3) $N_{N''} = 1 - N''/N_{cutoff}$ for $N'' \leqslant N_{cutoff}$ (negative linear ramp), and (4) $N_{N''} = \exp(-N''/K)$ (exponential). We conclude that the most probable N'' value of the experimental distribution must lie close to N''_{head} of the $Q_1 + R_{12}$ branch. The convolution procedure also provides a check on the method for determining band intensities. It is found that the band head areas of the computer-synthesized spectra ($T_{rot} \approx 800$ K) agree with the input vibrational populations to 2 %.

(*b*) Ba + HCl → BaCl + H

The BaCl $C^2\Pi_{\frac{1}{2}} - X^2\Sigma^+ \Delta v = 0$ and -1 sequences are well resolved (fig. 4) and are used for analysis. Vibrational band intensities are again estimated from areas of the band heads above a " background "; and populations, determined using the Franck–Condon factors in table 2, are shown in fig. 9a for the two sequences. The agreement of the $N_{v''}$ values deduced from the two sequences is encouraging, as it is a sensitive test of the consistency of the analysis. Because the $\Delta v = 0$ sequence is the more intense and because its Franck–Condon factors are expected to be more reliable, we present in table 3 and fig. 8 the $N_{v''}$ values from the $\Delta v = 0$ sequence.

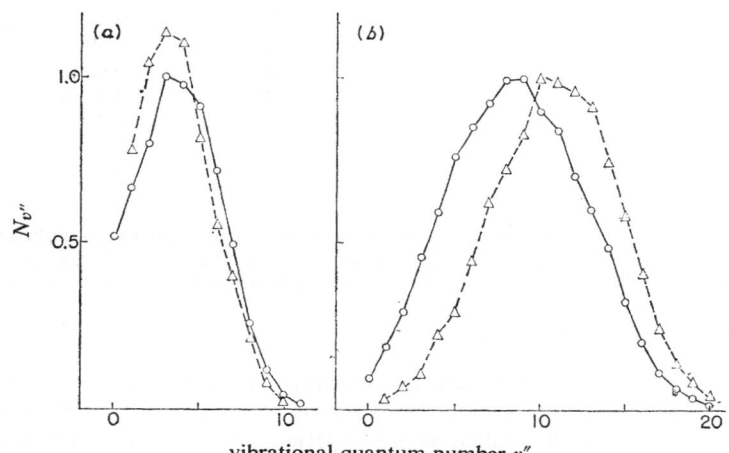

FIG. 9.—Relative vibrational populations derived from different sequences: solid line, $\Delta v = 0$ sequence; dashed line, $\Delta v = -1$ sequence. (*a*) The Ba + HCl reaction using the Franck–Condon factors given in table 2. Here the data are normalized so that the total population is equal for both sequences. (*b*) The Ba + HBr reaction assuming constant Franck–Condon factors in a sequence. Here the data are normalized so that for the most probable v'', $N_{v''}$ equals unity.

A rough estimate of the rotational distribution can be obtained in the same manner as for BaF. A rotational temperature of 1000 K, independent of vibrational level, appears to be consistent with the observed fluorescence spectrum. The band head areas of the computer-synthesized spectra agree with the input data to 10 % for the stronger bands.

The reaction Ba + DCl has also been studied. The population distributions for both vibration and rotation are very similar to the Ba + HCl results.

(c) $Ba + HBr \rightarrow BaBr + H$

The $\Delta v = 0$ and -1 sequences of the $C^2\Pi_{\frac{3}{2}}-X^2\Sigma^+$ subband of BaBr are moderately well resolved (see fig. 5), but the $\Delta v = 0$ sequence comprises only 45 % of the total intensity of the subband. Because Franck–Condon factors are not available, the conversion of vibrational band intensities to vibrational populations is impeded. An attempt to extract $N_{v''}$ is made by assuming that the Franck–Condon factors in a sequence are constant. The results for the $\Delta v = 0$ and -1 sequences are shown in fig. 9b. Since the Franck–Condon factors are expected to decrease with v'' in the $\Delta v = 0$ sequence but increase with v'' in the $\Delta v = -1$ sequence (see table 2), the actual vibrational population distribution is expected to be intermediate between those given in fig. 9b. We have chosen to present in table 3 and fig. 8 the values of $N_{v''}$ as deduced from the $\Delta v = 0$ band intensities with the above assumption.

(d) $Ba + HI \rightarrow BaI + H$

In the BaI C–X band system, the $\Delta v = 0$ sequence in each subband comprises about 95 % of the total intensity. Accordingly, we are well justified in assuming $q_{v'v''} = \delta_{v'v''}$. Table 3 and fig. 8 present the values of $N_{v''}$ as deduced from the $\Delta v = 0$ sequence of the $C^2\Pi_{\frac{3}{2}}-X^2\Sigma^+$ subband. This subband proved preferable because of the large variation of laser intensity over the $C^2\Pi_{\frac{1}{2}}-X^2\Sigma^+$ $\Delta v = 0$ sequence.

(e) RELATIVE TOTAL REACTIVE CROSS SECTIONS

For each reaction, the fluorescence from all bands of the BaX C–X system has been measured in order to estimate the relative total reactive cross sections σ_r. The ratio of the electronic transition moments, the factor in the proportionality constant k in eqn (2) needed to relate different molecules, is obtained from the estimated radiative lifetimes of the C states.[7] The σ_r are estimated using both the scattering chamber and

TABLE 4.—FRACTIONAL ENERGY DISPOSAL AND RELATIVE CROSS SECTIONS

reaction	exoergicity [a]/ eV	E_V/eV	E_R/eV	$\bar{f}_V$	$\bar{f}_R$	$\bar{f}_T$	σ_r [b] scatt. chamber	crossed beams
Ba+HF	0.55	0.06	0.07	0.12	0.13	0.75	0.12	
Ba+HCl	0.48	0.13	0.09	0.28	0.18	0.54	0.60	0.28
Ba+DCl	0.42	0.12	0.09	0.29	0.20	0.51	0.40	0.14
Ba+HBr	0.61	0.22 [c]		0.36			(1.00)	(1.00)
Ba+HI	1.26	0.22		0.18			0.67	0.41

[a] Taken as $D_0^\circ(BaX) - D_0^\circ(HX) + E + W$, where D_0° is the dissociation energy, E is the relative collision energy (using the most probable speeds, see fig. 7) and W is the reactant internal excitation energy. [b] The cross sections are normalized to the Ba+HBr reaction. [c] Average for the $\Delta v = 0$ and -1 sequences (see fig. 9b).

crossed beam arrangements. In the former, an ionization gauge is used to measure HX pressure, and no correction for the change in gauge calibration for different gases has been made. In the crossed-beam experiments, it is assumed that the beam densities for the different HX reactants are equal when the upstream nozzle pressures are the same. HX upstream pressures p_0 of <20 Torr are used, under which conditions the fluorescence signals are linearly proportional to p_0. Pressures of HCl, DCl, HBr, and HI are measured with a stainless steel Bourdon gauge (Matheson). No data have been taken for HF since a suitable pressure gauge was not available. Results are presented in table 4.

DISCUSSION

LASER-INDUCED FLUORESCENCE AS A MOLECULAR BEAM DETECTOR

The technique of laser-induced fluorescence as set forth in this Discussion is applicable to many molecules and reaction systems, whose number will increase as the wavelength range of tunable lasers is extended. Although this technique would encounter the same difficulties as conventional molecular beam methods in determining angular distributions for kinematically unfavourable reactions, nevertheless it is possible to obtain information on the translational energy of the products from a knowledge of the internal state distribution and the reaction exoergicity. This is demonstrated here for the reactions Ba + HX.

The laser source that we have used is capable of exciting a high percentage ($> 10\ \%$) of molecules absorbing at the laser wavelength and present in the observation zone during the pulse. However, since the laser pulse rate (10 pps) is less than the rate at which molecules are replenished in the observation zone ($\sim 5 \times 10^4\ \text{s}^{-1}$), only a small percentage of the total product flux is sampled. The laser pulse rate could ideally be increased to match the replenishment rate. Lasers with pulse rates on the order of 10^3 pps are presently available. Alternatively c.w. dye lasers can be used with similar gains in detection efficiency. The present detection system could be further improved by increasing the light collection efficiency (10^{-3}) by perhaps a factor of 10. With the present arrangement it has been possible to detect a density of 5×10^4 molecule cm^{-3} of BaO in an individual (v'', J'') level for the reaction [4] $\text{Ba} + \text{O}_2 \rightarrow \text{BaO} + \text{O}$, corresponding to an absorbance of only 10^{-8}. The potential sensitivity of the laser-induced fluorescence technique promises to make feasible the measurement of the angular distribution of a product in an individual (v'', J'') state and conceivably the velocity distribution as well by measurement of the absorption line Doppler profile.

THE Ba + HX REACTIONS

Table 4 summarizes the results for these reactions, where $\bar{E}_V$ and $\bar{E}_R$ denote the average BaX vibrational and rotational energies and $\bar{f}_V$, $\bar{f}_R$ and $\bar{f}_T$ denote the fractions of the total reaction exoergicity that appear as vibration, rotation and translation, respectively. The Ba + HX reactions are characterised by little vibrational and rotational excitation of the products, and there is a positive correlation of $\bar{f}_V$ with σ_r. We also note that the cross section for the reaction Ba + DCl is only $\sim 60\ \%$ of that for Ba + HCl. A similar result has been obtained for the K + HBr and K + DBr reactions.[24]

In general, it appears that exothermic reactions lead to the efficient formation of vibrationally excited products. This is expected for attractive reaction surfaces, in which most of the energy is released as the reactants approach. Trajectory calculations on repulsive surfaces, in which only a small amount of energy is released as the reactants approach, also show in general that a large fraction of the reaction exoergicity appears as vibrational excitation (mixed energy release).[25] However, for the exothermic Ba + HX reactions, the average product vibrational excitation does not exceed 36 % and in the case of Ba + HF is only 12 %. This suggests that these reactions proceed on highly repulsive surfaces where mixed energy release plays little role.

It is plausible that the variation in σ_r in this homologous series of reactions is caused by the existence of a barrier along the reaction coordinate, the barrier height being negatively correlated with σ_r. Mok and Polanyi[26] have found using LEPS surfaces and the BEBO method that as the barrier height decreases, the location of the barrier occurs earlier along the reaction coordinate. They have further found that earlier barrier locations lead to increased release of energy into vibrational excitation of the products. Thus, the positive correlation between σ_r and $\bar{f}_V$ observed in these reactions is consistent with the existence of a variable-height barrier. However, it is interesting to note that the Ba + HCl and Ba + DCl reactions, which proceed on the same potential energy surface and hence have the same barrier height, yield approximately equal $\bar{f}_V$ but different σ_r.

From fig. 8, it is seen that the shapes of the vibrational distributions are smooth and decidedly non-thermal in character. In the case of BaCl, BaBr and BaI, there is definate vibrational population inversion. The rotational distributions are less well characterized but the BaF distribution appears to be described quite satisfactorily by a temperature.

Hofacker and Levine [27] have proposed a non-adiabatic model in which the product vibrational distributions are described by a single parameter g, taken as a measure of the attractive character of the potential energy surface.[28] In this model, $g^2/2$ is equal to $\bar{E}_V/hc\omega_e$. We find that the BaX vibrational distributions are fit moderately well to the functional form given by Hofacker and Levine and that the parameter g increases monotonically along the series. Such a model shows promise as a means of relating the details of the product internal state distribution to the potential energy surface of the reaction.

We thank R. R. Herm, R. D. Levine, and J. C. Polanyi for helpful correspondence, and we are grateful to A. Schultz for his contributions to the construction of the experimental apparatus. This work has been supported by the U.S. Air Force Office of Scientific Research under Grant AFOSR-72-2275 and by the U.S. Office of Naval Research under Contract N00014-67-A-0108-0035.

[1] J. L. Kinsey, *Molecular Beam Reactions*, in *International Review of Science: Reaction Kinetics* (MTP, Oxford, 1972).

[2] D. H. Maylotte, J. C. Polanyi and K. B. Woodall, *J. Chem. Phys.*, 1972, **57**, 1547, and references therein.

[3] C. A. Mims, S.-M. Lin and R. R. Herm, *J. Chem. Phys.*, 1972, **57**, 3099.

[4] A. Schultz, H. W. Cruse and R. N. Zare, *J. Chem. Phys.*, 1972, **57**, 1354.

[5] T. R. Dyke, B. J. Howard and W. Klemperer, *J. Chem. Phys.*, 1972, **56**, 2442.

[6] T. W. Hänsch, *Appl. Opt.*, 1972, **11**, 895.

[7] P. J. Dagdigian, H. W. Cruse and R. N. Zare, *J. Chem. Phys.*, 1974.

[8] F. A. Jenkins and A. Harvey, *Phys. Rev.*, 1932, **39**, 922.

[9] A. E. Parker, *Phys. Rev.*, 1934, **46**, 301. The vibrational assignment of the $R_2 + Q_{21}$ band heads (denoted by Q_2 by Parker) of the $\Delta v = -1$ sequence is erroneous. The vibrational quantum numbers should be reduced by one.

[10] K. Hedfeld, *Z. Phys.*, 1931, **68**, 610.

[11] R. E. Harrington, *Ph.D. Thesis* (University of California, 1942).

[12] P. Mesnage, *Ann. Phys.* (Paris), 1939, **11** (12), 5.

[13] With the exception of the $A^2\Pi_i$ state of BeF, all known A parameters of the $^2\Pi$ states of the alkaline earth monohalides are positive. See T. E. H. Walker and W. G. Richards, *J. Phys. B*, 1970, **3**, 271.

[14] See G. Herzberg, *Spectra of Diatomic Molecules* (Van Nostrand, Princeton, N.J., 1950), p. 261, for the line positions of the branches of a $^2\Pi(a)$–$^2\Sigma^+$ system.

[15] Ref. (14), pp. 101 ff.

[16] R. F. Barrow, M. W. Bastin and B. Longborough, *Proc. Phys. Soc.*, 1967, **92**, 518.

[17] P. A. Akishin, V. P. Spiridonov, G. A. Sobolev and V. A. Naumov, *Zhur. Fiz. Khim.*, 1958, **32**, 58.

[18] E. Morgan and R. F. Barrow, *Nature*, 1960, **185**, 754.

[19] B. R. K. Reddy and P. T. Rao, *J. Phys. B*, 1970, **3**, 1008. The vibrational assignment of the first entry in their table 1 is erroneous, and this band was omitted from the fit.

292   INTERNAL STATE DISTRIBUTIONS

[20] J. L. Gole, C. D. Jonah and R. N. Zare, unpublished results.
[21] B. de B. Darwent, *Bond Dissociation Energies in Simple Molecules*, NSRDS-NBS 31 (U.S. Government Printing Office, 1970).
[22] R. E. Drullinger and R. N. Zare, *J. Chem. Phys.*, 1969, **51**, 5532.
[23] See R. N. Zare, *Rotational Line Strengths: The O_2^+ $b^4\Sigma_g^- - a^4\Pi_u$ Band System* in *Molecular Spectroscopy: Modern Research*, K. N. Rao and C. W. Mathews, eds. (Academic Press, New York, 1972).
[24] (a) K. T. Gillen, C. Riley and R. B. Bernstein, *J. Chem. Phys.*, 1969, **50**, 4019; (b) J. R. Airey, E. F. Greene, K. Kodera, G. P. Peck and J. Ross, *J. Chem. Phys.*, 1967, **46**, 3287.
[25] J. C. Polanyi, *Acc. Chem. Res.*, 1972, **5**, 161; B. A. Hodgson and J. C. Polanyi, *J. Chem. Phys.*, 1971, **55**, 4745.
[26] M. H. Mok and J. C. Polanyi, *J. Chem. Phys.*, 1969, **51**, 1451.
[27] G. L. Hofacker and R. D. Levine, *Chem. Phys. Letters*, 1971, **9**, 617.
[28] (a) R. D. Levine, *Chem. Phys. Letters*, 1971, **10**, 510; (b) G. L. Hofacker and R. D. Levine, *Chem. Phys. Letters*, 1972, **15**, 165.

Note added in proof: The exoergicities listed in column 2 of table 4 were calculated using the following value of $D_0^0(\text{BaX})$: 6.33 eV for BaF, 4.84 eV for BaCl, 4.28 eV for BaBr, and 4.22 eV for BaI. These values of $D_0^0(\text{BaX})$ were based on the preliminary analysis of the chemiluminescence data of Gole *et al.*[20] We have listed these values to facilitate any corrections which may be necessary when accurate dissociation energies are known.

Atmospheric chemistry

M. J. Pilling

Department of Chemistry, University of Leeds, Leeds, UK LS2 9JT

Commentary on: **Rate measurements of reactions of OH by resonance absorption. Part 1. Reactions of OH with NO$_2$ and NO,** C. Morley and I. W. M. Smith, *J. Chem. Soc., Faraday Trans.* 2, 1971, **68**, 1016.

Atmospheric models incorporating chemical processes are widely used on global, regional and local distance scales. Applications include climate change prediction and regional air pollution. The models incorporate rate coefficients for the elementary reactions that comprise the chemical mechanism and laboratory measurements still provide the most reliable route to rate coefficient data. Two techniques, discharge flow and flash photolysis, have contributed substantially, over the last 30 or so years, to the provision of rate coefficients for key reactions in atmospheric chemistry. In return, atmospheric applications have provided a focus for the continuing development of experimental techniques in reactions kinetics since the 1970s, when the initial impetus was provided by the recognition of the importance of catalytic cycles in the stratosphere.

The Morley and Smith (MS) paper preceded that era and was not motivated by atmospheric applications. Nevertheless it has had a profound impact. It is one of a group of what might be called the second phase of flash photolysis experiments.[p1–p5] The first phase primarily relied on the use of a second 'spectroscopic flash', which followed the photolysis flash after a known time delay and allowed the spectrum of the radical under study to be recorded on a photographic plate. This was the approach used in the famous Porter and Wright experiments on ClO.[1] Morley and Smith used resonance absorption to record the whole time decay in a single experiment, and with much higher sensitivity. Only four years earlier,[p16] the photographic plate technique required [OH] ~ 3 × 10^{14} molecule cm^{-3}, while this new resonance absorption method used [OH] ≤ 1.6 × 10^{12} molecule cm^{-3}, thereby considerably reducing potential interferences from secondary, and especially radical + radical reactions. The quality of the MS data is clearly superb, especially when it is recalled that signal averaging was not feasible at this stage and single shot oscilloscope traces were necessarily used. Nowadays, OH is detected almost exclusively by laser induced fluorescence; typical working concentrations are 10^{10} molecule cm^{-3} while 10^{6} molecule cm^{-3} are routinely measured in atmospheric field experiments.

The reactions studied by MS involve association to form HONO and HONO$_2$ and so are pressure dependent; MS provided the first measurements for such reactions in the 'fall-off' regime between the low pressure limit, previously studied by Porter for I + I, and the high pressure limit, which is unattainable for these reactions except at very high pressures (~1000 bar). MS analysed the pressure dependence using a RRKM model, based on the many developments of this fundamental theory by Rabinovitch and co-workers,[p28–p30] and obtained estimates of the high pressure limit and a rationalisation of the pressure dependence. Many reactions encountered in atmospheric chemistry show this type of behaviour and the MS experiments were undoubtedly seminal. In the mid-1970s, a series of papers by Troe[2] provided a simpler approach to fitting the fall-off curves that continues to be used by today's kineticists.

The OH + NO$_2$ reaction still attracts considerable attention. Fulle *et al.*[3] have studied the reaction under very high pressure conditions (~1–1000 bar) allowing them to approach the high pressure limiting rate coefficient, k_∞. Their extrapolation to lower pressures with He as the diluent gas relied heavily on the MS results obtained nearly 30 years earlier. Smith and Williams[4] also measured the high pressure limit using a technique that Smith pioneered, based on the rate of reaction of $OH(v = 1)$ + NO_2. The adduct will either re-dissociate to generate $OH(v = 0)$, or will be stabilised. Thus, no matter what the pressure is, this method provides a measure of k_∞. The two techniques give good agreement.[5]

The mechanism of the reaction provides a further focus, ever since it was recognised that it can proceed via two isomers HOONO and HONO$_2$.[6] The importance of the less stable isomer was demonstrated through the use of [18]OH,[7] which indicated that the rate of forming HONO$_2$ is a factor of five slower than the total rate, the rest being made up by HOONO formation. Hippler *et al.*[5] showed a biexponential loss of OH, provided the temperature was sufficiently high that the weakly bound HOONO isomer decomposes on the experimental timescale. The effect of HOONO formation of the rate of OH + NO$_2$ under atmospheric conditions is still not fully resolved.

The atmospheric implications of OH reactions with both NO and NO$_2$ were not appreciated in 1971. They certainly did not provide the justification of the MS experiments, which were driven more by the need to demonstrate the effectiveness of the new experimental technique and by high temperature applications. The final paragraph of the paper, however, provides a clear insight into the previously unrecognised role of these reactions in the troposphere. The so-called 'oxidising capacity' of the atmosphere refers to its ability to oxidise volatile organic compounds (VOCs). The hydroxyl radical plays a central role in this process. The current interpretation of atmospheric oxidation under moderately to highly polluted conditions is based on the dependence of the length of the short oxidation chain on the competition between the propagation reaction, OH + VOC, and the termination reaction, OH + NO$_2$. Morley and Smith recognised the significance of OH + NO$_2$ termination for the first time and correctly observed that '[it] will exert a profound effect on radical concentrations and the overall chemistry.'

References
References labelled p, above, refer to those in the original Morley and Smith paper.

1 G. Porter and F. J. Wright, *Discuss. Faraday Soc.*, 1953, **14**, 23.
2 J. Troe, *J. Chem. Phys.*, 1981, **75**, 226; 1983, **79**, 6017.
3 D. Fulle, H. F.Hamann, H. Hippler and J. Troe, *J. Chem. Phys.*, 1998, **108**, 5391.
4 I. W. M. Smith and M. D. Williams, *J. Chem. Soc., Faraday Trans. 2*, 1985, **81**, 1849
5 H. Hippler, S. Nasterlack and F. Striebel, *Phys. Chem. Chem. Phys.*, 2002, **4**, 2959
6 D. M. Golden and G. P. Smith, *J. Phys. Chem. A*, 2000, **104**, 3991.
7 N. M. Donahue, R. Mohrschladt, T. J. Dransfield, J. G. Anderson and M. K. Dubey, *J. Phys. Chem.*, 2001, **105**, 1515

Rate Measurements of Reactions of OH
by Resonance Absorption

Part 1.—Reactions of OH with NO_2 and NO

By C. Morley * and I. W. M. Smith

Dept. of Physical Chemistry, University Chemical Laboratories,
Lensfield Road, Cambridge CB2 1EP

Received 13*th December*, 1971

Some reactions of the OH radical, which can be initiated by flash photolysis, have been studied using resonance absorption to follow how the OH concentration varies with time. The reaction

$$OH + NO_2(+M) \rightarrow HNO_3 + M \qquad (4)$$

has been investigated at 300 K and 416 K, and at total pressures (of He) between 20 and 300 Torr. In this range, it is in the transition region between second and third order kinetics and the variation of the rate with total pressure is compared to the behaviour predicted by RRKM theory. Association of OH with NO,

$$OH + NO + M \rightarrow HNO_2 + M, \qquad (5)$$

has been studied, but only in its third order range, where $k_5 = 4.1 \ (\pm 0.6) \times 10^{-31} \ cm^6 \ molecule^{-2} \ s^{-1}$ at 300 K and with M = He; the activation energy $= -6.7 \ (\pm 2.0) \ kJ/mol$. Some measurements have also been made on the bimolecular reactions,

$$OH + HNO_3 \rightarrow H_2O + NO_3 \qquad (11)$$

and

$$O + HNO_3 \rightarrow OH + NO_3. \qquad (12)$$

At 300 K, $k_{11} = 1.3 \ (\pm 0.5) \times 10^{-13} \ cm^3 \ molecule^{-1} \ s^{-1}$ and $k_{12} < 1.3 \times 10^{-14} \ cm^3 \ molecule^{-1} \ s^{-1}$.

The importance of these results for several complex chemical systems is discussed and work on the steady-state photolysis of NO_2—HNO_3 mixtures is re-evaluated.

In flash photolysis experiments, the absorption spectra of transient species are normally photographed using a second flash lamp to provide a background continuum. Where the spectrum is discrete, high sensitivity can only be obtained by using a spectrograph of extremely high resolving power; because of this limitation other more sensitive methods of kinetic spectroscopy, which employ specific light sources and continuous photo-electric monitoring, have recently been developed.[1-5] In this paper, we describe how one of these techniques—resonance absorption—has been used to study the kinetics of some reactions of the OH radical.

Resonance absorption is most sensitive and its application is most straightforward when the emission line is not self-reversed and when it has a Doppler profile corresponding to a low translational temperature.[6-9] For molecules, the strength of an electronic transition is dissipated over many individual lines with the result that the emission and absorption intensity of any single line is generally less than for atoms. Although this lowers the sensitivity of the method, it also tends to reduce the problems which can arise from self-reversal of the lines emitted by the source. The OH radical is well suited to a study by resonance absorption because the intensity from a simple lamp [10] emitting the $A^2\Sigma^+ - X^2\Pi$ resonance system is concentrated in a few lines of the (0,0) band with low values of K, which are quite widely separated.

* present address : Thornton Research Centre, Shell Research Ltd., P.O. Box 1, Chester.

C. MORLEY AND I. W. M. SMITH   1017

Kinetic studies of OH using resonance absorption have been performed in discharge-flow experiments,[11, 12] but photographic techniques have been employed in previous flash photolysis experiments.[13-16]

In our work, OH radicals were produced in processes (1) to (3) which occur rapidly when mixtures of NO_2 and H_2 are flash photolyzed:

$$NO_2 + h\nu \rightarrow NO + O(^1D);\tag{1}$$

$$O(^1D) + H_2 \rightarrow OH + H;\tag{2}$$

$$H + NO_2 \rightarrow OH + NO.\tag{3}$$

Resonance absorption was used to follow the rate at which OH radicals were then removed either by

$$OH + NO_2(+M) \rightarrow HNO_3(+M);\tag{4}$$

or, in the presence of added NO, by

$$OH + NO + M \rightarrow HNO_2 + M.\tag{5}$$

Both reactions occur in a number of complex chemical environments, but their rates have not been determined previously. Reaction (4) was studied at total pressures between 20 Torr and 300 Torr (1 Torr $\equiv$ 133 N m^{-2}) at both 300 and 416 K. The kinetics are in the transition region between second and third order kinetics in this pressure range and the "fall-off" behaviour is compared with RRKM theory. Reaction (5) could be investigated only in its third order range.

EXPERIMENTAL

Gases were purified and mixtures prepared in a conventional high vacuum system. The reaction vessel was 80 cm long and constructed from quartz tubing. In order to minimize the scattered light from the photolysis lamp, the vessel was provided with a 1 cm thick outer jacket which was filled with 1 atm pressure of Cl_2. Without this filter, radiation of the same wavelength as the OH emission "blinded" the photomultiplier and measurements could not be made for 300 μs after the onset of the flash. Its use also reduced the number of $O(^3P)$ atoms produced by absorption in the long wavelength system of NO_2.

The reaction vessel and photolysis lamp lay within a cylindrical MgO reflector and the system could be heated to 420 K. The lamp was filled with a mixture of a few Torr of N_2 in 100 Torr of Kr. Flash energies of 500-1000 J were provided by charging a 5 μF capacitor to the appropriate voltage. The light output fell to half its peak intensity in less than 10 μs and was effectively zero after 30 μs.

The OH resonance lamp, constructed from 0.25 m of 13 mm int. diam. Pyrex tubing, was fitted with a quartz end-window. A mixture of ~ 0.5 % H_2O in Ar was passed through the lamp at ~ 170 cm^3 s^{-1} and a total pressure of 1 Torr. The tube was positioned in a type 5 microwave discharge cavity and 140 W of power was supplied from a Microtron Mk II (Electromedical Supplies Ltd.) 2450 MHz power unit, which was modified by running the magnetron heater on D.C. to reduce mains ripple on the output. The resultant discharge filled the whole of the lamp. By placing the centre of the lamp at the focus of a quartz lens a roughly parallel beam of light was sent through the reaction vessel. A second quartz lens focussed this radiation onto the slit of a Hilger medium quartz spectrograph, which had the plate-holder replaced by a unit incorporating a moveable slit and an R.C.A. C7045E photomultiplier.

A spectrum of the OH emission is shown in fig. 1; it is similar to that described by Carrington and Broida.[10] For time-resolved measurements, the spectrometer was centred on the Q_13 line at 308.15 nm[17] with the entrance and exit slits 0.09 mm wide. The spectrometer then accepted not only the Q_13 line but also the P_11 line at 308.17 nm, some contributions from the Q_12 (308.00 nm) and Q_14 (308.33 nm) lines, and the satellites of all four lines. These various components were not absorbed to the same extent but calculations based on the thermal populations in each level and the transition probabilities show that

1018   REACTIONS OF OH WITH NO₂ AND NO

the deviation from Beer's law was <6 % at the highest OH concentrations, and any error
in the derived rate constants was less than from other sources.

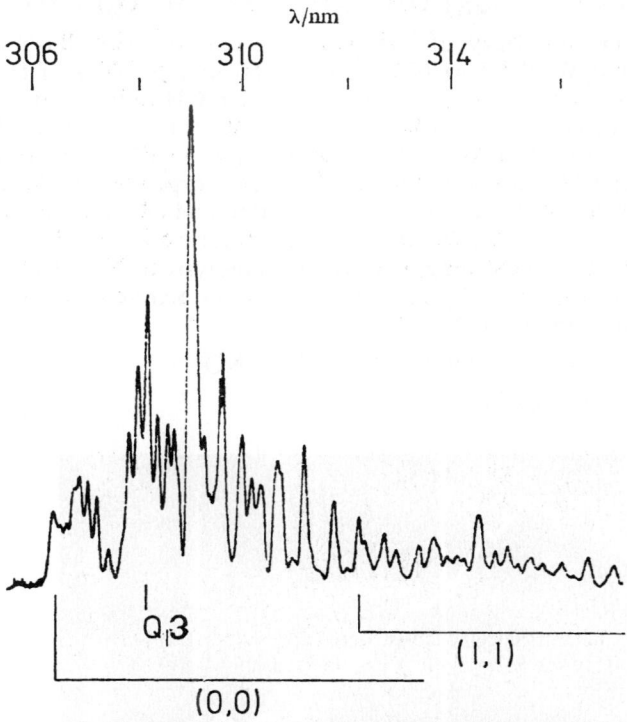

FIG. 1.—OH($A^2\Sigma^+ - X^2\Pi$) emission spectrum from the microwave resonance lamp.

The anode load of the photomultiplier possessed logarithmic current-voltage characteristics so that light intensities were converted to optical intensities directly. Besides
saving time, this made the signal independent of the standing anode current which varied
slowly due to fluctuations in the lamp output and to photomultiplier fatigue. The signal
was displayed on a 564B Tektronix storage oscilloscope. Constants for the experimental
decay of the absorption were generally determined by generating a low intensity exponential
curve and varying its decay constant to fit the stored experimental trace.

The circuit supplying the exponential was powered by the square wave generator of the
oscilloscope and the value of the decay constant was controlled by a capacitor and variable
resistors in the circuit. Rate coefficients determined by this method agreed with those
found by photographing the experimental trace and analyzing it graphically.

MATERIALS

NITROGEN DIOXIDE.—NO₂ was taken from a cylinder (Matheson Ltd.,) and degassed at
196 K. HYDROGEN.—H₂ was taken from a cylinder (British Oxygen Co.,) and passed
slowly through a DeOxo unit, a glass wool packed trap at 77 K and grade 5A molecular
sieve at 77 K. HELIUM.—He was taken from a grade A cylinder (B.O.C.) and passed slowly
through grade 5A molecular sieve at 77 K. NITRIC OXIDE.—NO was taken from a cylinder
(Matheson Ltd.,) and purified by a distillation from 140 K (pentane slush) to 77 K.
SULPHUR HEXAFLUORIDE.—SF₆ was taken from a cylinder (Matheson Ltd.,) and degassed
at 77 K. NITRIC ACID.—HNO₃ was distilled from a mixture of concentrated HNO₃
and a large excess of concentrated H₂SO₄.

RESULTS

THE REACTION: $OH + NO_2(+M) \rightarrow HNO_3(+M)$

The formation and decay of OH was observed following flash photolysis of mixtures containing 0.1-0.4 Torr NO_2, 2.5 Torr H_2 and 20-300 Torr He. A typical oscilloscope trace is shown in fig. 2. In this system OH radicals are produced as a result of processes (1) to (3). The rate coefficients for reactions (2)[18] and (3)[19] are $k_2 = 2 \times 10^{-10}$ cm^3 molecule^{-1} s^{-1} and $k_3 = 5 \times 10^{-11}$ cm^3 molecule^{-1} s^{-1}. With the NO_2 and H_2 concentrations used in these experiments, the half-life of H atoms was ~ 6 μs, that of $O(^1D)$ atoms much shorter, and OH radicals were essentially formed completely within the life-time of the photolytic flash. The chlorine filter almost completely prevented long wavelength photolysis of NO_2 and even if $O(^3P)$ atoms were produced, either by photolysis or by quenching of $O(^1D)$ atoms, they would react quite rapidly with NO_2

$$O(^3P) + NO_2 \rightarrow O_2 + NO; \tag{6}$$

where $k_6 = 6.0 \times 10^{-12}$ cm^3 molecule^{-1} s^{-1}.[20]

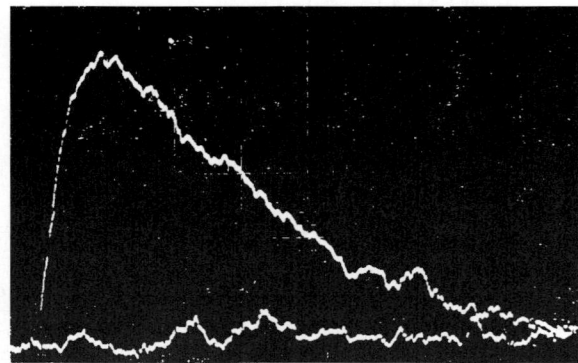

FIG. 2.—Decay of OH resonance absorption with time. 0.2 Torr NO_2 + 2.5 Torr H_2 + 40 Torr He; horizontal scale 50 μs/div.

Absolute concentrations of OH could be estimated from the extent of absorption. They never exceeded 1.6×10^{12} molecule cm^{-3}, which illustrates the sensitivity of resonance absorption. In a photographic study of the same OH absorption system, Horne and Norrish[16] estimated concentrations of approximately 3×10^{14} molecule cm^{-3}. The extent of NO_2 photodecomposition was investigated by observing its absorption at ~ 410 nm using a quartz-iodine bulb as background. No significant change was observed after flashing the gas 6 times, which meant that <0.7 % was decomposed per flash and that the maximum concentration of O atoms produced $(^3P + ^1D)$ was $\sim 5 \times 10^{13}$ molecule cm^{-3}. As a consequence of these very low concentrations, reactions which are second order in radicals, such as

$$OH + OH \rightarrow H_2O + O \tag{7}$$

where $k_7 = 2.5 \times 10^{-12}$ cm^3 molecule^{-1} s^{-1},[21, 22]

$$OH + O(^3P) \rightarrow O_2 + H \tag{8}$$

where $k_8 = 2.0 \times 10^{-11}$ cm^3 molecule^{-1} s^{-1},[22, 23] and

$$OH + OH + M \rightarrow H_2O_2 + M \tag{9}$$

1020 REACTIONS OF OH WITH NO_2 AND NO

where $k_9 = 8.6 \times 10^{-31}$ cm^6 molecule^{-2} s^{-1},[23] or those which involve OH and the product of a previous reaction were quite negligible. OH does react—rather slowly —with H_2

$$OH + H_2 \rightarrow H_2O + H, \qquad (10)$$

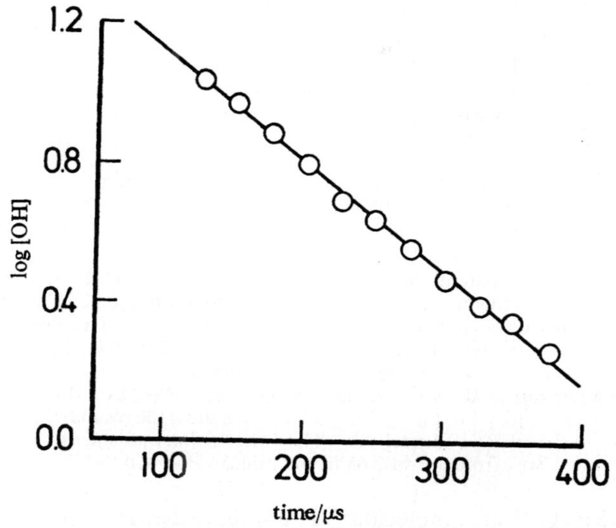

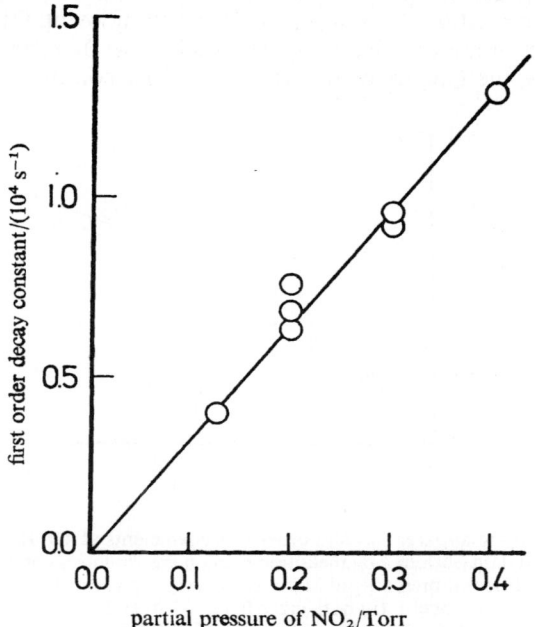

FIG. 3.—First order decay plots for [OH] in mixture containing 0.2 Torr NO_2+2.5 Torr H_2+40 Torr He.

FIG. 4.—Variation of the observed first order decay constants with partial pressure of NO_2. The gas mixtures contained 2.5 Torr H_2 and were made up to a total pressure of 30 Torr by adding He.

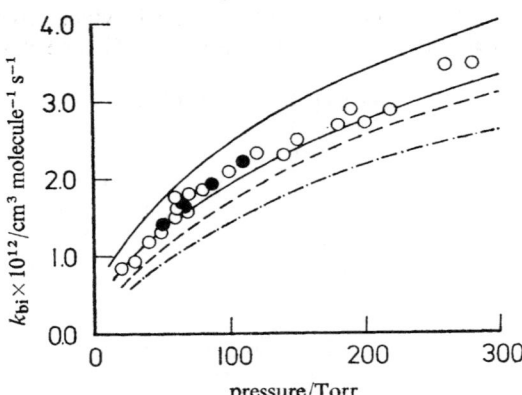

FIG. 5.—Dependence of the apparent second order rate coefficient at 300 K on total pressure. The open symbols refer to measurements made on gas mixtures containing 0.125 Torr NO_2, 2.5 Torr H_2 and different pressures of He; k_{bi} is plotted against the total pressure. The closed symbols were the results of experiments on mixtures which again contained 0.125 Torr NO_2 but in which the partial pressure of H_2 was varied. To bring these results into agreement with those from the first series of experiments, it was necessary to allow for the greater efficiency of H_2 as a third body relative to He and k_{bi} was plotted against $p_{He} + 4.0$ ($p_{H_2} - 2.5$). The curves show the results of RRKM calculations: — · — model A, – – – model B, — model A′. The lower solid line shows the result of lowering β_{SF_6} from 1.0 to 0.65 in the calculations with A′.

where $k_{10} = 6.5 \times 10^{-15}$ cm³ molecule⁻¹ s⁻¹,[21] but the H atom produced would regenerate OH through reaction (3) and no net loss of OH would occur.

The rate of removal of OH was shown to be first-order in OH (fig. 3), proportional to the NO_2 concentration (fig. 4) and dependent on P, the total pressure (fig. 5 and 6). It is clear that reaction (4) was responsible for removing OH, and it is evident from the discussion in the two previous paragraphs that this constitutes a very good system for studying the kinetics of this reaction. The non-linear dependence of the

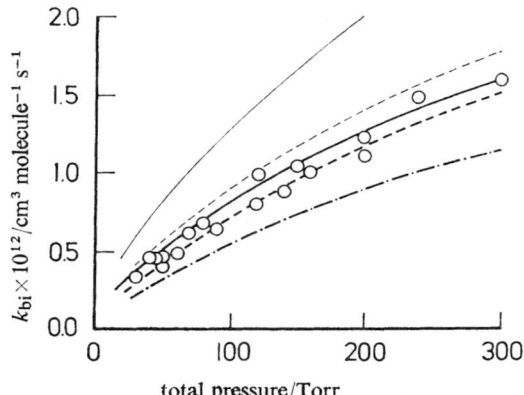

FIG. 6.—Dependence of the apparent second order rate coefficients at 416 K on total pressure. The points are experimental data obtained by measurements on gas mixtures containing 0.19 or 0.38 Torr NO_2, 3.8 Torr H_2 and different pressures of He. The curves represent the results of RRKM calculations; — · — model A, – – – model B (in both cases $\beta_{SF_6} = 1.0$) and — model A′ ($\beta_{SF_6} = 0.65$). The thin upper lines were calculated with values of k_{4a} predicted from the model { for A′, $= 1.6 \times k_{4a}$ (300 K) and for B, $= 1.2 \times k_{4a}$ (300 K)}, the thicker lower lines were calculated assuming that k_{4a} has the same value at 416 K as at 300 K.

1022 REACTIONS OF OH WITH NO$_2$ AND NO

apparent second order rate constant, $k_{bi} = (d[OH]/dt)/[OH][NO_2]$, on total pressure indicates that the reaction is in its transition region between second and third order kinetics between 20 and 300 Torr. Unfortunately, a closer approach to the high pressure limit was impossible. Above 300 Torr total pressure, the rate with 0.125 Torr NO$_2$ present became too fast to measure accurately, and reducing the concentration of NO$_2$ to compensate for this lowered the absorption signal and also extended the period over which OH was being formed in reaction (3).

Rate measurements were made at 300 (± 8) K and 416 (± 10) K. The quoted errors include uncertainties in estimating the temperature rise resulting from photochemical and chemical processes. As expected, the rate of reaction (4) at the higher temperature was slower; results plotted in fig. 6 show that the reaction was more nearly in its third-order region.

To find the efficiency of H$_2$ as a third body relative to He, rate measurements were made on mixtures containing different concentrations of H$_2$. The values of k_{bi} were plotted against $P_{He} + \delta_{H_2} P_{H_2}$, δ being adjusted until all the results fell on the the same line. This procedure gave $\delta_{H_2} = 4.0(\pm 1.0)$. As RRKM theory was to be applied to the reaction, it was desirable to measure the rate with a polyatomic molecule which would deactivate the internally excited HNO$_3$ molecule with high efficiency. Because it does not deactivate or react with O(^1D) atoms,[18] SF$_6$ was chosen, when δ_{SF_6} was found to be 4.8(± 0.8).

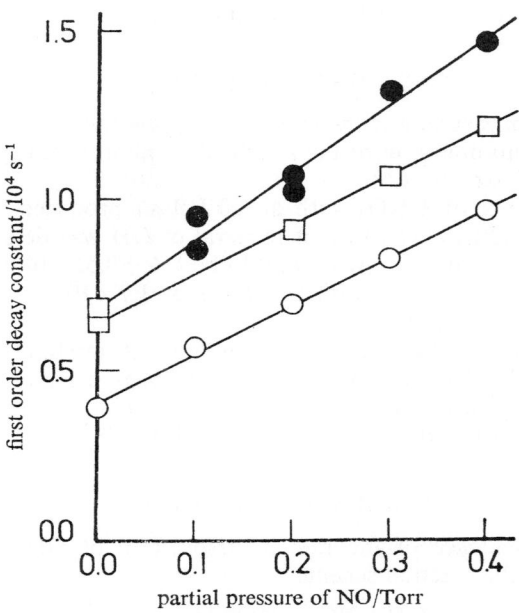

FIG. 7.—Variation of the observed first order decay constants partial pressure of NO. The gas mixtures contained: ○, −0.125 Torr NO$_2$+2.5 Torr H$_2$; □, −0.20 Torr NO$_2$+2.5 Torr H$_2$; ●, −0.125 Torr NO$_2$+15 Torr H$_2$. In all cases the total pressure was made up to 30 Torr by adding He.

Because of the possibility of forming an HONO$_2$ complex in which energy is distributed among several degrees of freedom before redissociation, NO$_2$ is probably peculiarly efficient at deactivating vibrationally excited OH. This point is discussed in greater detail later. He and H_2 are also likely to be quite efficient because of their light mass.[24] There was no indication of any complications arising from the formation of excited OH in reactions (2) and (3), and we conclude that the rates measured refer to reactions of thermally equilibrated OH.

THE REACTION: $OH + NO + M \rightarrow HNO_2 + M$

The association of OH radicals with NO was studied by carrying out experiments on mixtures containing added NO. The data shown in fig. 7 illustrate that the decay constants—which remained first order—were proportional to [NO]; they also depend on the nature and pressure of the diluent gas. With NO added, the rate could not be determined over as wide a range of total pressure, but since HNO_2 has three fewer internal modes than HNO_3, reaction (5) should certainly be in its third-order region at 30 Torr. This conclusion is confirmed by RRKM calculations presented below. From results like those shown in fig. 7, the rate coefficient for $OH + NO + He \rightarrow HNO_2 + He$ was found to be $4.1(\pm 0.6) \times 10^{-31}$ cm^6 molecule^{-2} s^{-1} at 300 K and $1.9(\pm 0.3) \times 10^{-31}$ cm^6 molecule^{-2} s^{-1} at 416 K. The rate was $2.3(\pm 0.9)$ times greater when H_2 replaced He as the third body.

OTHER REACTIONS

A few experiments were performed on the reactions

$$OH + HNO_3 \rightarrow H_2O + NO_3, \tag{11}$$

and

$$O + HNO_3 \rightarrow OH + NO_3. \tag{12}$$

The purpose of these measurements was to obtain approximate rate coefficients which would assist in a reinterpretation of the steady state photolysis experiments of Jaffe and Ford,[25] and of Bérces *et al.*[26]

Photolysis of 1 Torr of HNO_3 with a 120 J flash produced sufficient OH for reaction (11) to be studied. The rate of removal of OH was determined at several concentrations of HNO_3, and k_{11} was found to be $1.3(\pm 0.5) \times 10^{-13}$ cm^3 molecule^{-1} s^{-1} at 300 K, in good agreement with the value of 1.7×10^{-13} cm^3 molecule^{-1} s^{-1} obtained by Husain and Norrish.[27]

Only an upper limit could be estimated for k_{12}. A mixture of 0.1 Torr NO_2, 1.8 Torr HNO_3 and 20 Torr He was photolyzed through a plate glass filter so that only NO_2 was decomposed and only $O(^3P)$ atoms produced. No OH absorption could be detected and it was deduced that k_{12} is $< 1.3 \times 10^{-14}$ cm^3 molecule^{-1} s^{-1}.

RRKM CALCULATIONS

The pressure dependence of the rate of association of OH and NO_2 can be discussed in terms of the reaction scheme:

$$OH + NO_2 \rightarrow HONO_2^*, \tag{4a}$$

$$HONO_2^* \rightarrow OH + NO_2, \tag{-4a}$$

$$HONO_2^* + M \rightarrow HONO_2 + M. \tag{4b}$$

According to RRKM theory,[28] the apparent second order rate coefficient is given by

$$k_{bi} = k_{4a} \int_0^\infty \frac{\omega}{k(\varepsilon) + \omega} f(\varepsilon^+) d\varepsilon^+; \tag{I}$$

1024 REACTIONS OF OH WITH NO₂ AND NO

$k(\varepsilon)$ is the first order rate coefficient for decomposition of $HONO_2^*$ possessing a total energy ε in its internal degrees of freedom and is given by

$$k(\varepsilon) = \Sigma P(\varepsilon^+)/hN^*(\varepsilon), \qquad \text{(II)}$$

where $\Sigma P(\varepsilon^+)$ is the total number of ways in which the internal energy of the complex † may be distributed among its internal degrees of freedom, and $N^*(\varepsilon)$ is the number of states per unit energy of $HONO_2^*$ with a total internal energy ε. The thermal distribution function is given by

$$f(\varepsilon) = \Sigma P(\varepsilon^+) \exp\left(\frac{-\varepsilon^+}{kT}\right) \bigg/ \int_0^\infty \Sigma P(\varepsilon^+) \exp\left(\frac{-\varepsilon^+}{kT}\right) d\varepsilon^+, \qquad \text{(III)}$$

and ω is the frequency at which $HONO_2^*$ is deactivated by collision.

The first stage in the calculations is to select a transition complex so that the rate coefficient in the high pressure limit is equal to that predicted by transition state theory. This requires that Q^+, the total partition function per cm³ for the complex with the contribution from motion along the reaction coordinate removed, is given by

$$Q^+ = k_{4a}(h/kT)Q_{OH}Q_{NO_2}. \qquad \text{(IV)}$$

Since neither the rate of recombination of $OH + NO_2$ nor that of the thermal decomposition of HNO_3 has been determined close to the high pressure limit, it was necessary to estimate k_{4a} from the present results. A $(1/k_{bi}, 1/P)$ plot was quite linear and extrapolated to $k_{bi} = 5.8 \times 10^{-12}$ cm³ molecule⁻¹ s⁻¹ at $1/P = 0$, but as this kind of plot is known to curve sharply downwards at small values of $1/P$, k_{4a} was taken to be twice this value. Q_{OH} and Q_{NO_2} were calculated by normal methods from spectroscopic data.[31, 32]

Calculations were carried out on two, slightly different, models of the complex, each " loosened " versions of the $HONO_2$ molecule. Their structure and vibrational frequencies were chosen so that the product $q_{vib}^+ q_{rot}^+ q_{trans}^+$, calculated using statistical mechanics, equalled the value determined for $Q^+(q_{elec}^+$ was assumed to be unity). The magnitude of k_{4a}, similar to the high-pressure rate coefficient for methyl radical recombination,[33] indicated a considerable extension of the HO—NO₂ " bond " in the complex, so this distance was taken to be 4.2 Å, three times the bond length in the molecule,[34] and the internal rotation about this bond, which is hindered in the molecule,[35, 36] was treated as free. The other internuclear distances were taken to be the same as in HNO_3,[34] where they differ only very slightly from the corresponding bond lengths in OH and NO₂.[31, 32] The vibrational frequencies assigned to the complexes are compared in table 1 to those in HNO_3,[36] OH^{31} and NO₂.[32] In model A the two estimates of the partition function of the complex were finally matched by lowering the frequencies of the three bending modes, which appear as the radicals approach, by an appropriate, and identical, factor from their values in the HNO_3 molecule. In model B one of these degrees of freedom was regarded as a rotation, thereby allowing the OH group to " tumble ", and the other two were made equal and adjusted to achieve the necessary matching.

Next $\Sigma P(\varepsilon^+)$ and $N^*(\varepsilon)$ were evaluated by the method of Whitten and Rabinovitch.[37, 38] Calculations were performed for models A and B and also for a complex (A′) which was identical in every way to A except that the overall rotation about the HO—NO₂ axis was also considered to be " active ". The OH torsion was treated

---

† ε^+ denotes the energy in the complex and is equal to $(\varepsilon - \varepsilon_0)$, where ε_0 is the threshold energy. In this case ε_0 is the bond dissociation energy lowered by a factor which allows for flow into, or out of, two overall rotations, resulting from the complex and the molecule having different moments of inertia.[29, 30]

as a free rotor in the molecule, as well as in the complex. This should cause very little error in $N^*(\varepsilon)$ since there is ~ 210 kJ/mol in the energized molecules. $\Sigma P(\varepsilon^+)$ was also determined at a number of energies by direct count; the results were in good agreement with those obtained from Whitten and Rabinovitch's formula.

TABLE 1.—VIBRATION FREQUENCIES FOR HNO_3 AND COMPLEXES

vibration	molecule	radicals	complex A	complex B
OH stretch	3550.0	3589	3550	3550
NO antisymmetric stretch	1708.2	1618	1650	1650
HON bend	1330.7		121	free rotor
NO symmetric stretch	1324.9	1320	1325	1325
NO_2 deformation	878.6	750	800	800
NO_2 out-of-plane bend	762.2		69	109
NO' stretch	646.6		reaction coordinate	
ONO' in-plane bend	579.0		53	109
OH torsion	$(455.8)^a$		free rotor	
moments of inertia ^{b}				
internal	1.5		1.5	1.5, 1.5
overall	65, 73, 136		65,^{c} 430, 530	65, 430, 530

Frequencies in cm^{-1}. ^{a} Treated as free rotor. ^{b} In units of 10^{-40} g cm^2. ^{c} In A' this rotation was considered " active ".

Collision frequencies were calculated using the adjusted hard-sphere expression given by Curtiss, Hirschfelder and Bird.[39]   The collision diameter of HNO_3 has not apparently been measured and was assumed to be 6.7 Å, as for NO_2Cl; the values for He, H_2 and SF_6 were taken to be 2.6, 2.9 and 5.0 Å.   It was then assumed that SF_6 deactivates $HONO_2^*$ at every collision so that $\omega = Z[SF_6]$ but that $\omega = \beta Z[M]$ for other gases.   From the ratio of experimentally determined efficiencies, β was found to be 0.15 for He.

Now eqn (I) for k_{bi} could be integrated at various pressures using Simpson's rule.   The results of these calculations are shown in fig. 5.   A plot of the reciprocal of the calculated values of k_{bi} against $1/P$ only showed marked curvature at $P > 300$ Torr, and at $(1/P) = 0$, $1/k_{bi}$ was approximately half the value obtained by a linear extrapolation of the data at lower pressures.   This behaviour is entirely consistent with the experimental findings and supports the value of k_{4a} which had been chosen previously.

In order to compare the experimental results at 416 K with theoretical predictions, k_{4a} at the higher temperature was calculated from the transition state theory expression. For model A and A' this indicated that k_{4a} should increase by a factor of 1.6 and for model B by 1.2 times.   The positive temperature coefficient is associated with the low bending frequencies in the complex and decreases if these are replaced by rotations.   Where the radicals are completely free to rotate in the complex then the high pressure rate coefficient becomes proportional to $T^{-\frac{1}{2}}$.   A comparison of the curves in fig. 5 and 6 indicates that the temperature dependence of the rate in the transition range is well represented only when it is assumed that k_{4a} is approximately the same at 416 K as at 300 K.

Calculations were performed for the $OH + NO$ recombination with models of the activated complex (B″ and A″) which were exactly equivalent to the models B and A' used in the $OH + NO_2$ calculations.   Values of the high pressure rate coefficient, k_{5a}, at 300 K and 416 K were calculated using eqn (IV), and the low pressure rate coefficient deduced from the expression

$$k_5 = k_{5a}\omega N_{Av}^* kT \bigg/ \int_0^\infty \sum P(\varepsilon^+) \exp\left(\frac{-\varepsilon^+}{kT}\right) d\varepsilon^+ \, ;$$

N_{Av}^*, the value of $N^*(\varepsilon)$ when $\varepsilon^+ = kT$, was used since $N^*(\varepsilon)$ is not a strongly varying function of ε^+. The collisional efficiencies of He and SF₆ were assumed to be 0.15 and 1.0, as for HNO₃, and the collision diameter of HNO₂ was taken to be 5.2 Å. The experimental and predicted values of k_5 and k_{5a} are shown in table 2. The rate coefficients calculated with model B″ agree more closely with the experimental values than those computed with A″. At 80 Torr He, k_5[He] $= 1.2 \times 10^{-12}$ cm³ molecule⁻¹ s⁻¹ (model B″) or 0.47×10^{-12} cm³ molecule⁻¹ s⁻¹ (model A″). Clearly the assumption that this recombination reaction is in its third order region when $P_{He} < 80$ Torr is fully justified.

TABLE 2.—THEORETICAL AND EXPERIMENTAL RATE COEFFICIENTS FOR THE RECOMBINATION OF OH AND NO

	high pressure limit k_{5a}/cm³ molecule⁻¹ s⁻¹	low pressure limit k_5/cm⁶ molecule⁻² s⁻¹	
		300 K	416 K
experimental	—	4.1×10^{-31}	1.9×10^{-31}
calculated			
model B″	1.7×10^{-10}	4.7×10^{-31}	2.5×10^{-31}
model A′	1.7×10^{-10}	1.8×10^{-30}	7.8×10^{-31}

Within the uncertainties of experiment and theory, the behaviour of reactions (4) and (5) predicted by RRKM theory agrees with that observed experimentally. These uncertainties would be reduced if the limiting high pressure rate of the OH + NO₂ recombination could be determined, and if the OH + NO reaction could be followed over a wider range of pressures than in the present experiments.

DISCUSSION

The rates of reactions (4) and (5) have not been directly measured previously. They have been recognized [40, 41] as terminating the chain reaction between NO₂ and H₂, and Rosser and White [41] concluded that at 600-700 K with H₂ as third body, the reactions had " essentially equal " rate constants; this is consistent with our results. The thermal decomposition of HNO₃ has been studied by Johnston and coworkers [42-44] at temperatures up to 1000 K. They proposed that the first stages of the mechanism are

$$\text{HNO}_3 + \text{M} \rightleftharpoons \text{OH} + \text{NO}_2 + \text{M} \qquad (-4', 4')$$

$$\text{OH} + \text{HNO}_3 \rightarrow \text{H}_2\text{O} + \text{NO}_3, \qquad (11)$$

and that at 650 K the decomposition is in its second order range at 10 Torr but in its transition range at 1 atm (of N₂) when $k_{11} \sim k_{-4'}$[M]. Godfrey *et al.* agreed with this mechanism but found that at 622 K a pressure of 230 Torr N₂ was sufficient for the reaction to be in its high pressure region. Even allowing for N₂ being a more efficient third body than He, it seems impossible to reconcile our results with this conclusion whereas our results do agree with Johnston's findings if reasonable estimates are made for the temperature dependences of the rates of reactions (4) and (11).

C. MORLEY AND I. W. M. SMITH   1027

There have been two studies of the photolysis of mixtures of HNO_3 and NO_2. Jaffe and Ford (JF)[25] used light at 366 nm where only NO_2 absorbs and suggested the following mechanism to explain their results:

$$NO_2 + h\nu \rightarrow NO + O(^3P), \qquad (13)$$

$$O + NO_2 \rightarrow O_2 + NO, \qquad (6)$$

$$O + HNO_3 \rightarrow OH + NO_3, \qquad (12)$$

$$OH + HNO_3 \rightarrow H_2O + NO_3, \qquad (11)$$

$$NO + NO_3 \rightarrow 2NO_2, \qquad (14)$$

$$NO_2 + NO_3 \rightarrow NO_2 + NO + O_2, \qquad (15)$$

$$2NO + O_2 \rightarrow 2NO_2. \qquad (16)$$

Bérces and Förgeteg (BF)[26] irradiated at 265 nm, where HNO_3 also absorbs, and included some other reactions, including $OH + NO_2 + M \rightarrow HNO_3 + M$, in their reaction mechanism. Both groups of authors explained the increase in NO_2 concentration—in JF's experiments, after it had initially decreased—by proposing that reaction (12) could compete with (6); values of $k_{12} \sim 2 \times 10^{-11}$ cm^3 molecule^{-1} s^{-1} (JF) and 2×10^{-10} cm^3 molecule^{-1} s^{-1} (BF) were then deduced. BF assumed that reaction (4) remains in its third-order region with up to 1 atm of Kr present, and deduced that $k_{4'}/k_{11} = 2.2 \times 10^{-20}$ cm^3 molecule^{-1}. Using their value for k_{11}, this gave $k_{4'} = 5 \times 10^{-34}$ cm^6 molecule^{-2} s^{-1}, compared to our low pressure value of 1.1×10^{-30} cm^6 molecule^{-2} s^{-1}. Their failure to take into account the fall-off behaviour would result in some error but cannot explain the discrepancy of a factor of ~ 2000.

We believe that the disagreement between our results and the rate coefficients derived from the steady-state photolysis experiments occurs because the mechanism proposed by JF and then adopted by BF is incorrect. Their values of k_{12} would be remarkably high for a reaction of an atom with a complex molecule which is approximately thermoneutral; our own flash photolysis experiments show unequivocally that k_{12} is $< 1.3 \times 10^{-14}$ cm^3 molecule^{-1} s^{-1} at 300 K, and reaction (12) could not therefore compete with reaction (6) in JF's and BF's experiments. The results can be explained without requiring any reaction between O atoms and HNO_3.

Although JF and BF noted that NO and HNO_3 react in the dark, neither included the reaction in their analysis. According to Smith,[46] the reaction is fast, partly heterogeneous, and catalyzed by H_2O and NO_2. Assuming the mechanism

$$NO + HNO_3 \rightarrow NO_2 + HNO_2, \qquad (17)$$

$$2HNO_2 \rightarrow H_2O + NO + NO_2, \qquad (18)$$

JF found a rate coefficient for (17) of 2×10^{-19} cm^3 molecule^{-1} s^{-1}. Their subsequent neglect of these reactions in the photochemical system seems unjustified since in their experiments $[HNO_3]_{t=0} \sim 5 \times 10^{16}$ molecules cm^{-3}, so that the half-life of NO with respect to reaction (17) was initially ~ 100 s compared to periods of illumination ≥ 1 h. If (17) and (18) are simply added to reactions (13) and (6), and it is assumed that (18) is much faster than (17), then one obtains the rate expression:

$$\frac{d[NO_2]}{dt} = -2I_a + 1.5k_{17}[NO][HNO_3],$$

where I_a is the rate of light absorption. The quantum yield $\Phi = I_a^{-1}d[NO_2]/dt$

and will equal -2 at $t = 0$ when $[NO] = 0$, approach 4 if $[NO]$ reaches its steady-state value, and then diminish as HNO$_3$ is used up. If NO is added initially, then the period when Φ is negative or small will be absent. These effects were observed by JF and fair quantitative agreement with their results can be obtained using their value of k_{17}. This explanation does not depend vitally on the mechanism of the reaction between NO and HNO$_3$ provided that the stoichiometry is NO$+2$HNO$_3\rightarrow$ 3NO$_2+$H$_2$O.

We obtained further strong evidence to support this mechanism by performing some simple flash photolysis experiments. Mixtures of NO$_2$ and HNO$_3$ were flashed with light restricted to wavelengths $\gtrsim 320$ nm so that only the NO$_2$ was photolyzed. By monitoring the NO$_2$ absorption at 410 nm, it was observed that its concentration decreased by an amount c just after the flash but then rose over about 15 min until it was greater by $2c$ than the original concentration. The species reacting during this period cannot be O, NO$_3$ or HNO$_2$ because they are too reactive, and the rate is entirely consistent with NO undergoing reaction (17) with HNO$_3$.

Although the rate of light absorption was lower in BF's experiments than in JF's, there seems every reason to suppose that reaction (17) is again a crucial one in the NO$_2$-sensitized reaction. BF observed a lowering in the quantum yield at high total pressures when NO$_2$ was present and deduced from this effect the rate which was quoted earlier for OH$+$NO$_2+$M. In a revised scheme this pressure effect can be explained by the inclusion of the reactions

$$O+NO_2+M\rightarrow NO_3+M, \tag{19}$$

and

$$NO+NO_3'\rightarrow 2NO_2. \tag{14}$$

At 1 atm, $k_{19}[M] \simeq 2k_6$ and the observed inhibition of the reaction is to be expected. To conclude: the lack of agreement between our rate coefficients for reactions (4), (11) and (12) and those deduced by JF and BF can be ascribed to their incorrect mechanism; the accuracy of other rate coefficients derived by BF must therefore be regarded as doubtful.

The reaction

$$H+NO_2\rightarrow OH+NO \tag{20}$$

is frequently used as a source of OH radicals in discharge-flow systems.[47] At the relatively low total pressures generally used in the experiments, the association of OH with NO and NO$_2$ will be too slow to cause appreciable removal of OH. However, our results do cast some doubt on Del Greco and Kaufman's [11] claim to have shown that reaction (20) leads predominantly to OH$(v = 0)$. If the bimolecular rate coefficients at the high pressure limits of reactions (4) and (5) refer to the formation of energized HNO$_3$ and HNO$_2$ complexes in which randomization of the energy is possible, and therefore correspond approximately to the rate coefficients for de-excitation of vibrationally excited OH by NO$_2$ and NO, then in Del Greco and Kaufman's experiments, where the NO concentration was typically 10^{-2} Torr, the relaxation time of vibrationally excited OH would be only about 15 μs. Their first observations were made 300 μs after mixing the reagents, so it is scarcely surprising that excited vibrational states of OH could not be detected.

Finally, we consider briefly the possible role of reactions (4) and (5) in the chemistry of the atmosphere. OH has been recognized as an important intermediate in both polluted and unpolluted environments [48-51] and it is thought to take part in chain processes resulting in oxidation of NO to NO$_2$ and CO to CO$_2$. Levy [52] has proposed that OH is removed mainly by reaction with CO

$$OH+CO\rightarrow CO_2+H, \tag{21}$$

where $k_{21} = 1.5 \times 10^{-13}$ cm^3 molecule^{-1} s^{-1},[47] but recombination of OH with NO$_2$ and NO seems to have been ignored. N$_2$ and O$_2$ are both likely to act as more efficient third bodies than He for these reactions, and in air at atmospheric pressure the second order rate coefficients for both reactions must be $\gtrsim 10^{-11}$ cm^3 molecule^{-1} s^{-1}. In an " unpolluted " atmosphere, the concentrations of CO and of NO+NO$_2$ are respectively 3×10^{12} and 7.5×10^{10} molecules cm^{-3},[52] so OH will be removed at least as fast by combination with the oxides of nitrogen, as by reaction with CO. Moreover, since reactions (4) and (5) terminate reaction chains and are orders of magnitude faster than steps such as

$$OH+OH+M \rightarrow H_2O_2+M, \qquad\qquad (9)$$

which are bimolecular in radicals, they will exert a profound effect on radical concentrations and the overall chemistry.

We thank S.R.C. for a maintenance grant (C. M.).

[1] R. J. Donovan, D. Husain and L. J. Kirsch, *Trans. Faraday Soc.*, 1970, **66**, 2551.
[2] D. Husain and L. J. Kirsch, *Trans. Faraday Soc.*, 1971, **67**, 2025.
[3] W. Braun and M. Lenzi, *Disc. Faraday Soc.*, 1967, **44**, 252.
[4] D. D. Davies, W. Braun and A. M. Bass, *Int. J. Chem. Kinetics*, 1970, **2**, 101.
[5] G. Hancock, C. Morley and I. W. M. Smith, *Chem. Phys. Letters*, 1971, **12**, 193.
[6] A. C. G. Mitchell and M. W. Zemansky, *Resonance Radiation and Excited Atoms* (Cambridge University Press, Cambridge, 1934).
[7] W. Braun and T. Carrington, *J. Quant. Spec. Rad. Transfer*, 1969, **9**, 1133.
[8] F. Kaufman and D. A. Parkes, *Trans. Faraday Soc.*, 1970, **66**, 1579.
[9] T. T. Kikuchi, *Appl. Optics*, 1971, **10**, 1288.
[10] T. Carrington and H. P. Broida, *J. Mol. Spectr.*, 1958, **2**, 273.
[11] F. P. Del Greco and F. Kaufman, *Disc. Faraday Soc.*, 1962, **33**, 128.
[12] R. V. Poirier and R. W. Carr, *J. Phys. Chem.*, 1971, **75**, 1593.
[13] G. Black and G. Porter, *Proc. Roy. Soc. A*, 1962, **266**, 185.
[14] N. R. Greiner, *J. Chem. Phys.*, 1967, **46**, 2795.
[15] N. R. Greiner, *J. Chem. Phys.*, 1970, **53**, 1070.
[16] D. G. Horne and R. G. W. Norrish, *Nature*, 1967, **215**, 1373.
[17] G. H. Duke and H. M. Crosswhite, *J. Quant. Spec. Rad. Transfer*, 1962, **2**, 97.
[18] R. J. Donovan and D. Husain, *Chem. Revs.*, 1970, **70**, 489.
[19] L. F. Phillips and H. I. Schiff, *J. Chem. Phys.*, 1962, **37**, 1233.
[20] D. L. Baulch, D. D. Drysdale and D. G. Horne, *High Temperature Reaction Rate Data, No. 5* (The University, Leeds, 1970).
[21] D. L. Baulch, D. D. Drysdale and A. C. Lloyd, *High Temperature Reaction Rate Data, No. 2* (The University, Leeds, 1968).
[22] J. E. Breen and G. P. Glass, *J. Chem. Phys.*, 1970, **52**, 1082.
[23] D. L. Baulch, D. D. Drysdale and A. C. Lloyd, *High Temperature Reaction Rate Data, No. 3* (The University, Leeds, 1969).
[24] D. Rapp and T. Kassel, *Chem. Rev.*, 1969, **69**, 61.
[25] S. Jaffe and H. W. Ford, *J. Phys. Chem.*, 1967, **71**, 1832.
[26] T. Bérces and S. Förgeteg, *Trans. Faraday Soc.*, 1970, **66**, 633, 640.
[27] D. Husain and R. G. W. Norrish, *Proc. Roy. Soc. A*, 1963, **273**, 165.
[28] For a recent review of the theory see B. S. Rabinovitch and D. W. Setser, *Adv. Photochem.*, 1964, **3**, 1.
[29] R. A. Marcus, *J. Chem. Phys.*, 1965, **43**, 2658.
[30] E. V. Waage and B. S. Rabinovitch, *Chem. Revs.*, 1970, **70**, 377.
[31] G. Herzberg, *Molecular Spectra and Molecular Structure. I. Spectra of Diatomic Molecules* (Van Nostrand, New York, 1950).
[32] G. Herzberg, *Molecular Spectra and Molecular Structure. III. Electronic Spectra of Polyatomic Molecules* (Van Nostrand, New York, 1966).
[33] H. E. Van den Berg, A. B. Callear and R. J. Norstrom, *Chem. Phys. Letters*, 1969, **4**, 101.
[34] A. P. Cox and J. M. Riveros, *J. Chem. Phys.*, 1965, **42**, 3106.
[35] W. R. Forsythe and W. F. Giauque, *J. Amer. Chem. Soc.*, 1942, **64**, 48.
[36] G. E. McGraw, D. L. Bernitt and I. C. Hisatsune, *J. Chem. Phys.*, 1965, **42**, 237.

1030 REACTIONS OF OH WITH NO₂ AND NO

[37] G. Z. Whitten and B. S. Rabinovitch, *J. Chem. Phys.*, 1963, **38**, 2466.

[38] G. Z. Whitten and B. S. Rabinovitch, *J. Chem. Phys.*, 1964, **41**, 1883.

[39] J. O. Hirschfelder, C. F. Curtiss and R. B. Bird, *Molecular Theory of Gases and Liquids* (Wiley, New York, 1954).

[40] P. G. Ashmore and B. P. Levitt, *Trans. Faraday Soc.*, 1957, **53**, 945.

[41] W. A. Rosser and H. Wise, *J. Chem. Phys.*, 1957, **26**, 571.

[42] H. S. Johnston, L. Foering and R. J. Thompson, *J. Phys. Chem.*, 1953, **57**, 390.

[43] H. S. Johnston, L. Foering and J. R. White, *J. Amer. Chem. Soc.*, 1955, **77**, 4208.

[44] H. Harrison, H. S. Johnston and E. R. Hardwick, *J. Amer. Chem. Soc.*, 1962, **84**, 2478.

[45] T. S. Godfrey, E. D. Hughes and C. Ingold, *J. Chem. Soc.*, 1965, 1063.

[46] J. H. Smith, *J. Amer. Chem. Soc.*, 1947, **69**, 1741.

[47] D. D. Drysdale and A. C. Lloyd, *Oxid. Comb. Rev.*, 1970, **4**, 157.

[48] P. A. Leighton, *Photochemistry of Air Pollution* (Academic Press, New York, 1961).

[49] K. Westberg, N. Cohen and K. W. Wilson, *Science*, 1971, **171**, 1013.

[50] R. D. Cadle and E. R. Allen, *Science*, 1970, **167**, 243.

[51] H. S. Johnston, *Science*, 1971, **173**, 517.

[52] H. Levy, *Science*, 1971, **173**, 141.

Astrophysical Chemistry
The H_3^+ Molecule

P. J. Sarre

School of Chemistry, The University of Nottingham, UK NG7 2RD

Commentary on: **Infrared Spectrum of H_3^+ as an Astronomical Probe**, T. Oka and M.-F. Jagod, *J. Chem. Soc., Faraday. Trans*, 1993, **89**, 2147.

When the Faraday Society was formed in 1903 the field of Astrophysical Chemistry had not yet been born. Even in 1929, at what must surely have been a truly exhilarating General Discussion of the Society on 'Molecular Spectra and Molecular Structure', there was only one very brief exchange that touched on astrophysical matters.[1] The title of that 1929 Discussion is well known as it embraces the four volumes by Gerhard Herzberg who was a participant in that meeting in Bristol. He became keenly interested in molecular astrophysics and it was through a combination of laboratory spectroscopy and astronomical observation that identification of interstellar CH and CN molecules and, notably, the CH^+ molecular ion, was accomplished in the late 1930s and 1940s.[2] Such were the contributions of Herzberg that the Institute in Ottawa where he worked bore his name, and it was within that building that Takeshi Oka designed and carried out a truly outstanding experiment to record the first spectrum of H_3^+ in 1980.[3] My selected paper is from the first Faraday Discussion, a Faraday Symposium to be exact, that was devoted to chemistry in the interstellar medium and held in 1992. Entitled 'Infrared spectrum of H_3^+ as an astronomical probe', the paper bridges laboratory chemical physics and astrophysical research, including the detection by then of H_3^+ in the atmospheres of Jupiter, Uranus and Saturn, but not yet in interstellar space.

Using an electric discharge in hydrogen, Oka employed tunable infrared laser radiation to record fifteen absorption lines of the doubly degenerate v_2 infrared active band in the 2950–2450 cm^{-1} region and which, for an apparently straightforward molecule, are arranged in anything but a simple spectral pattern as illustrated in Fig. 3. The work established the equilibrium geometry of H_3^+ to be an equilateral triangle. Obtaining the spectrum was a major triumph and resulted from four years of commitment and skill. When I visited Ottawa in 1979, Oka told me that this was a unique experiment as the principal components of the hydrogen discharge were non-polar and this would permit, in principle, infrared absorption by H_3^+ to be seen. In other discharge systems there would be strong infrared absorption from neutrals and radicals that would be expected to obscure weak signals from ions. While ways of circumventing this problem have been developed, perhaps most notably velocity modulation and techniques for controlling ions in beams and traps, there is no doubt that the Ottawa experiment laid the foundation for laboratory infrared studies of gaseous molecular ions. There have been numerous subsequent spectroscopic and theoretical studies of H_3^+ to which this brief commentary focussed on astrophysical aspects cannot do justice; I refer the reader to the proceedings of a Royal Society meeting devoted entirely to H_3^+.[4]

It had been suspected since the 1960s that H_3^+ might play a pivotal role in the gas-phase chemistry of interstellar clouds. The ion–molecule reaction

$$H_2^+ + H_2 \rightarrow H_3^+ + H$$

has an extremely high rate and proceeds readily at the low temperatures of interstellar clouds. Its importance in astrochemistry arises from the fact that it acts as proton donor, thereby allowing larger molecules to be built up. The definitive detection of H_3^+ in molecular clouds required much observational effort but this was finally rewarded with success in 1996.[5]

There is a further fascinating chemistry associated with 'isotopic fractionation'. At low temperatures, the difference in zero-point energies of the reactants and products in the reaction

$$H_3^+ + HD \rightarrow H_2D^+ + H_2$$

means that the reaction proceeds to the right yielding relatively high abundances of H_2D^+ and consequent enhanced incorporation of D into molecules through deuteron-transfer reactions of H_2D^+.

H_3^+ has also been detected towards the Galactic Centre and along lines of sight containing diffuse interstellar clouds.[6] The second of these is a particular surprise and shows that H_3^+ is ubiquitous in the diffuse interstellar medium, a possible explanation for which is a larger than expected rate of H_2 ionization.[7] Related to this topic is the long-standing controversial issue of the recombination rate for H_3^+ with electrons. This is discussed in a recent experimental paper and an enhanced cosmic-ray flux is indeed inferred.[8] While there remain unresolved issues, this illustrates that H_3^+ is not only at the heart of interstellar chemistry but is also playing a role in elucidating the fundamental physical characteristics of the Universe. This is analogous to the 3 K background temperature deduced from observations of the CN molecule over 50 years ago. Finally, although the first extraterrestrial detection of H_3^+ was made in planets of our own solar system, a recent paper has reported H_3^+ emission from a protoplanetary disk at a distance of 320 ly. If confirmed this will be another landmark in the short history of the simplest polyatomic molecule as an astronomical probe.[9,10]

One of the greatest strengths of the Faraday Society and now the Faraday Division is the encouragement of the sciences lying between chemistry, physics and biology. To these three one is tempted to add the field of astronomy as the exciting interdisciplinary research emanating from the discovery of the first spectrum of H_3^+ continues to grow.

References

1   Faraday Society General Discussion on 'Molecular Spectra and Molecular Structure', *Trans. Faraday Soc.*, 1929, **25**, 611–949
2   See discussion in G. Herzberg, *Molecular Spectra and Molecular Structure*. I. Spectra of Diatomic Molecules, Van Nostrand Reinhold, New York, 1950, p. 496, and references therein.
3   T. Oka, *Phys. Rev. Lett.*, 1980, **45**, 531
4   *Philos. Trans. R. Soc. London, Ser. A* 2000, **358**, 2363–2559 and references therein.
5   T. R. Geballe and T. Oka, *Nature*, 1996, **384**, 334
6   B. J. McCall, K. H. Hinkle, T. R. Geballe and T. Oka, *Faraday.Discuss.*, 1998, **109**, 267
7   B. J. McCall, K. H. Hinkle, T. R. Geballe, G. H. Moriarty-Schieven, N. J. Evans II, K. Kawaguchi, S. Takano, V. V. Smith and T. Oka, *Astrophys. J.* 2002, **567**, 391
8   B. J. McCall, A. J. Huneycutt, R. J. Saykally, T. R. Geballe, N. Djuric, G. H. Dunn, J. Semaniak, O. Novotny, A. Al-Khalili, A. Ehlerding, F. Hellberg, S. Kalhori, A. Neau, R. Thomas, F. Österdahl and M. Larsson, *Nature*, 2003, **422**, 500
9   S. D. Brittain and T.W. Rettig, *Nature*, 2002, **418**, 57
10 T. Oka, *Nature*, 2002, **418**, 31

J. CHEM. SOC. FARADAY TRANS., 1993, **89**(13), 2147–2154   2147

Infrared Spectrum of H_3^+ as an Astronomical Probe

Takeshi Oka and Mary-Frances Jagod
Department of Chemistry and Department of Astronomy and Astrophysics, The University of Chicago, Chicago, IL 60637-1403, USA

According to the current theory of interstellar chemistry advanced in the early 1970s, the protonated hydrogen molecular ion H_3^+ plays the crucial role in initiating a chain of ion–neutral reactions. The infrared spectrum of H_3^+ was first observed in the laboratory in 1980 and has been extended greatly since. The spectrum has been observed as intense emission in polar regions of Jupiter, and very recently in Uranus and Saturn. Its detection has also been claimed in Supernova 1987A, and intense searching for the spectrum in interstellar space is in progress. We summarize the laboratory observation of the H_3^+ spectrum and its use as an astronomical probe.

Hydrogenic Species

Since hydrogen is the most abundant element in the universe, spectra of hydrogenic species provide the most general probe with which astronomers can study a wide variety of gaseous objects in the cosmos. The electronic transitions of atomic hydrogen, initially observed by Fraunhofer in his solar spectrum[1] (the C and the F lines), identified as due to hydrogen by Ångström,[2] and explained by Bohr,[3] are used over the wide spectral region from the vacuum ultraviolet (Lyman series) all the way through to the radiofrequency (recombination lines). The 21 cm magnetic transition between the hyperfine levels of the hydrogen atom, initially observed in the laboratory by Rabi and co-workers[4] and detected in our galaxy by Ewen and Purcell[5] and by Muller and Oort,[6] is the most universally intense radio signal with which we can observe deep into the galaxy. The spectrum of the negative ion of atomic hydrogen H^- causes the opacity of stars, as initially observed by Milne,[7] suggested by Wildt,[8] and explained by Chandrasekhar,[9] and is a general probe for studies of circumstellar atmospheres.

With the advent of molecular astrophysics, the spectrum of the hydrogen molecule has come to play increasingly important roles in astronomical observations. The quadrupole-induced vibration–rotation spectrum of H_2, initially observed in the laboratory by Herzberg,[10] has now been observed not only in planetary atmospheres,[11] circumstellar space[12] and interstellar space,[13] in our galaxy, but also very strongly in extragalactic objects.[14] The infrared emissions for the $v = 1 \rightarrow 0$ fundamental band from superluminous galaxies are extremely intense [$\sim 10^8$ $L_\odot$† for the $S(1)$ line in NGC 6240],[15–17] and pure rotational emissions show an extremely high temperature of gaseous hydrogen in certain regions. (Up to $J = 17 \rightarrow 15$ has been observed in OMC;[18] the energy of the upper state corresponds to a temperature of 21 600 K.) The electronic transition of H_2 in the vacuum ultraviolet region, initially studied in the laboratory by Lyman[19,20] and Werner,[21] has been observed in interstellar space as strong absorption lines by a rocket spectrometer of Carruthers[22] and the Copernicus satellite spectrometer of Spitzer and his colleagues.[23] Even the VUV spectrum of HD is sufficiently strong to be clearly detected.[24] Just as the presence of abundant H^- is detected through broad bound–free and free–free absorption, the abundant H_2 in Uranus and Neptune was first identified by Herzberg[25] in the spectrum taken by Kuiper[26] through the collision-induced, near-infrared spectrum of H_2. Note that this was the first detection of extraterrestrial H_2. More recently the abundant H_2 in Jupiter and Saturn was identified by McKellar[27,28] in the spectrum by

Gautier *et al.*[29] through the H_2 dimer spectrum. The spectrum was also suggested in the mid-infrared region by Trafton and Watson.[30]

In all these cases hydrogenic species introduced something fundamentally new theoretically, experimentally and observationally. The impact of hydrogenic species on physics, astronomy and chemistry has been more thoroughly discussed in the two inspiring reviews of hydrogenic species by Herzberg.[31,32]

The purpose of this paper is to discuss yet another hydrogenic species, protonated hydrogen H_3^+, whose spectrum was observed in the laboratory in 1980[33] and recently has been observed in space. It is likely that this spectrum plays a unique role in the study of the universe.

Thumbnail Sketch of H_3^+

H_3^+ is the simplest stable polyatomic system, in which three protons are bound by two electrons. It was discovered by J. J. Thomson[34] in the early days of mass spectrometry, and since then a great many experimental and theoretical studies have been reported. Readers are referred to a review article for detail.[35] We summarize below some fundamental properties of H_3^+ needed for this discussion. (1) The H_3^+ system is well bound. The proton affinity of H_2 (*i.e.*, the dissociation energy of H_3^+ into $H_2 + H^+$) is 4.4 eV,‡ equal to the dissociation energy of H_2. (2) The three protons are equivalent, and the two electrons are paired. The equilibrium structure is an equilateral triangle. There is neither a permanent dipole nor an electronic magnetic moment. (3) While its ground state is stable, H_3^+ has no stable electronic excited states, except for a triplet state very close to its second dissociation limit. Therefore its spectroscopic detection must be through either vibration–rotation transitions or rotational transitions induced by intramolecular interaction or isotopic substitution. (4) H_3^+ is the most abundant ionic species in molecular hydrogen plasmas both in the laboratory and in space. This is because of the extremely efficient ion–neutral reaction

$$H_2 + H_2^+ \rightarrow H_3^+ + H \tag{1}$$

which has a large Langevin rate ($\sim 10^{-9}$ cm³ s⁻¹) and exothermicity (~ 1.7 eV). (5) H_3^+ is destroyed by wall collisions (in the laboratory), electronic recombination, and the proton-hop reactions,

$$H_3^+ + X \rightarrow H_2 + HX^+ \tag{2}$$

This reaction is efficient for almost all molecules and atoms because of the relatively low proton affinity of H_2. Thus H_3^+

---

† 1 $L_\odot$ = 3.8 × 10²⁶ W; solar luminosity.

‡ 1 eV ≈ 1.602 18 × 10⁻¹⁹ J.

acts as the universal protonator in space and initiates inter-stellar chemistry. Once protonated, the reactions

$$HX^+ + Y \rightarrow XY^+ + H$$

proceed rapidly as opposed to the direct reactions

$$X + Y \xrightarrow{\hspace{2cm}} XY + h\nu$$

The great many papers that led to these conclusions are found in ref. 35.

Spectrum in the Laboratory

Fundamentals

The normal vibrations of H_3^+ are shown in Fig. 1. The totally symmetric v_1 mode is infrared inactive, and the doubly degenerate v_2 mode is infrared active. The doubly degenerate mode has unit vibrational angular momentum ($\zeta_2 = -1$) as initially shown by Teller.[36] The vibrational states of H_3^+ are specified by two vibrational quantum numbers v_1 and v_2 and the vibrational angular momentum quantum number l. The vibrational energy structure of H_3^+ relevant for astronomical observations is shown in Fig. 2. The transitions observed in the laboratory are shown by upward-pointing arrows, while the emissions observed in astronomical objects are shown by bold downward-pointing arrows. The numbers in parentheses are $|l|$.

The rotational energy levels are uniquely specified by three quantum numbers: the rotational angular momentum J and its projection on the molecular axis k, and the vibrational angular momentum l. The last enters through the vibration–rotation interaction. The approximate rotational energy

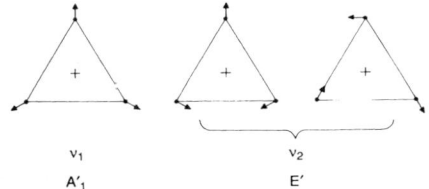

$$v_1 \qquad\qquad v_2$$
$$A'_1 \qquad\qquad E'$$

Fig. 1 Normal vibrational modes for H_3^+. The totally symmetric v_1 mode (3178.3 cm^{-1}) is Raman active and infrared inactive. The degenerate v_2 mode (2521.3 cm^{-1}) is infrared active.

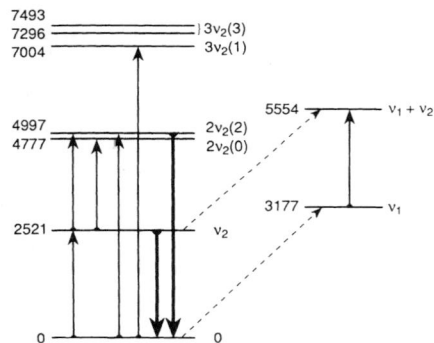

Fig. 2 Vibrational states of H_3^+. Transitions observed in the laboratory are shown by upward-pointing arrows. The transitions so far observed in astronomical objects are shown by bold downward-pointing arrows.

levels of H_3^+ are given by[37]

$$E_r(J, K, l) = BJ(J + 1) + (C - B)k^2 - 2\zeta Ckl + E_c(J, k, l) \quad (3)$$

where B and C are the rotational constants for the specified vibrational levels and ζ is the Coriolis coupling constant. $E_c(J, k, l)$ represents the higher-order centrifugal correction to the rotational energy which is treated in a variety of ways. There are many vibration–rotation coupling terms which have been treated by Watson[38] in the traditional contact transformation formalism. The largest of them is the term

$$\hat{H}_l = q[q_+^2 J_-^2 + q_-^2 J_+^2]/4 \quad (4)$$

where q is the l-doubling constant, and $q_\pm$ and $J_\mp$ are the vibrational and the rotational ladder operators for l and k, respectively. This term mixes levels $|J, k + 1, l = 1\rangle$ and $|J, k - 1, l = -1\rangle$ and causes l-type doubling when the levels are degenerate and the l-type resonance otherwise. The effect of this term is universal and dominating in H_3^+ for the following reasons: (1) The value of q is exceptionally large (*ca.* -5.38 cm^{-1}) because of the large vibrational anharmonicity and small mass. (2) The relations $B \approx 2C$ and $C\zeta \approx -C$ cause the two interacting levels to always be nearly degenerate. For these reasons the two levels are completely mixed; k and l individually are no longer good quantum numbers, but $k - l$ is. It is for this reason that $|k - l|$ has been used to specify the levels of H_3^+. In order to discriminate the two mixed levels with the same G, the labels I and II were used in the earlier papers[33,39] in analogy with Fermi resonance. Watson[38,40] introduced the $U = \pm l$ label ($+l$ for the upper level), and this notation is currently used.

The analysis of the v_2 fundamental band was done using the traditional formalism for an ever increasing number of observed transitions.[33,39–41] This formalism has met difficulties, however, in interpreting spectral lines related to higher energy levels because of the unusually large vibration–rotation interaction in H_3^+. Brute force variational calculations using supercomputers have taken its place.[42] Readers are referred to ref. 43. It is nevertheless necessary to be versed in the traditional formalism to have insight of, and insight into, the general energy structure, assignment of quantum numbers and selection rules.

In order to understand the selection rules, it is useful to identify three quantum numbers that stay good even after the levels are badly mixed; they are J, the parity $\pm$, and the total nuclear spin angular momentum I ($I = I_1 + I_2 + I_3$). These quantum numbers are related to the invariance of the Hamiltonian with respect to rotation in space, to space inversion, and to (123) permutation. If we ignore the very small hyperfine interaction, they are good quantum numbers and follow the rigorous selection rules

$$\Delta J = 0, \pm 1 \quad (5a)$$
$$+ \leftrightarrow - \quad (5b)$$

and

$$\Delta I = 0 \quad (5c)$$

for electric single-photon transitions. In the H_3^+-type molecules, parity is related to the evenness or oddness of k as parity $= (-1)^k$, and the $I = 3/2$ (*ortho*) and the $I = 1/2$ (*para*) spin wavefunctions combine with the $G = 3n$ and the $G = 3n \pm 1$ coordinate wavefunctions, respectively.[44] Thus the last two selection rules are equivalent to

$$\Delta k = \text{odd} \quad (6b)$$

and

$$\text{ortho} \nleftrightarrow \text{para} \quad (6c)$$

J. CHEM. SOC. FARADAY TRANS., 1993, VOL. 89

In discussing the intensities of forbidden transitions, eqn. (6c) is more conveniently written as[45]

$$\Delta g = 3n \quad (n \text{ integer}) \qquad (6d)$$

where $g \equiv k - l$. Eqn. (5a) and eqn. (6) give the selection rules which are always followed. The strongest transitions are those with the smallest value of Δk (*i.e.*, $\Delta k = \pm 1$) and n. U is not a quantum number, and it has no place in the selection rules.

v_2 Fundamental Band

The rovibrational transitions between the v_2 state and the ground state are the strongest and are best suited for astronomical observations. The laboratory observation of this band has been substantially extended[33,39,40,41,46-50] since its initial discovery. Effort has continually been made to observe weaker transitions corresponding to higher rotational levels, in anticipation of their detection in hot astronomical objects. To date, 190 rovibrational transitions have been identified within the spectral range 3579–1798 cm^{-1}. The highest rotational level so far observed is $J = K = 15$ with the energy of 5092 cm^{-1} ($=7327$ K, 0.63 eV). A computer-generated stick diagram of the fundamental band is shown in Fig. 3 for assumed temperatures of $T = 250$, 500, 1000 and 2000 K. The first two temperatures correspond approximately to the rotational temperatures of H$_3^+$ in pure H$_2$ discharges in liquid-N$_2$-cooled and water-cooled plasma tubes, respectively. Likewise, 500 and 1000 K correspond approximately to the rotational temperatures of H$_3^+$ in He-dominated H$_2$ discharges in the respective plasmas. The experimental details may be found in the original papers.[33,40,41,46] The values of individual frequencies, intensities, Einstein coefficients of spontaneous emission, *etc.*, are listed in a table in ref. 51. This table has been revised considerably; new tables are available upon request.

Fig. 4 shows the energy structure and individual transitions for low rotational levels. The transitions are best specified by giving all labels $(J, G, U) \leftarrow (J, K)$. Nevertheless, the traditional spectroscopist's shorthand is sometimes useful. This may be done by looking at the actual energy level structure in the v_2 state. Thus, for example, $(J, G, U) \leftarrow (J, K)$: (3, 2, +1) $\leftarrow$ (2, 2), (3, 1, +1) $\leftarrow$ (2, 1), (3, 0, +1) $\leftarrow$ (2, 0) and (3, 1, -1) $\leftarrow$ (2, 1) correspond to R(2, 2)^1, R(2, 1)^1, R(2, 0)^1 and R(2, 1)$^{-1}$, respectively. Note that the $K = 0$ levels in the ground state with odd J are not allowed by the Pauli principle.[44]

Hot Bands

All three hot bands $2v_2(2) \leftarrow v_2$, $2v_2(0) \leftarrow v_2$ and $v_1 + v_2 \leftarrow v_1$ which are allowed from the first excited states of v_1 and v_2 have been identified.[46] They are weaker than the fundamental bands, typically by a factor of *ca.* 50, owing to the Boltzmann factors for the vibrationally excited states in laboratory plasmas ($T_v \approx 1000$ K).[46] So far 112, 29 and 44 rovibrational transitions have been identified for each respective band.[46,49]

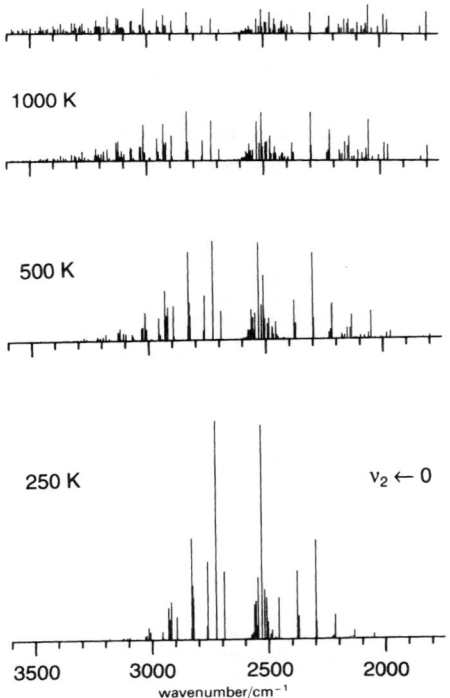

Fig. 3 Computer-generated stick diagrams of the absorption spectrum of the v_2 fundamental band of H$_3^+$ at four different temperatures. Note the large number of lines at *ca.* 2500 cm^{-1} and Q branch transitions with the band origin at 2521.3 cm^{-1}.

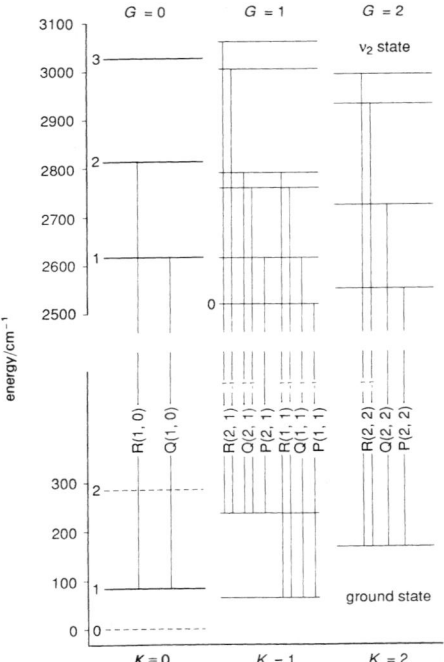

Fig. 4 Lower rotational energy levels of H$_3^+$ in the v_2 state and the ground state. They are specified by J, G, U and J, K, respectively; see text. The allowed rovibrational transitions are shown by vertical lines.

2150

J. CHEM. SOC. FARADAY TRANS., 1993, VOL. 89

Computer-generated stick diagrams of the observed transitions are given in Fig. 5.

During the preparation of this manuscript, we identified a hot-band transition in the Jupiter spectrum (see below). We believe the sensitivity of modern spectrometers is sufficient to observe many of them. These spectral lines will serve as a useful thermometer for astronomical objects.

Overtone Bands

Because of the small mass of the proton, the decrease of the transition dipole moment as we move to higher overtone bands of H_3^+ is not as drastic as in ordinary molecules. The band origins, transition moments, relative intensities and Einstein's spontaneous emission probabilities theoretically calculated by Dinelli, Miller and Tennyson[52] are listed in Table 1. Note that the value of A_{ij} is larger for the $2v_2(2)$ overtone band than for the v_2 fundamental band because the v^3 factor in the Einstein formula overrides the reduction of $|\mu|^2$. This explains the strong 2 μm overtone emission observed in Jupiter.[53,54]

In the laboratory, the $2v_2(2)$ first overtone band[48] and the $3v_2(1)$ second overtone band[55,56] have thus far been observed. The selection rules (6b) and (6d) require for this band $\Delta k = \pm 1$, $\Delta l = \mp 2$, *i.e.*, $\Delta g = \pm 3$. Thus combinations

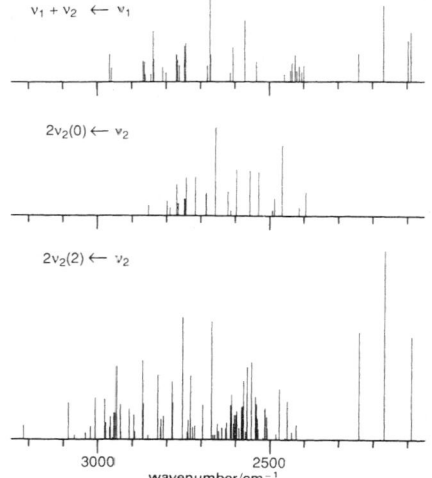

$v_1 + v_2 \leftarrow v_1$

$2v_2(0) \leftarrow v_2$

$2v_2(2) \leftarrow v_2$

3000   2500
wavenumber/cm⁻¹

Fig. 5 Computer-generated stick diagrams of the three observed hot bands $2v_2(2) \leftarrow v_2$, $2v_2(0) \leftarrow v_2$ and $v_1 + v_2 \leftarrow v_1$ at a temperature of 1000 K

Table 1 Intensities of fundamental, overtone and combination bands predicted by Dinelli, Miller and Tennyson[52]

bands	band origin	transition moment/D[a]	relative intensity[b]	A_{ij}/s^{-1}
$v_2(1)$	2521.3	0.160	1	129
$2v_2(2)$	4997.4	0.0609	1/7	145
$v_1 + v_2(1)$	5553.7	0.0026	1/3900	0.35
$3v_2(1)$	7003.5	0.0121	1/180	15.7
$v_1 + 2v_2(2)$	7868.7	0.0097	1/270	14.4
$2v_1 + v_2(1)$	8487.0	0.0028	1/3300	1.5
$4v_2(2)$	9107.6	0.0054	1/890	6.8
$5v_2(1)^c$	10870.0	0.0026	1/3800	2.7

[a] 1 D $\approx 3.335\,64 \times 10^{-30}$ C m. [b] This applies to laboratory absorption spectroscopy. [c] J. K. G. Watson, personal communication.

of the sum of the v_2 fundamental and the $2v_2(2)$-v_2 hot band ($\Delta g = 0$ for both) and the $2v_2(2)$ overtone band yield combination differences in the ground state between levels with different k ($\Delta k = \pm 3$) and allow us to determine their absolute energy values.

Forbidden Transitions

Weaker rovibrational transitions of $v_1 \leftarrow 0$ and of $v_1 + v_2 \leftarrow v_2$ have been observed in the laboratory[49] based on the theoretical calculation of Miller *et al.*[42] The $v_1 \leftarrow v_2$ transition with the band origin at 600 cm⁻¹ has yet to be observed. These transitions, induced by various vibration–rotation mixing terms, are much weaker than the other transitions discussed so far and are not likely to be convenient astronomical probes. However, they will play important roles in the radiative thermalization of H_3^+ in interstellar space. For example, H_3^+ molecules produced in the v_1 state will relax to the ground state *via* the cascading transitions $v_1 \rightarrow v_2$ and $v_2 \rightarrow 0$.

Forbidden pure rotational transitions of H_3^+, following the selection rules $\Delta k = \pm 3$, occur in the wide region from millimetre wave to mid-infrared.[57,58] These transitions are caused by centrifugal distortions of the symmetric structure.[45,59,60] No laboratory observation of them has been reported so far. These transitions are much weaker than the usual dipole-allowed rotational transitions in polar molecules, and their spontaneous emission rates range from *ca.* 10^{-9} s⁻¹ to *ca.* 10^{-2} s⁻¹. Nevertheless, such weak transitions may be observable in low-density regions just like the H_2 quadrupole transitions. Also, the spontaneous emission lifetimes are short compared with the collisional time in low-density areas, making the forbidden rotational transitions important processes for cooling the rotational temperature of H_3^+.

H_3^+ Spectrum in Astronomical Objects

Because of the efficient reaction (1), H_3^+ is expected to be produced in abundance in areas with (*a*) a large amount of H_2 and (*b*) a significant degree of ionization. Since both of these conditions are met by a great variety of gaseous astronomical objects, the H_3^+ spectrum may become a general means to probe weakly ionized areas. Searches for the spectrum in interstellar space have been attempted so far without success. In the meantime, strong infrared emission of H_3^+ has been observed in the planetary ionospheres of Jupiter, Uranus and Saturn. Its detection in Supernova 1987A has been claimed. Here we summarize these observations and possible future observations in various objects. Readers are referred to a review[61] for a more informal summary.

H_3^+ in Interstellar Space

Recent work on ion–molecule reactions indicates that the molecular ion H_3^+ may also be expected in interstellar space.... It now appears desirable to consider the possibilities for detecting H_3^+ because this molecular ion may be present under some circumstances to the virtual exclusion of H_2^+.

D. W. Martin, E. W. McDaniel and M. L. Meeks,[62] 1961.

The H_3^+ molecular ion plays the pivotal role in the ion–neutral reactions scheme now generally believed to be the major mechanism for the chemical evolution of dense molecular clouds.[63-67] H_3^+ is produced through cosmic ray ionization of H_2 followed by the ion–neutral reaction (1). Since the latter reaction is extremely fast, the rate-determining process for the production of H_3^+ is the cosmic ray ionization, whose rate is generally taken to be $\zeta \approx 10^{-17}$ s⁻¹.[63,68] The main destruction mechanism of H_3^+ is the proton-hop reaction (2).

J. CHEM. SOC. FARADAY TRANS., 1993, VOL. 89 2151

Equating the rates of production and destruction, we obtain for the study-state number densities of H_3^+, H_2 and X.

$$n(H_3^+)/n(H_2) = \zeta/kn(X) \qquad (7)$$

where k is the rate constant for reaction (2) which is the Langevin rate (*ca.* 10^{-9} cm^3 s^{-1}) or a little higher if X is polar.[69] Thus $n(H_3^+)/n(H_2) \approx 10^{-8}$ cm$^{-3}/n(X)$. For $n(X) \approx 1$–100 cm^{-3}, we obtain $n(H_3^+)/n(H_2) \approx 10^{-8}$–$10^{-10}$. More detailed chemical model calculations gave $n(H_3^+)/n(H_2) = 1.1 \times 10^{-9}$, 9×10^{-11} (Prasad and Huntress[70]), 5.3×10^{-10} (Millar and Freeman[71]), 1.3×10^{-10}, 5×10^{-11} (Watt[72]), 6.3×10^{-9} (Brown and Rice[73]), 1.8×10^{-9} (Herbst and Leung[69]) and 3×10^{-9} (Langer and Graedel[68]). For the H_2 column density of 2×10^{23} cm^{-2} of Orion A(KL) and Sgr B2(OH), the last number gives the H_3^+ column density of 6×10^{14} cm^{-2}.

Owing to the low temperature and quiescent conditions in dense clouds, the infrared spectrum of H_3^+ is expected in absorption. Thus some bright infrared object is required behind the cloud containing H_3^+. Since $n(H_3^+)$ is inversely proportional to the concentration of the proton acceptor X (which is mostly CO), de Jong, Dalgarno and Boland,[74] and Lepp, Dalgarno and Sternberg[75] recommended searching for H_3^+ in carbon-depleted clouds. Thus obscured infrared objects, such as W33A where a strong solid CO spectrum was reported by Geballe,[76] are more promising sources.

Because of the low temperature in dense molecular clouds and the high rotational constants of H_3^+, only the lowest few rotational levels are significantly populated. The lowest rotational level is $J = 1$, $K = 1$ which is *para*-H_3^+ and has the lower weight of the 2 : 1 spin statistics. (Note that the $J = 0$, $K = 0$ level is not allowed by the Pauli principle.[44,61]) The lowest *ortho* level $J = 1$, $K = 0$ is higher than the $J = K = 1$ level by 22.84 cm^{-1}. Thus, in spite of the spin statistical weight, the $J = K = 1$ level is more populated for $T < 48$ K. The next lowest level $J = K = 2$ is higher by 105.18 cm^{-1}, and its population is negligible at the low temperatures of dense clouds. A computer-generated stick diagram for the H_3^+ spectrum at 30 K is given in Fig. 6. The assignments, frequencies and the squares of the transition moments are listed in Table 2. Each of these spectral lines has some nearby atmospheric interference which may be avoided by choosing the time of optimum Doppler shift. Oka[39] used lines 1 and 2,

$v_2 \leftarrow 0$

30 K

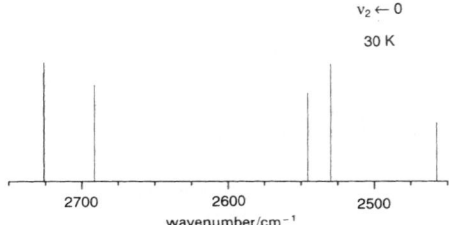

| 2700 | 2600 | 2500 |

wavenumber/cm^{-1}

Fig. 6 A computer-generated stick diagram of the v_2 fundamental band for $T = 30$ K. This simple spectrum should represent absorption in dense clouds.

Table 2 Detail of the spectral lines in Fig. 6

line no.	transition	frequency/cm^{-1}	$\lvert\mu_{ij}\rvert^2/10^{-2}$ D^2
1	$2, 1, +1 \leftarrow 1, 1$	2726.219	1.58
2	$2, 0, +1 \leftarrow 1, 0$	2725.898	2.59
3	$2, 1, -1 \leftarrow 1, 1$	2691.444	1.41
4	$1, 1, +1 \leftarrow 1, 1$	2545.418	1.28
5	$1, 0, -1 \leftarrow 1, 0$	2529.724	2.54
6	$0, 1, +1 \leftarrow 1, 1$	2457.960	0.86

Geballe and Oka[77] 5 and 6, and Black *et al.*[78] 3. The standard formula[39,79] gives the peak absorption as

$$\alpha \equiv \Delta I/I = 5.86 \times 10^{-14} N(H_3^+) f_{JK}(T) \, \lvert\mu_{ij}\rvert^2/\Delta v \qquad (8)$$

where $N(H_3^+)$ is the column density of H_3^+ in cm^{-2}, $f_{JK}(T)$ is the temperature-dependent fraction of molecules in the given J, K level, $\lvert\mu_{ij}\rvert^2$ is the square of the transition dipole moment in D^2, and Δv is the linewidth (HWHM) in km s^{-1}. If we apply this formula for line 5 at 30 K ($f_{JK} = 0.398$), and for a spectrometer with $\Delta v = 10$ km s^{-1} and the minimum detectable absorption α_{min} of 5%, we find the column density of H_3^+ needed for the observation to be 8×10^{14} cm^{-2}, a value comparable to the predicted values given earlier. Compared with the recent infrared detection of more abundant neutral molecules such as CH_4,[80] C_2H_2 and HCN,[81] the expected column density of H_3^+ is two to three orders of magnitude lower. However, the H_3^+ absorption should be comparable to those of $^{13}C_2H_2$ and OCS which were also reported[81] because of the smaller rotational partition function and larger transition dipole moment.

So far, unsuccessful searches for H_3^+ in BN,[39,77] GL 2591, LkH$_\alpha^\#$ 101, NGC 2024/IRS, W33IR,[77] NGC 2264 and AFGL 2591[78] have been reported. Black *et al.*[78] observed spectral lines of CO simultaneously with their search for H_3^+. The abundance of CO thus obtained together with the upper limit of the H_3^+ column density set a limit on the rate of the cosmic ray ionization ζ through eqn. (7). van Dishoeck and Black have proposed on chemical grounds that the abundance of H_3^+ may be equally high in diffuse interstellar clouds.[82]

Further increases in sensitivity and resolution of observational spectrometers will lead us to the detection of H_3^+ in interstellar space. If detected, the spectrum will give crucial information on the working of the ion–neutral reaction scheme for the chemical evolution of molecular clouds.

H_3^+ in Planetary Ionospheres

While the conscious search for the absorption spectrum of H_3^+ was being made in interstellar space over several years without success, the H_3^+ spectrum was found by chance as intense infrared emission in the auroral regions of Jupiter by two groups of astronomers. In 1989 Drossart *et al.* reported[53] rovibrational transitions in the region 5000–4557 cm^{-1} that were assigned to the $2v_2(2) \rightarrow 0$ first overtone of H_3^+, and Trafton *et al.*[54] reported five features comprising six transitions of the same band. Until then the overtone band was unknown to laboratory spectroscopists. For us (the authors), the intense spectrum from Jupiter, the strong overtone band and emission rather than absorption were all unanticipated and were revelations. Readers are referred to ref. 61 for more information. A computer-generated stick diagram of the $2v_2(2) \rightarrow 0$ emission band is shown in Fig. 7. The full range of the spectrum which extends over 1500 cm^{-1} may some day be employed for useful purposes.

After the observation of the 2 μm overtone emission band, even stronger emission for the 4 μm fundamental v_2 band was observed.[83–85] The extensive Fourier-transform infrared emission spectrum reported by Maillard *et al.*[85] is shown in Fig. 8. This spectrum matches very well with the laboratory spectrum for $T = 1000$ K given in Fig. 3 (except for the intensity of the line corresponding to the $4, 3, -1 \rightarrow 3, 3$ transition at 2829.923 cm^{-1} which is stronger in Fig. 3; Telluric line absorption was mentioned by the authors as a possible cause for the anomaly[85]). The observed emission spectrum given in Fig. 7 is remarkably free from Jovian background emission. Especially in the region 2900–2600 cm^{-1}, the background is almost non-existent. The infrared radiation from Jupiter is all absorbed by the pressure-broadened spectrum of CH_4 in the

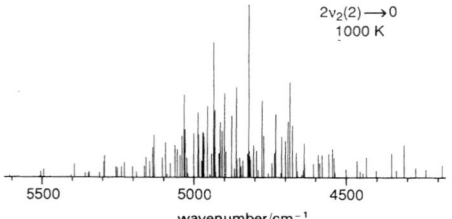

Fig. 7 A computer-generated stick diagram of the theoretical $2v_2(2) \to 0$ emission spectrum at 1000 K. We used theoretical calculations communicated to us by J. K. G. Watson.

same region (the v_3 band with $v_0 = 3019$ cm^{-1} and the $v_2 + v_4$ band with $v_0 = 2845$ cm^{-1}). Only the H_3^+ emission occurring at the high altitude of the ionosphere reaches us unhindered. The purity of the spectrum has allowed two groups of astronomers to study the morphology of the Jovian ionosphere by using infrared cameras.[86,87]

With the benefit of hindsight, the observation of the intense H_3^+ emission in the Jovian ionosphere was to be expected. The atmospheres of the outer planets are rich in hydrogen, and the colossal and tempestuous Jovian magnetosphere was well documented by its decametric activities and by the observations of the Pioneer and Voyager spacecraft.[88] According to Trafton, 'Jovian H_3^+ [emission] was overly ripe for discovery.'[89] The observed results showed: (a) The H_3^+ emission is observable only in the polar auroral regions, and its intensity has good correlation with the auroral ultraviolet intensity measured by the International Ultraviolet Explorer.[53] (b) The distribution of H_3^+ over rovibrational levels is approximately thermal and corresponds to temperatures of 1100 ± 100 K,[53,84] 670 ± 100 K[83] and 1000 ± 40 K (south), and 830 ± 50 (north).[85] This means H_3^+ is collisionally pumped to the excited vibrational states. The variation of temperatures for different observations is believed to be genuine due to temporal variation of Jovian plasma activities. The H_3^+ emission will serve as a ground-based monitor of the dynamical Jovian plasmas. (c) The total column density of H_3^+ is estimated to be *ca.* 10^{12}–10^{13} cm^{-2}, under the assumption of thermal equilibrium.

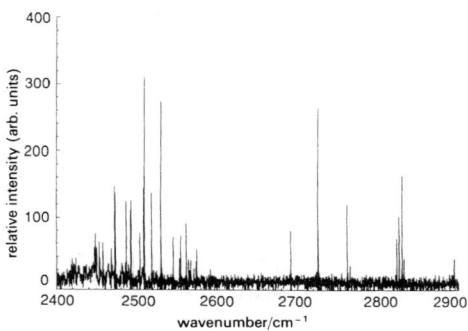

Fig. 8 The apodized H_3^+ emission spectrum from the southern auroral zone of Jupiter recorded by Maillard *et al.*[85] using the Canada–France–Hawaii telescope Fourier-transform spectrometer. The spectral lines match well with the laboratory spectrum at $T = 1000$ K given in Fig. 3. Reprinted with permission from The Astrophysical Journal (1990).

Our most recent observation[90] has added a new chapter to this already rich field. First we found that the H_3^+ emission is not limited to the polar regions but is observable over the entire planet. A spectral image of Jupiter is shown in Fig. 9. As we move from the polar regions to the equator, the emission decreases by one order of magnitude but is still clearly visible. The observation of H_3^+ emission will provide a useful means to study the Jovian ionosphere including its temporal variation. Secondly, we have detected a hot band transition $2v_2(0) \to v_2$ (9, $9 \to 8$, 9, 1), with an intensity of *ca.* 1/50 of neighbouring fundamental band lines. Unlike H_3^+ in the $2v_2(2)$ state which can radiatively relax directly to the ground state through $\Delta v = 2$ (overtone) transitions, H_3^+ in the $2v_2(0)$ state relaxes only through cascading $\Delta v = 1$ transitions. More observations of hot band lines including $2v_2(2) \to v_2$ and $v_1 + v_2 \to v_1$ will enable us to determine temperatures directly and to acquire more knowledge about the dynamics of the Jovian ionosphere.

On April 1 1992 the 4 µm emission spectrum of H_3^+ was observed in Uranus by Trafton *et al.*[91] Twelve emission features comprising 21 Q branch transitions of the fundamental v_2 band have been detected between 2570 and 2545 cm^{-1} with the resolution of *ca.* 1300. The emission lines are weaker than those from Jupiter by two orders of magnitude but were observed with a good signal-to-noise ratio because of the small background and of the high sensitivity of the spectrometer (CGS4). Considering the fact that Uranus has a smaller diameter (*ca.* 1/3) and much smaller magnetic moment (*ca.* 1/410) than Jupiter, and is four times further away from Earth, the emission is amazingly intense and suggests great plasma activity in this planet whose magnetic axis is tilted from the axis of rotation by 58.6°.

On July 18 1992, the H_3^+ emission spectrum was observed in Saturn. The emission is weaker than that of Jupiter by three orders of magnitude, but the doublet at 2725.898 cm^{-1} (2, 0, $1 \to 1$, 0) and 2726.219 cm^{-1} (2, 1, $1 \to 1$, 1) and the singlet at 2829.428 cm^{-1} (4, 3, $-1 \to 3$, 3) have been clearly identified. Observation of Q branch lines was hampered by a stronger background.

On the whole observation of H_3^+ emission has emerged as a useful ground-based observational method for the study of plasma activities in the outer planets.

Other Astronomical Objects

A most surprising paper appeared early this year in which Miller *et al.*[92] claimed detection of H_3^+ emission in the infrared spectrum of Supernova 1987A reported by Meikle *et al.*[93] The spectral features, that are most clearly noted in a spectrum 192 days after the event, are a blend of H_3^+ rovibrational transitions which are Doppler-broadened owing to the rapid expansion of the gas. Since the detailed fingerprint of the individual rovibrational lines is absent, the detection in Supernova 1987A is not very definitive, but their theoretically convoluted spectrum matches well with two observed strong emission features which had remained unidentified. Formation of molecular species within such a short period after the violent catastrophe seems difficult, but there is convincing spectroscopic evidence[94] and a theoretical argument[95] that CO exists abundantly after *ca.* 100 days. From a chemical-model calculation, Miller *et al.*[92] indeed show that formation of H_3^+ with the total mass of *ca.* 10^{-7} M$_\odot$† is possible. (This is approximately the amount of H_3^+ needed to explain the observed strong features.)

All those previous astrophysical studies related to H_3^+ suggest that no gaseous object should be neglected as a pos-

† 1 M$_\odot$ = 2 × 10^{30} kg.

J. CHEM. SOC. FARADAY TRANS., 1993, VOL. 89

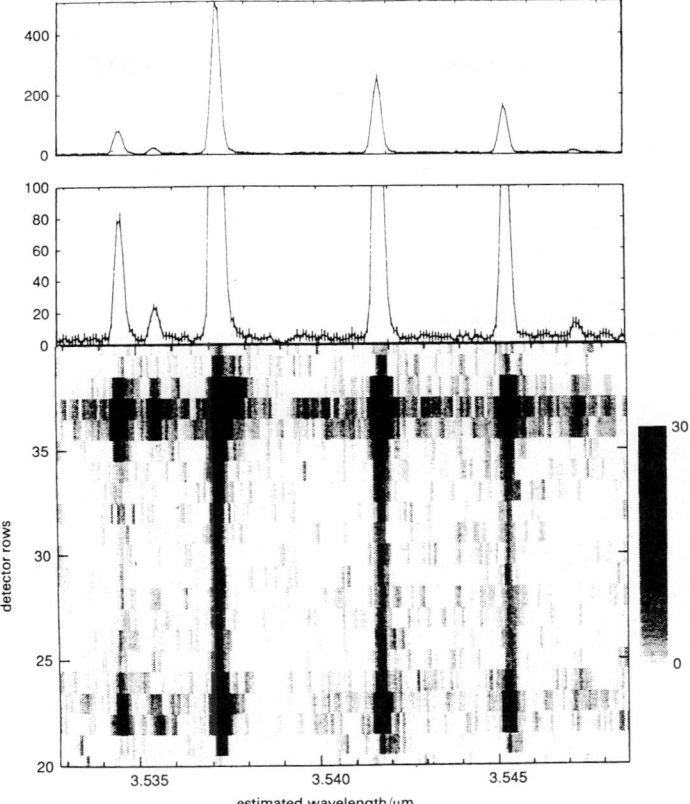

Fig. 9 A spectral image of Jupiter from 2833 to 2821 cm^{-1}

sible source of the H$_3^+$ spectrum. The dense and diffuse clouds, planetary ionospheres and supernovae are widely different astronomical objects. The universal nature of the reaction (1) indicates that there always is the possibility that abundant H$_3^+$ exists as long as there are plenty of hydrogen gas and active plasmas. One guiding principle from the experience of detecting the H$_3^+$ emission in the Jovian ionosphere is to search for objects in which intense H$_2$ quadrupole emission is seen. Thus hot infrared objects,[13] planetary nebulae,[96] superluminous galaxies,[15-17] and H–H objects,[97] are all possible candidates. The vibrational excitation of H$_2$ in these areas proceeds through collisional excitation due to hot gas in shockwave-heated areas,[98] ultraviolet pumping by stellar radiation,[99] or plasma excitation by charged particles in local magnetospheres. Stars, which are assumed to provide the vacuum ultraviolet radiation for the second mechanism in the wavelengths below 914 Å which excites the Lyman and Werner band of H$_2$, must also provide radiation for ionization. The existence of many vibrationally excited H$_2$ molecules also accelerates the ionization. However, such radiation will also photodissociate H$_3^+$ efficiently. H$_3^+$ will exist most abundantly in the areas with high magnetohydrodynamic activity where such UV photons do not penetrate.

It is our pleasure to acknowledge collaboration over many years with J. K. G. Watson, T. R. Geballe and S. Miller. Many of their ideas have entered in this paper. This work has been supported by N.S.F. grant PHY-90-22647 and US Air Force Contract F33615-90-2035.

References

1 J. Fraunhofer, *Denkschiften der königlichen Akademie der Wissenschaften zu München*, 1817, **5**, 193.
2 A. J. Ångström, *Ann. Phys. Chem.*, 1855, **94**, 141.
3 N. Bohr, *Philos. Mag.*, 1913, **26**, 1.
4 J. E. Nafe, E. B. Nelson and I. I. Rabi, *Phys. Rev.*, 1947, **71**, 914.
5 H. I. Ewen and E. M. Purcell, *Nature* (London), 1951, **168**, 356.
6 C. A. Muller and J. H. Oort, *Nature* (London), 1951, **168**, 357.
7 E. A. Milne, *Philos. Trans. R. Soc. London, A*, 1922, **223**, 201.
8 R. Wildt, *Astrophys. J.*, 1939, **89**, 295.
9 S. Chandrasekhar, *Astrophys. J.*, 1945, **102**, 223; 395.
10 G. Herzberg, *Nature* (London), 1949, **163**, 170.
11 C. C. Kiess, C. H. Corliss and H. K. Kiess, *Astrophys. J.*, 1960, **132**, 221.
12 H. Spinrad, *Astrophys. J.*, 1964, **140**, 1639; 1965, **145**, 195.
13 T. N. Gautier III, U. Fink, R. R. Treffers and H. P. Larson, *Astrophys. J.*, 1976, **207**, L129.
14 R. I. Thompson, M. J. Lebofsky and G. H. Rieke, *Astrophys. J.*, 1978, **222**, L49.

2154

J. CHEM. SOC. FARADAY TRANS., 1993, VOL. 89

15 R. D. Joseph, G. S. Wright and R. Wade, *Nature (London)*, 1984, **311**, 132.
16 G. H. Rieke, R. M. Cutri, J. H. Black, W. F. Kailey, C. W. McAlary, M. J. Lebofsky and R. Elston, *Astrophys. J.*, 1985, **290**, 116.
17 D. L. DePoy, E. E. Becklin and C. G. Wynn-Williams, *Astrophys. J.*, 1986, **307**, 116.
18 R. F. Knacke and E. T. Young, *Astrophys. J.*, 1981, **249**, L65.
19 T. Lyman, *The Spectroscopy of the Extreme Ultra-violet*, Longmans, Green and Co., London, 1914.
20 T. Lyman, *Astrophys. J.*, 1916, **43**, 89.
21 S. Werner, *Proc. R. Soc.*, 1926, **113**, 107.
22 G. Carruthers, *Astrophys. J.*, 1970, **161**, L81.
23 L. Spitzer, J. F. Drake, E. B. Jenkins, D. C. Morton, J. B. Rogerson and D. G. York, *Astrophys. J.*, 1973, **181**, L116.
24 J. F. Drake, *Proc. R. Soc. London, A*, 1974, **340**, 457.
25 G. Herzberg, *Astrophys. J.*, 1952, **115**, 337.
26 G. P. Kuiper, *Astrophys. J.*, 1949, **109**, 540.
27 A. R. W. McKellar, *Can. J. Phys.*, 1984, **62**, 760.
28 A. R. W. McKellar, *Astrophys. J.*, 1988, **326**, L75.
29 D. Gautier, A. Marten, J. P. Blateau and G. Bachet, *Can. J. Phys.*, 1983, **61**, 1455.
30 L. M. Trafton and J. K. G. Watson, *Astrophys. J.*, 1992, **385**, 320.
31 G. Herzberg, *Trans. R. Soc. Can.*, 1967, **5**, 3.
32 G. Herzberg, *Trans. R. Soc. Can.*, 1982, **20**, 151.
33 T. Oka, *Phys. Rev. Lett.*, 1980, **45**, 531.
34 J. J. Thomson, *Philos. Mag.*, 1912, **24**, 209.
35 T. Oka, in *Molecular Ions: Spectroscopy, Structure and Chemistry*, ed. T. A. Miller and V. E. Bondybey, North-Holland, New York, 1983, pp. 73–90.
36 E. Teller, in *Hand. Jahrb. Chem. Phys.: Molekü Kristallgitterspektren*, ed. A. Euecken and K. L. Wolf, Akademische Verlag, Leipzig, 1934, vol. 9, pt. II, pp. 43–160.
37 G. Herzberg, *Molecular Spectra and Molecular Structure. II. Infrared and Raman Spectra of Polyatomic Molecules*, Van Nostrand Reinhold, New York, 1945; reprint Kreiger Publishing, Malabar, FL, 1991.
38 J. K. G. Watson, *J. Mol. Spectrosc.*, 1984, **103**, 350.
39 T. Oka, *Philos. Trans. R. Soc. London, A*, 1981, **303**, 543.
40 J. K. G. Watson, S. C. Foster, A. R. W. McKellar, P. Bernath, T. Amano, F. S. Pan, M. W. Crofton, R. S. Altman and T. Oka, *Can. J. Phys.*, 1984, **62**, 1875.
41 W. A. Majewski, M. D. Marshall, A. R. W. McKellar, J. W. C. Johns and J. K. G. Watson, *J. Mol. Spectrosc.*, 1987, **122**, 341.
42 S. Miller, J. Tennyson and B. T. Sutcliffe, *J. Mol. Spectrosc.*, 1990, **141**, 104, and references therein.
43 J. Tennyson, this symposium.
44 L. D. Landau and E. M. Lifshitz, *Quantum Mechanics (Non-Relativistic Theory)*, Pergamon Press, New York, 3rd edn., 1977.
45 T. Oka, in *Molecular Spectroscopy: Modern Research*, ed. K. Narahari Rao, Academic Press, New York, 1976, vol. 2, pp. 229–253.
46 M. G. Bawendi, B. D. Rehfuss and T. Oka, *J. Chem. Phys.*, 1990, **93**, 6200.
47 W. A. Majewski, P. A. Feldman, J. K. G. Watson, S. Miller and J. Tennyson, *Astrophys. J.*, 1989, **347**, L51.
48 L-W. Xu, C. M. Gabrys and T. Oka, *J. Chem. Phys.*, 1990, **93**, 6210.
49 L-W. Xu, M. Rösslein, C. M. Gabrys and T. Oka, *J. Mol. Spectrosc.*, 1992, **153**, 726.
50 D. Uy, C. M. Gabrys, M-F. Jagod and T. Oka, unpublished results.
51 L. Kao, T. Oka, S. Miller and J. Tennyson, *Astrophys. J. Suppl.*, 1991, **77**, 317.
52 B. M. Dinelli, S. Miller and J. Tennyson, *J. Mol. Spectrosc.*, 1992, **153**, 718.
53 P. Drossart, J-P. Maillard, J. Caldwell, S. J. Kim, J. K. G. Watson, W. A. Majewski, J. Tennyson, S. Miller, S. K. Atreya, J. T. Clarke, J. H. Waite Jr. and R. Wagener, *Nature (London)*, 1989, **340**, 539.
54 L. Trafton, D. F. Lester and K. L. Thompson, *Astrophys. J.*, 1989, **343**, L73.
55 S. S. Lee, B. F. Ventrudo, D. T. Cassidy, T. Oka, S. Miller and

J. Tennyson, *J. Mol. Spectrosc.*, 1991, **145**, 222.
56 B. F. Ventrudo, D. T. Cassidy, Z-Y. Guo, S. Joo and T. Oka, unpublished results.
57 F-S. Pan and T. Oka, *Astrophys. J.*, 1986, **305**, 518.
58 S. Miller and J. Tennyson, *Astrophys. J.*, 1988, **335**, 486.
59 T. Oka, F. O. Shimizu, T. Shimizu and J. K. G. Watson, *Astrophys. J.*, 1972, **165**, L15.
60 J. K. G. Watson, *J. Mol. Spectrosc.*, 1971, **40**, 536.
61 T. Oka, *Rev. Mod. Phys.*, 1992, **64**, 1141.
62 D. W. Martin, E. W. McDaniel and M. L. Meeks, *Astrophys. J.*, 1961, **134**, 1012.
63 E. Herbst and W. Klemperer, *Astrophys. J.*, 1973, **185**, 505.
64 W. D. Watson, *Astrophys. J.*, 1973, **183**, L17.
65 W. D. Watson, *Rev. Mod. Phys.*, 1976, **48**, 513.
66 A. Dalgarno and J. H. Black, *Rep. Prog. Phys.*, 1976, **39**, 573.
67 H. Suzuki, *Prog. Theor. Phys.*, 1979, **62**, 936.
68 W. D. Langer and T. E. Graedel, *Astrophys. J. Suppl.*, 1989, **69**, 241.
69 E. Herbst and C. M. Leung, *Astrophys. J.*, 1986, **310**, 378.
70 S. S. Prasad and W. T. Huntress Jr., *Astrophys. J.*, 1980, **239**, L151.
71 T. J. Millar and A. Freeman, *Mon. Not. R. Astron. Soc.*, 1984, **207**, 405; 425.
72 D. G. Watt, *Mon. Not. R. Astron. Soc.*, 1985, **212**, 93.
73 R. D. Brown and E. H. N. Rice, *Mon. Not. R. Astron. Soc.*, 1986, **223**, 405.
74 T. de Jong, A. Dalgarno and W. Boland, *Astron. Astrophys.*, 1980, **91**, 68.
75 S. Lepp, A. Dalgarno and A. Sternberg, *Astrophys. J.*, 1987, **321**, 383.
76 T. R. Geballe, *Astron. Astrophys.*, 1986, **162**, 248.
77 T. R. Gaballe and T. Oka, *Astrophys. J.*, 1989, **342**, 855.
78 J. H. Black, E. F. van Dishoeck, S. P. Willner and R. C. Woods, *Astrophys. J.*, 1990, **358**, 459.
79 L. A. Pugh and K. N. Rao, in *Molecular Spectroscopy: Modern Research*, ed. K. Narahari Rao, Academic Press, New York, 1976, vol. 2, pp. 165–227.
80 J. H. Lacy, J. S. Carr, N. J. Evans II, F. Baas, J. M. Achtermann and J. F. Arens, *Astrophys. J.*, 1991, **376**, 556.
81 N. J. Evans II, J. H. Lacy and J. S. Carr, *Astrophys. J.*, 1991, **383**, 674.
82 E. F. van Dishoeck and J. H. Black, *Astrophys. J. Suppl.*, 1982, **62**, 109.
83 T. Oka and T. R. Geballe, *Astrophys. J.*, 1990, **351**, L53.
84 S. Miller, R. D. Joseph and J. Tennyson, *Astrophys. J.*, 1990, **360**, L55.
85 J-P. Maillard, P. Drossart, J. K. G. Watson, S. J. Kim and J. Caldwell, *Astrophys. J.*, 1990, **363**, L47.
86 S. J. Kim, P. Drossart, J. Caldwell, J-P. Maillard, T. Herbst and M. Shure, *Nature (London)*, 1991, **353**, 536.
87 R. Baron, R. D. Joseph, T. Owen, J. Tennyson, S. Miller and G. E. Ballester, *Nature (London)*, 1991, **353**, 539.
88 *Physics of the Jovian Magnetosphere*, ed. A. J. Dessler, Cambridge University Press, Cambridge 1983.
89 L. M. Trafton, personal communication.
90 T. R. Gabelle, M-F. Jagod and T. Oka, *Astrophys. J.*, submitted.
91 L. M. Trafton, T. R. Geballe, S. Miller, J. Tennyson and G. E. Ballester, *Astrophys. J.*, in the press.
92 S. Miller, J. Tennyson, S. Leep and A. Dalgarno, *Nature (London)*, 1992, **355**, 420.
93 W. P. S. Meikle, D. A. Allen, J. Spyromilio and G. F. Varani, *Mon. Not. R. Astron. Soc.*, 1989, **238**, 193.
94 J. Spyromilio, W. P. S. Meikle, R. C. M. Learner and D. A. Miller, *Nature (London)*, 1988, **334**, 327.
95 W. Liu, A. Dalgarno and S. Lepp, *Astrophys. J.*, 1992, **396**, 679.
96 R. J. Treffers, U. Fink, H. P. Larson and T. N. Gautier III, *Astrophys. J.*, 1976, **209**, 793.
97 H. Zinnecher, R. Mundt, T. R. Geballe and W. Zealey, *Astrophys. J.*, 1989, **342**, 377.
98 D. J. Hollenbach and A. Dalgarno, *Astrophys. J.*, 1977, **216**, 419.
99 J. H. Black and A. Dalgarno, *Astrophys. J.*, 1976, **203**, 132.

Paper 2/04740C; Received 3rd September, 1992

Theoretical Dynamics

M. S. Child

Physical and Theoretical Chemistry Laboratory, University of Oxford, UK OX1 3QZ

Commentary on: **The transition state method**, E. Wigner, *Trans. Faraday Soc.*, 1938, **34**, 29–41.

The paper chosen for its impact on theoretical dynamics is Eugene Wigner's contribution to the Faraday Discussion on Reaction Kinetics, held appropriately in Manchester in September 1937. Although couched in purely classical terms, its influence lies in the emphasis on dynamical as distinct from statistical aspects of transition state theory. The first important point is that the transition state, marked by the line F in Fig. 1 is seen as a $3n - 1$ dimensional dividing surface between the reactant and product regions, not as some pre-equilibrium intermediate. Secondly Wigner shows that knowledge of the transition state and reactant profiles, which determine the integrands of eqns. (1) and (3), is sufficient to determine the classical reaction rate, provided that no trajectory that crosses F ever returns to reactants. Finally, as a corollary, the resulting transition state theory estimate is, in general, an upper bound on the true rate constant, because any re-crossings must reduce the rate.

Readers will find the discussion remarkably brief, and may not even recognize Wigner's transition state formula as a classical partition function ratio, because terms involving the classical momenta and kinetic energy are factored out of the discussion. A detailed commentary by Pechukas[1] on Wigner's approach and the status of classical and quantum transition state theory is therefore strongly recommended as further reading. Wigner himself clearly recognized the importance of quantum effects in statistical sampling of the reactants, in handling tunneling and in the extent to which the velocity at a point, required by eqn. (1) can be properly defined within the uncertainty principle. It is interesting to see what has grown out of Wigner's ideas over the past 65 years, and what can now be said about these final questions.

Starting with classical studies, there has been a considerable interest in the dynamics of low dimensional systems, based on recognition that the exact dividing surface between one class of trajectories and another, is itself a classical trajectory permanently trapped in the interaction region of the potential surface: a periodic orbit dividing surface, or PODS. One finds in particular that such PODS typically change their locations as the energy increases and then bifurcate to produce multiple PODS, and associated quantum mechanical resonances, as the energy becomes too high for the trajectories to pass smoothly round the corner in the potential surface.[2,3] Extensions of these ideas have been applied, in the non-linear dynamics literature, to problems ranging from the ionization of hydrogen atoms in crossed electric and magnetic fields to the rates of escape from asteroid belts.[4]

Changes in the position of the transition state, as signaled by the above migration of the PODS, were anticipated in the literature, first by Keck[5] who recognized that Wigner's upper bound implied a variational version of transition state theory, according to which the 'best' transition state surface is the one with the lowest quantum mechanical partition function. The adiabatic channel model of loose transition states due to Quack and Troe[6] is based on the same idea, while Truhlar, Garrett and coworkers[7] have paid particular attention to incorporating tunneling corrections in variational transition state theory.

The final and most fundamental impact of Wigner's paper lies in the impetus that it gave to formulating a rigorous quantum theory of the chemical rate constant, without the need to solve the full state to state scattering problem. Early work by McLafferty and Pechukas[1,8] concluded that can be no direct quantum analogue of transition state theory, because the quantum flux operator, which is the analogue of the classical velocity in Wigner's eqn. (1), is actually singular in any representation that is sharply localized on the dividing surface, a problem that becomes intractable at energies below the barrier because any proper tunneling correction depends on the nature of the quantum dynamics over a finite region of space. A further difficulty is that the traditional use of a multiplicative tunneling correction is strictly valid only if motion along the reaction coordinate is separable from motion within the transition state, whereas the interesting systems commonly involve tunneling in the vicinity of the corner of the potential surface between the reactant and product valleys. Special methods have evolved to optimize the tunneling correction, by use of complex-time corner-cutting trajectories,[9,10] but nothing of this nature has been proved to be exact. Major progress has however come from Miller's recognition that the exact quantum rate constant can be expressed in the form[11]

$$k(T) = [2\pi\hbar Q_r(T)]^{-1} \int_{-\infty}^{\infty} dE \, e^{-E/kT} N(E)$$

where $N(E)$ is the cumulative reaction probability, summed over all reactant and product quantum states. The important point is that $N(E)$ need not be evaluated by this sum; it can be rigorously expressed as the quantum mechanical trace (sum over diagonal matrix elements) of an operator involving the correlation between localized flux operators at different times. Thus $N(E)$ is manifestly independent of any reference to the asymptotic reactant and product states. Moreover the required thermal average of $N(E)$ is readily evaluated on a coordinate grid, with absorbing boundary conditions in the reactant and product zones, which suppress the quantum mechanical equivalent of re-crossing trajectories. The precise width of the zone must depend on the quantum characteristics of the system, because by Wigner's arguments it must shrink to a simple dividing surface when classical conditions apply.

References

1 P. Pechukas in *Dynamics of Molecular Collisions*, Part B, ed. W. H. Miller, Plenum Press, New York, 1976.
2 E. Pollak, in *Theory of Chemical Reactions*, ed. M. Baer., CRC Press, Boca Raton, FL, 1985, vol. III.
3 R. Sadeghi and R. T. Skodje, *J. Chem. Phys*, 1995, **102**, 193.
4 E. Lee, A. F. Brunello, C. Cerjan, T. Uzer and D. Farrelly, in *The Physics and Chemistry of Wave Packets*, ed. J. A. Yeazell and T. Uzer, John Wiley, New York, 2000
5 J. Keck, *Adv. Chem. Phys.*, 1967, **13**, 85.
6 M. Quack and J. Troe, in *Theoretical Chemistry, Advances and Perspectives*, ed. D. Henderson, Academic Press, 1981, vol. 6B.
7 D. G. Truhlar, B. C. Garrett and S. J. Klippenstein, *J. Phys. Chem*, 1996, **100**, 12771.
8 F. J. McLafferty and P. Pechukas, *Chem. Phys. Lett.*, 1974, **27**, 511.
9 R. A. Marcus and M. E. Coltrin, *J. Chem. Phys.*, 1977, **67**, 2609.
10 N. Makri and W. H. Miller, *J. Chem. Phys.*, 1989, **91**, 4026.
11 W. H. Miller, *Acc. Chem. Res.*, 1993, **26**, 174.

THE TRANSITION STATE METHOD.

By E. Wigner.

Received 29th June, 1937.

1.

According to our present notions, the theory of reaction rates involves three steps. First, one should know the behaviour of all molecules present in the system during the reaction, how they will move, and which products they will yield when colliding with definite velocities, etc. Practically, this amounts in most cases to the construction of the energy surface for the reacting system. Professor Eyring told us about the results which can be obtained by the application of quantum mechanics to molecular systems for this part of the theory. The second step in the theory I would call the statistical part. It endeavours to solve the problem of the rate of elementary reactions. Assuming only the material on the left side of a chemical equation to be present in a vessel, and the molecules of these to have the Maxwell-Boltzman energy and velocity distribution, one wants to know how many molecules corresponding to the right side of the equation will be formed in unit time. The elementary properties of the molecules are supposed to be known in this second step and one wants to express the reaction rate of elementary reactions in terms of these. The present paper will be devoted entirely

to this second step.   The third step is the consideration of the co-opera-
tion of the various elementary reactions, which may occur beside and
must occur after each other in order to complete a real reaction.   In
especially favourable cases there is only one important chain of reactions
leading to the final products and this has one link which is so much slower
than all the others, that it is made responsible for the observed rate.
The others are then assumed to be so much faster that one has practically
equilibrium between the two sides of their chemical equations.

The fact that one has to subdivide the calculation of the reaction
rates into three different steps shows that it has not acquired the neatness
and finality of the theory of equilibria.   I should like, on this occasion,
to point again to one " elementary reaction," which, though often mentioned, never has been properly taken into account.   It is the production
of fast molecules, the establishing of the Maxwell-Boltzman velocity and
energy distribution.   This was tacitly assumed to be a fast reaction and
in complete equilibrium, before I began the discussion of the second step.
In many cases this will not be true.

Naturally, the treatment of the statistical step will be different for
the different kinds of elementary reactions.   It seems that one can
divide the elementary reactions into three groups.   I would class into
the first group those which do not change either the electronic quantum
state of the colliding particles nor their chemical formula and effect an
interchange of kinetic, vibrational and rotational energy only.   The
consideration of these is, as I just pointed out, a rather neglected dis-
cipline in chemistry.[1]   The physicist is strongly interested in them for
the so-called problems of mean free path (viscosity, heat conduction, etc.).
Into the second group, one could class all reactions which involve no
jump in the electronic quantum numbers but change the chemical constitution.   These are the most common types of chemical reactions [2] and
the transition state method applies to these alone.   The remaining third
class [3] deals with reactions which involve a jump in the electronic structure.   This type of reactions has been dealt with in several papers of
Professor Polanyi ;   it is clearly the most general type and probably the
most difficult of all.

2.

The rest of this paper will be devoted to the consideration of the
second kind of reactions.   I shall endeavour rather to emphasise the
basic assumptions of the theory than to derive ready formulas.
Especially on account of some discrepancies with experiment, I think
that it may be useful to see that the transition state method is based, in
addition to well-established principles of statistical mechanics, on only
three assumptions, two of which are generally accepted.

The first assumption is that the comparatively slow motion of the
nuclei is followed by the rapid motion of the electrons to such an extent

[1] C. Zener, *Physic. Rev.*, 1931, **37**, 556, and O. K. Rice, *J. Am. Chem. Soc.*,
1932, **54**, 4558 have studied the excitation of vibrations and the changing of
rotational energy through collisions.   *Cf.* also J. M. Jackson and N. F. Mott,
Proc. Roy. Soc., A, 1932, **137**, 703.

[2] According to quantum mechanics, chemical reactions can occur without
a jump in the quantum state of the electrons.   This fact was first recognised
by F. London, *Sommerfeld Festschrift*, S. Hirzel, 1928, p. 104.

[3] Quantum mechanics was first applied to these by F. London, *Z. Physik*,
1932, **74**, 143, and by L. Landau, *Physik. Z. Sowjetunion*, 1932, **1**, 88 and 1932,
2, 146.   *Cf.* also O. K. Rice, *J. Amer. Chem. Soc.*, 1933, **1**, 375.

E. WIGNER 31

that they are, for every position of the nuclei, in the lowest quantum state [2] (adiabatic assumption). The energy of this lowest state, together with the electrostatic energy of the nuclei, will be denoted by $E_1(X_1, X_2, \ldots, X_{3n})$ where $X_1, X_2, \ldots, X_{3n}$ are the co-ordinates of the nuclei. Under the adiabatic assumption the nuclei will move as if they were acted upon only by the potential $E_1(X_1, X_2, \ldots, X_{3n})$.

The second assumption is that the motion of the nuclei under this potential can be described sufficiently accurately by classical mechanics (or, if this is not the case, the usual way of taking the quantum effects into account is correct).

The last assumption can be best given at the end of the description of the transition state method, which will be the subject of the next section.

3.

For the calculation of the reaction rate it is sufficient to consider a system with as many atoms as occur in the chemical equation of the elementary reaction, since for a system with many atoms, the reaction rate will be simply proportional to the concentrations. Following Gibbs'

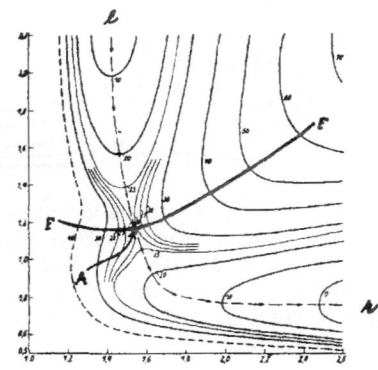

Fig. 1.—The regions (*l*) and (*r*), the activation point A and the surface F, in the section of the configuration space, corresponding to a linear configuration of the atoms H, H, Br. The ordinate is the distance of the two H atoms in A, the abscissa the distance of the middle H from the Br. The numbered lines are contour lines of the energy surface, constructed by Eyring and Polanyi for the reaction

$$H + HBr = H_2 + Br.$$

procedure, one will consider very many such systems, forming a macrocanonic ensemble. The configuration space of the system will have $3n$ co-ordinates and one can plot the function $E_1(X_1, X_2, \ldots, X_{3n})$ along a $(3n + 1)$th direction. Examples of such energy surfaces, for two-dimensional sections of the configuration space, are given in Figs. 1, 2, and 3. (The legends of Figs. 2 and 3 refer to questions to be brought up in section 5).

The configuration of every system of the macrocanonic ensemble will be represented by a point in the configuration space. At ordinary temperature the points of most systems will be in regions of low energy. One such low region will correspond to the left side of the chemical equation, another low region to the right side of the equation. In the first region (*l*), the co-ordinate differences are small for those atoms which are in the same molecule on the left side of the equation. In the second region (*r*), the co-ordinate differences are small for those atoms which form a molecule on the right side of the chemical equation. The transition between the two regions forms the reaction.

We assume now that the two low regions are separated by a higher region, the height of which is at least great compared with the ordinary thermal energy. Then, all paths leading from the first valley into the second will have a highest point. This highest point will be different for different paths. The paths for which the highest point is as low as possible are said to pass through an activation point A, which indeed is the highest point of these paths. The situation is better illustrated in the figures than can be done in words (Figs. 1, 2, and 3).

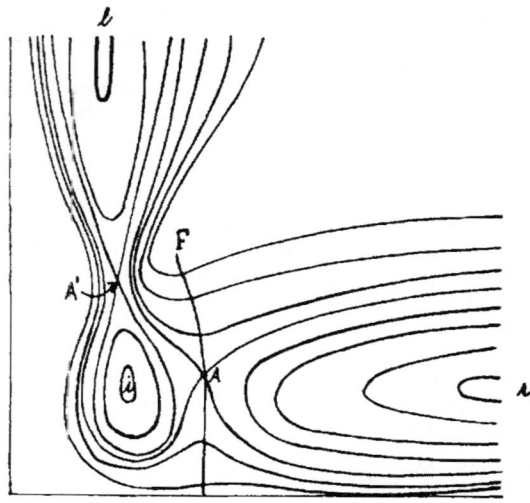

Fig. 2. — Reaction with an intermediate region (i). At not too low temperatures, most points coming from (r) and entering (i) will return to (r), rather than to go through the narrow channel A to (l). In the reverse reaction, the points coming from (l) and passing A′, will soon go over to (r), so that the concentration of points in (i) will not correspond to equilibrium.

Most reactions will have not only one activation point, but a whole super-surface A of activation points (activation surface). Indeed, if we displace or rotate a system in the activated state, the state thus obtained will still be an activated state. All activation points have the same energy Q_0 and no system can react which has less energy than this amount.

Fig. 3. — The section of the activation surface with the plane of the Figure is the contour line 6, from A to its forking point. A path shows a point crossing the activation surface and returning to (l) again.

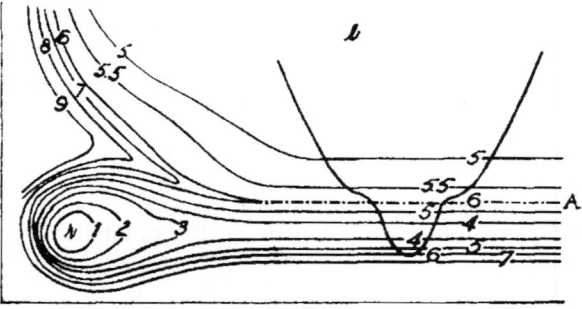

If we move on the activation surface A, the energy E_1 will remain Q_0. If we move in the direction from which the path defining the activated state comes, or if we move in the direction of this path, the energy will decrease, because the activation point is the highest one of this path. In all directions perpendicular to the direction of steepest descent, however, E_1 will *increase*. Otherwise, the highest point of an adjoining

E. WIGNER 33

path would be *lower* than Q_0. The activation surface A forms a saddle-region; it is a maximum in one direction (in the direction of the path) but it is a minimum in all other directions.

We can now draw a $3n - 1$ dimensional surface F through the activated points, perpendicular always to the direction of steepest descent. A will have, in general, less than $3n - 1$ dimensions and lie entirely in F (*cf.* Fig. 1). For a reaction in a system to occur, it is necessary that the representative point shall cross this $3n - 1$ dimensional surface. Most crossings will occur, of course, in the neighbourhood of the activation points, at any rate at such points the energy of which is not many kT greater than the activation energy Q_0. It is easy to calculate the number of systems of our assembly which cross F in one direction in unit time. It is simply equal to the density of points on the surface multiplied by the mean velocity perpendicular to the surface, and this integrated all over the surface:

$$n = \int_F \exp. \left(- E_1/kT\right) \bar{v}_p d\sigma \qquad . \qquad . \qquad . \quad (1)$$

For the practical evaluation of this integral one can expand E_1 into a power series of the co-ordinates which represent the distance from the activation surface A:

$$E_1 = Q_0 + \sum_{i=1}^{3n-s-1} Q_i(\xi_1, \ldots, \xi_s)\eta_i^2 . \qquad . \qquad . \quad (2)$$

where $\xi_1, \ldots, \xi_s$ are s co-ordinates within A and $\eta_1, \ldots, \eta_{3n-s-1}$ are the other $3n - s - 1$ co-ordinates within F. All $a_i(\xi_1, \ldots, \xi_s)$ are necessarily positive and (1) can be evaluated, if (2) is sufficiently accurate, by elementary methods.

We can formulate now the third assumption for the validity of the formulas of the transition state method. If all paths crossing the surface F in one direction originate in the low region l and, crossing F only once, end in the low region r, the reaction rate is simply given by the ratio of (1) to the number of systems in the region l, which is

$$\int_l \exp. \left(- E_1/kT\right) dX_1 \ldots dX_{3n} \qquad . \qquad . \qquad . \quad (3)$$

Here again, a development of the form (2) is possible in most cases and this allows a simple evaluation of (3).

An especially simple way to evaluate the ratio of (1) and (3), *i.e.* the rate constant, has been developed by Eyring and by Evans and Polanyi.[4] Their considerations can be summarized as follows. Clearly, the ratio of (1) and (3) will be the same for all systems in which the potential in (l) and F is the same, while it may be different in the intermediate region. One can replace, therefore, the real potential which is decreasing in the direction perdendicular to F, by a potential which increases in this direction. The region of F will represent then a metastable molecule, the "activated molecule," the probability C of which can be calculated by the well-known formulas for chemical equilibria. The activated molecule will have a vibrational frequency ν corresponding to the motion perpendicular to F, so that C will contain a factor kT/ν.

[4] M. C. Evans and M. Polanyi, *Trans. Faraday Soc.*, 1935, **31**, 875 ; H. Eyring, *J. Chem. Physics*, 1935, **3**, 107.

2

34   THE TRANSITION STATE METHOD

The activated molecule will pass through F in one direction ν times per second, so that the rate constant becomes

$$k = C\nu \qquad . \qquad . \qquad . \qquad . \qquad . \qquad (4)$$

This, of course, in reality does not depend on ν, *i.e.*, on the steepness of the increase of the artificial potential perpendicular to F. Equation (4) is very practical for actual calculations. It is, however, quite easy to evaluate (1) and (3) as they stand also.[5]

Most formulæ for C take quantum effects into account also. Strictly speaking, it should be remembered that it is only in classical statistics permissible to make the above change in the potential.

Before discussing the theoretical foundations of the theory, I should like to quote a few characteristic instances in which it was possible to compare the theory with experiment.

The difficulty of this comparison lies mainly in the difficulty of obtaining accurate energy surfaces. I think, therefore, that at the present stage of the theory one must derive the most important quantity in (4), the activation heat Q_0, from the experiments themselves. The frequencies of the activated state can be derived from the calculated surface with less danger. If one determines these from the observed rate also, very few data may be left as an actual check of the theory. This procedure can be used only for reactions for which rich experimental material is available, especially if the isotopic effect is known.

The experimental difficulties in measuring reaction rates are only too well known. Since at least one quantity for (4), namely Q_0, is derived from the experiments, the values of the rate at at least two temperatures must be known. If these temperatures are too close, a relatively small error in the measurement of the rate may lead to an incorrect determination of Q_0 and a very great inaccuracy in the remaining factors.

The oldest example of the application of the transition state method is the reaction $H + H_2 = H_2 + H$ and the similar reactions with deuterium.[6] They have been measured partly directly, partly by the *ortho-para*-conversion.[7] The calculation of the corresponding energy surface[8] is the simplest example of its kind. Although in my opinion, the agreement between experiment and theory is not complete, the disagreement amounts to less than 50 per cent. with the most suitably chosen set of constants. A similar example is $Cl + H_2 = ClH + H$, where the agreement with the measurement of Rodebush[9] is within[10]

[5] In this way the transition state method was first applied to actual energy surfaces. *Cf.* H. Pelzer and E. Wigner, *Z. physik. Chem.*, B, 1932, **15**, 445. The quantum corrections were included *ibid.*, 1932, **19**, 203. Formulæ similar to those resulting from the transition state method[5] were obtained not much later by W. H. Rodebush, *J. Chem. Physics*, 1933, **1**, 440 and by O. K. Rice and H. Gershinowitz, *ibid.*, 1934, **2**, 853. However, these are very ingenious guesses rather than real derivations based on statistical mechanics. Consequently, the results are not quite identical with those of the transition state method.

[6] *Cf.* H. Pelzer and E. Wigner, and E. Wigner[4]; J. Hirschfelder, H. Eyring and B. Topley, *J. Chem. Physics*, 1936, **4**, 170; L. Farkas and E. Wigner, *Trans. Faraday Soc.*, 1936, **32**, 1.

[7] A. Farkas, *Z. Electrochem.*, 1930, **36**, 782; *Z. physik. Chem.*, B, 1931, **10**, 419; K. H. Geib and P. Harteck, *ibid.*, 1931; *Bodensteinband*, 849; A. Farkas and L. Farkas, *Proc. Roy. Soc.*, A, 1935, **152**, 124.

[8] H. Eyring and M. Polanyi, *Z. physik. Chem.*, B, 1931, **12**, 279; J. Hirschfelder, H. Eyring and N. Rosen, *J. Chem. Physics*, 1936, **4**, 121.

[9] W. H. Rodebush and Klingelhoeffer, *J. Amer. Chem. Soc.*, 1933, **55**, 130; M. Bodenstein, *Trans. Faraday Soc.*, 1931, **27**, 8, 413.

[10] A. Wheeler, B. Topley and H. Eyring, *J. Chem. Physics*, 1936, **4**, 178.

E. WIGNER

the experimental error. The two temperatures at which the rate is measured are very close in this case (0° and 25°) and the limits of error very large.

Two four-body reactions have been investigated : $H_2 + I_2 = 2HI$ and $H_2 + ICl = HI + HCl$. The discrepancy in these cases is a factor of 25 and 300 respectively.[11] The construction of potential surfaces is quite difficult in these cases.

Perhaps the most interesting example is that of the reaction $2NO + O_2 = 2NO_2$ studied theoretically by H. Gershinowitz and H. Eyring.[12] The energy surface is not known for this reaction. It is clear, however, that in the activated state vibrational degrees of freedom take the place of several rotational degrees of freedom of the left side of the chemical equation. Thus the phase space available is strongly diminished in the activated state and the reaction rate unusually low. In spite of the absence of a detailed energy surface, Gershinowitz and Eyring could calculate the reaction rate with a surprising accuracy and also explain the negative temperature coefficient of this reaction.

I have given only the simplest examples of the comparison of theory and experiment, omitting entirely the highly interesting work of Polanyi and Evans on liquids,[13] of Bawn on " slow " reactions, and the extension of the transition state method to the ionogenic reactions by Ogg, Polanyi, and A. G. and M. G. Evans.[14] This subject will be taken up probably by other speakers.

On the whole, I feel that the agreements are not as good as could be expected. Doubtless the experiments are partly to blame for this, and the way of comparison I adopted here. On the other hand, the disagreements call for a review of the theoretical foundations of the transition state method.

In the next two sections I will, therefore, discuss the validity of the three conditions heretofore stated.

4.

(a) **Adiabatic Condition.**—The electronic motion will be able to follow the motion of the nuclei, if that latter motion is not too rapid and if the electronic wave function does not change too radically for a small change in the nuclear positions. The first of these is fulfilled practically always. In all cases in which a non-adiabatic reaction is assumed, the rapid change of the wave function in the neighbourhood of an approximate crossing of energy levels is made responsible for it. I think that for an energy surface which does not show singularities in abnormal curvatures, etc., the adiabatic condition cannot cause serious discrepancies.

[11] For $H_2 + I_2 = 2HI$ cf.[10], for $H_2 + ICl = HCl + HI$, Wm. Altar and H. Eyring, *J. Chem. Physics*, 1936, **4**, 661.

[12] H. Gershinowitz and H. Eyring, *J. Amer. Chem. Soc.*, 1935, **57**, 985. *Cf.* also O. K. Rice, *J. Chem. Physics*, 1936, **4**, 53. The measurements are due to M. Bodenstein, *Z. Electrochem.*, 1918, **24**, 183 ; M. Bodenstein and Lindner, *Z. physik. Chem.*, 1922, **100**, 87 ; M. Bodenstein and Ramstätter, *ibid.*, 106 ; Briner, Pfeiffer and Malet, *J. chim. physique*, 1924, **21**, 25.

[13] M. G. Evans and M. Polanyi, *Trans. Faraday Soc.*, 1935, **31**, 875 ; M. G. Evans and M. Polanyi, *ibid.*, 1936, **32**, 1333 ; W. F. K. Wynne-Jones and H. Eyring, *J. Chem. Physics*, 1935, **3**, 492. For the theory of slow reactions C. E. H. Bawn, *Trans. Faraday Soc.*, 1935, **31**, 1.

[14] R. A. Ogg and M. Polanyi, *Trans. Faraday Soc.*, 1935, **31**, 1375 ; A. G. Evans and M. G. Evans, *ibid.*, 1400. *Cf.* also C. F. Goodeve, *ibid.*, 1934, **30**, 60.

(b) Quantum Effects.—In the argument leading to (4) it was necessary to assume the validity of classical mechanics only for the activated state. In most cases, however, some vibrations of the activated molecule will not be in the classical region. Even this will not matter for those vibrations which either are too far from the bond to be changed, or, for some other reason, do not interfere with the motion of the important degree of freedom. It appears indeed natural to use the usual quantum expressions for C in (4), even for those degrees of freedom which interact strongly with the critical bond. A reasonably rigorous argument shows [15] that this expression is (when modified by an additional factor to take into account tunnelling) valid up to the fourth power of h. This may render (4) probable, but we must admit that it does not prove it beyond the third power terms in h.

Indeed, one can point out two cases in which (4) with the usual quantum expression for C breaks down badly. The first one is given in Fig. 4. The zero-point energy corresponding to the vibration along OX will appear in (4), not, however, in reality, since the potential barrier is very thin.

In the second case, tunnelling under the high region of E_1 plays an important rôle. A negative curvature of E_1 in the activated state can be described as imaginary ν. It may be interesting to note that (4) still holds up to the third power of h, if this imaginary frequency is used as one of the ν of the activated state. Of course, the higher power terms of (4) are not correct in this case, partly, indeed, because the rate depends on the further continuation of the series of (2).

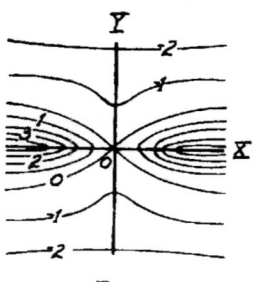

FIG. 4.

The reason that it is so much more difficult to apply the transition state method in quantum theory than it is in classical theory is twofold. Firstly, there is no corresponding simple expression to Boltzmann's exp.$(- E_1/kT)$ for the probability of a configuration. But an even greater difficulty is that one cannot speak about the mean velocity $\bar{v}$ at the activation point. (Heisenberg's indetermination principle.) There is thus, strictly speaking, no way to define the activated complex and its lifetime. All that appears to be possible, in general, is to start out from the classical expression and develop the quantum corrections into a power series of h (only even powers will occur). This has been done to the second power, but the incentive to continue this procedure is not very great, partly because of the laborious mathematical work, partly because three terms will not be very much better than two are, and finally and mainly because our knowledge of energy surfaces hardly warrants an elaborate work.

One can form an idea of the magnitude of the corrections from the one dimensional case where, for several potentials, one has exact solutions of the Schrödinger equation which make a power series development unnecessary. Bell [16] has calculated the penetration of Eckart's one dimensional barrier [17] (Fig. 5a) as function of the temperature. His results for the reaction rate, the classical values, for the same conditions

[15] E. Wigner, *Z. physik. Chem.*, B, 1932, **19**, 203.
[16] R. P. Bell, *Proc. Roy. Soc.*, A, 1933, **139**, 466.
[17] C. Eckart, *Physic. Rev.*, 1930, **35**, 1303.

and the values of A and Q derived from Bell's results in the usual way by representing them as $Ae^{-Q/RT}$ are given in Table I. The values of A and Q become 1 and 14·5 Cal. for all temperatures in classical theory.

One sees that Q and A both decrease with decreasing temperature. This behaviour can be seen qualitatively without calculation : at very low temperature, the tunnelling rate becomes independent of temperature (which means $Q = 0$) and very slow (*i.e.*, A is very small). One

TABLE I.

T.	k_{Bell}.	k_{Class}.	A_{Bell}.	Q_{Bell}.
273	$7 \cdot 10^{-10}$	$2 \cdot 5 \cdot 10^{-12}$	10^{-7}	3·5
323	$4 \cdot 2 \cdot 10^{-9}$	$1 \cdot 5 \cdot 10^{-10}$	$1 \cdot 5 \cdot 10^{-3}$	8·5
373	$3 \cdot 1 \cdot 10^{-8}$	$3 \cdot 10^{-9}$	$3 \cdot 10^{-2}$	10
473	$7 \cdot 3 \cdot 10^{-7}$	$2 \cdot 10^{-7}$	0·15	11·5
573	$6 \cdot 4 \cdot 10^{-6}$	$2 \cdot 9 \cdot 10^{-6}$	0·30	12·5
673	$3 \cdot 4 \cdot 10^{-5}$	$2 \cdot 10^{-5}$	0·55	13·2

is tempted to use this behaviour for the explanation of the very small A values, occasionally observed. It must be remembered, however, that even a potential curve like that of Bell gives an appreciable tunnelling for very light atoms only (Bell's calculation is for hydrogen).

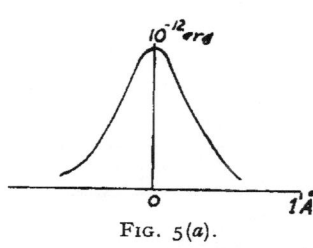

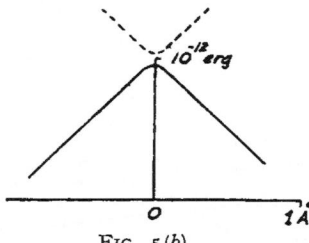

FIG. 5(a).   FIG. 5(b).

A potential curve which may be realised more frequently is given as a full line in Fig. 5b. Such a shape would occur always if two energy surfaces nearly cross (the other energy surface is dotted in Fig. 5b). A non-adiabatic reaction is possible in such cases, but we shall disregard this. For such a potential curve one may expect some tunnelling near the top even for somewhat heavier atoms.

The penetration probability can be exactly calculated if one neglects the small rounding off of the top of the curve and makes it quite sharp-edged. For energies W which are not too near the classical activation energy Q_0, the penetration probability is [18]

$$R = 4W^{1/2}(Q_0 - W)^{1/2}Q_0^{-1} \exp. (- \alpha(Q_0 - W)^{3/2}), \quad \alpha = 4a\sqrt{2M}/3\hbar \quad (5)$$

where M is the mass of the penetrating particle, a is the sum of the reciprocal slopes of the two sides of the potential curve. At the temperature $T = 1/k\beta$ the rate of penetration, divided by the classical rate $e^{-\beta Q_0}/\beta$ is, assuming (5) correct throughout,

$$\frac{k}{k_{\text{class.}}} = \int_0^\infty R\beta e^{\beta(Q_0 - W)}dW . \quad . \quad . \quad . \quad (6)$$

[18] R. H. Fowler and L. Nordheim, *Proc. Roy. Soc.*, A, 1928, **119, 173**.

38   THE TRANSITION STATE METHOD

For $1 \ll \beta^3/\alpha^2 \ll \beta Q_0$ this can be evaluated and gives

$$\frac{k}{k_{\text{class.}}} = \frac{32\sqrt{\pi}\beta^{5/2}}{9\alpha^2\sqrt{Q_0}}e^{4\beta^3/27\alpha^2} \qquad . \qquad . \qquad . \qquad . \qquad (7)$$

A plot of (7) against T shows that for an atomic weight 10 and the potential curve of Fig. 5b, the quantum correction becomes insignificant above 0° C. It seems, therefore, that apart from reactions involving H, the tunnelling effect cannot be made responsible at ordinary temperatures for any large decrease of the temperature independent factor.

These considerations have a particular reference to the dissociation of N_2O in which the temperature independent factor was found [19] to be about 10^{-3} times the value obtained by estimating (4). This was taken, for awhile, as an indication of the non-adiabatic nature of the reaction,[20] which was suggested by the fact that it violates the spin conservation law. However, it has been shown by Zener [21] on the basis of the interaction integrals obtained from the intensities of forbidden

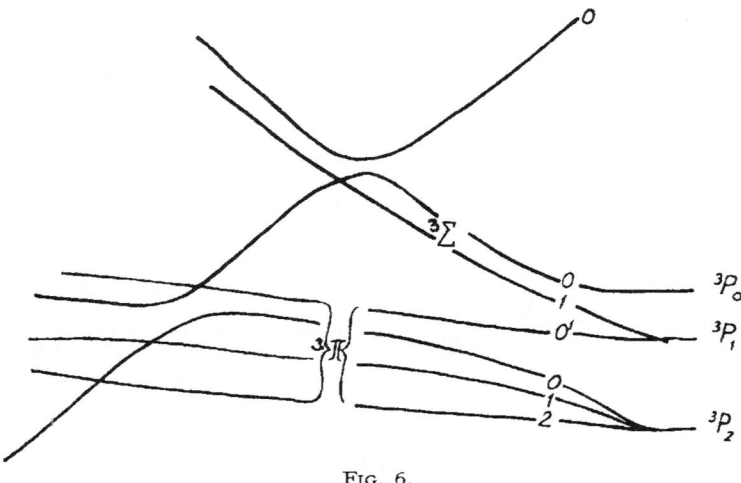

FIG. 6.

lines, that the non-adiabatic character of the reaction cannot be made responsible for a factor of the order of 10^{-3}. We saw that the same holds of the tunnelling effect, so that it must be admitted that the experimental finding is unexplained at present. Perhaps some explanation can be obtained from a more detailed level scheme, like the one drawn schematically and very tentatively, in Fig. 6. If the lower level arising from the ^3P term of oxygen is a $^3\Pi_2$ level, this will strictly cross the $^1\Sigma_0$ level of the ground state for a straight configuration of the N_2O molecule. Then, the reaction will certainly not be adiabatic for a linear

[19] M. Volmer and H. Kummerow, *Z. physik. Chem.*, B, 1930, **9**, 141 ; Nagasako and Volmer, *ibid.*, 1930, **10**, 414 ; Hunter, *Proc. Roy. Soc., A*, 1934, **144**, 386.
[20] H Eyring, *Chem. Rev.*, 1932, **10**, 103 ; H. Pelzer and E. Wigner,[5] G. Herzberg, *Z. physik. Chem.*, B, 1932, **17**, 68 ; A. E. Stearn and H. Eyring, *J. Chem. Physics*, 1935, **3**, 778.
[21] C. Zener, *private communication*. The spectroscopic data were interpreted by A. F. Stevenson, *Proc. Roy. Soc., A*, 1932, **137**, 298 ; E. U. Condon, *Astrophys. J.*, 1934, **79**, 217.

dissociation—not on account of the spin conservation law but because of the conservation law for angular momentum around the molecular axis. The reaction still can go adiabatically for a bent molecule, but it seems that the energy surface becomes so complicated that a larger effect of the tunnelling cannot be *a priori* excluded.

This example again emphasises how every calculation of reaction rates is bound to founder unless we have more detailed energy surfaces than are available at present for N_2O. One is also led to believe that the energy surfaces show more often a complicated structure, as exemplified in Fig. 6, than has been assumed hitherto. If this is true, the deviations from (4) become readily understandable, not only on account of the uncertainty in the quantum theoretical expression for C, but also because the last condition for the validity of (4) will be strongly violated.

5.

The last assumption of the transition state method was that all systems crossing the potential barrier are reacting systems. This appears to be quite natural for a simple energy surface of the kind illustrated in Fig. 1. Figs. 2 and 3 show, however, that especially for more complicated potential surfaces the system is even most likely to cross the activated state and to return then to the original configuration.

It appears thus that the formula (4) will lead, in general, to too high values of the reaction rate and should be corrected by a factor γ, smaller than one, which is the ratio of the number of the crossings through F which end in r, coming from l (or end in l, coming from r), to the number of all the crossings. The problem is then to calculate γ.

There is at least one case in which γ is certainly 1. If the dimensionality s of the activation surface A is *less* than $3n - 1$ and if the temperature is very low, then most of the crossings through F will go through a very narrow strip of F around A. Even if the system does not go straight through into the final state, after having crossed F, but is reflected as in [21a] Fig. 2, the probability that it will find that very narrow strip through which it can return into the region from where it started, is very small. It will be the smaller, the thinner the strip is, *i.e.* the less the energy of the system surpasses Q_0. Since the mean energy excess of the molecules, which have an energy larger than Q_0, over this energy Q_0 is proportional to T, the probability of a return will go to zero with decreasing temperature (unless quantum effects should interfere), and γ goes to 1. However, at higher temperatures γ will decrease, and, apart from special cases (like that of an intermediate valley), it may be quite difficult [22] to find its accurate value. The complication indicated in Fig. 2 is not quite an adequate picture of the situation, since in the $3n$ dimensional configuration space, the picture may be even more involved.

[21a] On account of a mistake in drawing Fig. 2, the original substances correspond to r in this case and the reaction products to l. (In all other examples it is the opposite way.)

[22] Mathematically, the problem is the same as that attacked by G. Lemaitre and M. S. Vallarta (*Physic. Rev.*, 1933, **43**, 87) and by G. Lemaitre, M. S. Vallarta and L. P. Bouckaert (*ibid.*, 1935, **47**, 434) for the calculation of the allowed cone of cosmic radiation. A complete solution involves in every case very much labour.

THE TRANSITION STATE METHOD

6.

Since one cannot calculate Q_0 at present sufficiently accurately, the comparison of (4) with experiment requires the experimental determination of the rate constant at at least two temperatures. This procedure tends to multiply the experimental errors and this may be the cause of a good part of the discrepancies mentioned in section 3. This difficulty would not arise in reactions without a heat of activation, like

$$A + B + C = AB + C \qquad . \qquad . \qquad . \qquad (8)$$

However, the transition state method cannot be applied directly to this type.

Eyring, Gershinowitz and Sun [23] calculated the reaction rate of (8) if A, B, C are atoms by a very ingenious method for the case that the energy surface consists of two parts, both of which are developable. This is the case to a large extent for the $H + H + H = H_2 + H$ reaction. I should like to show how to calculate the rate constant without using this assumption, but using a symbolic application of the transition state method, as first considered by Polanyi.[24]

The energy surface of two attracting atoms shows clearly that one cannot define an activation surface in the usual way. If one puts it at a short distance, the atoms which cross it will not necessarily dissociate, if one puts it a large distance, most of the atoms crossing it are coming from the dissociated state to begin with (Fig. 7). One obtains the impression that it is not the passing through a surface in space which is important, but the passing of the relative energy of the two atoms (kinetic and potential)

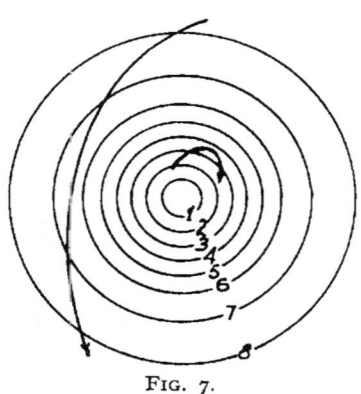

FIG. 7.

through the value 0. One is thus led to consider the energy (or something equivalent) as a co-ordinate instead of the usual space co-ordinates.

Following this line of thought I derived the formula for the velocity constant

$$k = \frac{2(2\pi\beta)^{3/2}}{\sqrt{m_r}} \iiint (-V_0)e^{-\beta(V - V_0)} g r_1 r_2{}^2 \sin \epsilon \, dr_1 \, dr_2 \, d\epsilon \qquad (9)$$
$$g = [(\partial V/\partial \epsilon)^2 + (r_1 \partial (V - V_0)/\partial r_1)^2]^{1/2}.$$

Here $\beta = 1/kT$; the relative mass of the two atoms A and B is m_r; the potential energy of these at the distance r_1 is V_0 if C is very distant. If C is near, the potential energy of all three is $V(r_1 r_2 \epsilon)$, where r_1 is the distance from A to B, the distance of their centre of mass C' from C is r_2 and ϵ is the angle between the lines C'C and AC'. The integration has to be extended over all the region in which V_0 is negative. The details of the derivation of (9) will be given elsewhere.[24a] $V - V_0$ is assumed to

[23] H. Eyring, H. Gershinowitz and C. E. Sun, *J. Chem. Physics*, 1935, **3**, 786, H. Gershinowitz, *ibid.*, 1936, **4**, 363.
[24] M. Polanyi and E. Wigner, *Z. physik. Chem.*, 1928; *Haber Band*, 439.
[24a] *J. Chem. Physics*, 1937, **5**, September issue.

be positive throughout ; (9) will be slightly modified otherwise. A factor $1/2$ enters if A and B are identical atoms.

The formula (9) for the association reaction is much more complicated than that for exchange reactions, *e.g.*, the reason for this is that the latter could be simplified by an assumption like (2), since only the neighbourhood of the activation surface was important for these reactions. This is not true for the type (8).

Formula (9) tells us how often the relative energy of two atoms A and B decreases in unit time below o, if there is a third atom C together with them in a box, assuming complete equilibrium. Again, this will be only an upper limit for the reaction rate, because the relative energy of A and B may increase above zero immediately after it has fallen below, thus necessitating the introduction of a factor γ similar to that for exchange reactions. It may be again expected, however, that γ will be of the order 1 if the interaction between C and the A-B attraction is not very great. There is one important difference between this and the γ of Fig. 2 : the γ for (9) will not go to 1 with decreasing temperature, and will be in general further from 1 than that for (4).

Equation (9) shows the same dependence on the mass of the reacting particles as that of Eyring, Gershinowitz and Sun. This is in agreement with measurements on the association of hydrogen atoms.[25] It also shows that the " third body " C will be more efficient which exerts greater forces on A and B. For equal chemical properties, the efficiency will be roughly proportional to the surface of C. Even so far as numerical values are concerned, the disagreement between (9) and experiment is not very great, for the $2I + A = I_2 + A$ reaction, (9) yields $k = 2 \cdot 0 \times 10^{-32}$ cm.^{6}/sec., the experimental value of E. Rabinowitch and W. C. Wood [26] being $3 \cdot 6 \times 10^{-32}$. The values with He as the third body, instead of A, are $1 \cdot 4 \times 10^{-32}$ and $1 \cdot 8 \times 10^{-32}$, respectively.

Department of Physics,
University of Wisconsin,
Madison, Wisconsin, U.S.A.

[25] I. Amdur, *J. Amer. Chem. Soc.*, 1935, **57**, 856 ; W. Steiner, *Trans. Faraday Soc.*, 1935, **31**, 623.
[26] E. Rabinowitch and W. C. Wood, *J. Chem. Physics*, 1936, **4**, 497.

Statistical Thermodynamics

J. S. Rowlinson

Physical and Theoretical Chemistry Laboratory, University of Oxford, UK OX1 3QZ

Commentary on: **The statistical mechanics of systems with non-central force fields,**
J. A. Pople, *Discuss. Faraday Soc.* 1953, **15**, 35-43

In the first half of the 20th century statistical mechanics was a branch of science cultivated by physicists but rarely by chemists. The physicists were interested mainly in gases and solids whereas the typical physical chemist of the time usually worked on liquid systems. The few who worked on the theory of liquids, such as Guggenheim in Britain and Eyring and Hirschfelder in America, subscribed to the standard opinion of the time that liquids were closer to solids than to gases in their properties, and so also in their theoretical interpretation. A typical theory was the lattice model of Lennard-Jones and Devonshire in which it was supposed that molecules were confined by their neighbours to move around the lattice sites of a close-packed crystal. This solid-like picture had not been the earlier view of van der Waals and the Dutch school who had believed that the structure of a liquid was determined by the space-filling properties of the molecules and that the whole assembly was held together by a relatively unstructured attractive field. Such a picture is what we now call a *mean-field approximation*, in which the attractive potential is treated as a perturbation of the structure of an essentially hard-sphere fluid.

The realisation that lattice theories of liquids were getting nowhere came only slowly from about 1950 onwards. A key paper for chemists was that of Longuet-Higgins on what he called *conformal solutions* in 1951. In this[1] he avoided the assumption that a liquid had a lattice (or any other particular) structure but treated the different strengths of the intermolecular potentials in a mixture as a first-order perturbation of the physical properties of one of the components. In practice, if not formally in principle, his treatment was restricted to molecules that could be assumed to be spherical, but it was so successful for many mixtures of non-polar liquids that this and later derivatives[2] drove lattice theories of liquid mixtures from the field.

The power of perturbation theory in statistical thermodynamics was consolidated when it was extended to polar and other non-spherical molecules. Here an important advance was the treatment that John Pople put before the Faraday Society at the April Discussion held in its Jubilee year of 1953, which I have chosen as my text. An independent perturbation treatment of the dipole–dipole potential in the Proceedings of the Royal Society[3] was soon followed by Pople's definitive treatment in the same journal[4]. His Faraday paper is nicely balanced between the old and the new. The basic treatment is quite general but when he wants to apply the theory to a liquid mixture he resorts to a lattice model for his unperturbed system. In 1953 this was the natural choice, one that was still favoured by Prigogine[5] in his monograph of 1957 and by Guggenheim in his Baker Lectures[6] at Cornell in 1963, but thereafter it soon fell into disuse, and perturbation theory of one kind or another came to dominate statistical thermodynamics in the second half of the century. It is central to the first quantitatively successful theory of liquids, Barker and Henderson's treatment of simple liquids which was put before the Faraday Discussion[7] at Exeter in April 1967. It is one of the techniques used by Buckingham[8] in his treatment of the optical and electrical properties of fluids composed of non-spherical molecules, it is the basis of much of the work summarised in the 1984 monograph of Gray and Gubbins,[9] and has now made its way into chemical engineering practice. A recent example, based in part on the extension of Barker and Henderson's treatment to mixtures, has appeared in the new embodiment of the *Faraday Transactions*.[10]

In placing this emphasis on perturbation theory it must not be implied that other statistical methods were not also used, as, for example, the development, mainly by physicists, of integral equations for determining the basic or hard-sphere structure of liquids. The most striking advance in what might be called experimental statistical mechanics was the use of computer simulation of the equilibrium and transport properties of fluid systems from the 1950s onwards. Moreover some systems, such as aqueous solutions, have structures that are determined by the attractive forces and so fall outside the hard-sphere or argon-like structures that are the basic systems over which the perturbing forces of interest are usually averaged, but perturbation theory has nevertheless been central to the application of statistical thermodynamics to physical chemical problems in the second half of the 20th century.

References

1 H. C. Longuet-Higgins, *Proc. R. Soc. London, Ser. A*, 1951, **205**, 247

2 W. B. Brown, *Philos. Trans. R. Soc. London, Ser. A*, 1957, **250**, 175, 221; T.W. Leland, J. S. Rowlinson and G.A. Sather, *Trans. Faraday Soc.*, 1968, **64**, 1447

3 D. Cook and J. S. Rowlinson, *Proc. R. Soc. London, Ser. A*, 1953, **219**, 405

4 J. A. Pople, *Proc. R. Soc. London, Ser. A*, 1954, **221**, 498

5 I. Prigogine, *The Molecular Theory of Solutions,* North-Holland, Amsterdam, 1957.

6 E. A. Guggenheim, *Applications of Statistical Mechanics,* Clarendon Press, Oxford, 1966.

7 J. A. Barker and D. Henderson, *Discuss. Faraday Soc.*, 1967, **43**, 50; J. A. Barker and D. Henderson, *J. Chem.Phys.*, 1967, **47**, 2856, 4714

8 A. D. Buckingham and J.A. Pople, *Trans. Faraday Soc.*, 1955, **51**, 1029, 1173, 1179; A. D. Buckingham, *Trans. Faraday Soc.*, 1956, **52**, 747 and later papers.

9 C. G. Gray and K. E. Gubbins, *Theory of Molecular Fluids*, vol.1: *Fundamentals*, Clarendon Press, Oxford, 1984; vol. 2, in preparation

10 P. J. Leonard, D. Henderson and J. A. Barker, *Trans. Faraday Soc.* 1970, **66**, 2439; C. McCabe and G. Jackson, *Phys. Chem. Chem. Phys.* 1999, **1**, 2057.

THE STATISTICAL MECHANICS OF SYSTEMS WITH NON-CENTRAL FORCE FIELDS

By J. A. Pople

Department of Theoretical Chemistry, University of Cambridge

Received 2nd February, 1953

A general method for investigating the thermodynamic effects of the dependence of the intermolecular potential on orientation is described. A systematic expansion of the angular dependent part of the intermolecular energy forms the basis of an approximate method of evaluating the partition function which is then used to discuss the thermodynamic properties of liquids and liquid mixtures. Although complete expressions for the intermolecular energies are not available, it is possible to draw certain general conclusions applicable to all types of directional forces. A simple relation is found connecting the additional cohesive energy due to directional forces with the consequent loss of entropy due to restricted rotation. A similar treatment of liquid mixtures leads to a relation between the heat, free energy and entropy of mixing which differs significantly from corresponding theories based on central forces. The theory is used to interpret some experimental data on mixtures in terms of intermolecular forces.

---

1. Introduction.—Most theoretical work on the relation between intermolecular forces and the thermodynamic properties of liquids and liquid mixtures has been limited to potential fields which are independent of the orientation of the particles. This condition, however, is only strictly satisfied by monoatomic substances. For a great many molecular substances directional intermolecular forces are likely to be important and will have a significant effect on the thermodynamic properties of liquids both because of the additional cohesive energy and because of the loss of entropy associated with hindrance to free rotation. Although many of the observed properties of liquids have been attributed to directional forces in a qualitative manner,[1] there has been little in the way of general quantitative theory.

The development of a general theory of systems with non-central force fields can be divided into two parts. First the many types of directional interaction that may occur have to be classified within a general mathematical framework and then approximate methods of evaluating the partition function have to be devised. This paper summarizes some of the results of a method developed by the author[2] with particular reference to its application to the properties of liquid mixtures.

In order to simplify the presentation only axially symmetric molecules are considered in detail. This restriction is not altogether necessary as other systems can be treated by similar methods with very similar results. In the next section a general method of separating the intermolecular field into a central-force part and directional terms of various angular symmetries is described. Simple models such as dipole-dipole forces to represent interactions between polar molecules correspond to particular terms of this expansion. The additional free energy due to the directional part of the field can then be estimated by a perturbation method, provided that the additional field is not too large. The method is applicable at any density and enables approximate theories of monatomic systems to be extended so as to apply to more realistic intermolecular fields.

36   NON-CENTRAL FORCE FIELDS

In the two final sections an approximate version of the theory, based on a lattice distribution, is used to discuss the thermodynamic properties of liquids and liquid mixtures.

2. THE INTERMOLECULAR ENERGY OF AXIALLY SYMMETRIC MOLECULES.—In order to discuss the thermodynamic properties of assemblies of axially symmetric molecules in a comparative manner, it is necessary to use a systematic expansion for the intermolecular field so that various types of angular dependence can be distinguished. To specify the general configuration of two axially symmetric molecules, five co-ordinates are needed, the distance between centres and two angular co-ordinates for each molecular axis. These angular co-ordinates will be taken as spherical polar angles as in fig. 1.

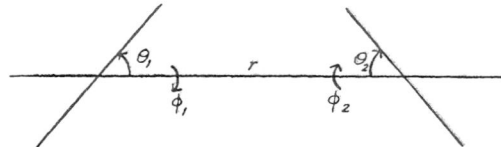

FIG. 1.—Angular co-ordinates of interacting molecules.

The intermolecular energy u_{st} of a molecule 1 of species s and another molecule 2 of species t can be written in the form

$$u_{st}(r_1, \theta_1, \phi_1, \theta_2, \phi_2) = \sum_{l_1=0}^{\infty} \sum_{l_2=0}^{\infty} \sum_m \zeta_{st}^{(l_1 l_2 \,:\, m)}(r)\, S_{l_1 m}(\theta_1, \phi_1) S_{l_2 m}(\theta_2, \phi_2), \quad (2.1)$$

where $\zeta_{st}^{(l_1 l_2 :\, m)}(r)$ is a function of r only (zero if $|m|$ is greater than l_1 or l_2) and S_{lm} are surface harmonics defined by

$$S_{lm}(\theta, \phi) = \left[(2l + 1) \frac{(l - |m|)!}{(l + |m|)!} \right]^{\frac{1}{2}} P_l^{|m|}(\cos \theta)\, e^{im\phi}, \quad (2.2)$$

$P_l^m(x)$ being an associated Legendre function

$$P_l^m(x) = (1 - x^2)^{m/2} \frac{d^m}{dx^m} P_l(x). \quad (2.3)$$

Since $S_{00} = 1$, (2.1) reduces to its first term $\zeta_{st}^{(00 :\, 0)}(r)$ if there are only central forces. The other terms represent the various types of directional field that can occur.

This expansion can be applied to any intermolecular field, but it is particularly useful since many simple models of directional interaction lead to simple expressions for the functions $\zeta_{st}^{(l_1 l_2 :\, m)}(r)$. Further it will be shown in the next section that (2.1) is a very convenient basis for a general method of evaluating the free energy.

The expansion (2.1) is particularly useful if the intermolecular forces are well represented by the interaction of permanent electrostatic moments. Attempts have been made, for example, to discuss the interaction of simple polar molecules in terms of a central force field representing the dispersion and repulsion forces, together with the interaction of point dipoles situated at the molecular centres.[3-5] If the moments of the dipoles are μ_s and μ_t, this directional field can be written in the form (2.1) where

$$\zeta_{st}^{(11 :\, -1)}(r) = \tfrac{1}{2} \zeta_{st}^{(11 :\, 0)}(r) = \zeta_{st}^{(11 :\, 1)}(r) = \tfrac{1}{3} \mu_s \mu_t\, r^{-3}. \quad (2.4)$$

When considering short-range interaction, however, the point dipole model may be inadequate because the dimensions of the molecules are not small compared with the intermolecular distance. It then becomes necessary to include interactions due to higher order moments. If the molecules are represented by dipole moments μ_s, μ_t and quadrupole moments Θ_s, Θ_t the complete directional field is given by (2.4) together with

$$\zeta_{st}^{(12:-1)} = 3^{-\frac{1}{2}} \zeta_{st}^{(12:0)} = \zeta_{st}^{(12:1)} = 5^{-\frac{1}{2}} \mu_s \Theta_t r^{-4},$$

$$\zeta_{st}^{(21:-1)} = 3^{-\frac{1}{2}} \zeta_{st}^{(21:0)} = \zeta_{st}^{(21:1)} = 5^{-\frac{1}{2}} \mu_t \Theta_s r^{-4},$$

$$\zeta_{st}^{(22:-2)} = \tfrac{1}{4} \zeta_{st}^{(22:-1)} = \tfrac{1}{6} \zeta_{st}^{(22:0)} = \tfrac{1}{4} \zeta_{st}^{(22:1)} = \zeta_{st}^{(22:2)} = \tfrac{1}{6} \Theta_s \Theta_t r^{-5}. \quad (2.5)$$

3. STATISTICAL MECHANICS OF ASSEMBLIES OF AXIALLY SYMMETRIC MOLECULES.— To illustrate the way in which the contributions of the directional forces to the thermodynamic functions can be estimated by statistical mechanics, we shall limit ourselves to pure substances in this section. The extension to mixtures will be discussed in the last section. The Helmholtz free energy of an assembly of N identical molecules occupying a volume V at temperature T is given by

$$e^{-F/kT} = \{\phi(T)\}^N (N!)^{-1} \int \ldots \int e^{-U/kT} \, dv d\omega, \quad (3.1)$$

where $\phi(T)$ is a molecular partition function (independent of V) allowing for the kinetic and vibrational energies, U is the total potential energy for any configuration of particles and $\int dv$ and $\int d\omega$ have been written for integration over the positional and angular co-ordinates of all molecules. If U is written as the sum of contributions from all pairs of molecules

$$U = \sum_{i>j} u(i, j) \quad (3.2)$$

and $u(i, j)$ is expanded in the form (2.1), we have a complete expression for the free energy F and consequently for all other thermodynamic functions in terms of the ζ-functions of (2.1).

In practice, the complete integration of (3.1) is very difficult, particularly because of the dependence of U on orientation. The quantity in which we are interested, however, is the *extra* free energy that arises from the orientational forces, that is the contribution of all terms but the first in the expansion (2.1). This can be estimated by treating these orientational components in the intermolecular energy as small quantities compared with the central force energy $\zeta^{(00:0)}(r)$ and retaining only the leading term in the expansion. If the orientational forces are not small, this procedure is only approximate, but in any case it should lead to results of the correct order of magnitude.

The mathematical details of the method are given elsewhere [2] and only the results will be quoted here. These are simplest if the only orientational terms in (2.1) are those with *both* l_1 and l_2 greater than zero. This applies to many types of field including those described in the last section and we shall limit the discussion to such systems. It then follows that the total free energy can be written as the sum of two contributions

$$F(T, V) = F^{(0)}(T, V) + F^{(or)}(T, V), \quad (3.3)$$

where $F^{(0)}(T, V)$ if the free energy of a similar assembly of molecules interacting according to the central force term $\zeta^{(00:0)}(r)$ only, and $F^{(or)}(T, V)$ is the additional free energy due to the orientational forces. $F^{(or)}(T, V)$ is given explicitly by

$$F^{(or)} = -N(4kT)^{-1} \sum_{l_1=1}^{\infty} \sum_{l_2=1}^{\infty} \sum_{m} \int \left[\zeta^{(l_1 l_2 : m)}(r) \right]^2 n_2^{(0)}(r) dr, \quad (3.4)$$

where $n_2^{(0)}(r)dr$ is the probability of a particle *in the central force system* being in a volume element dr at r given that there is one at the origin. Once the pair distribution function $n_2^{(0)}$ for the central force system is known, therefore, the orientational free energy $F^{(or)}$ can be calculated immediately. The theory of central force assemblies has not yet reached a stage where the pair distribution function can be calculated accurately from the intermolecular field, but eqn. (3.4) does enable approximate theories of monatomic systems to be extended to systems where there are orientational forces. It has the further advantages of not being restricted to any particular statistical model and of being applicable at any density.

4. THERMODYNAMIC EFFECTS OF DIRECTIONAL FORCES IN PURE LIQUIDS.—At high densities characteristic of the liquid phase well below the critical temperature, the pair distribution function $n_2^{(0)}(r)$ for the central force system is dominated by considerations of packing. The simplest approximation to use in (3.4) is to assume that $n_2^{(0)}(r)$ can be replaced by the corresponding function for a close-packed lattice, the dimensions of which are chosen to give the correct total volume. This means that only certain discrete values of the intermolecular distance are allowed instead of a continuous distribution, so that the integral in (3.4) has to be replaced by a sum over lattice sites. If the sites of the lattice are R_i, a typical site being chosen as origin, the expression for the orientational free energy becomes

$$F^{(or)} = - N(4kT)^{-1} \sum_{l_1=1}^{\infty} \sum_{l_2=1}^{\infty} \sum_{m} \sum_{i>0} \left[\zeta^{(l_1 l_2 : m)}(R_i) \right]^2. \tag{4.1}$$

If only nearest neighbour interactions are taken into account this can be written

$$F^{(or)} = - zN(4kT)^{-1} \sum_{l_1=1}^{\infty} \sum_{l_2=1}^{\infty} \sum_{m} \left[\zeta^{(l_1 l_2 : m)}(a) \right]^2, \tag{4.2}$$

where z is the lattice co-ordination number (the number of nearest neighbours of a given site) and a is the nearest neighbour distance.

Eqn. (4.1) and (4.2) lead to simple expressions for the extra contribution to the thermodynamic functions if the intermolecular field can be represented by dipole or quadrupole interactions as discussed in section 2. Using a face-centred cubic lattice for which $z = 12$ and

$$a^3 = \sqrt{2}\, V/N \tag{4.3}$$

the contribution of dipole-dipole interactions is found to be

$$F^{(or)}/NkT = - (N\mu^2/VkT)^2 \tag{4.4}$$

from (4.2). If more distant interactions are taken into account (4.1) must be used and (4.4) is multiplied by a factor 1·2045.[7] Corresponding expressions involving quadrupole moments are easily derived using (2.5).

Even if simple expressions for the directional field are not available, it is possible to draw certain conclusions about the other thermodynamic functions from (4.2). The orientational free energy $F^{(or)}$ leads to the following additional contributions to the entropy, internal energy and heat capacity at constant volume

$$S^{(or)} = \tfrac{1}{2} E^{(or)}/T = - \tfrac{1}{2} C_V^{(or)} = - \frac{zN}{4kT^2} \sum_{l_1=1}^{\infty} \sum_{l_2=1}^{\infty} \sum_{m} \left[\zeta^{(l_1 l_2 : m)}(a) \right]^2. \tag{4.5}$$

The extra entropy $S^{(or)}$, which is always negative, arises because, in order to take advantage of the additional energy of certain orientations, the molecules have to restrict their freedom of rotation. The heat capacity $C_V^{(or)}$ which, according to (4.5) is twice $(- S^{(or)})$ is due to the loosening-up of the rotational degrees of freedom with rising temperature.

It should be possible to determine whether directional forces are important in a liquid by examining its thermodynamic properties. The most marked effect will be probably on the heat capacity. A direct test of this sort can only be applied to substances for which there are no other complicating factors such as internal rotation about single bonds. Accurate molar heat capacities at constant volume are not available for many liquids but the additional term given by (4.5) should also contribute to the molar heat capacity at constant pressure. As only the configurational heat capacity due to intermolecular forces is required, the corresponding heat capacities of a perfect gas (at constant volume) must be subtracted from the observed C_P. The gas heat capacities can be estimated theoretically from the observed vibrational frequencies. The values of $\{(C_P)_{\text{liq}} - (C_V)_{\text{gas}}\}$ for some simple liquids near their normal boiling points are given in table 1.

TABLE 1.—CONFIGURATIONAL HEAT CAPACITIES OF SOME LIQUIDS AT CONSTANT PRESSURE (cal mole^{-1} deg^{-1})

substance	T (°K)	$\{(C_p)_{\text{liq}} - (C_v)_{\text{gas}}\}$	substance	T (°K)	$\{(C_p)_{\text{liq}} - (C_v)_{\text{gas}}\}$
A	87	7·1 [7]	Cl_2	240	9·9 [13]
Kr	120	7·8 [8]	H_2S	210	10·3 [14]
Xe	165	7·7 [9]	CH_3Cl	249	11·1 [15]
N_2	78	8·7 [10]	NH_3	240	11·9 [16]
CS_2	310	9·0 [11]	C_6H_6	353	14·0 [17]
HCl	185	9·2 [12]	CCl_4	330	14·1 [18]

As expected, the configurational heat capacities of molecular liquids are larger than those of the monatomic substances. Further, most of the values increase in a reasonable order, those for symmetrical non-polar molecules being rather less than those for simple polar molecules. The heat capacity of liquid carbon tetrachloride is surprisingly high and is rather difficult to reconcile with Hildebrand's suggestion [19] that almost free rotation occurs. If the extra heat capacities shown in table 1 are enturely due to the increasing importance of directional forces, there will be corresponding extra entropies according to (4.5). For benzene and carbon tetrachloride the negative entropy due to hindered rotation would correspond to a value of $(-TS)$ as large as 1 kcal mole^{-1}.

5. THERMODYNAMIC EFFECTS OF DIRECTIONAL FORCES IN LIQUID MIXTURES.— The theory applied to pure liquids in the last two sections can be generalized to liquid mixtures and can be used to discuss the effects of directional forces on the thermodynamic functions of mixing. Classical statistical mechanics leads to a complete expression for the free energy of a multicomponent system in terms of the intermolecular energies u_{st} for all pairs of components s and t. Each u_{st} can be expanded in the general manner (2.1), so that it is separated into a spherically symmetric part and various directional terms.

The general method is to suppose that all the intermolecular energies u_{st} differ only slightly from the corresponding energy u_{00} for some reference substance, u_{00} being a central force energy and consequently a function of r alone. If all the u_{st} were *equal* to u_{00}, the mixture would be ideal and there would be no excess thermodynamic functions. The aim of the present method is to express the excess functions in terms of the differences between the intermolecular fields u_{st} and u_{00}, these differences being assumed small.

The difference $(u_{st} - u_{00})$, which is to be treated as a perturbation, can be written

$$u_{st}(r_1\theta_1, \phi_1, \theta_2, \phi_2) - u_{00}(r) = \left\{ \zeta_{st}^{(00:0)}(r) - u_{00}(r) \right\}$$

$$+ \sum_{l_1=1}^{\infty} \sum_{l_2=1}^{\infty} \sum_{m} \zeta_{st}^{(l_1 l_2 : m)}(r) S_{l_1 m}(\theta_1, \phi_1) S_{l_2 m}(\theta_2, \phi_2). \quad (5.1)$$

We are again considering only directional terms for which both l_1 and l_2 are greater than zero. The two parts of (5.1) representing central force and directional force differences respectively lead to distinct contributions to the free energy. If only leading terms are retained these two contributions are additive and can

40 NON-CENTRAL FORCE FIELDS

be considered separately. The excess Helmholtz free energy of mixing at constant volume, therefore, can be written

$$(\Delta^* F)_{T, V} = (\Delta^* F^{(cent)})_{T, V} + (\Delta^* F^{(or)})_{T, V} \qquad (5.2)$$

where the two contributions arise from the first and second parts of (5.1). For binary mixtures the explicit expressions for these two quantities are.[2]

$(\Delta^* F^{(cent)})_{T, V}$

$$= N x_A x_B \int \left[\zeta_{AB}^{(00:0)}(r) - \tfrac{1}{2} \zeta_{AA}^{(00:0)}(r) - \tfrac{1}{2} \zeta_{BB}^{(00:0)}(r) \right] n_2^{(0)}(r) dr, \qquad (5.3)$$

and

$$(\Delta^* F^{(or)})_{T, V} = - N(2kT)^{-1} x_A x_B \sum_{l_1 = 1}^{\infty} \sum_{l_2 = 1}^{\infty} \sum_{m}$$

$$\int \left\{ \left[\zeta_{AB}^{(l_1 l_2 : m)} \right]^2 - \tfrac{1}{2} \left[\zeta_{AA}^{(l_1 l_2 : m)}(r) \right]^2 - \tfrac{1}{2} \left[\zeta_{BB}^{(l_1 l_2 : m)}(r) \right]^2 \right\} n_2^{(0)}(r) dr, \qquad (5.4)$$

where x_A, x_B are the mole fractions of the components and $n_2^{(0)}(r)$ is the pair distribution function for the reference liquid.

This paper is mainly concerned with the excess free energy $(\Delta^* F^{(or)})_{T, V}$ but we shall first give a short discussion of theories of the central force term $(\Delta^* F^{(cent)})_{T, V}$ which has been evaluated by a variety of methods. If the continuous distribution function $n_2^{(0)}$ is replaced by a lattice distribution as in the previous section and if interactions between non-neighbouring sites are neglected, (5.3) becomes

$$(\Delta^* F^{(cent)})_{T, V} = N w x_A x_B, \qquad (5.5)$$

where

$$w = z \left\{ \zeta_{AB}^{(00:0)}(a) - \tfrac{1}{2} \zeta_{AA}^{(00:0)}(a) - \tfrac{1}{2} \zeta_{BB}^{(00:0)}(a) \right\}, \qquad (5.6)$$

a being the distance between neighbouring sites and z the co-ordination number of the lattice. This is the zeroth approximation in the theory of strictly regular solutions.[20] One important consequence of (5.5) is that $(\Delta^* F^{(cent)})_{T, V}$ is independent of temperature so that there is no entropy of mixing at constant volume. A lattice model will only give an excess entropy if higher order calculations are carried out and then $(\Delta^* S)_{T, V}$ is small and always negative.

An improvement that has recently been introduced [21, 22] is to treat the particles as moving in cells rather than restrict them to lattice sites. This improved model does lead to an excess entropy of mixing because the vibrational motions of particles in their cells may differ in the mixture and pure liquids. Quantitative estimates of this entropy vary according to the type of cell field used, but the more realistic fields lead to a value of $\Delta^* S$ which is of the same sign as $\Delta^* F$.

The relative contributions of the excess entropy predicted by the various theories are shown in table 2, together with some experimental data on equimolar non-polar mixtures. It is important that the theoretical predictions should be compared with measurements of mixing functions at constant volume, for a considerable part of the measured excess entropy of mixing at constant pressure can be attributed directly to the volume change. The figures of table 2 show that, even after the entropy due to the volume change has been allowed for, the remaining excess entropy is considerably larger than that predicted by any of the models of the reference liquid. The cell model of Lennard-Jones and Devonshire, in particular, has proved very successful in other respects, so that we have some justification for concluding that the observed entropies cannot be explained in terms of central forces, however good the statistical model.

Longuet-Higgins [30] has shown how, under certain circumstances, it is possible to express $(\Delta^* F^{(cent)})_{T, V}$ and other thermodynamic functions in terms of experimentally measurable properties of the reference liquid, thereby eliminating appeal to any statistical model. In particular it is found that

$$(\Delta^* G)_{T, P} : (\Delta^* S)_{T, P} = RT - Q_0 : dQ_0/dT - R, \qquad (5.7)$$

TABLE 2.—EXCESS FREE ENERGIES AND ENTROPIES OF MIXING FOR EQUIMOLAR MIXTURES

	$(\Delta^*F)_{T,\,V}$ (cal mole^{-1})	$T(\Delta^*S)_{T,\,P}$ (cal mole^{-1})	$T(\Delta^*S)_{T,\,V}$ (cal mole^{-1})	$\dfrac{T(\Delta^*S)_{T,\,V}}{(\Delta^*F)_{T,\,V}}$
central forces, lattice theory	—	—	—	0·00
central forces, harmonic oscillator cell theory [21]	—	—	—	0·08
central forces, smoothed potential cell theory [21]	—	—	—	0·00
central forces, Lennard-Jones and Devonshire cell theory [22]	—	—	—	0·10
benzene + cyclohexane (298° K)	74·4 [23]	119·1*	66·2*	0·89
benzene + carbon tetrachloride (298° K)	19·5 [24]	11·2*	10·5*	0·54
cyclohexane + carbon tetrachloride (298° K)	16·7 [25]	21·3*	8·2*	0·49

* calculated using heats of mixing given by Scatchard *et al.*[26] and volume changes measured by Wood *et al.*[27-29]

where Q_0 is the latent heat of conversion of a mole of reference liquid to its vapour at the same temperature and zero pressure. If either of the components is used as reference liquid and the quantities on the right-hand side of (5.7) are obtained experimentally, this equation predicts a larger excess entropy than any of the completely theoretical treatments quoted in table 1 and consequently rather better agreement with experiment. It should be noted, however, that this only means that the observed properties of mixtures can be interpreted in terms of central forces *if the properties of the pure liquids are also interpretable in this way.* In fact dQ_0/dT is closely related to the heat capacity at constant pressure and we have seen in the previous section that this shows a systematic variation which is possibly due to directional forces.

If we now consider the expression (5.4) for the excess free energy due to differences in directional forces, it is immediately clear that $(\Delta^*F^{(or)})_{T,\,V}$ is more temperature dependent than $(\Delta^*F^{(cent)})_{T,\,V}$ because of the extra factor T^{-1}. If the simple lattice distribution function is used for $n_2^{(0)}$, $(\Delta^*F^{(or)})_{T,\,V}$ is given by

$$(\Delta^*F^{(or)})_{T,\,V} = - N(2kT)^{-1} x_A x_B \sum_{l_1=1}^{\infty} \sum_{l_2=1}^{\infty} \sum_m \sum_{i>0}$$
$$\left\{ \left[\zeta_{AB}^{(l_1 l_2:\,m)} (R_i) \right]^2 - \tfrac{1}{2} \left[\zeta_{AA}^{(l_1 l_2:\,m)} (R_i) \right]^2 - \tfrac{1}{2} \left[\zeta_{BB}^{(l_1 l_2:\,m)} (R_i) \right]^2 \right\}, \quad (5.8)$$

or, if only interactions between nearest neighbours are considered,

$$(\Delta^*F^{(or)})_{T,\,V} = - zN(2kT)^{-1} x_A x_B \sum_{l_1=1}^{\infty} \sum_{l_2=1}^{\infty} \sum_m$$
$$\left\{ \left[\zeta_{AB}^{(l_1 l_2:\,m)} (a) \right]^2 - \tfrac{1}{2} \left[\zeta_{AA}^{(l_1 l_2:\,m)} (a) \right]^2 - \tfrac{1}{2} \left[\zeta_{BB}^{(l_1 l_2:\,m)} (a) \right]^2 \right\}, \quad (5.9)$$

a being directly related to the total available volume.

Both these expressions lead to an excess entropy of mixing which has the same sign as the excess free energy and which is such that

$$(\Delta^*F^{(or)})_{T,\,V} = T(\Delta^*S^{(or)})_{T,\,V} = \tfrac{1}{2}(\Delta^*E^{(or)})_{T,\,V}. \quad (5.10)$$

If more accurate statistical models were used allowing for the temperature dependence of $n_2^{(0)}$, these ratios would no doubt be modified, but as these refinements only make small differences to the central force theory, (5.10) is probably substantially correct.

The most important qualitative conclusion to be drawn is that the excess entropy of mixing due to directional force differences is approximately equal to the excess free energy divided by the temperature. This is more in accord with the experimental data of table 2. The observed values of the ratio $T(\Delta^*S)_{T,\,v}/(\Delta^*F)_{T,\,v}$ lie between 0·5 and 1·0 as would be expected if both effects are operative. The entropy change due to directional forces is to be interpreted physically in terms of hindrance to free molecular rotation. If the directional forces between molecules of different species are stronger than those in the pure components, for example, there will be less random orientation in the mixture and consequently less entropy. This has often been suggested as the qualitative explanation of observed excess entropies, but simple quantitative expressions for the effect in terms of intermolecular forces have not previously been put forward. In the remainder of this section, we shall examine the expressions (5.8) or (5.9) in some particular cases to see whether observed data can be interpreted in terms of reasonable intermolecular fields.

(a) *Benzene + cyclohexane.*—This system has been extensively studied [23, 26, 27] and is found to show positive deviations from Raoult's law together with a considerable excess entropy of mixing (table 2). Scatchard, Wood and Möchel [23] suggested that the excess entropy might be due to incomplete randomness of orientation in either of the pure liquids. This might be expected on the grounds that the plane or puckered hexagons fit together better among themselves than with each other.

Neither molecule is truly axially symmetric, but we may consider them to be approximately so about axes perpendicular to the hexagons. Suitable approximate forms for the intermolecular fields are

$$u_{st} = \zeta_{st}^{(00:0)}(r) + 5\,\zeta_{st}^{(22:0)}(r)P_2(\cos\theta_1)P_2(\cos\theta_2) \quad (s,\,t = \text{A or B}), \quad (5.11)$$

where $P_2(x)$ is $\frac{3}{2}x^2 - \frac{1}{2}$ and $\zeta_{st}^{(22:0)}(r)$ are negative so that face-to-face configurations are preferred.

If it is now assumed that the central forces are all the same, that the directional forces only operate between like particles and that the additional intermolecular energy of the face-to-face configurations in both pure liquids is $-q$ at the nearest neighbour separation a so that

$$\zeta_{AA}^{(22:0)}(a) = \zeta_{BB}^{(22:0)}(a) = -\tfrac{1}{5}q, \qquad (5.12)$$

then the expression (5.9) for the excess free energy becomes

$$(\Delta^*F)_{T,\,v}/NkT = 0 \cdot 02z(q/kT)^2\,x_A x_B. \qquad (5.13)$$

Taking $z = 8$ and using the observed value of $(\Delta^*F)_{T,\,v}$ (table 2), we find that $q/kT = 1 \cdot 77$ corresponding to $q = 1050$ cal mole^{-1}. According to (5.10), $(\Delta^*S)_{T,\,v}$ will be equal to $(\Delta^*F)_{T,\,v}/T$ in approximate agreement with observation. These figures are, of course, derived from special assumptions and may not represent the actual interactions at all well, but the calculation does show how the qualitative suggestion of Scatchard, Wood and Möchel can be put in approximate quantitative form.

It should be mentioned that in a later paper Scatchard, Wood and Möchel [25] withdrew their explanation on the grounds that the excess entropy of $C_6H_{12} + C_6H_6$ is considerably greater than the sum of excess entropies for the mixtures $C_6H_6 + CCl_4$ and $C_6H_{12} + CCl_4$. Equality is only to be expected, however, if there are no orientation effects either in pure CCl_4 or in the other two mixtures. As has been pointed out in the previous section, it is by no means certain that CCl_4 acts as a spherically symmetric molecule.

(b) *Simple polar molecules.*—If the directional forces between simple polar molecules can be represented by the interactions of dipoles and quadrupoles, then eqn. (5.8) or (5.9) lead to simple expressions for $(\Delta^*F^{(or)})_{T,\,v}$. As in the corresponding treatment of pure liquids, we shall use a face-centred cubic lattice. If only the dipole moments μ_A and μ_B are important, we find, using (2.4),

$$(\Delta^*F^{(or)})_{T,\,v}/NkT = 1 \cdot 2045[N(\mu_A^2 - \mu_B^2)/VkT]^2 x_A x_B. \qquad (5.14)$$

the corresponding energy and entropy of mixing being given by (5.10). If the quadrupole moments Θ_A and Θ_B also have to be taken into account, (5.14) has to be replaced by

$$(\Delta^* F^{(or)})_{T,\,V}/NkT = \{1\cdot2045[N(\mu_A^2 - \mu_B^2)/VkT]^2$$
$$+ 2\cdot540\,N^{8/3}(\mu_A^2 - \mu_B^2)(\Theta_A^2 - \Theta_B^2)/V^{8/3}k^2T^2$$
$$+ 2\cdot716[N^{5/3}(\Theta_A^2 - \Theta_B^2)/V^{5/3}kT]^2\}x_A x_B. \qquad (5.15)$$

Eqn. (5·14) can be tested by calculating the free energy of mixing from observed vapour phase dipole moments. The results for two simple mixtures are compared with the observed values of $(\Delta^* G)_{T,\,P}$ [31] (which should be approximately equal to $(\Delta^* F)_{T,\,V}$) in table 3. (V is taken as the mean of the molar volumes of the pure components.)

TABLE 3.—EXCESS FREE ENERGY FOR POLAR-NON-POLAR MIXTURES
(EQUIMOLAR COMPOSITION)

mixture	μ_A (debyes)	V (cm³)	$(\Delta^* F)_{T,\,V}$ (calc) (cal mole⁻¹)	$(\Delta^* G)_{T,\,P}$ (obs) (cal mole⁻¹)
$CHCl_3 + CS_2$	1·05	73·4	9	76
$(CH_3)_2CO + CS_2$	2·74	68	460	267

The large difference between the two calculated values is due to the fact that the expression (5.14) varies as the fourth power of μ. It would appear reasonable to attribute the mixing properties of acetone + carbon disulphide mainly to the dipolar interaction, but other types of intermolecular force must play a significant part in the system chloroform + carbon disulphide.

1 Pitzer, *J. Chem. Physics*, 1939, **7**, 583.
2 Pople (to be published).
3 Stockmayer, *J. Chem. Physics*, 1941, **9**, 398.
4 Rowlinson, *Trans. Faraday Soc.*, 1949, **45**, 984.
5 Pople, *Proc. Roy. Soc. A*, 1952, **215**, 67.
6 Lennard-Jones and Ingham, *Proc. Roy. Soc. A*, 1925, **107**, 636.
7 Frank and Clusius, *Z. physik. Chem. B*, 1939, **42**, 395.
8 Clusius, *Z. physik. Chem. B*, 1936, **31**, 459.
9 Clusius and Riccoboni, *Z. physik. Chem. B*, 1937, **38**, 81.
10 Clayton and Giauque, *J. Amer. Chem. Soc.*, 1933, **55**, 4875.
11 Brown and Manov, *J. Amer. Chem. Soc.*, 1937, **59**, 500.
12 Giauque and Wiebe, *J. Amer. Chem. Soc.*, 1928, **50**, 101.
13 Giauque and Powell, *J. Amer. Chem. Soc.*, 1939, **61**, 1970.
14 Giauque and Blue, *J. Amer. Chem. Soc.*, 1936, **58**, 831.
15 Messerly and Aston, *J. Amer. Chem. Soc.*, 1940, **62**, 886.
16 Overstreet and Giauque, *J. Amer. Chem. Soc.*, 1937, **59**, 254.
17 Oliver, Eaton and Huffman, *J. Amer. Chem. Soc.*, 1948, **70**, 1502.
18 Lord and Blanchard, *J. Chem. Physics*, 1936, **4**, 707.
19 Hildebrand, *J. Chem. Physics*, 1947, **17**, 727.
20 Guggenheim, *Mixtures* (Oxford University Press, 1952).
21 Prigogine and Mathot, *J. Chem. Physics*, 1952, **20**, 49.
22 Pople, *Trans. Faraday Soc.* (in press).
23 Scatchard, Wood and Möchel, *J. Physic. Chem.*, 1939, **43**, 124.
24 Scatchard, Wood and Möchel, *J. Amer. Chem. Soc.*, 1940, **62**, 713.
25 Scatchard, Wood and Möchel, *J. Amer. Chem. Soc.*, 1939, **61**, 3206.
26 Scatchard, Ticknor, Goates and McCartney, *J. Amer. Chem. Soc.*, 1952, **74**, 3721.
27 Wood and Austin, *J. Amer. Chem. Soc.*, 1945, **67**, 480.
28 Wood and Brusie, *J. Amer. Chem. Soc.*, 1943, **65**, 1891.
29 Wood and Gray, *J. Amer. Chem. Soc.*, 1952, **74**, 3729.
30 Longuet-Higgins, *Proc. Roy. Soc. A*, 1951, **205**, 247.
31 Hirschberg, *Bull. Soc. chim. Belg.*, 1932, **41**, 163.

Polymer Science

Anthony J. Ryan

University of Sheffield, Sheffield, UK S3 7HF

Commentary on: **Organization of macromolecules in the condensed phase,** F. C. Frank, *Faraday Discuss. R. Soc. Chem.*, 1979, **68**, 7–13.

Polymer science is a fortunate strand of physical chemistry in that it also lies in the domain of condensed matter physics, materials science, molecular biology and chemical engineering, to name but a few other disciplines. *Faraday Discussions* in this area attract a wide range of contributors from the diverse population interested in synthetic and natural polymers and the announcement of a new *Faraday Discussion* is always greeted with great excitement, just in case "it lives up to '79". The field has a tendency to be combative and the confrontations provoked by *Faraday Discussions* are often seized upon with great relish. The only other conferences that come close are the Gordon Research Conferences, but, because there are no written papers distributed beforehand and they rely on 40 minute talks, the axes that are ground tend to be old and dull. Moreover, they particularly suffer from the recent tendency to political correctness which has even made some *Faraday Discussions* rather bland.

The best example of a *Faraday Discussion* providing a stage for a great fight was FD68 held in 1979. The field of polymer structure, particularly polymer crystallisation, had been very active up to this point, and there were a number of issues to be resolved. All the most influential people of the time were assembled and a look through the attendance list reveals a large number of those active today. Unfortunately, I was taking my A-levels at the time and was more interested in rugby and The Clash than polymer morphology, but I have felt the repercussions of this meeting throughout my research career and this has prompted my choice.

There are many great papers from FD68 that stand the test of time, Andrew Keller provided an excellent and even-handed review of the field[1] that set the scene for the contributions by Hoffman *et al.*,[2] and Yoon and Flory,[3] that are mutually opposed in their conclusions, but provide the basis for our understanding of chain folding. Both models have aspects that are unphysical, and in the case of Flory's paper this was first pointed out in the discussion by Bob Stepto. This was my first link to this meeting, as Stepto became my undergraduate tutor the following year! The ubiquity of the reptation model for polymer dynamics was only starting to be appreciated at the time and a number of papers in the volume introduced them to the industrially relevant area of crystallisation from the melt, notably those by Hoffman *et al.*[2] and Klein *et al.*[4] The recorded discussion in this volume is an illuminating read, and I only hope that the passion with which these scientists held their views will continue to be a feature of our work for the foreseeable future.

The contentious issues discussed here relied on new experimental techniques, particularly the development of small angle neutron scattering (SANS) for measuring polymer dimensions.[2,3,5] The theories and models discussed were founded on data from careful scattering experiments on specially synthesised, deuterium-labelled polymers and demonstrate a breadth of skills that Faraday himself would have admired. The paper I choose that exemplifies the contribution of the Faraday Society to the development of the field, however, is none of the above. It is the "General Introduction" by Sir Charles Frank[6] because it has all the unique qualities of *Faraday Discussions* and highlights certain problems that are still under discussion to this day. The opening paragraph sets the scene and the author certainly doesn't "adopt a neutral stance as between right and wrong". In fact he takes it upon himself to point out the ridiculous nature of some of the models that were about to be presented. This was "getting your retaliation in first" and exactly the behaviour adopted by the British Lions rugby teams on tour in the southern hemisphere around this time. The polymer crystallisation field has undergone a resurgence in activity in recent times and the issues now are still some of those highlighted by Frank. How does the polymer disentangle as it forms crystallites? How is the process affected by flow? What is the nature of the density change from crystalline to amorphous regions? How do these crystals nucleate? The partially crystalline state of polymer materials will continue to engage scientists in theory and experiment for quite some time.

This account of the significance of the paper has dwelled upon its sociological context, the role of arguments and the nature of dialogue. It should be emphasised, however, that throughout the paper there are very clear theoretical arguments and beautiful analogies with delightful connections to everyday life, moderated by the physicists renormalisation through time-scales. Finally, there is a magnificent use of language, demonstrating the precision of German compared to the subtleties of English and, as far as I am aware, the only use of the word "perspicacious" in a scientific publication. I present to you a scholarly piece of work that stands as a landmark in polymer science. Furthermore it is important, both as the introduction to a volume, and in the context of the history and philosophy of science. As Sir Charles tells us "Only Gibbs and God made no mistakes…Let the arguments begin!"

References

1 A. Keller, *Faraday Discuss. R. Soc. Chem.*, 1979, **68**, 145–66.
2 J. D. Hoffman, M. Guttman and E. A. DiMarzio, *Faraday Discuss. R. Soc. Chem.*, 1979, **68**, 177–197; C. M. Guttman, J. D. Hoffman and E. A. DiMarzio, *Faraday Discuss. R. Soc. Chem.*, 1979, **68**, 297–309.
3 D. Y. Yoon and P. J. Flory, *Faraday Discuss. R. Soc. Chem.*, 1979, **68**, 288–296.
4 J. Klein and R. C. Ball, *Faraday Discuss. R. Soc. Chem.*, 1979, **68**, 198-209.
5 M. Stamm, E. W. Fischer, M. Dettenmaier and P. Convert, *Faraday Discuss. R. Soc. Chem.*, 1979, **68**, 263–278.
6 F. C. Frank, *Faraday Discuss. R. Soc. Chem.*, 1979, **68**, 7–13.

General Introduction

By F. C. Frank

H. H. Wills Physics Laboratory, University of Bristol,
Royal Fort, Tyndall Avenue, Bristol BS8 1TL

Received 25th September, 1979

In this general introduction to the Discussion I look on it as my task to highlight those points where it seems to me that we can reach decision, or at least clarify the lines of enquiry which can lead us to decision. In doing this, I see it as no part of my task to adopt a neutral stance as between right and wrong.

Basically, this is a discussion on certain aspects of the fascinating general topic of restricted randomness, particularly in application to long-chain polymers.

In the first section, up to Uhlmann's paper, we are concerned with polymer melts in equilibrium. In dilute solution, the dominant restriction on randomness, apart from the very fact that the polymer molecule *is* a chain, is *self-exclusion*, and Paul Flory taught us how to cope with that many years ago. In the interior (and I emphasize interior) of the pure amorphous phase, *mutual exclusion* has a large effect on the total entropy, but its effect on molecular conformations is the relatively minor one of virtually cancelling the effects of self-exclusion. That knowledge we also owe to Flory. Hence a limited number of parameters suffice both to describe and explain the conformations in this case. Uhlmann's paper dismisses for us the aberrant nodular structures which have been proposed: there only remains to ask how in certain circumstances the *appearance* of nodular structure can be produced.

What I have said about the pure amorphous phase is not, logically speaking, contradicted by the paper by Pechhold and Grossman, because according to their theory the polymer melt is not an amorphous phase but rather a remarkable kind of cubic mesophase. To justify that they have to use a modification of the rules of statistical mechanics, the cluster-entropy-hypothesis, and how they can justify that I do not know. No doubt we shall come to that in the discussion. For myself I will take one valuable point from that paper, a warning to show what bizarrely different models can be deemed consistent with the same diffraction evidence.

Many more subtleties arise when we have to deal with phase boundaries, and particularly the boundary between crystalline and disordered polymer, whether the latter is the phase from which a crystal grows, or the disordered layer at the surface of crystal lamellae in the final state. This, in one way or another, is what we are concerned with in much the largest section of our Discussion, from Point's paper almost to the end. I would like to emphasize that the most important restriction on randomness in these situations, whether we are considering statistical equilibrium or kinetic restrictions on the attainment of equilibrium, is not self-exclusion but *mutual exclusion*.

In this connection we have become accustomed to the famous switchboard analogy. I always have a great deal of sympathy for that poor girl, the telephone operator trying to get all the sockets plugged up on Flory's switchboard, when the plugs and cables have the same thickness as the centre-spacing between plug-holes which are in close-packed array, and the cables are required to run in a way which is in some sense random, and she is told she may use only a very few double plugs connecting adjacent holes. Of course she'll never do it, poor girl, and I'm surprised that after twenty years or so of trying she hasn't gone on strike.

8 GENERAL INTRODUCTION

I said that was an impossible model at the Cooperstown Conference on Crystal Growth in 1958.[1] Explicitly, we were talking about the nucleus for polymer crystallization, but the point at issue was the same. I called it an impossible model, and I quote, " because if these fringing chains are not in crystalline packing they need more cross-sectional area per chain than they do in the crystal ". Flory made the curious response that all models which had been presented could be reasoned to be impossible. I think that is an inadequate excuse for going on presenting impossible models. Fig. 3 of the paper by Flory and Yoon in this Discussion is a classic example of the impossible pictures which have continued to appear in the polymer literature, presented by a wide variety of authors, from then to now. Guttman *et al.* show, in the next paper, that the computer-generated model of Yoon and Flory, which fits the neutron-scattering evidence, is likewise sterically impossible.

Let me attempt a quantitative estimate of the overcrowding factor which makes these models impossible. In doing that I shall be covering much the same ground as DiMarzio, but I prefer to do it in my own words; after all, I did get my foot into this doorway first, only I failed to perceive the necessity of belabouring what seemed an obvious point.

I start with the assumption that, at an interface normal to the chain directions in the crystal, all chains enter the disordered region, as represented in fig. 3 of Yoon's paper. Then I see the overcrowding as the product of three factors, D, A and B, (fig. 1).

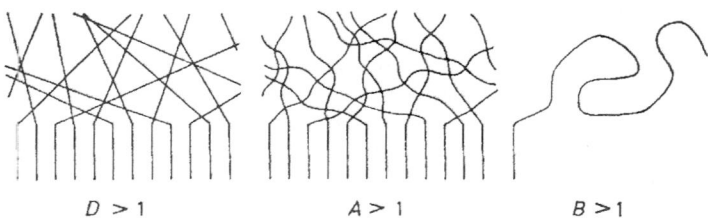

$D > 1$ $A > 1$ $B > 1$

Fig. 1.—Density factors at a crystal–amorphous boundary with through-going chains. $D > 1$: directional randomization; for anisotropy, $D = 2$. $A > 1$: avoidance; say, $A \approx 1.5$. $B > 1$: backtracking; $B \approx N^{\frac{1}{2}}$ (what is N?). Product $BAD \approx 3$.

$D > 1$: *the factor from randomization of direction.* If the chains all parallel in the crystal are still straight in the amorphous region, but have an isotropic distribution of directions, then $D = 2$, a result first obtained in the polymer context by Flory, I believe. Note in passing that if the emerging chains randomized their directions over the hemisphere we should have $D = \infty$, but the distribution would not be isotropic: to produce isotropy the probability of changing direction by θ from the normal must be weighted by a factor $\cos \theta$; that's how the sun comes to look about equally bright across the whole disc.

$A > 1$: *the avoidance factor.* Straight chains in random directions will intercept each other: to avoid this they must make deviations increasing the required length by a factor A which I will estimate as ≈ 1.5.

$B > 1$: *the back-tracking factor.* If the path is anything like a random walk it makes decreases as well as increases in all coordinates. This increases the chain density by a factor B, which for truly random walks would be of order $N^{\frac{1}{2}}$ where N is the number of persistence lengths in a loop or tie-chain. We could estimate $N^{\frac{1}{2}}$ as the thickness of the disordered layer measured in persistence lengths, but I will just leave it as $B > 1$.

Hence on the initial assumption, the density of the disordered layer exceeds that of the crystal by the product BAD, in which all three factors are certainly >1, and the product is >3 if it is remotely reasonable to apply the word " amorphous " in description of that disordered layer. I think there will be universal consent that the density of the disordered region, so far from being more than 3 times greater than that of the crystal should instead be something like 10% less. Where do we find a countervailing factor of at least 10/3?

The answer is very simple. Putting aside crystal chains which terminate at the interface (only significantly available in low molecular weight material) there are two possibilities (fig. 2).

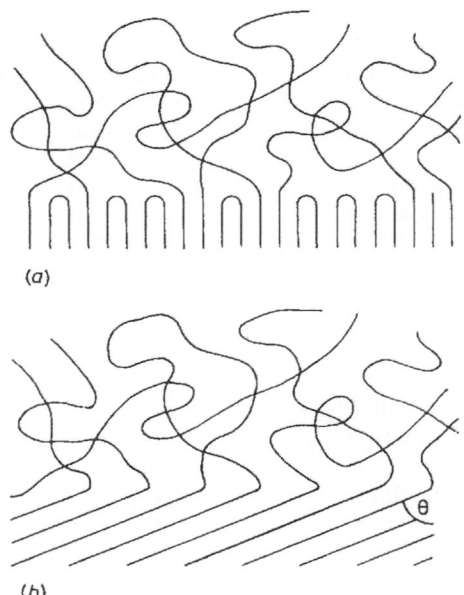

(a)

(b)

Fig. 2.—Alternative resolutions of the density paradox
$(1 - p - 2l/L) \cos \theta = 0.9/BAD \leqslant 3/10$.

One, if the interface is normal to the crystalline chains, is that at least $(3 - 0.9)/3$ of them, *i.e.* 70%, fold back immediately, at, or even before, the interface. If the fold occurs before the interface the resulting two-chain space near the surface will draw in chains from the disordered region, but I think that makes an energetic defect which will not be very common.

The alternative, if there is no backfolding, as I said at Cooperstown in 1958, is to have an interface oblique to the chain axes. The required obliquity is at least arc sec $(3/0.9) = 72.5°$. Lamellae with tilted chains are familiar, but never with a tilt nearly as large as this.

The general formula, with a back-folding probability p and an obliquity θ is

$$(1 - p - 2l/L) \cos \theta = 0.9/BAD \leqslant 3/10$$

where the term $2l/L$, with l the crystalline stem-length, L the full length of a chain, makes the greatest possible allowance for chain ends at the interface.

One may enquire whether back-folding can be avoided, with a mean interface perpendicular to the crystal chains, by giving it a steep city-scape profile of roofs and steeples. The answer is No: overcrowding is relieved on the roofs, but is worse than ever in the valleys, the streets of the city.

That, I think, is about as much as purely steric arguments can tell us about our problem. No kind of microscopy can help us, at least until someone invents a neutron microscope: only for neutrons can we label individual molecules to make them distinguishable, without excessively modifying the intermolecular forces, and even in that case, using deuteration as the label, that difficulty is still with us. X-rays and electrons will distinguish crystalline and non-crystalline regions for us, and indeed tell us the structure in the crystalline regions, but to get at the configuration of single whole molecules we have to make use of neutron scattering with deuteration-labelling, despite its faults. To avoid segregation of labelled from unlabelled material one resorts to rapid quenching. Until we fully understand the mechanism of the crystallization process we cannot know how that affects the resulting conformations, and in any case it limits the range of conditions in which conformations can be studied.

The simplest mode of analysis of neutron scattering data gives us a radius of gyration. That is essentially the square root of the ratio of the second to the zeroth moment of a distribution function. If you have ever looked at the question how well is a function of any complexity defined by knowledge of its first few moments, you will be aware that that is very limited knowledge indeed. Nevertheless a result has emerged from these analyses, that for a variety of different polymers, of various molecular weights, the radius of gyration is nearly the same in the crystallized state as in the melt: a result of sufficient universality to indicate that it must mean something. But the meaning is not necessarily, I think, that which it suggests to Fischer, his *Erstarrungsmodell*. He calls it *solidification model* here, but I think the German word gives me a slightly more definite idea of what he means, which seems to accord also with Flory's beliefs: but in either language that is a very brief specification of what has to be a very complicated process. I would like to see a much fuller analysis of the necessary details of its mechanism. The flat uniformity of the experimental result is enlivened by the results reported by Guenet *et al.* in their paper for isotactic polystyrene which is found fortunately free from the complication of segregation of the deuterated species. The radius of gyration increases on crystallization to only a small extent when the molecular weight is 2.5×10^5, increases by 40% for molecular weight 5×10^5, both in a matrix of molecular weight 4×10^5, and decreases about 10% for molecules of molecular weight 5×10^5 in a matrix of molecular weight 1.75×10^6. Something more interesting than simple *rigor mortis* is happening there.

When we use the neutron-scattering data more completely, what we can obtain is the mean square Fourier transform of the distribution functions for individual molecules: and there is no uniqueness theorem for the problem of inverting that. Even if there was, knowing the distribution functions would not tell us the conformations. All we can do is to make models and see whether they will fit the scattering data within experimental error. If they don't, they are wrong. If they do, they are not necessarily right. You must call in all aids you can to limit the models to be tested. It is essential that they should pass tests of steric acceptability, as everyone who uses the corresponding trial-and-error method for X-ray crystal structure determination knows.

Keller and Sadler's model fits the shape of the scattering curve, but fails by a factor of two in absolute intensity: but how reliable are the absolute intensities in neutron-scattering measurements? A factor of two is quite a lot to laugh off. Yoon and Flory produce a computer-generated stochastic model which buys agreement in shape and intensity at the expense of unacceptably large variation in real space density, as

Guttman shows. It must be rejected. We have a good many other models before us in this Discussion, and I must leave it to the experts in the subject to try and thrash out the question of which of them are not demonstrably wrong.

In making models, we must respect the principles of equilibrium statistical mechanics, but cannot wholly rely on them, since we have every reason to believe that in polymer crystallization we have only a frustrated approach towards thermodynamic equilibrium. Most of the models are in some way or another founded on their authors' conceptions of the nature of the process of high-polymer crystallization from the melt. That is a process on which direct detailed information is hard to get, and some imaginative extrapolation from what one knows about related problems is almost unavoidable.

In the work of Kovacs on low molecular weight poly(ethylene oxide) the combination of a favourable material and brilliant technique tells us how complicated a real polymer crystallization process can be, and gives us much insight into what, at any rate, chains of modest length can do. It would be helpful for the problem of melt crystallization if we fully understood crystallization of polymers from dilute solution, but we don't. I think we still don't understand why the rate of crystal growth is proportional to a fractional power of the concentration, as shown by Keller and co-workers.[23] Mandelkern showed more than twenty years ago that the growth kinetics implied repeated surface nucleation. Lauritzen showed that that implied a transition between what he called regime I and regime II kinetics, and there I myself finished off the solution of the problem which Lauritzen had started using tricks first taught me by Burton and Cabrera (and, *pace* Hoffman, my result is *not* unwieldy).[4] Hoffman *et al.* appear to have found that transition, but the puzzle then is to identify the defects which by implication dissect the growth front into half-micron segments. I feel that there is some key idea still missing from the picture, and perhaps Point's new look at the problem of crystallization from dilute solution will point the way forward.

These, however, are all laboratory problems. What matters practically is crystallization from the melt. Flory has put forward a kinetic argument to show that close folding is impossible. If it were true, one would wonder how crystallization of any kind is possible, but his argument is refuted both by Hoffman *et al.* and by Klein, pointing out that he has overlooked de Gennes' process of reptation: the ability of a chain to worm its way longitudinally through the tangle. I like pictorial analogues which aid thought, like the telephone switchboard which helps me to perceive the impossibility of some of Flory's models. My analogue for the polymer melt is a pan of spaghetti. If I can shake the spaghetti pan three times a second, and a characteristic frequency of thermal agitation is 3×10^{11} Hz, then a millisecond for the polymer melt is equivalent to 3 years for the spaghetti. I doubt whether 3 years shaking is long enough for standard length unbroken spaghetti, which corresponds to a medium high molecular weight polymer, in length to diameter ratio, to reach an equilibrium degree of entanglement: but if I get one piece of spaghettti between my lips, and gently suck while I go on shaking, I think I shall have it out within a minute: and the free energy of crystallization provides just that suction. The resistance to extraction rises initially as we pull out un-pinned loops further and further along the chain, and if the molecular weight is too high the crystal may lose patience, stop pulling on that chain and go to work on another. The first one remains attached to the crystal, the tension in it relaxes, but it cannot start crystallizing again except by a surface nucleation event or the arrival of another growth step. If it utilizes the latter it leaves another chain dangling, and so on. Klein's estimates put the typical experiment in polymer melt crystallization rather nearer to his transition to high molecular weight behaviour than Hoffman's estimates do. Guenet's remarkable change in behaviour between isotactic polystyrene matrices of 4×10^5 and 1.75×10^6 molecular weight may indicate passage beyond this transition limit.

Equilibrium statistical mechanics may still have something useful to teach us in this problem. At that same discussion at Cooperstown in 1958 Flory made the perspicacious remark " (the crystal) surface presents an impenetrable barrier to the random coil, and this restricts the statistical possibilities of the coil. The problem resembles that encountered in treating the surface tension of a dilute polymer solution ". I do not believe the consequences of that remark have been sufficiently followed up.

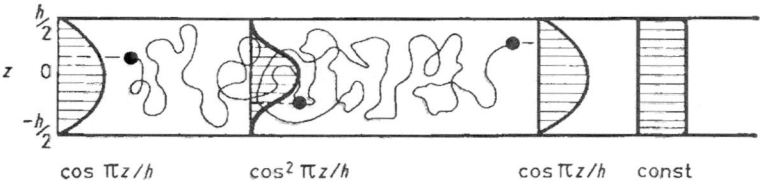

FIG. 3.—Distribution of end-points and of points not near ends, for a random chain between impenetrable walls; $h \ll (L\lambda)^{\frac{1}{2}}$.

Let me consider a simple problem which may have some relevance to the state of affairs in the disordered layer between crystal lamellae. Consider first a long-chain polymer molecule in dilute solution, confined in the gap of width h between impenetrable walls at $z = +h/2$. Let the persistence length be λ and the full length of the chain be $N\lambda$. Now, neglecting self-exclusion corrections, which should make no drastic difference, it is easy to show that the distribution function in z for either end point of the chain is proportional to cos $\pi z/h$. It is maximum in the middle of the gap, because configurations which end there suffer least rejection of phantom configurations, those calculated without regard for the presence of the boundary walls, which must be rejected for infringement of the non-crossing boundary conditions; and it is zero at the walls because the probability of such infringement, high everywhere, is virtually infinite for configurations terminating there. Now consider a point in the mid-range of the chain, further, along the chain, from either end than h^2/λ. Such a point is the end-point of two long part-chains, and the configuration number is the product of the configuration numbers of the two parts. The distribution of such points in z is therefore proportional to $\cos^2 \pi z/h$. By my initial assumption the greater part of the chain belongs to this mid-range and should have this distribution. Now, instead of dilute solution, let us have molten polymer in the gap. If chains still behaved in the same way that would be the full density of matter. We should have twice the mean density in the middle, and virtual emptiness near the walls. But intuition, which I strongly trust in this case, tells me the density should be much more nearly uniform. Evidently, where boundaries are present, mutual exclusion does something much more drastic than just to cancel the effects of self-exclusion. How does one modify the theory to make it talk sense?

Formally, the answer is to bring in Boltzmann factors: to assign a potential energy $V_i(r)$ to each segment of a chain and hence factors exp $[-V_i(r)/kT]$ for each segment in the statistical weight for any configuration. The consequences are not simple, especially as usually presented, and a nice little recent paper by DiMarzio and Guttman[5] attempts to set them out in simple terms. It is an instructive paper, but I think it is partly wrong. In explaining the use of a potential to represent mutual exclusion, they equate exp $(-V_i/kT)$ to the " fraction of emptiness ". I think that cannot be right. Space is equally occupied by separate molecules of solvent, or similar molecules tied together as polymer chains, but the consequences are entirely different, as my example shows. Statistically, of course, the effect of chain connection comes from the fact that if a chain is excluded from one point, it has a correlated partial exclusion from neighbouring points: but I would like to think of it from another point of view, namely from the fact that chains can transmit non-hydrostatic stress and so maintain pressure gradients. After all, I can discuss the free energy of deformation of rubber entirely in terms of constraints on the randomness of chain configuration, without ever mentioning stress: but stress is still a valid concept in rubber, and has to obey the laws of stress continuity known to us in elasticity theory. In my example there is certainly a z-wise compressive stress, because the walls constrain the chain configurations more the closer they are together. That stress, $-\sigma_{zz}$, should be uniform in z. I would like to think there are transverse tensile stresses σ_{xx} and σ_{yy} in thin layers near the walls sucking the segments down towards those regions. If I may think of it that way those potential energies V_i become more real, much less of a formal device. Will somebody tell me whether I may? What we actually need to flatten this density distribution is an attractive potential $V_i = -kT \ln 2$, for a monolayer of segments, adjacent to the wall. Di Marzio will recognize that.

Several other themes of considerable interest emerge from papers in this Discussion which I have not found time to mention but I hope I have said enough now to start a few arguments. I expect to be told where I have been mistaken. Only Gibbs and God made no mistakes, as the Russians say, though they don't like to be quoted when they say it. Let the arguments begin!

[1] *Growth and Perfection of Crystals* (Proceedings of an International Conference on Crystal Growth held at Cooperstown, New York, August 27–29, 1958), ed. Doremus, Roberts and Turnbull (John Wiley, New York, 1958), discussion remarks of F. C. Frank and P. J. Flory, pp. 529 and 530.
[2] D. J. Blundell and A. Keller, *Polymer Letters*, 1968, **6**, 433.
[3] A. Keller and E. Pedemonte, *J. Crystal Growth*, 1973, **18**, 111.
[4] F. C. Frank, *J. Crystal Growth*, 1974, **22**, 233.
[5] E. A. DiMarzio and C. M. Guttman, *J. Res. Nat. Bur. Standards*, 1978, **83**, 165.

Colloids

Brian Vincent

Bristol Colloid Centre, School of Chemistry, University of Bristol, UK BS8 1TS

Commentary on: **J. Th. G. Overbeek and M. J. Sparnaay**, "London-van der Waals Attraction between Macroscopic Objects", *Discuss. Faraday Soc.*, 1954, **18**, 12–24.

The foundations for the quantitative interpretation of the stability of colloidal dispersions were laid down, independently, by Derjaguin and Landau[1] in Russia, and by Verwey and Overbeek[2] in the Netherlands, during the years of the Second World War. However, because of the difficulties imposed by occupation, the Dutch work was not published (as a monograph*) until 1948. The "DLVO theory", as it is now popularly and universally known, marked a turning point in colloid science. Not only has the stability/aggregation behaviour of dispersions of charged particles (including mixtures of oppositely charged particles) been successfully interpreted, at least semi-quantitatively, in terms of its basic tenets, but also the stability of all types of thin liquid films, both symmetric and asymmetric. Indeed, a veritable "industry" of research publications was spawned by this theory, testing and extending its basic ideas, and this activity has still not ceased today. The basic concept is that the interaction between two (charged) like bodies, across an intervening medium, comprise two parts: (a) the attractive van der Waals interaction between those bodies (moderated in a liquid medium) and (b) the repulsive (screened in a liquid medium) Coulombic interaction between the charged surfaces. The calculation of the van der Waals attraction between two bodies, had its origins in the earlier work of de Boer[3] in 1936 and Hamaker[4] in 1937, following London's earlier, quantum mechanical derivation of the dispersion forces between molecules in 1930.[5] The evaluation of the electrostatic repulsive forces, between two adjacent charged surfaces, followed on naturally from the Gouy–Chapman theory of the electrical double-layer for isolated charged surfaces in a liquid (usually aqueous) medium.

The 18th Discussion of the Faraday Society, held in Sheffield in September 1954, entitled "Coagulation and Flocculation", was not the first to be held in the colloid area. This honour must go to the General Discussion on "Colloids and Their Viscosity" held in 1913 at Burlington House,[6] at which one of the "forefathers" of European Colloid Science, Dr Wolfgang Ostwald, gave the introductory lecture (James W. McBain, the first Leverhulme Professor of Physical Chemistry at Bristol, was also present). The 1954 meeting was seminal, and every 12 years since a similar Faraday Discussion Meeting has been held on a closely related topic, namely:

1966: "Colloid Stability in Aqueous and Non-Aqueous Media" (**42,** Nottingham)
1978: "Colloid Stability" (**65,** Lunteren, The Netherlands)
1990: "Colloidal Dispersions" (**90,** Bristol)
2002: "Non-Equilibrium Behaviour of Colloidal Dispersions" (**123,** Edinburgh).

Of course, there have been other Faraday Discussions on topics in the colloids area in between, but let us hope someone does remember to organise a colloid stability meeting in 2014: I am sure there will still be many new ideas to discuss !

One reason that the 1954 meeting was seminal is that it was the first occasion on which Derjaguin and Overbeek met. They presented the first two papers at the meeting. Overbeek's paper is the one reprinted here; Derjaguin's paper[7] is titled "Investigations of the Forces of Interaction of Surfaces in Different Media and their Application to the Problem of Colloid Stability", and is primarily concerned with the stability of soap films. Derjaguin was clearly annoyed with Verwey and Overbeek for seemingly "ignoring" his 1941 paper with Landau, and says as much in a footnote to his paper[7] on p. 26. Verwey and Overbeek proffered an apology in the ensuing "general discussion" (p. 180) and said that they were unaware of the Russian papers and state that "the work for our monograph was performed during the war when the German occupation of our country cut us off from all Allied information".

To return to Overbeek and Sparnaay's paper, I selected this because it describes the first experiments aimed at directly measuring the forces between macroscopic solid surfaces (optically flat, polished glass plates), in this case across a vacuum. The experiments were carried out in the basement of the famous van't Hoff Laboratory, at Sterrenbos 19, in Utrecht. When I visited there as a young postdoc in 1969 the apparatus was still there, and I distinctly remember Overbeek saying that they had to perform the measurements at night, when the trains at the nearby Utrecht Central Station had stopped running! The main aim of these experiments was to test de Boer's and Hamaker's expression for the van der Waals force (F) between two parallel flat plates, as a function of their separation (d),

$$F = \frac{AO}{6\pi d^n}$$

where $n = 3$, O is the area of the plates, and A is the "force constant", subsequently named the "Hamaker constant". Overbeek and Sparnaay wanted to check the predicted d^3-dependence of F. They found $n = 3 \pm 0.3$, seemingly in excellent agreement - although the experimental data (presented in their Fig. 3) are somewhat scattered. They also found that $A = 3.8 \times 10^{-18}$ J. They had predicted a value from the de Boer–Hamaker theory of 10^{-19} J. The explanation that Overbeek and Sparnaay suggested for this difference was that the de Boer–Hamaker theory underestimates A. They pointed to the weakness of the "additivity" assumption made in this theory, and set out to correct for this in Part 2 of their paper. However, a more recent estimate[8] for A for fused quartz, based on the Lifshitz theory, is 6.5×10^{-20} J, which is actually in reasonable agreement with the original de Boer–Hamaker theoretical value. So it would seem, after all, that the experimental value obtained by Overbeek and Sparnaay is in fact an overestimate, even allowing for the fact they used glass rather than quartz. Nevertheless their work was an outstanding, landmark achievement.

Other attempts at measuring van der Waals forces followed in the wake of the Overbeek and Sparnaay work. Derjaguin himself in the mid 1950's[9] measured the forces between a convex lens and a flat glass surface in a vacuum, over separations ~100 nm to 1 μm (*cf.* Overbeek and Sparnaay: 600 nm to 1.8 μm). In this work an electrobalance was used to measure the forces and an optical technique to determine the separation. They obtained an A value within 50% of the predicted Lifshitz value. Tabor and Winterton (in 1969)[10] and Israelachvili and Tabor (in 1972–3)[11] designed equipment for measuring the van der Waals forces between molecularly smooth mica surfaces in air or vacuum., down to separations of 1.5 nm; they confirmed the Lifshitz predictions. This led on to the famous "Surface Forces Apparatus" (SFA), designed and marketed by Israelachvili,[12] for measuring the forces between crossed mica surfaces immersed in a liquid medium. The distance sensitivity achieved was now ~0.1 nm, and the force sensitivity was ~10^{-8} N (in the Overbeek–Sparnaay apparatus the force sensitivity was ~10^{-6} N).

Other methods have appeared more recently for measuring forces between macroscopic surfaces or between a (large) colloidal particle and a surface, immersed in a liquid. These include the total internal reflection microscope in 1990 (TIRM)[13] and the atomic force microscope in 1991 (AFM).[14,15] With TIRM, incredibly weak forces can be measured (~10^{-15} N), whilst with AFM forces ~ 10^{-10} N can be determined. With the development of optical tweezers,[16] we now have the ability to measure forces directly between two colloidal particles. Using these latest techniques, not only may *interaction* forces between surfaces be measured but, by performing dynamic measurements, the *hydrodynamic* forces can also be examined. We are now at a stage surely undreamt of by Theo Overbeek 50 or so years ago when he made his own measurements. It is surely fitting that he has lived to witness all this, and that he has reached an age almost commensurate with that of the Faraday Society/Division itself!

* This monograph has recently been republished.[17]

References

1 B. V. Derjaguin and L. Landau, *Acta Phys. Chim. URSS* , 1941 **14** 633 [2]
2 E. J. W.Verwey and J. Th. G.Overbeek, *Theory of the Stability of Lyophobic Colloids*, Elsevier, Amsterdam, 1948.
3 J. H. de Boer, *Trans. Faraday Soc.*, 1936, **32,** 21.
4 H. C. Hamaker, *Physica*, 1937, **4,** 1058.
5 F. London, *Z. Phys.*, 1930, **63,** 245.
6 *Trans. Faraday Soc.*, 1918, **9,** 34.
7 B. V. Derjaguin, A. S. Titijevskala, I. I. Abricossova, and A. M. Malkina, *Discuss. Faraday Soc.*, 1954, **18,** 24.
8 D. B.Hough and L. R.White, *Adv. Colloid Interface Sci.*, 1980, **14,** 3.
9 B. V. Derjaguin, I. I. Abricossova and E. M. Lifschitz, *Q. Rev. Chem.*, 1956, **10,** 295.
10 D. Tabor and R. H. S. Winterton, *Proc. R. Soc., London, Ser. A*, 1969, **312,** 435.
11 J. N. Israelachvili and D. Tabor, *Proc. R. Soc., London, Ser. A*, 1972, **331,** 19.
12 J. N. Israelachvili and G. E. Adams, *J.Chem.Soc. Faraday Trans.*, 1978, **74,** 975.
13 D. C. Prieve and N. A. Frej, *Langmuir*, 1990, **6,** 396.
14 P. K. Hansma, V. B. Elings, O. Marti, and C. E. Bracker, *Science*, 1988, **242,** 209.
15 W. A. Ducker, T. J. Sendon and R. M. Pashley, *Nature*, 1991, **353,** 239.
16 J. C. Crocker, R. T. Valentine, E. R. Weeks, T. Gisler, R. P. Kaplan, A. G. York and D.A. Weitz, *Phys. Rev. Lett.*, 2000, **85,** 888.
17 E. J. W. Verwey and J. Th. G. Overbeek, *Theory of the Stability of Lyophobic Colloids*, Dover, New York, 1999.

I. CLASSICAL COAGULATION

LONDON-VAN DER WAALS ATTRACTION BETWEEN MACROSCOPIC OBJECTS

By J. Th. G. Overbeek and M. J. Sparnaay

van't Hoff Laboratory, Sterrenbos 19, Utrecht, Netherlands
Philips Research Laboratories, Eindhoven

Received 24th June, 1954

In the first part of this paper a description is given of an apparatus with which attractive forces between two flat glass plates have been measured. One of the glass plates was attached to a spring. The bending of this spring was directly proportional to the force between the glass plates and could be followed with an accuracy of 10-30 Å with the aid of an electrical capacity method. The distances between the glass plates were measured by means of Newton interference colours. A discussion of the errors is given. The force-distance relation found, an inverse third-power law, followed the London-Hamaker theory but the force constant was found to be about 40 times larger than predicted by their theory.

In the second part an extension of London's harmonic oscillator is discussed, leading to deviations from additivity of London-van der Waals forces which might be helpful in the understanding of our experimental results. Two *groups* of atoms a large distance apart instead of two atoms are considered, and a weak interaction is assumed between atoms of each group. It then appears that the polarizability is no longer the determining quantity in the force between the atoms such as given by London, but that large deviations from additivity can occur in the attractive force, whereas the polarizability remains practically unaffected.

PART 1. EXPERIMENTAL

1.1. Introduction

The role ascribed to London-van der Waals forces in the stability of colloids [1] and in the formation of aggregates between particles, and especially the long-range character of these forces made an independent proof of their existence very desirable. We investigated therefore the forces between optical flat glass (or quartz) plates in air. Preliminary publications [2, 3, 4] of the results have appeared.

The attraction between two (electrically neutral) parallel flat plates is given by an expression derived by de Boer [5] and Hamaker [6] from London's [7] theory on the attraction between two atoms :

$$F = \frac{AO}{6\pi d^3}. \qquad (1)$$

F is the force, d the distance between the plates, O is the area. The force constant A was predicted to be of the order of 10^{-12} erg.

It is easily seen that, to check this expression, taking for instance $O = 2$ cm², forces of the order of 1 dyne have to be measured with a distance d of about $\frac{1}{2}\mu$ between the plates. This condition proved to be a nearly insurmountable difficulty, probably because of dust particles and an irregularly shaped gel-layer present on the glass plates.

The force F was measured by the bending of a spring to which one of the plates was attached. The distance d was estimated with the aid of Newton interference colours. The results obtained confirmed the exponent 3 in expression (1) for the dependence of the force upon the distance but the force-constant A was about 40 times larger than that predicted.

1.2. THE VALUE OF THE FORCE-CONSTANT

The value $A = 10^{-12}$ erg was obtained from the following relations:

$$A = \pi^2 q^2 \lambda \tag{2}$$

$$\lambda = \tfrac{3}{4} h \nu \alpha^2. \tag{3}$$

Relation (2) is introduced in the theory of de Boer and Hamaker; q is the number of atoms per cm³ involved. λ is the energy constant in London's theory in which the two atoms are represented by three-dimensional harmonic oscillators with a characteristic frequency ν and a polarizability α. h is Planck's constant.

In the case of glass or quartz the main contribution to A is expected from oxygen, the polarizability for silicon being 40 times less than for oxygen.[8] Margenau[9] inserting $h\nu = 1\cdot37\,\text{eV} = 2\cdot05 \times 10^{-11}$ erg and $\alpha = 1\cdot57 \times 10^{-24}\,\text{cm}^3$ found $\lambda = 39\cdot8 \times 10^{-60}$ erg cm⁶. As q is about 5×10^{22} this leads to $A = 10^{-12}$ erg.

Expression (1) is derived on the basis of additivity, i.e. the attraction between two atoms is considered to be independent of the presence of a third atom. It is doubtful whether this procedure is allowed for glass or quartz. The oxygen atoms here do not have the same individuality as free atoms. In part 2 of this paper a certain type of deviation from additivity is considered theoretically.

1.3 THE APPARATUS.

The essential parts of the apparatus used are schematically shown in fig. 1.

A_1 and A_2 are the glass plates, their optically flat surfaces facing each other. A_1 is attached to a spring F of known resilience with the aid of frame B and holder D, both made of brass, A_2 rests on the pins T on top of three boxes K. These boxes were fixed to three micrometer movements regulating roughly the position of A_2 towards A_1. These micrometer screws are not shown in the figure. The fine adjustment of A_2 with

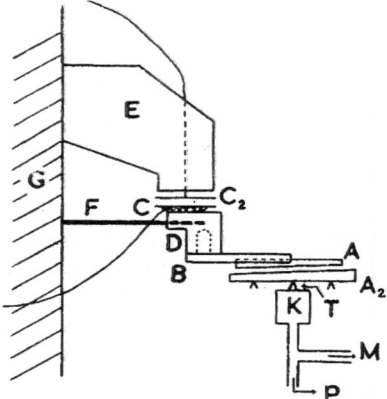

Fig. 1.—Essential parts of the apparatus.

respect to A_1 was obtained by changing the pressure in the boxes by means of a pump P. Changing the pressure in the three boxes simultaneously by 1 atmosphere resulted in a displacement of the glass plate A_2 of 4μ. The pressures were measured by the manometers M with an accuracy of 1 mm mercury.

If the distance between the glass plates was small enough the lower plate pulled the upper plate down over a certain distance. This distance was measured by means of the decrease of the capacity of the condenser formed by the two silvered microscope cover-glasses C_1 and C_2, C_1 being fixed to the holder D with insulating wax. C_2 was immovably fixed to the massive brass body E. The other brass body G carried both E and the fixed end of the spring F.

The parts shown in fig. 1 were mounted in a cylindrical box with flat top and bottom, that could be evacuated to 10^{-5} mm. The top plate carried two glass windows for observation. Leads to the condenser-plates C_1 and C_2, transmission to the micrometer screws and conductance to the pressure boxes K, were all vacuum-tight sealed through the cylinder wall or bottom. The evacuation was necessary to decrease the viscous resistance of the air between the glass plates (see 1.6).

The equilibrium condition for the system is (fig. 2):

$$F_1 + F_2 = \frac{AO}{6\pi d^3} - \beta (b - d) = 0 \tag{4}$$

β is the force-constant of the spring, b is the distance between the plates in absence of attractive forces, d is the actual distance. $(b - d)$ is thus the displacement of the upper glass plate and also that of the lower condenser plate.

14   LONDON–VAN DER WAALS ATTRACTION

By manipulating the micrometer screws and the pressure in the pressure-boxes, b can be changed. Due to the steepness of the attractive force stable equilibrium can only be obtained if d is larger than

$$d_{min} = \left(\frac{AO}{2\pi\beta}\right)^{\frac{1}{4}} \tag{5}$$

In fig. 2 where $A = 10^{-11}$ ergs, $O = 1$ cm^2 and $\beta = 1\cdot5 \times 10^5$ dyne/cm d_{min} is near to 5700 Å. Four values of b are given as an illustration, $b = 5000$ Å, 7500 Å, 10,000 Å and 12,500 Å respectively. The dotted curves indicated with 1s, 2s, 3s and 4s are the results of the addition of the attractive force-curve and the straight lines 1, 2, 3, 4 representing the force of the spring. The units on the abscissa vary with proportional variation of AO and β, the rest of the figure being unaltered.

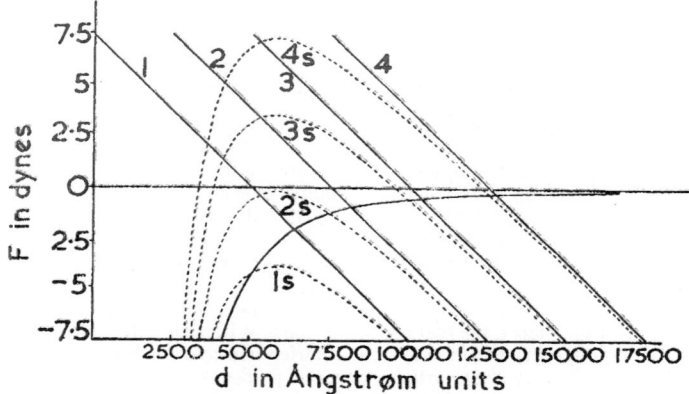

FIG. 2.—Force against distance in the system used.
$A = 10^{-11}$ erg; $\beta = 1\cdot5 \times 10^5$ dyne/cm.

Relatively stiff springs have to be used and this limits the displacements $(b - d)$ of the upper glass plates to less than 100 Å. Consequently the condenser plates C_1 and C_2 have to be very close together. If their distance is 4×10^{-3} cm a displacement of 10 Å changes the capacity by about 5×10^{-4} pF. This change could just be observed with a heterodyne set-up consisting of two oscillating circuits, one with a crystal stabilized frequency of 1500 kc, the other containing the condenser C_{12} in parallel with a precision condenser. The beat frequency between the two circuits was compared with a 1024 period tuning fork and could be reproduced within one beat per second corresponding to a variation in the capacity of 5×10^{-4} pF. The precision of the whole procedure, however, depended largely on the quality of the condenser C_{12}. The preparation of this condenser was one of the most delicate operations involved in the measurements. The relation between the distance d and the interference colours is given in table 1.[10]

TABLE 1.—RELATION BETWEEN INTERFERENCE COLOUR AND DISTANCE BETWEEN PLATES

first order		*second order*		*third order*	
d (Å)	colour	d (Å)	colour	d (Å)	colour
0	black	2850	violet	5750	indigo
1000	grey	2950	indigo	6300	green blue
1300	white	3350	blue	6687	bright green
1400	straw yellow	4125	green	7150	green yellow
1650	bright yellow	4300	green yellow	7500	rose
2200	orange yellow	4750	orange	7650	carmine
2750	red	550	dark violet	8100	violet
				8250	violet grey

It appeared after long experience that colours of the first order could be observed such that the distance d could be determined with a precision of 15 %, colours of the second order up to 3 %, then a decrease followed, the precision at $d = 15,000$ Å being about 5 %.

In measuring large distances (higher orders than the third one give alternating green and red bands) it was necessary to count the number of orders passed through upon increasing the distance from that at a colour of known order. This could be checked by the mano-meters M giving the position of the pins on top of the pressure boxes K and thus giving the position of the lower glass plate A_2. Vice-versa the method could be used to check the rate of displacement per atmosphere of the pins previously determined under a micro-scope.

There are no interference colours at distances smaller than 1000 Å, but the distance can be measured with the aid of the light reflected from the gap between the glass plates. If its intensity is I, then, according to Lord Rayleigh :[11]

$$\frac{I}{I_0} = \frac{16\pi^2 d^2}{\lambda(1 - e^2)}.$$ (6)

I_0 is the intensity of the incident light; e, a numerical factor, is 0.2; λ is the wavelength, $\approx 5 \times 10^{-5}$ cm.

1.4 CLEANING AND MOUNTING OF THE GLASS AND QUARTZ PLATES.

The plates used were slightly wedge-shaped in order to make the different reflection images more easily separable. The unevennesses on the optical flat surfaces were smaller than 400 Å which is probably the limit obtainable.[12] The area of the plates mainly used was 4 cm², their density at 15° C was $d_{15} = 2.556$, their refractive index $n_d = 1.5209$. Later on glass plates with area 1 cm², $d_{15} = 2.55$, $n_d = 1.515$ and quartz plates ($d_{15} = 2.66$; n_d (ord.) $= 1.544$ and n_d (extra-ord.) $= 1.539$) with an area of 1.7 cm² were obtained. The radius of curvature of the big glass plates was 300-500 metres, that of the smaller plates was too large to be measurable.

The glass plates were first cleaned with chamois-leather dipped in 3 % H_2O_2 and after-wards with alcohol. Then the plates, the plate A_1 being fixed in the frame B, were quickly but carefully brought together in such a way that interference colours became visible. If this could not be done very easily, dust particles were still present between the plates and cleaning was repeated.

The two plates with interference colours still visible were then mounted in the apparatus. The brass body G (see fig. 1) with spring F, holder D and condenser $C_{1,2}$ could be raised about 4 cm. The two glass plates were laid down on the three pins T. Part A was then carefully lowered until the upper glass plate A_1 was attached to the spring F by two pins on the frame B sliding into loosely fitting holes in D. The connection was made sufficiently solid by pouring molten wax between B and D. The whole system was then closed, evacuated to the desired pressure of about 0.01 mm (see (1.7)) and only then the plates were separated by lowering the lower plate A_2.

It required considerable experience to fix the plates into their proper places and to avoid them attracting so strongly that relative movements became too difficult. In these circumstances frictional movements between the plates can have a very bad influence upon the quality of the surfaces.[14] However, if the distance is too large and the attraction not strong enough, the plates may lose contact before the apparatus is evacuated and this almost certainly brings dust-particles between the plates and spoils the measurement. The viscosity of the air makes the manipulation more easy by providing an " air-cushion " between the plates that disappears only slowly.

It has never been possible to get the plates moving completely freely at a distance smaller than 7000 Å or perhaps 10,000 Å. The obstacles apparently present could occas-ionally be crushed if a force was exerted sufficiently large to obtain a distance smaller than d_{min} (see eqn. (5)). The smallest distance thus obtained was about 200 A.

These obstacles probably were silica gel particles. Their influence increased after exposure to a wet atmosphere in agreement with the hygroscopic properties of glass, and decreased after rubbing the surfaces with chamois-leather provided with some finest quality polishing rouge. This is in agreement with general ideas on poilshing glass, the role of polishing rouge being the removal of gel-layers formed by water on the uneven surface.[10]

1.5 CALIBRATION OF THE SPRING AND CONDENSER.

The condenser $C_{1,2}$ was calibrated before and after each series of measurements in terms of force on the spring F by putting small weights on the upper glass plate A_1, and observing the capacity. As the sensitivity of the condenser was not always constant,

16 LONDON–VAN DER WAALS ATTRACTION

frequent rechecking during measurements was desirable. This was done in the following way. When the glass plates were mounted and the apparatus evacuated, the glass plate A_2 was tilted using the micrometer screws such that it just touched the other plate A_1. Then the pressure in the three boxes K was uniformly varied. This resulted in a uniform displacement over a known distance of A_2 and thus of A_1 and the lower condenser plate C_1.

The corresponding change of the capacity was measured. In this way the relation between capacity and distance was obtained. It could be compared to the relation between capacity and force by means of the force constant β of the spring F. Three different springs were used with

$$\beta_1 = 1\cdot5 \times 10^5 \text{ dyne/cm}, \quad \beta_2 = 8 \times 10^5 \text{ dyne/cm}, \quad \beta_3 = 15 \times 10^5 \text{ dyne/cm}.$$

1.6 MEASUREMENT AND EVALUATION.

The glass plates were brought into a parallel position with the aid of a white light-source. This gave two images upon reflecting against the two optical flat surfaces facing each others These two images were brought to coincidence while the plates might still be separated by 0·1 mm. Then the distance was carefully decreased until interference-colours became visible. This procedure required some experience. Once interference-colours were visible the plates could be brought into the desired position and measurements made.

Two measurements of the capacity were needed at least, one at a large distance d_∞ such that attractions were practically absent and one at the distance where the attractive force was to be measured. A convenient value for d_∞ was 4μ. A measurement was repeated many times upon increasing and decreasing the distance. An (arbitrarily chosen) example is given in table 2.

One scale-division on the precision-condenser corresponds with a bending of 40 Å of the spring with $\beta = 1\cdot5 \times 10^5$ dyne/cm. The average force F is found to be 0·7 dyne.

In general more than one colour was visible at distances of the order of 1μ, either due to the curvature or to a deviation from parallel position. Each colour was seen over an area O_c belonging to an approximately constant distance d_c. The attractive force F_c for each distance was supposed to be given by (1). For the whole surface the force becomes

$$F = F_c = \frac{A}{6\pi} \sum \frac{O_c}{d_c^3} = \frac{A \sum O_c}{6\pi d_m^3} \left(1 + 3 \sum \frac{\delta_c}{d_c} \cdot \frac{O_c}{d_c^3}\right) \tag{7}$$

$d_m = d_c + \delta_c$ is a distance such that $\sum \frac{\delta_c}{d_c} \cdot \frac{O_c}{d_c^3} = 0$. Two criticisms can be made concerning

this interpretation : first, the force-distance law was not yet known : second, the summation should be replaced by an integral. However, after many experiments were carried out, the use of the force-distance law with an exponent near to 3 was justified, and replacement by an integral would hardly increase the precision, much larger sources of error originating from elsewhere. Table 3 gives an illustration of eqn. (7). The same experiment is chosen as in table 2. Table 3 is representative in that one term O_c/d_c^3 predominates in most cases. Terms smaller than 0·1 of the largest term were generally neglected.

TABLE 2.—REPEATED MEASUREMENTS OF THE FORCE OF ATTRACTION AT A DISTANCE BETWEEN THE PLATES OF 13,000 Å (see table 3)

scale-divisions	force (dynes)
18·5	1·11
10	0·60
15	0·90
20	1·20
17·5	1·05
11	0·66
14	0·84

TABLE 3.—EVALUATION OF TOTAL AREA AND AVERAGE DISTANCE IN AN EXPERIMENT

O_c in cm^2	d_c in Å	O_c/d_c^3
0·5	11,000	0·39
2	13,500	0·82
1	15,000	0·30

$d_m = 13,000$ Å $O_c = 3\cdot5$ cm^2

Using the force as given in table 2, average distance and total area resulting from table 3 and taking an exponent 3, A becomes $0\cdot8 \times 10^{-11}$ erg.

J. TH. G. OVERBEEK AND M. J. SPARNAAY   17

1.7 THE INFLUENCE OF THE VISCOSITY OF THE AIR.

Reynolds [16] gave an expression for the relation between the velocity $dd/dt = -\dot{q}$ under the influence of a force F of a flat circular plate (radius c) towards a second plate at a distance d in a parallel position in a medium with viscosity η:

$$-\dot{q} = \frac{dd}{dt} = \frac{2Fd^3}{3\pi\eta c^4}. \tag{8}$$

Taking for the viscosity of the air $\eta = 1\cdot8 \times 10^{-6}$ poise, assuming c to be 1 cm and assuming the van der Waals force (1) to be the driving force, it is found that for $d = 5000\,\text{Å}$ the velocity is only $0\cdot6\,\text{Å/sec}$. It goes asymptotically to zero if the equilibrium position is approached. Furthermore, if a measurement is started with a large distance $b \approx d$ (see eqn. (4)) which must be decreased, d the actual distance between the plates, stays considerably behind, say 2000 Å. It then requires 10-20 minutes before $b = d$ which is still not enough. These long periods were very undesirable because they allowed all kinds of mechanical and thermal disturbances to occur. Consequently the viscosity of the air was decreased by means of a decrease of the air-pressure P in the apparatus. As the mean free path of N_2 and O_2 molecules is about 600 Å at 1 atm the viscosity does not decrease until the pressure has become $600/d = P_0$ atm. Below this pressure (i.e. the Knudsen region) the viscosity is an approximately linear function of the pressure. It was found necessary to leave some air in the apparatus to damp vibrations of glass plate A_1. Eqn. (8) has to be extended with an inertia term in order to get insight into the magnitude and character of the vibrations. This leads to

$$m\ddot{q} + m\omega^2q + B\dot{q} = 0. \tag{9}$$

m is the mass of the vibrating glass plate A_1 together with frame B, holder D and condenser-plate C_1: $m\omega^2 = \beta$ is the force-constant of the spring F.

$$B = \frac{3\pi\eta c^4}{2d^3} \approx \frac{3\pi\eta c^4}{2b^3}; \quad \eta = \eta_0\frac{P}{P_0}, (P < P_0).$$

The attractive force was neglected in (9). It would make the vibrations only slightly less harmonic (see fig. 2). Equation (9) is the expression for a damped oscillator. Upon solving it appears that critical damping takes place at

$$B = 2m\omega \tag{10}$$

Inserting $m = 20$ (which gives $\omega = 6\cdot3$ if $\beta = 8 \times 10^5$ dyne/cm is taken), $d \approx h = 5000\,\text{Å}$, $c = 1$ cm one finds $\eta = 6\cdot25 \times 10^{-12}$ poise. This corresponds to $P = 0\cdot36 \times 10^{-2}$ mm. If $d \approx b = 1\mu$, then $P = 1\cdot4 \times 10^{-2}$ mm. If $d \approx b = 2\mu$, then $P = 5\cdot2 \times 10^{-2}$ mm. Under these conditions the time in which an oscillator is damped down to 2 % of its original value is calculated to be of the order of $\frac{1}{4}$ sec. This means that even deviations from equilibrium of for instance 500 Å occasionally occurring are not too harmful. It was preferred to take the pressure slightly less than that derived from (10) because the glass plate A_1 could then be seen " dancing " around its equilibrium-value indicating that it was moving freely. Due to their harmonic character the vibrations, although having an unfavourable influence upon the precision, permitted a reasonable estimate of the equilibrium-distance.

1.8. POSSIBLE ELECTROSTATIC EFFECTS.

In order to prevent electrostatic charges on the glass plates a radio-active preparation was always present. The attraction never decreased even after waiting a week and resuming the measurements. Neither did it decrease after ionization of the air in the cylinder-box ($P = 10^{-2}$ mm) or after evaporation of a silver-layer of 200 Å thickness upon the plates.

The distance between the evaporating silver-droplet and the glass plates was only 12 cm whereas Tolansky [17] recommends 30 cm. Therefore the silver-layer was rather poor and prevented the plates coming closer than $1\cdot2\,\mu$.

The amount of moisture had no influence upon the attraction. Lastly the strong dependence of the force measured upon the distance shows that an electrostatic interpretation must be ruled out Homogeneously distributed electric charges on the glass plates would result in a force not depending upon the distance at all.

1.9. DISCUSSION OF THE ERRORS.

As far as can be seen the main errors involved were due to :

        (a) obstacles between the plates,
        (b) vibrations of the glass-plate fixed to the spring,
        (c) unexpected alterations of the condenser $C_{1, 2}$.

18   LONDON–VAN DER WAALS ATTRACTION

(*a*) It is difficult to account quantitatively for the errors due to obstacles. They might play a role in any measurement. For instance in one case there was found a repulsion of 0·5 dyne at $d = 15{,}000$ Å, an attraction of 2·2 dyne at $d = 7500$ Å and again a repulsion at $d = 6500$ Å. This kind of irregular values, completely irreproducible, was often found. Such a series of measurements was discarded.

(*b*) Table 4 gives the influence of the vibrations as estimated after long experience.

TABLE 4.—INFLUENCE OF VIBRATIONS ON THE PRECISION OF THE FORCE MEASUREMENTS

d (Å)	ΣO_c	precision at d (in dyne/cm²)	precision at d_∞ (in dyne/cm²)
200	0·2	—*	0·15 †
6000	0·5	—*	0·15 †
10,000	1·5	0·1	0·4
15,000	3	0·15	0·2
18,000	3·6	0·1	0·15

* precision determined by obstacles.   † $d_\infty \approx 15{,}000$ Å.

At the large distance d_∞ the vibrations were damped to a smaller extent than at the distance d where an attraction was measured. The area O in the second column is equal to O_c in eqn. (7) and is averaged over many measurements.

(*c*) If no sudden change in the capacity was observed the difference before and after a series of measurements of the sensitivity was considered to be a linear function of the time.

1.10. RESULTS

The results for the glass-plates mainly used are given in fig. 3.

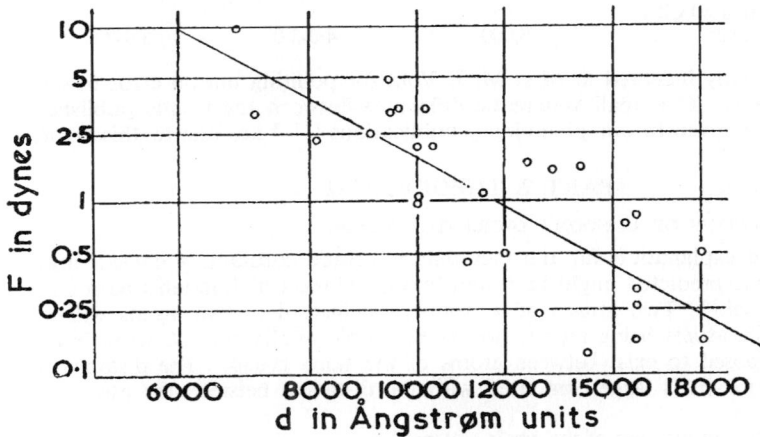

FIG. 3.—Force against distance for glass plates with $n_d = 1\cdot5209$; $d_{15} = 2\cdot556$.

The straight line fitting the measurements corresponds to

$$F = \frac{A}{6\pi d^n}. \tag{11}$$

F is the force/cm²; $A = 3\cdot8 \times 10^{-11}$ erg; $n = 3 \pm 0\cdot3$.[18]

Other results are given in table 5. These results are considered less reliable than those of fig. 3, but they still give the same order of magnitude for A.

The force F in row 1 and 3 was found upon tearing the plates apart. The rows 2, 4, 9 and 10 give forces measured with detectable obstacles between the plates. The force values in the rows 5, 6, 7 and 8 are highly inaccurate due to the small area of the plate A_1 allowing for strong vibrations. The thickness of the silver-layer, row 13, was 200 Å, thus making the quartz less transparent. This made the distance-value less accurate.

J. TH. G. OVERBEEK AND M. J. SPARNAAY   19

The values found with the glass plates $n_d = 1.5209$, $d_{15} = 2.556$ were divided between those given in fig 3 and those given in row 2, table 5, the first being values with the plates moving freely, the second being values with obstacles detected. This division is somewhat arbitrary and it might be that low values in fig. 3 are still due to obstacles.

It is seen, however, that the exponent in (11) is near to the predicted one. A possible explanation for the large deviations from the predicted values for the constant A will be given in part 2.

TABLE 5

no.	material	d (Å)	F (dyne/cm²)	$A \times 10^{11}$ (erg)
1	glass	200	$750 - 1.5 \times 10^5$	0.11-2.2
2	$n_d = 1.5209$	varying between	varying between	
	$d_{15} = 2.556$	2500 and 7000	0.6 and 50	0.11-3
3	glass plate A₁	200	10^2-10^5	0.015-1.5
4	$n_d = 1/1.515$	5,000	20	3.2
5	$d_{15} = 2.55$	7,000	10.0-20.0	6.5-13.0
6	area 1 cm²	9,000	5.0-10.0	6.7-13.5
7	glass plate A₂	11,000	2.5-4.5	6.3-11.1
8	as before	16,000	0.8-3.0	6.0-23.0
9	quartz	3,000	20	1.1
10	both A₁ and A₂	4,000	25	3.0
11	area 1.7 cm²	13,000	0.7-2.0	3.0-8.0
12		17,000	0.2	1.9
13	quartz treated with silver	8,000	4.0-8.0	3.9-7.8

We are greatly indebted to Mrs. M. J. Vold for pointing out an error in our original table 1. The small systematic difference between the results published here and those given in our preliminary publications [1, 2, 3] are due to this error.

PART 2. THEORETICAL

2.1. THE EXTENSION OF LONDON'S OSCILLATOR MODEL

A possible explanation for the fact that the force constant A considerably exceeds the one predicted might be found in an extension of London's harmonic oscillator model.[19] Two *groups* of n atoms each instead of two atoms will be considered, the atoms being represented as harmonic oscillators. A weak interaction is supposed to exist between atoms of the same group. The distance R between the groups is large compared with the distances between the atoms of the same group.

The Hamiltonian function of the whole system [20] is

$$
H = \sum_{i=1}^{n} G_{ii}(p_x^2 + p_y^2 + p_z^2)_i + \sum_{i=1}^{n} A_{ii}(x_i^2 + y_i^2 + z_i^2)
$$
$$
+ \sum_{i=1}^{n}\sum_{k=1}^{n} A_{ik}(x_i x_k + y_i y_k - 2 z_i z_k) + \sum_{i=1}^{n}\sum_{k=n+1}^{2n} B_{ik}(x_i x_k + y_i y_k - 2 z_i z_k)
$$
$$
+ \sum_{i=n+1}^{2n} G_{ii}(p_x^2 + p_y^2 + p_z^2)_i + \sum_{i=n+1}^{2n} A_{ii}(x_i^2 + y_i^2 + z_i^2)
$$
$$
+ \sum_{i=n+1}^{2n}\sum_{k=n+1}^{2n} A_{ik}(x_i x_k + y_i y_k - 2 z_i z_k)
$$
$$
+ \sum_{i=n+1}^{2n}\sum_{k=1}^{n} B_{ik}(x_i x_k + y_i y_k - 2 z_i z_k). \quad (12)
$$

$G_{ii} = \dfrac{1}{2m}$; $A_{ii} = \tfrac{1}{2}m\omega_0^2$; A_{ik} $(i \neq k)$ represent the interactions between atoms of the same

group; $B_{ik} = \dfrac{e^2}{2R^3}$ represent the interactions between atoms of different groups. All

the B_{ik} will have the same value throughout the whole treatment. (e = electronic charge). $(p_x)_i$, $(p_y)_i$, $(p_z)_i$ are the x-, y- and z-components of the momenta of the i-th oscillator, x_i, y_i, and z_i the same for the co-ordinates.

Generalized co-ordinates and momenta can be introduced upon transforming (12) with

$$s_1 = \sum_{k=1}^{2n} U_{lk} z_k \,;\; (p_z)_l = \sum_{k=1}^{2n} U_{lk}(p_z)_k. \tag{13}$$

We write down only the equations for the z co-ordinates. For instance taking $n = 2$ the transformation becomes:

$$s_1 = \tfrac{1}{2}(z_1 + z_2 + z_3 + z_4), \qquad s_3 = \tfrac{1}{2}(z_1 - z_2 + z_3 - z_4),$$
$$s_2 = \tfrac{1}{2}(z_1 + z_2 - z_3 - z_4), \qquad s_4 = \tfrac{1}{2}(z_1 - z_2 - z_3 + z_4), \tag{14}$$

and analogous equations for the momenta.

The generalized co-ordinates will be of secondary interest for the calculation of the energy required and will not be considered further.

The transformation means that matrix (A, B) formed by the A_{ii}, A_{ik} and B_{ik} elements must be brought into diagonal form.[21] The diagonal elements are the roots of det $(A, B) = 0$. The transformed Hamiltonian function is the sum of the Hamiltonian functions of $2n$ unperturbed harmonic oscillators with identical masses but with different frequencies. The $2n$ energy values obtained can be expanded in inverse powers of the distance R, thus giving the interaction-energy between the two groups, distance R apart. Jehle [22] gave a wave-mechanical treatment of the problem. Our case (1) (see 2.2) will correspond with his and the same result concerning the energy will be obtained.

As is done in London's theory it will also be important in this case to compare the attraction to the polarizability.

The polarizability α_n of the groups can be found upon introduction of generalized co-ordinates and momenta of the Hamilton function.

$$H = \sum_{i=1}^{n} G_{ii}(p_z)_i^2 + \sum_{i=1}^{n} A_{ii} z_i^2 - 2 \sum_{i=1}^{n} \sum_{k=1}^{n} A_{ik} z_i z_k \tag{15}$$

and calculating the corresponding frequencies.

The frequencies found must be inserted in

$$\alpha_n = \sum_{l=1}^{n} \frac{e^2}{m\omega_l^2}. \tag{16}$$

It will appear that in general the relation between polarizability and van der Waals' constant is not of the simple type given in eqn. (3) for a single pair of atoms.

2.2. THREE CASES.

It is evident that the procedure described above can be carried out for special cases only. Three types of interaction are chosen:

(1) $A_{ik} = c_z$ for all values of i and k given by (12).
   are zero.
(2) $A_{ik}(i = k + 1$ and $k = i + 1$, except when i, $k = n$ or $2n) = c_z$; all other A_{ik}
(3) The same as (2) but now the four elements A_{ik}: $i = 1$, $k = n$; $i = n$, $k = 1$; $i = n + 1$, $k = 2n$; $i = 2n$, $k = n + 1$ also have the value c_z.

The first case means that the interaction between any two atoms of the same group is equally strong. This seems unlikely on geometrical grounds. The calculation is simple, however, and the final results give an upper limit for the deviation from additivity.

The second case is based upon the picture of a string of atoms (to be taken in the z-direction) each having interactions with its nearest neighbours only. The picture underlying the third case differs from that in the second case in that the first and the last atom are considered as nearest neighbours.

J. TH. G. OVERBEEK AND M. J. SPARNAAY   21

2.3 CALCULATION OF THE THREE CASES

CASE 1

It can be seen upon suitable addition and subtraction of rows [22] in det (1) that there are $2n-2$ roots of det (1) $= 0$, $a = -c_z$, one root $a + (n-1) c_z - nb = 0$ and one root $a - (n-1) c_z + nb = 0$.

$$\det (1) = \begin{vmatrix} \begin{array}{ccc|ccc} a\ c_z & - - - & c_z & b & - - - - & b \\ c_z\ a & & & & & \\ & & \searrow & & & \\ & & a\ c_z & & & \\ c_z & - - - & c_z\ a & b & - - - - & b \\ \hline b & - - - - & b & a\ c_z & - - - & c_z \\ & & & c_z\ a & & \\ & & & & \searrow & \\ & & & & & a\ c_z \\ b & - - - - & b & c_z & - - - & c_z\ a \end{array} \end{vmatrix}$$

$\underbrace{\qquad\qquad}_{n} \underbrace{\qquad\qquad}_{n}$

$b = -2B_{ik} = -\dfrac{e^2}{R^3}$

$\begin{matrix} i = n+1 \ldots 2n \\ i \neq k = 1 \ldots n \end{matrix}$ and $\begin{matrix} i = 1 \ldots n \\ i \neq k = n+1 \ldots 2n \end{matrix}$

$a = A_{ii}$ $\qquad\qquad$ $i = 1 \ldots 2n$

$c_z = -2A_{lk}$

$\begin{matrix} i = 1 \ldots n \\ i \neq k = 1 \ldots n \end{matrix}$ and $\begin{matrix} i = n+1 \ldots 2n \\ i \neq k = n+1 \ldots 2n \end{matrix}$

The latter two roots are of interest only since they are the only ones containing terms depending upon R.

The potential energies of two harmonic oscillators with frequencies ω_+ and ω_- can be deduced from:

$$\omega_+ = \omega_0 \sqrt{1 + \frac{2(n-1)}{m\ \omega_0{}^2} c_z + \frac{2ne^2}{m\omega_0{}^2 R^3}}, \qquad (17)$$

$$\omega_- = \omega_0 \sqrt{1 + \frac{2(n-1)}{m\omega_0{}^2} c_z - \frac{2ne^2}{m\omega_0{}^2 R^3}}. \qquad (18)$$

Just as London did, these frequencies are considered as frequencies of two quantum mechanical harmonic oscillators with lowest eigen values $\frac{1}{2}\hbar\omega_+$ and $\frac{1}{2}\hbar\omega_-$ where $\hbar = \dfrac{h}{2\pi}$.

Expanding ω_+ and ω_- one finds for the interaction energy:

$$E_R = -n^2 \frac{\hbar\omega_c\alpha_c{}^2}{2R^6}; \quad \omega_c = \omega_0 \sqrt{1 + \frac{2(n-1)}{m\omega_0{}^2} c_z};$$

$$\alpha_c = \frac{e^2}{m\omega_c{}^2} = -n^2 \frac{\hbar\omega_0\alpha_0{}^2}{2R^6} \left(1 - \frac{3(n-1)}{m\omega_0{}^2} c_z + \ldots \right);$$

$$\alpha_o = \frac{e^2}{m\omega_0{}^2}. \qquad (19)$$

If ω_0 is low enough to allow for the classical limit, the attraction free energy becomes

$$F_R = -n^2 \frac{4e^4kT}{m^2\omega_c{}^2 R^6}. \qquad (20)$$

This can be found after transforming the partition function into a Gauss integral.[23] Then the value of det (1) is required, not its roots. The classical limit might be important in physiology.

$$\det (1A) = \begin{vmatrix} \begin{array}{ccc} a\ c_z & - - - & c_z \\ c_z\ a & & \\ & \searrow & \\ & & \\ & & a\ c_z \\ c_z & - - - & c_z\ a \end{array} \end{vmatrix}$$

The roots of det (1A) $= 0$ give the generalized frequencies of the n oscillators of one group. There are $(n-1)$ roots, $\qquad (a - c_z) = 0$ and one root $\qquad a + (n-1) c_z = 0$.

22   LONDON–VAN DER WAALS ATTRACTION

Inserting the corresponding frequencies into (16) the polarizability for case 1 becomes

$$\alpha_n = n\alpha_0 \left\{ 1 + (n - 1) \frac{4c_z^2}{m^2\omega_0^4} + \ldots \right\}. \tag{21}$$

Case 2

The matrix (A, B) for the second case can be easily constructed. Adding the $2n$th row to the first, the $(2n - 1)$th to the second and so on, then subtracting the $2n$th column from the first, the $(2n - 1)$th from the second and so on, its determinant value is the product of the determinants $(A + bU)$ and $(A - bU)$.

$$\det (A + bU) = \begin{vmatrix} a+b & c_z+b & b & — & — & — & — & — & — & — & — & b \\ c_z+b & a+b & c_z+b & — & — & — & — & — & — & — & b \\ b & & & & & & & & & & & \\ & & & & & & & & & & & \\ & & & & & & & & & & & b \\ & & & & & c_z+b & a+b & c_z+b \\ b & — & — & — & — & — & — & b & c_z+b & a+b \end{vmatrix}$$

This is generally true for the type of matrices involved here. The matrices must be symmetrical with respect to the main diagonal. Generally the transformation

$$z_i(i = 1 \ldots n) = \frac{1}{\sqrt{2}} (s_i - t_i)$$

$$z_k(k = n + 1 \ldots 2n) = \frac{1}{\sqrt{2}} (s_i + t_i) \tag{22}$$

will give the desired separation as was kindly pointed out to us by Dr. Bouwkamp of the Philips Research Laboratories.

The same procedure (adding and subtracting of rows and columns) will serve to simplify $\det (A + bU)$ and $\det (A - bU)$. For $n = 3$ and $n = 4$ there are four roots containing b; for $n = 5$ and $n = 6$ there are six. From these the total attraction can be calculated in the same way as in case 1, and appears to be

$$E_R = n^2 \frac{\hbar\omega_0\alpha_0^2}{R^6} \left(1 - \frac{n - 1}{n} \frac{6c_z}{m\omega_0^2} + \ldots \right), \quad n = 2, 3, 4, 5, 6. \tag{23}$$

No calculations for other n-values were made.

The polarizability can be found upon solving the determinant with diagonal elements a and elements c_z immediately bordering them, all other elements being zero. The roots a_r have been found by Coulson [24] to be

$$a_r = 2c_z \cos \frac{r\pi}{n + 1} \quad r = 1 \ldots n. \tag{24}$$

The polarizability then becomes

$$\alpha_n = n\alpha_0 \left(1 + \frac{4c_z^2}{(m\omega_0^2)} \sum_{r=1}^{n} \cos^2 \frac{r\pi}{n + 1} - \ldots \right). \tag{24}$$

Case 3

It can be proved that the determinant for this case has only two roots containing b. First add all the rows to the first. This gives a root $(a - 2c_z - nb) = 0$. Then add columns $2 \ldots n$ to the first and subtract the sum of columns $n + 1 \ldots 2n$. This gives $(a - 2c_z + nb) = 0$. It is seen upon suitable subtraction of rows and columns, preferably such that only two elements b symmetrical to the main diagonal remain, that all terms containing b cancel in the further development of the determinant. One could of course as well have started with determinants $(a + bU)$ and $(a - bU)$. Considering the two roots containing b the attraction becomes:

$$E_R = - n^2 \frac{\hbar \omega_2 \alpha_2{}^2}{2R^6}; \quad \omega_2 = \omega_0 \sqrt{1 + \frac{4c_z}{m\omega_0{}^2}};$$

$$\alpha_2 = \frac{e^2}{m\omega_2{}^2} = - n^2 \frac{\hbar \omega_0 \alpha_0{}^2}{2R^6} \left(1 - \frac{6}{m\omega_0{}^2} c_z + \dots \right). \tag{26}$$

The determinant leading to the polarizability has also been treated by Coulson [24] and has roots

$$a_s = 2c_z \cos \frac{2\pi s}{n} \quad s = 0 \dots n - 1. \tag{27}$$

The polarizability becomes

$$\alpha_n = n\alpha_0 \left(1 + \frac{4c_z{}^2}{(m\omega_0{}^2)^2} \sum_{s=0}^{n-1} \cos^2 \frac{2\pi s}{n} - \dots \right). \tag{28}$$

Although there is no direct proof case 3 will probably be the limit of case 2 for $n = \infty$.

2.4. DISCUSSION

It appears that in all cases treated the weak interactions between atoms in the same group give only second-order effects in the polarizability whereas the total interaction energy which is itself a second-order effect will be strongly affected.

If, however, the interactions are so strong that the oscillators are always in phase, the two oscillators would behave as one oscillator with charge $2e$. Additivity is then restored on a new basis and a London-type expression applies again. Such an expression is found by Coulson and Davies [25] considering extended oscillators as a model for large chain molecules.

A negative value of c_z means that there is a tendency of the oscillators to be in phase, a positive value means a tendency to be in counterphase.

If there is dipole-interaction only between atoms of one group, then

$$c_z = - \frac{e^2}{R_0{}^3}; \quad c_x = c_y = \frac{e^2}{2R_0{}^3}. \tag{29}$$

R_0 is the distance between the atoms involved. A reasonable value of c_z is [26] $- 0 \cdot 1 \ m\omega^2$.

Considering then a chain along the z-axis there is an increased attraction of about 50 %. If, however, a chain is taken along the x- or y-axis the attraction is decreased. In these cases there is a tendency of two neighbouring oscillators to be in counterphase as far as the z-direction is concerned, giving the largest contribution to the total attraction. If $n \leqslant 10$ there is in all these cases only an increase of a few % of the polarizability.

London's additivity theorem applies to the case of varying distance between *all* the atoms involved. This means that it is often misinterpreted in colloid chemistry where one is concerned with the case of two *groups* of atoms with a variable distance, the distance between the atoms in one group being constant.

The models treated here are, of course, oversimplifications. The A_{ik}-terms may depend upon the direction in space and they certainly will assume different values between different pairs of atoms instead of having either the value $c_{x,y,z}$ or zero as assumed in the models. One might think of a mixture of the cases treated here in a given physical situation. However, the models, although not giving a quantitative explanation of the strong forces experimentally found, may serve to demonstrate the possibility of a strongly increased van der Waals interaction between groups of atoms combined with a normal value of the polarizability.

24   FORCES OF INTERACTION OF SURFACES

[1] Verwey and Overbeek, *Theory of the stability of lyophobic colloids* (Elsevier, Amsterdam, 1948).

[2] Overbeek and Sparnaay, *Proc. K. Akad. Wetensch., B*, 1951, **54**, 387.

[3] Overbeek and Sparnaay, *J. Colloid Sci.*, 1952, **7**, 343.

[4] Sparnaay, *Thesis* (Utrecht, 1952).

[5] de Boer, *Trans. Faraday Soc.*, 1936, **32**, 21.

[6] Hamaker, *Physica*, 1937, **4**, 1058.

[7] London, *Z. Physik*, 1930, **63**, 245 ; *Z. physik. Chem.*, 1931, **11**, 222. Eisenschitz and London, *Z. Physik*, 1930, **60**, 491.

[8] Stevels, *Progress in the theory of the physical properties of glass* (Elsevier, Amsterdam, 1948), p. 94. Fajans and Kreidl, *J. Amer. Ceram. Soc.*, 1948, **31**, 105.

[9] Margenau, *Rev. Mod. Physics*, 1939, **11**, 1.

[10] Van Heel, *Inleiding in de optica* (Nijhoff, s'-Gravenhage, 3rd edn., 1950), p. 86.

[11] Lord Rayleigh, *Proc. Roy. Soc. A*, 1936, **156**, 343.

[12] Mayer, *Physik dünner Schichten, Wissenschaftliche Verlagsgesellschaft* (Stuttgart. 1950).

[13] Tolansky, *Multiple beam interferometry* (Clarendon Press, Oxford, 1948), p. 24.

[14] Bowden and Tabor, *The friction and lubrication of solids* (Clarendon Press, Oxford, 1950), p. 168.

[15] Grebenschikov, *Keramika i Stekle*, 1936, **7**, 36 ; *Sotsialisticheskaya Reconstruktsuya i Nauka*, 1935, **2**, 22.   Bowden and Tabor, *The friction and lubrication of solids* (Clarendon Press, Oxford, 1950), p. 302.

[16] Reynolds, *Phil Trans.*, 1885, **177**, 157.

[17] Tolansky, *Multiple beam interferometry* (Clarendon Press, Oxford, 1948), p. 25.

[18] Weatherburn, *A first course in mathematical statistics* (Cambridge University Press, Cambridge, 1947) (" t-test ").

[19] London, *Z. physik. Chem. B*, 1931, **11**, 222.

[20] Margenau, *Rev. Mod. Physics*, 1939, **11**, 1.

[21] Born and Jordan, *Elementare Quantenmechanik* (Springer, Berlin, 1930), especially chapters II and III.

[22] Jehle, *J. Chem. Physics*, 1950, **18**, 1150.

[23] Eidinoff and Aston, *J. Chem. Physics*, 1935, **3**, 379.

[24] Coulson, *Proc. Roy. Soc. A*, 1938, **164**, 393.

[25] Coulson and Davies, *Trans. Faraday Soc.*, 1952, **48**, 777. Davies, *Trans. Faraday Soc.*, 1952, **48**, 790.

[26] London, *Z. physik. Chem., B*, 1931, **11**, 250.

Liquid Crystals

G. R. Luckhurst

School of Chemistry, University of Southampton, UK SO17 1BJ

Commentary on: **On the Theory of Liquid Crystals,** F. C. Frank, *Discuss. Faraday Soc.,* 1958, **25**, 19–28.

The Faraday Society has done much to further the development of liquid crystal science, through the organisation of two Discussions, in 1933 and 1958, and a Symposium in 1971, as well as by the publication of numerous and diverse papers in the *Transactions*.

Of all these contributions, many seminal, that by Sir Charles Frank clearly stands out. This unique paper was presented at the 1958 Discussion, entitled *Configurations and Interactions of Macromolecules and Liquid Crystals,* and it earned him his enviable international reputation in the field. The paper describes what is known as the continuum theory of liquid crystals and is now widely used to design and optimise display devices for the LCD industry through the prediction of the macroscopic organisation of the liquid crystal when subject to competing external constraints, such as surface interactions and electric fields.

The term *liquid crystal* was aptly chosen to describe this fascinating state of matter because it is especially informative about its macroscopic characteristics. Thus, as a *liquid* it is unable to support a shear stress but like a *crystal* its properties are anisotropic. Accordingly the liquid crystal has a unique axis, known as the director, and it is the factors that control the spatial distribution of the director's orientation which attracted Frank's attention. Curiously at the 1933 Discussion papers by Oseen[1] and by Zocher[2] had begun to attack this problem of how the director would be distributed, using an elastic-like approach. From these beginnings, Sir Charles, in an elegant and remarkably accessible paper, developed the modern form of continuum theory that describes how the elastic free energy of the nematic phase increases as the director deviates from its state of uniform alignment. This energy is written as an expansion in terms of the director gradients or curvature strains. The coefficients of the terms are the elastic constants and in a deceptively straight forward analysis he demonstrated how the symmetry of the system can be used to reduce the large number of such constants to just four and under many conditions three. He then went further and showed in a qualitative sense how the theory could be applied to other liquid crystals, namely the smectic and chiral nematic phases. He also touched on the hypothetical ferroelectric nematic in a way which hints at the flexoelectric effect discovered some years later.[3] The name *nematic* has its origins in the thread-like structures which the phase exhibits under the polarising microscope and are known to stem from defects in the director field. He used his expression for the elastic free energy to determine the director distributions around a line defect and found optical behaviour in accord with experiment. His considerable knowledge of the dislocations found in crystals prompted him to refer to the defects in nematics as *disinclinations* although over the years this term has metamorphosed to disclinations.

The paper by Sir Charles is also a joy to read both for its style and his digressions. For example, he notes that the long standing controversy between the swarm and continuum theories was illusory and that the properties could be understood with continuum theory as could be clearly seen from his paper! He also exhorted experimentalists to renew their interests in liquid crystals since the conclusion that they might have drawn after the 1933 Discussion that there were no major puzzles remaining was manifestly false. Indeed, similar conclusions appear to have been drawn throughout the history of liquid crystal research but as soon as it is thought that there are no new challenges, discoveries, often unexpected, have revitalised the subject and demonstrated that the pessimism in such claims is misplaced. Thus, Sir Charles urged the measurement of the elastic constants, which now bear his name, because they were clearly defined and because they provide a unique insight into the nature of anisotropic intermolecular forces. In fact, his appeal went largely unheeded for the next ten years until, stimulated by the clear applications of nematic liquid crystals in electro-optic display devices, there was a resurgence of interest in their elastic behaviour. For example, in 1970 the response of a helical director distribution within a thin nematic slab subject to a magnetic field was investigated by Leslie using continuum theory and shown to be unwound above a threshold field.[4] This helical configuration is identical to the structure in the twisted nematic display device that was described just a few months later by Schadt and Helfrich,[5] which lead to a revolution in the display industry. It was then appreciated that the Frank (or perhaps more correctly the Frank–Oseen–Zocher) theory was an invaluable tool with which to model the static performance of display devices based on nematics.

Indeed, it was this fact together with the extension to continuum dynamics[6] that was responsible, in part, for the rapid development of displays. However, to be able to model a device the elastic constants must be known and this has led, at last, to the measurements[7] that Sir Charles had called for at the 1958 Discussion.

References

1 C. W. Oseen, *Trans. Faraday Soc.,* 1933, **29**, 883.
2 H. Zocher, *Trans. Faraday Soc.*, 1933, **29**, 945.
3 R. B. Meyer, *Phys. Rev. Lett.*, 1969, **22**, 918.
4 F. M. Leslie, *Mol. Cryst. Liq. Cryst.*, 1970, **12**, 57.
5 M. Schadt and W. Helfrich, *Appl. Phys. Lett.*, 1971, **18**, 127.
6 J. L. Ericksen, *Trans. Soc. Rheol.*, 1961, **5**, 23; F. M. Leslie, *Arch. Ration. Mech. Anal.*, 1968, **28**, 265.
7 D. A. Dunmur in *Physical Properties of Liquid Crystals: Nematics*, ed. D.A. Dunmur, A. Fukuda and G. R. Luckhurst, INSPEC, IEE, London, 2001, p. 216.

I. LIQUID CRYSTALS

ON THE THEORY OF LIQUID CRYSTALS

By F. C. Frank

H. H. Wills Physics Laboratory, University of Bristol

Received 19th February, 1958

A general theory of curvature-elasticity in the molecularly uniaxial liquid crystals, similar to that of Oseen, is established on a revised basis. There are certain significant differences : in particular one of his coefficients is shown to be zero in the classical liquid crystals. Another, which he did not recognize, does not interfere with the determination of the three principal coefficients. The way is therefore open for exact experimental determination of these coefficients, giving unusually direct information regarding the mutual orienting effect of molecules.

---

1. INTRODUCTION

One of the principal purposes of this paper is to urge the revival of experimental interest in its subject. After the Society's successful Discussion on liquid crystals in 1933, too many people, perhaps, drew the conclusion that the major puzzles were eliminated, and too few the equally valid conclusion that quantitative experimental work on liquid crystals offers powerfully direct information about molecular interactions in condensed phases.

The first paper in the 1933 Discussion was one by Oseen,[1] offering a general structural theory of the classical liquid crystals, i.e. the three types, smectic, nematic and cholesteric, recognized by Friedel[2] (1922). In the present paper Oseen's theory (with slight modification) is refounded on a securer basis. As with Oseen, this is a theory of the molecularly uniaxial liquid crystals, that is to say, those in which the long-range order governs the orientation of only one molecular axis. This certainly embraces the classical types, though in the smectic class one translational degree of freedom is also crystalline. Other types of liquid crystal may exist, but are at least relatively rare : presumably because ordering of one kind promotes ordering of another—it is already exceptional for *one* orientational degree of freedom to crystallize without simultaneous crystallization of the translational degrees of freedom. Fluidity (in the sense that no shear stress can persist in the absence of flow) is in principle compatible with biaxial orientational order, with or without translational order in one dimension. It is very unlikely that translational order in two dimensions, and not in the third, can occur as an equilibrium situation. The existence of a three-dimensional lattice is not compatible with true fluidity. A dilute solution with lattice order, which appeared to be fluid, would not be considered a liquid crystal from the present viewpoint, but rather a solid with a very low plastic yield stress.

The Oseen theory embraces smectic mesophases, but is not really required for this case. The interpretation of the equilibrium structures assumed by smectic substances under a particular system of external influences may be carried out by essentially geometric arguments alone. The structures are conditioned by the existence of layers of uniform thickness, which may be freely curved, but in ways which do not require a breach of the layering in regions of greater extension than lines. These conditions automatically require the layers to be Dupin cyclides and the singular lines to be focal conics. Nothing, essentially, has been added

19

to the account of these given by Friedel, and not much appears to be needed, though a few minor features (the scalloped edges of Grandjean terraces, and scalloped frills of battonets) have not been fully interpreted.

The case is quite different for the nematic and cholesteric liquid crystals. This is particularly clear for the former. In a thin film, say, of a nematic substance, particular orientations are imposed at the surfaces, depending on the nature or prior treatment of the materials at these surfaces; if the imposed orientations are not parallel some curved transition from one orientation to the other is required. Curvature may also be introduced when, say, the orienting effect of a magnetic field conflicts with orientations imposed by surface contacts. Something analogous to elasticity theory is required to define the equilibrium form of such curvatures. It is, however, essentially different from the elasticity theory of a solid. In the latter theory, when we calculate equilibrium curvatures in bending, we treat the material as having undergone homogeneous strains in small elements: restoring forces are considered to oppose the change of distance between neighbouring points in the material. In a liquid, there are no permanent forces opposing the change of distance between points: in a bent liquid crystal, we must look for restoring torques which directly oppose the curvature. We may refer to these as torque-stresses, and assume an equivalent of Hooke's law, making them proportional to the curvature-strains, appropriately defined, when these are sufficiently small. It is an equivalent procedure to assume that the free-energy density is a quadratic function of the curvature-strains, in which the analogues of elastic moduli appear as coefficients: this is the procedure we shall actually adopt.

Oseen likewise proceeded by setting up an expression for energy density, in terms of chosen measures of curvature. However, he based his argument on the postulate that the energy is expressible as a sum of energies between molecules taken in pairs. This is analogous to the way in which Cauchy set up the theory of elasticity for solids, and in that case it is known that the theory predicted fewer independent elastic constants than actually exist, and we may anticipate a similar consequence with Oseen's theory.

It is worth remarking that the controversial conflict between the " swarm theory " and the " continuum theory " of liquid crystals is illusory. The swarm theory was a particular hypothetical and approximative approach to the statistical mechanical problem of interpreting properties which can be well defined in terms of a continuum theory. This point is seen less clearly from Oseen's point of departure than from that of the present paper.

2. BASIC THEORY

We first require to define the components of curvature. Let L be a unit vector representing the direction of the preferred orientation in the neighbourhood of any point. The sign of this vector is without physical significance, at least in most cases. If so, it must be chosen arbitrarily at some point and defined by continuity from that point throughout the region in which L varies slowly with position. In multiply-connected regions it may be necessary to introduce arbitrary surfaces of mathematical discontinuity, where this sign changes without any physical discontinuity. At any point we introduce a local system of Cartesian co-ordinates, x, y, z, with z parallel to L at the origin, x chosen arbitrarily perpendicular to z, and y perpendicular to x so that x, y, z form a right-handed system. Referred to these axes, the six components of curvature at this point are (see fig. 1):

$$\left.\begin{array}{lll}
\text{" splay ":} & s_1 = \partial L_x/\partial x, & s_2 = \partial L_y/\partial y; \\
\text{" twist ":} & t_1 = -\partial L_y/\partial x, & t_2 = \partial L_x/\partial y; \\
\text{" bend ":} & b_1 = \partial L_x/\partial z, & b_2 = \partial L_y/\partial z.
\end{array}\right\} \quad (1)$$

Then, putting

$$
\left.
\begin{aligned}
L_x &= a_1 x + a_2 y + a_3 z + 0(r^2), \\
L_y &= a_4 x + a_5 y + a_6 z + 0(r^2), \\
L_z &= 1 + 0(r^2), \ (r^2 = x^2 + y^2 + z^1),
\end{aligned}
\right\}
\tag{2}
$$

we have

$$
s_1 = a_1, \ t_2 = a_2, \ b_1 = a_3, \ -t_1 = a_4, \ s_2 = a_5, \ b_2 = a_6. \tag{3}
$$

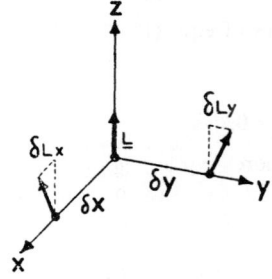

Splays.

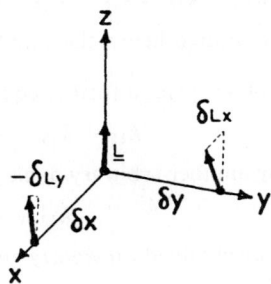

Twists.

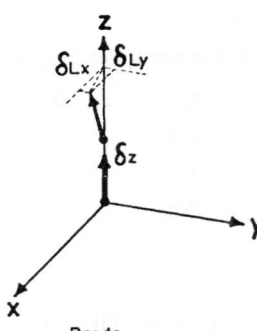

Bends.

Fig. 1.

We postulate that the free energy G of a liquid crystal specimen in a particular configuration, relative to its energy in the state of uniform orientation, is expressible as the volume integral of a free-energy density g which is a quadratic function of the six differential coefficients which measure the curvature :

$$
G = \int_v g\,d\tau, \tag{4}
$$

$$
g = k_i a_i + \tfrac{1}{2} k_{ij} a_i a_j, \ (i, j = 1 \ldots 6, \ k_{ij} = k_{ji}), \tag{5}
$$

where summation over repeated suffixes is implied.

In so far as there was arbitrariness in our choice of the local co-ordinate system (x, y, z), we require that when we replace this by another equally permissible one (x', y', z') in which we have new curvature components a'_i, g shall be the same function as before of these curvature components :

$$
g = k_i a'_i + \tfrac{1}{2} k_{ij} a'_i a'_j. \tag{6}
$$

This requirement will impose restrictions on the moduli, k_i, k_{ij}.

The choice of the x-direction was arbitrary, apart from the requirement that it should be normal to z, which is parallel to the physically significant direction $\mathbf{L}$:

hence any rotation of the co-ordinate system around z is a permissible one. Putting $z' = y$, $y' = -x$, $z' = z$, gives us the equations:

$$
\left.
\begin{aligned}
&k_1 = k_5,\ k_2 = -k_4,\ k_3 = k_6 = 0,\\
&k_{11} = k_{55},\ k_{22} = k_{44},\ k_{33} = k_{66},\\
&k_{12} = -k_{45},\ k_{14} = -k_{25},\\
&k_{13} = k_{16} = k_{23} = k_{26} = k_{34} = k_{35} = k_{36} = k_{46} = k_{56} = 0
\end{aligned}
\right\}
\tag{7}
$$

The working is omitted here: the simpler examples of eqn. (13-17) below exhibit the principle.

A rotation of 45° gives a further equation

$$
k_{11} - k_{15} - k_{22} - k_{24} = 0,
\tag{8}
$$

and rotation by another arbitrary angle just one more

$$
k_{12} + k_{14} = 0,
\tag{9}
$$

which with those obtained previously gives

$$
k_{12} = -k_{14} = k_{25} = -k_{45}.
\tag{10}
$$

Thus, of the six hypothetical moduli k_i two are zero and only two are independent:

$$
k_i = (k_1\ k_2\ 0\ -k_2\ k_1\ 0),
\tag{11}
$$

while of the thirty-six k_{ij} eighteen are zero and only five are independent:

$$
k_{ij} =
\left\{
\begin{array}{cccccc}
k_{11} & k_{12} & 0 & -k_{12} & (k_{11} - k_{22} - k_{24}) & 0\\
k_{12} & k_{22} & 0 & k_{24} & k_{12} & 0\\
0 & 0 & k_{33} & 0 & 0 & 0\\
-k_{12} & k_{24} & 0 & k_{22} & -k_{12} & 0\\
(k_{11} - k_{22} - k_{24}) & k_{12} & 0 & -k_{12} & k_{11} & 0\\
0 & 0 & 0 & 0 & 0 & k_{33}
\end{array}
\right\}
\tag{12}
$$

If the molecules are non-polar with respect to the preferentially oriented axis, or, if polar, are distributed with equal likelihood in both directions, the choice of sign of $\mathbf{L}$ is arbitrary. It is a significant convention in our definition of curvature components that z is positive in the positive direction of $\mathbf{L}$: and if z changes sign, one of x and y should change sign also to retain right-handed co-ordinates. Hence a permissible transformation *in the absence of physical polarity* is $\mathbf{L}' = -\mathbf{L}$, $x' = x$, $y' = -y$, $z' = -z$. This gives us

$$
\left.
\begin{aligned}
L'_{x'} &= -a_1 x' + a_2 y' + a_3 z' + 0(r^2),\\
L'_{y'} &= \ \ a_4 x' - a_5 y' - a_6 z' + 0(r^2).
\end{aligned}
\right\}
\tag{13}
$$

Since, compared with (2), the coefficients with indices 1, 5 and 6 have changed sign, the required invariance of (6) gives us the equations

$$
k_1 = 0,\ k_5 = 0,\ k_6 = 0,
\tag{14}
$$

and (from the second order terms in which only one factor has changed sign):

$$
k_{12} = k_{13} = k_{14} = k_{25} = k_{26} = k_{35} = k_{36} = k_{45} = k_{46} = 0.
\tag{15}
$$

Some of this information is already contained in eqn. (7-10). The effect upon (11) and (12) is that k_1 and k_{12} vanish.

There is a further element of arbitrariness in our insistence on right-handed co-ordinates, unless the molecules are enantiomorphic, or enantiomorphically arranged. Empirically, it appears that enantiomorphy does not occur in liquid crystals unless the molecules are themselves distinguishable from their mirror images, and that it also vanishes in racemic mixtures. *In the absence of enantiomorphy*, a permissible transformation is $x' = x$, $y' = -y$, $z' = z$, giving

$$\left.\begin{array}{l} L_{x'} = a_1 x' - a_2 y' + a_3 z' + 0(r^2), \\ L_{y'} = - a_4 x' + a_5 y' - a_6 z' + 0(r^2), \end{array}\right\} \quad (16)$$

whence, by the same argument as before (omitting the redundant information),

$$k_2 = 0 \text{ and } k_{12} = 0. \quad (17)$$

Hence, while (15) with (11) and (12) expresses the most general dependence of free energy density on curvature in molecularly uniaxial liquid crystals, k_1 vanishes in the absence of polarity, k_2 vanishes in the absence of enantiomorphy, and k_{12} vanishes unless both polarity and enantiomorphy occur together.

The general expression for energy density in terms of the notation of eqn. (1) is

$$g = k_1(s_1 + s_2) + k_2(t_1 + t_2) + \tfrac{1}{2}k_{11}(s_1 + s_2)^2 + \tfrac{1}{2}k_{22}(t_1 + t_2)^2 + \tfrac{1}{2}k_{33}(b_1{}^2 + b_2{}^2)$$
$$+ k_{12}(s_1 + s_2)(t_1 + t_2) - (k_{22} + k_{24})(s_1 s_2 + t_1 t_2). \quad (18)$$

By introducing

$$s_0 = -k_1/k_{11}, \quad t_0 = -k_2/k_{22}, \quad (19)$$

and

$$g' = g + \tfrac{1}{2}k_{11}s_0{}^2 + \tfrac{1}{2}k_{22}t_0{}^2, \quad (20)$$

i.e. by adopting a new and (in the general case) lower zero for the free-energy density, corresponding not to the state of uniform orientation but to that with the optimum degree of splay and twist, we obtain the more compact expression

$$g' = \tfrac{1}{2}k_{11}(s_1 + s_2 - s_0)^2 + \tfrac{1}{2}k_{22}(t_1 + t_2 - t_0)^2 + \tfrac{1}{2}k_{33}(b_1{}^2 + b_2{}^2)$$
$$+ k_{12}(s_1 + s_2)(t_1 + t_2) - (k_{22} + k_{24})(s_1 s_2 + t_1 t_2). \quad (21)$$

An alternative form of this expression is given as eqn. (25) below.

2.2 COMPARISON WITH OSEEN'S THEORY

According to Oseen, the energy is expressed by

$$\frac{1}{2m^2} \int \int \rho_1 \rho_2 Q(\xi_1, \xi_2) d\tau_1 d\tau_2$$

$$+ \frac{1}{2m^2} \int \rho_2 \{ K_1 \mathbf{L} \cdot \nabla \times \mathbf{L} + K_{11}(\mathbf{L} \cdot \nabla \times \mathbf{L})^2 + K_{22}(\nabla \cdot \mathbf{L})^2$$

$$+ K_{33}((\mathbf{L} \cdot \nabla)\mathbf{L})^2 + 2K_{12}(\nabla \cdot \mathbf{L})(\mathbf{L} \cdot \nabla \times \mathbf{L}) \} d\tau. \quad (22)$$

We are not concerned with the first integral, which is not related to the dependence of energy on curvature (and which plays only a minor role in Oseen's theory). It is the integrand of the second integral which should be compared with our free-energy density g. Noting that

$$\mathbf{L} \cdot \nabla \times \mathbf{L} = \partial L_y/\partial x - \partial L_x/\partial y \quad = -(t_1 + t_2),$$
$$\nabla \cdot \mathbf{L} = \partial L_x/\partial x + \partial L_y/\partial y \quad = (s_1 + s_2), \quad (23)$$
$$((\mathbf{L} \cdot \nabla)\mathbf{L})^2 = (\partial L_x/\partial z)^2 + (\partial L_y/\partial z)^2 = (b_1{}^2 + b_2{}^2),$$

we see that with

$$- (\rho^2/2m^2)K_1 = k_2, \ (\rho^2/m^2)K_{11} = k_{22}, \ (\rho^2/m^2)K_{22} = k_{11},$$
$$- (\rho^2/m^2)K_{12} = k_{12}, \ (\rho^2/m^2)K_{33} = k_{33}, \quad (24)$$

Oseen's expression is equivalent to (18), except that the latter contains the additional terms,

$$k_1(s_1 + s_2) - (k_{22} + k_{24})(s_1 s_2 + t_1 t_2).$$

This accords with the anticipation that Oseen was in danger of missing terms by adopting a Cauchy-like approach to the problem.

The first omission is not very important, since it is virtually certain that there is no physical polarity along the direction $\mathbf{L}$ in any of the normal liquid crystal substances which were discussed by Oseen; and k_1 is then zero. The second omission is of more general significance: but $(s_1s_2 + t_1t_2)$ relates to an essentially three-dimensional kind of curvature. It occurs in pure form (with $(s_1 + s_2)$ and $(t_1 + t_2)$ equal to zero) in what we may call " saddle-splay ", when the preferred directions $\mathbf{L}$ are normal to a saddle-surface; and then contributes a positive term to the energy if $(k_{22} + k_{24})$ is positive. It is zero if $\mathbf{L}$ is either constant in a plane or parallel to a plane. It may be disregarded in all the simpler configurations which would be employed for the determination of moduli other than k_{24}, provided only that $(k_{22} + k_{24})$ is non-negative.

The most gratifying result of the comparison is to notice that there is one term in Oseen's expression which can be omitted, namely the last, since K_{12} is always zero under the conditions which justify omitting the term $k_1(s_1 + s_2)$. For many purposes we actually have a simpler result than Oseen's. In detailed application, he in fact assumed $K_{12} = 0$, supposing this to be an approximation.

We may conveniently use relations (23) to cast eqn. (21) into co-ordinate-free notation:

$$g' = \tfrac{1}{2}k_{11}(\nabla \cdot \mathbf{L} - s_0)^2 + \tfrac{1}{2}k_{22}(\mathbf{L} \cdot \nabla \times \mathbf{L} + t_0)^2 + \tfrac{1}{2}k_{33}((\mathbf{L} \cdot \nabla)\mathbf{L})^2$$
$$- k_{12}(\nabla \cdot \mathbf{L})(\mathbf{L} \cdot \nabla \times \mathbf{L}) - \tfrac{1}{2}(k_{22} + k_{24})((\nabla \cdot \mathbf{L})^2$$
$$+ (\nabla \times \mathbf{L})^2 - \nabla\mathbf{L} : \nabla\mathbf{L}), \tag{25}$$

where

$$\nabla\mathbf{L} : \Delta\mathbf{L} = \left(\frac{\partial L_x}{\partial x}\right)^2 + \left(\frac{\partial L_y}{\partial y}\right)^2 + \left(\frac{\partial L_z}{\partial z}\right)^2 + \left(\frac{\partial L_y}{\partial x}\right)^2 + \left(\frac{\partial L_y}{\partial y}\right)^2 + \left(\frac{\partial L_y}{\partial z}\right)^2$$
$$+ \left(\frac{\partial L_z}{\partial x}\right)^2 + \left(\frac{\partial L_z}{\partial y}\right)^2 + \left(\frac{\partial L_z}{\partial z}\right)^2,$$

or $(\nabla \cdot \mathbf{L})^2 + (\nabla \times \mathbf{L})^2 - (\nabla\mathbf{L} : \Delta\mathbf{L})$

$$= 2\left\{\frac{\partial L_x}{\partial x}\frac{\partial L_y}{\partial y} + \frac{\partial L_y}{\partial y}\frac{\partial L_z}{\partial z} + \frac{\partial L_z}{\partial z}\frac{\partial L_x}{\partial x} - \frac{\partial L_y}{\partial x}\frac{\partial L_x}{\partial y} - \frac{\partial L_z}{\partial y}\frac{\partial L_y}{\partial z} - \frac{\partial L_z}{\partial z}\frac{\partial L_z}{\partial x}\right\},$$

in a fully arbitrary system of co-ordinates.

3. PARTICULAR CASES

3.1. THE SMECTIC STATE

According to Oseen, the smectic state corresponds to the vanishing of all the moduli except k_{22} and k_{33} (our notation). The (free) energy is then minimized when

$$\mathbf{L} \cdot \nabla \times \mathbf{L} = 0, \quad (\mathbf{L} \cdot \nabla)\mathbf{L} = 0. \tag{26}$$

The second of these equations states that a line following the preferred direction of molecular axes is straight: the first, that the family of such straight lines is normal to a family of parallel surfaces (defining the surface parallel to a given curved surface as the envelope of spheres of uniform radius centred at all points on the given surface). Hence he formally predicts the geometry explicable by molecular layering without apparently appealing to the existence of layers: their real existence he explains separately by use of his Q integral. The present writer considers this a perverse approach. We need to explain why k_{22} and k_{33} are so large that the other moduli are negligible, and can do so from the existence of the layering. There is no real need to employ the theory of curvature strains to interpret smectic structures, until one requires to deal with small departures from the geometrically interpretable structures, which will be permitted if the other moduli are merely small, instead of vanishing, compared with k_{22} and k_{33}.

Before leaving the subject of the smectic state, we may remark that to explain deformations in which the area of individual molecular layers does not remain constant, it is necessary to invoke dislocations of these layers. It is likely that these dislocations are usually combined with the focal conic singularity lines, being then essentially screw dislocations.

3.2. THE NEMATIC STATE

The nematic state is characterized by $s_0 = 0$, $t_0 = 0$, $k_{12} = 0$, corresponding to the absence both of polarity and enantiomorphy. There are four non-vanishing moduli, k_{11}, k_{22}, k_{33} and k_{24}. The last has no effect in " planar " structures. Hence Zocher's [3] three-constant theory is justified (his k_1, k_2, k_t are the same as k_{11}, k_{33}, k_{22}). Formerly, this appeared to be only an approximate theory, neglecting k_{12} which appeared in the theory of Oseen. The simplest way to measure the moduli is to impose body torques by imposing magnetic fields. k_{22} can be determined straightforwardly; k_{11} and k_{33} are more difficult to separate from each other. There is evidence that they are about equal. Using the experimental data of Fréederickz and of Foëx and Royer, Zocher shows that if they are equal the value for *p*-azoxyanisole is $1 \cdot 0 \times 10^{-6}$ dynes. Information about the relative magnitude of these moduli is obtainable from the detailed geometry of the " disinclination " structures described in § 4. The fact that there is not another unknown in k_{12} ought to encourage a complete experimental determination of the moduli for some nematic substances.

3.3. THE CHOLESTERIC STATE

Enantiomorphy, either in the molecules of the liquid-crystal-forming substance, or in added solutes, converts the nematic into the cholesteric state. $k_2 \neq 0$, and the state of lowest free energy has a finite twist, $t_0 = k_2/k_{22}$. In the absence of other curvature components, there is only one structure of uniform twist, in which **L** is uniform in each of a family of parallel planes, and twists uniformly about the normal to these planes. The torsion has a full pitch of $2\pi/t_0$: but since **L** and $-$ **L** are physically indistinguishable, the physical period of repetition is π/t_0. When, as is usually the case, this is of the order of magnitude of a wavelength of light or greater, it can be measured with precision by optical methods. Some additional information should be obtainable by perturbing this structure with a magnetic field. Uniform twist can also exist throughout a volume when the repetition surfaces are not planes, but curved surfaces : for example, spheres, though in this case there has to be at least one singular radius on which the uniformity breaks down. Since the cholesteric substances, like the smectic substances (though for entirely different reasons) give rise to structures containing families of equidistant curved surfaces, their structures show considerable geometric similarity to those of the smectic substances. This was appreciated by Friedel, who also realized that it was misleading, and that the cholesteric phases are in reality thermodynamically equivalent to nematic phases.

3.4. THE CASE $k_1 \neq 0$

If $k_1 \neq 0$, the state of lowest free-energy density has a finite splay, $s_0 = k_1/k_{11}$. This can only exist when the molecules are distinguishable end from end, and there is polarity along **L** in their preferred orientation. Then, almost inevitably, the molecules have an electric dipole moment, and therefore, unless the material is an electric conductor, the condition $\nabla \cdot \mathbf{L} \neq 0$ implies $\nabla \cdot \mathbf{P} \neq 0$ (where **P** is the electric polarization), so that finite splay produces a space-charge. As a second consideration, it is not geometrically possible to have uniform splay in a three-dimensionally extended region. The simple cases of uniform splay are those in

26   LIQUID CRYSTALS

which L is radial in a thin spherical shell of radius $2/s_0$, or a thin cylindrical shell of radius $1/s_0$. These considerations relate this hypothetical polar class of liquid crystals to the substances which produce " myelin figures " : but since the only allowed structures for this class have a high surface-to-volume ratio, a theory of their configurations which does not pay explicit attention to interfacial tensions will be seriously incomplete.

4. " Disclinations "

The nematic state is named for the apparent threads seen within the fluid under the microscope. Their nature was appreciated by Lehmann and by Friedel. In thin films they may be seen end on, crossing the specimen from slide to cover-slip, and their nature deduced from observation in polarized light. In this position they were named *Kerne* and *Konvergenzpunkte* by Lehmann, positive and negative nuclei (*noyaux*) by Friedel. They are line singularities such that the cardinal direction of the preferred axis changes by a multiple of π on a circuit taken round one of the lines. They thus provide examples of the configurations excluded (under the name of " Moebius crystals ") from consideration in the establishment of a general definition of crystal dislocations.[4] In analogy with dislocations, they might be named " disclinations ". It is the motion of these disclination lines which provides one of the mechanisms for change of configuration of a nematic specimen under an orienting influence, in the same way as motion of a domain boundary performs this function for a ferromagnetic substance. It is the lack of a crystal lattice which allows the discontinuities of orientation to have a line topology, instead of a topology of surfaces dividing the material into domains, in the present case. Disclination lines occur in cholesteric as well as in nematic liquid crystals.

The actual configuration around disclination lines was calculated by Oseen for the case $k_{11} = k_{33}$, $k_{12} = 0$, the latter assumption actually being exact. Then if L is parallel to a plane and ϕ is the azimuth of L in this plane, in which x_1, x_2 are Cartesian co-ordinates, the free energy is minimized in the absence of body torques when

$$\partial^2\phi/\partial x^2 + \partial^2\phi/\partial y^2 = 0. \qquad (27)$$

The solutions of this equation representing disclinations are

$$\phi = \tfrac{1}{2}n\psi + \phi_0, \tan \psi = x_2/x_1, \qquad (28)$$

n being an integer. These configurations are sketched in fig. 2. Changing ϕ_0, merely rotates the figure in all cases except $n = +2$, for which three examples are shown. Of these, the first or the third will be stable, for a nematic substance according as k_{33} or k_{11} is the larger. In the other cases, non-equality of k_{11} and k_{33} should not make drastic changes, but only changes of curvature in patterns of the same topology. Thus, for $n = -2$ or -1, if k_{33} is larger than k_{11} the bends will be sharpened, and conversely. This would be observed as a non-uniform rotation of the extinction arms with uniform rotation of polarizer and analyzer. The ratio of k_{11} and k_{33} can thus be determined from a simple optical experiment.

In cholesteric substances ϕ_0 is not constant, but a linear function of the co-ordinate x_3, normal to the x_1, x_2 plane : except for the case $n = 2$, with $k_{11} \neq k_{33}$, which should show a periodic departure from linearity near the core, from which the relative magnitudes of k_{22} and $(k_{11} - k_{33})$ could be deduced, though the observations would not be simple.

Oseen was puzzled at the non-occurrence of configurations corresponding to

$$\phi = c \ln r + \text{const.}, r = (x_1^2 + x_2^2)^{\frac{1}{2}}. \qquad (29)$$

The reason is obvious : unlike n in (28), c is not restricted to integral values, and can relax continuously to zero. Alternatively stated, this configuration requires an impressed torque at the core for its maintenance.

At the time Friedel and Oseen wrote their papers, the values of the disclination-tion strength n which had been observed were -2, -1, 1, 2. Since then the case $n = 4$ has been observed by Robinson [5] in the radial singularity of a cholesteric " spherulite". The non-occurrence of high values of $|n|$ is explained by the fact that the energy is proportional to n^2. The fact that higher values than one occur indicates a relatively high energy in the disorderly core of the disclination-ation line, which must be as large as its field energy so that it becomes profitable for a pair of disclinations to share the same core.

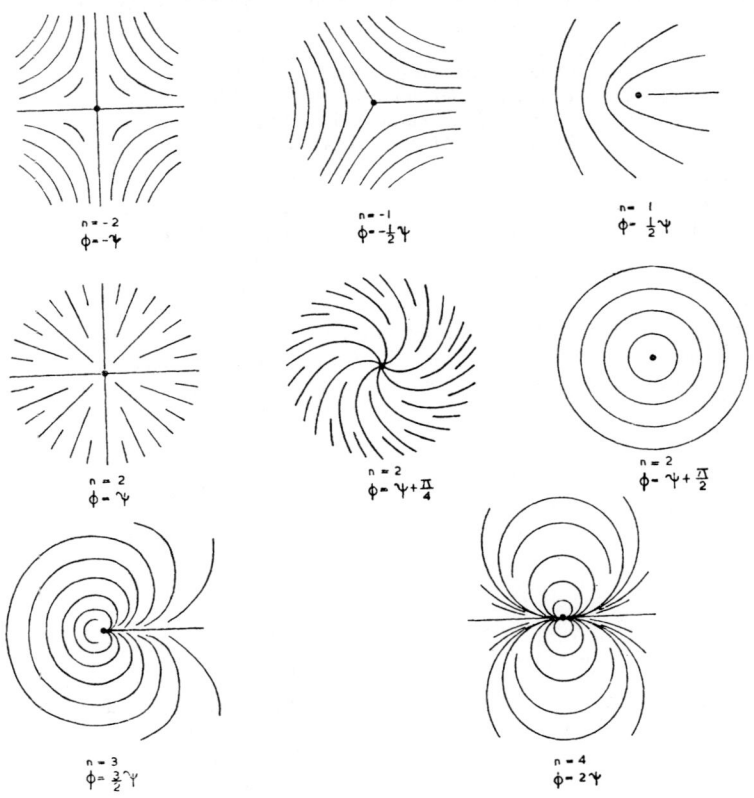

FIG. 2.

5. k_{24}

Let us leave aside the question of how to determine k_{24}, necessarily involving the observation of three-dimensional curvatures, until we have better information about the moduli of plane curvature, k_{11}, k_{22} and k_{33}.

6. RELATIONSHIP TO THE ORDINARY ELASTIC CONSTANTS

Let us take note that the molecular interactions giving rise to liquid crystal properties must also be present in solids. This indicates that conventional elastic theory is incomplete: the direct curvature-strain moduli should also be included. This is true, but does not seriously invalidate the accepted theory of elasticity. Consider the bending of a beam, of thickness $2a$, to a radius R. Then the stored free energy according to ordinary elasticity theory is $g_E = Ea^2/24R^2$, where E

28   LIQUID CRYSTALS

is Young's modulus. The stored free energy arising from a curvature modulus k (corresponding to k_{11} or k_{33} according to the molecular orientation in the beam) would be $g_k = k/2R^2$. Their ratio is

$$g_k/g_E = 12k/Ea^2.$$

Taking k as 10^{-6} dyne, and E as 10^{10} dyne/cm^2, this ratio is unity when a is about $3 \cdot 5 \times 10^{-8}$ cm, and negligible for beams of thickness as large as a micron. Thus, on the visible scale, the curvature-elastic constants are always negligible compared with the ordinary elastic constants, unless the latter are zero.

[1] Oseen, *Trans. Faraday Soc.*, 1933, **29**, 883.
[2] Friedel, *Anales Physique*, 1922, **18**, 273.
[3] Zocher, *Trans. Faraday Soc.*, 1933, **29**, 945.
[4] Frank, *Phil. Mag.*, 1951, **42**, 809.
[5] Robinson, this Discussion.

Liquid–Solid Interfaces

R. K. Thomas

Physical and Theoretical Chemistry Laboratory, University of Oxford, UK OX1 3QZ

Commentary on: **Theory of Self-Assembly of Hydrocarbon Amphiphiles into Micelles and Bilayers,** J.N.Israelachvili, D.J.Mitchell and B.W.Ninham, *J. Chem. Soc., Faraday Trans. 2*, 1976, **72**, 1525.

Amphiphilic molecules in water spontaneously self-assemble into a wonderful range of aggregates. When the solution is relatively dilute the basic shapes of these aggregates may be spherical, oblate or prolate ellipsoids, rods, or vesicles, and, at any given concentration, the aggregates will be very well defined in both shape and size. In a unit of the smallest of these aggregates, a spherical micelle, there will typically be upwards of 50 molecules. As the concentration of amphiphile increases the aggregates coalesce to form another great variety of periodic structures. It is almost a matter of common sense that the size and shape of amphiphilic aggregates should in some way depend on the architecture of the individual molecules and many of those who have devised industrial formulations of surfactants will have utilized some sort of intuitive relation between the performance of the surfactant, its aggregation characteristics, and the molecular architecture. The importance of the paper by Israelachvili, Mitchell and Ninham is that it was the first paper to establish a quantitative basis for assessing the stability of different amphiphilic aggregates in dilute solution.

The simplest structure of a surfactant aggregate is the spherical micelle. In 1941 G. S. Hartley, in a paper in the *Faraday Transactions*,[1] recognized that the combination of the spherical structure with the requirement that the ionic head groups cannot penetrate into the liquid hydrocarbon-like interior of the micelle must impose some geometric constraints on the packing. His idea was that in a double chained surfactant the chains would already be "huddled together" with less exposure of the individual hydrophobic groups to water and hence the tendency of the surfactant to self-assemble would be reduced in comparison with a single-chained surfactant. Since this should not affect adsorption at a flat interface it would relatively enhance the performance of a double chain surfactant as a surface active agent at a flat interface because there would be less competition from aggregation in the bulk. He made a series of double chained surfactants to test this idea and found some experimental support for it, but the work was not finished. In his words, "It seems, however, opportune to publish this short note, although my own experimental work has been interrupted while still far from complete". It was wartime.

After a long gap two other authors, Tartar in 1955[2] and Tanford in 1972,[3] considered the relation between molecular size and micellar shape. Tartar put forward the idea that the maximum extent of the hydrocarbon core is determined by the fully extended length of the hydrocarbon chain, which became a key idea in the work of Israelachvili *et al.*, and he then showed that the surface area per molecule must therefore increase as the micelle changes from spherical to ellipsoidal. He also examined the effects of the ionic atmosphere on the electrostatic screening at the surface of the micelle. Although Tartar's contribution has been neglected, Tanford's paper on the subject is commonly cited but not because of his contribution to the ideas of surfactant packing, which make a small extension to the ideas of Tartar, but because it contains his widely accepted empirical estimates of the relation between the volume and length of a hydrocarbon chain and the number of CH_2 units in it. Ironically, these are the quantities that are essential for calculating the packing factor of Israelachvili *et al.*

The packing factor derived by Israelachvili *et al.* is $V/a_0 l_c$ where V is the volume of the hydrophobic chain, a_0 is the mean cross sectional area of the head group in the aggregate, and l_c is the length of the fully extended chain. V and l_c are empirical quantities, usually taken from the paper by Tanford, but a_0 is less accessible, although trends in a_0 are easily identified. Given that aggregation is always associated with a loss of entropy and hence the most stable aggregate is always the smallest one, simple geometrical considerations lead to the rules that spherical micelles are formed when the packing factor is less than 1/3, ellipsoids when it is between 1/3 and 1/2, and lamellar structures (vesicles *etc.*) when it is between 1/2 and 1. While the rules in the paper extend further than these three basic statements, they are still relatively short and simple, both to understand and apply, yet the paper is very long. This is because the use of simple geometric criteria for micellar packing raises a number of difficult and subtle issues. Among these issues are questions about the most efficient packing in non-spherical aggregates, the role of polydispersity, whether it is valid to use a mean area per molecule and the location of this area in the aggregate, the existence or non-existence of tubular vesicles, and the optimum size of vesicles. It is in addressing these more subtle issues that the authors found it necessary to delve deeper, and at length, into the underlying basis of self-assembly. Amphiphiles are of an awkward size for statistical mechanics, too large for liquid state theories, and too small for the scaling arguments that can be applied to polymers. By the standards of these fields the paper by Israelachvili *et al.* is both crude and oversimple. However, as stated in the authoritative book on colloid science by Hunter,[4] "This deceptively simple packing model allows many physical properties of micelles and vesicles such as size, shape, polydispersity, etc. to be predicted."

References

1 G. S. Hartley, *Trans. Faraday Soc.*, 1941, **37**, 130.
2 H. V. Tartar, *J. Phys. Chem.*, 1955, **59**, 1195.
3 C. Tanford, *J. Phys. Chem.*, 1972, **76**, 3020.
4 R. J. Hunter, *Foundations of Colloid Science*, Clarendon Press, Oxford, 1987, vol. 1.

Theory of Self-Assembly of Hydrocarbon Amphiphiles into Micelles and Bilayers

By Jacob N. Israelachvili,† D. John Mitchell and
Barry W. Ninham*

Department of Applied Mathematics, Institute of Advanced Studies,
Research School of Physical Sciences, The Australian National University,
Canberra, A.C.T. 2600, Australia

Received 7th November, 1975

A simple theory is developed that accounts for many of the observed physical properties of micelles, both globular and rod-like, and of bilayer vesicles composed of ionic or zwitterionic amphiphiles. The main point of departure from previous theories lies in the recognition and elucidation of the role of geometric constraints in self-assembly. The linking together of thermodynamics, interaction free energies and geometry results in a general framework which permits extension to more complicated self-assembly problems.

† And Department of Neurobiology, Research School of Biological Sciences, The Australian National University, Canberra.

1526   ASSEMBLIES OF HYDROCARBON AMPHIPHILES

1. INTRODUCTION

" Despite enormous progress in understanding the genetics and biochemistry of molecular synthesis we still have only primitive ideas of how linearly synthesised molecules form the multimolecular aggregates that are cellular structures. We assume that the physical forces acting between aggregates of molecules and between individual molecules should explain many of their associative properties; but available physical methods have been inadequate for measuring or computing these forces in solids or liquids." These few succinct opening sentences from the review of V. A. Parsegian [1] embrace and define a whole grey area bridging chemistry, physics and biology which is as yet but little explored. They imply a formidable injunction. For while it is axiomatic to the physicist or chemist that structural changes in any system should be reduced to a consideration of forces or free energies which cause those changes, the burden of proof lies with the proponent. The axioms of physics do not always receive so ready an acceptance from biologists whose whole thinking in the past has been centred on the role of geometry to the almost complete exclusion of forces and entropy. The burden of proof becomes especially great if one considers the increasing sophistication of those few successful theoretical advances in our understanding of condensed matter. To be convincing, and to have any hope whatever of reducing to some semblance of order the vast complexity of those intricate multimolecular structures that are the subject of biology, any successful theories of self-assembly must have as a minimal requirement extreme simplicity— to make them accessible to the biologist who has enough concerns of his own not to be dragged into the subtleties of modern physics.

There is merit in the view that forces and entropy are important. There is merit in the view that geometry is a determining factor in self-assembly. But there have been few attempts at model self-assembly problems which embrace both views. The aim of this paper is to develop a simple theory dealing with one area of self-assembly, the spontaneous aggregation of one-component lipid suspensions, where the fusion of both notions results in a unification of a multiplicity of diverse observations.

We shall deal with ionic and zwitterionic amphiphiles. These can take on a confusing variety of different shapes and sizes : some aggregate into small spherical or globular micelles, others appear to form long cylindrical micelles, while others coalesce spontaneously into vesicular or lamellar bilayers.

The outline of the paper is as follows : In section 2 we delineate the necessary

J. N. ISRAELACHVILI, D. J. MITCHELL AND B. W. NINHAM 1527

thermodynamics which must underlie any description of self-assembly. Section 3 is concerned with the formulation of a simple model, with minimal assumptions, for the free energies of amphiphiles in both aggregated and dispersed states. In section 4 we carry out a preliminary application of the model to spherical micelles, and show that while the agreement between prediction and observation is qualitatively correct, the missing component can come only when geometric constraints are built into the model. Section 5 is devoted to a study of these geometric constraints. In section 6 thermodynamic consequences of these constraints are examined in detail, and the refined model is shown to give remarkable agreement with observation. This confidence in our model having been established, we proceed in section 7 to study its consequences for the assembly of biological phospholipids which aggregate into bilayer structures. Finally we summarise our conclusions.

2. THERMODYNAMICS OF SELF-ASSEMBLY

(a) SOME GENERAL RESULTS

To set notation and develop a framework for our later studies we briefly recapitulate the thermodynamics of micelle formation. The literature is voluminous and confusing; the most detailed statement of the problem being that of Hall and Pethica [2] based on Hill's small system thermodynamics.[3] We follow the approach of Tanford [4] in which micelles (here meaning any lipid aggregate) made up of N amphiphiles are considered to be distinct species each characterised by its own self free energy $N\mu_N^0$. N will be called the aggregation number. Assume that dilute solution theory holds. Then equilibrium thermodynamics demands that the mole fraction X_N of amphiphile incorporated into micelles of aggregation number N is given by

$$\mu_N^0 + \frac{kT}{N} \ln (X_N/N) = \mu_1^0 + kT \ln X_1, \tag{2.1}$$

where k is Boltzmann's constant, T is temperature, and the suffix 1 denotes isolated amphiphiles. Alternatively

$$X_N = NX_1^N \exp [N(\mu_1^0 - \mu_N^0)/kT]. \tag{2.2}$$

For a prescribed total amphiphile concentration S, measured in mole fraction units, provided μ_N^0 are known functions of N and T, the X_N may be determined from eqn (2.2) together with the supplementary relation

$$S = \sum_{N=1}^{\infty} X_N. \tag{2.3}$$

Two quantities which we shall require are the mean micelle aggregation number $\overline{N}$ and the standard deviation σ of the distribution of sizes which measures polydispersity. These quantities are defined by

$$\overline{N} \equiv \sum_{N>1} NX_N \Big/ \sum_{N>1} X_N \tag{2.4}$$

$$\sigma^2 = \langle N - \overline{N} \rangle^2 = \sum_{N>1} N^2 X_N \Big/ \sum_{N>1} X_N - \overline{N}^2. \tag{2.5}$$

It follows from eqn (2.2)–(2.4) that

$$\overline{N} = \frac{\partial \ln (S - X_1)}{\partial \ln X_1}. \tag{2.6}$$

This equation relates the rate of change of micelle concentration with monomer concentration to the mean micelle aggregation number. Observed aggregation numbers $\overline{N}$ tend to be large (typically > 40) since amphiphiles can decrease their free energies significantly by forming, and only by forming, large aggregates. Therefore, an immediate consequence of eqn (2.6) is that the micelle concentrations $S - X_1$ will be rapidly varying functions of monomer concentration X_1. Define a critical micelle concentration (cmc) to be that value of X_1 for which $S - X_1^c = X_1^c$. For X_1 slightly less than the cmc X_1^c the micelle concentration will be much less than the monomer concentration, and for $X_1 > X_1^c$, the micelle concentration will be much greater than X_1. Another useful expression which follows from eqn (2.3)–(2.5) is

$$\sigma^2 = \frac{\partial \overline{N}}{\partial \ln X_1} = \overline{N} \frac{\partial \overline{N}}{\partial \ln (S - X_1)}. \tag{2.7}$$

This equation relates the standard deviation to the rate of change of mean aggregation number with respect to micelle concentration $(S - X_1)$. Clearly, a rapid change of $\overline{N}$ with concentration is evidence of a large distribution of polydispersity in micelle size (at a given concentration).

(b) CONSEQUENCES OF END EFFECTS

The preceding relations are standard. If $\mu_N^0 - \mu_1^0$ is negative, i.e., aggregates are energetically favoured, the transition from disaggregated to aggregated states comes about due to competition between entropy and energy, and little more can be said without spelling out the detailed form and magnitude of $\mu_N^0 - \mu_1^0$. This will be reserved for later sections, but we anticipate here the forms which will emerge and make some general remarks. Mathematically only two possibilities exist for μ_N^0 which will allow the formation of large aggregates. We consider these in turn.

(1) μ_N^0 may have a minimum value attained for some finite value of N, say $N = M$. This is the usual case discussed in the literature.[4] Physically, such a minimum or optimal value for μ_N^0 comes about due to the competing effects, an increased hydrophobic free energy of the hydrocarbon tails for $N \lesssim M$, and increased head group interaction due to electrostatic or geometric constraints for $N \gtrsim M$. Very little can be said without a specific form of μ_N^0 and a particular example of this case will be discussed extensively in section 4. A more convenient form for eqn (2.2) is here

$$\frac{X_N}{N} = \left(\frac{X_M}{M} \exp \left[M(\mu_M^0 - \mu_N^0)/kT \right] \right)^{N/M} \tag{2.8}$$

Two points can be made. First, even if $\mu_N^0 = \mu_M^0$ for all $N > M$, as will be shown below $\overline{N}$ cannot exceed M by very much for dilute lipid solutions. In fact $\overline{N}$ will be found in section 4 to be somewhat smaller than the optimal value M. Second, if the spread of micelle size is not large, the cmc will be given to a good approximation by neglecting polydispersity and assuming $\overline{N} = M$. Whence (solving $X_1 = X_M$ for X_1^c), we have

$$\ln (\text{cmc}) = \left(\frac{M}{M-1} \right)(\mu_M^0 - \mu_1^0)/kT - \left(\frac{1}{M-1} \right) \ln M$$
$$\simeq (\mu_M^0 - \mu_1^0)/kT. \tag{2.9}$$

This observation greatly facilitates analysis of experimental data.

(2) The other possible form for μ_N^0 is that μ_N^0 decreases with increasing N tending to a finite limit as $N \to \infty$. This form will be seen later to arise from contributions to μ_N^0 from end effects, e.g., hemispherical-like ends on rod-shaped micelles, and hemicylindrical-like ends at the extremities of planar aggregates.

J. N. ISRAELACHVILI, D. J. MITCHELL AND B. W. NINHAM   1529

One functional form for μ_N^0 which will occur in connection with rod-shaped micelles is

$$\mu_N^0 = \mu_\infty^0 + \alpha\, kT/N \qquad N \geqslant m$$
$$= \infty \qquad\qquad 1 < N < m \qquad\qquad (2.10)$$

where α is a constant, and m is a finite number below which micelles are energetically disfavoured, *i.e.*, $\mu_N^0 \gg \mu_\infty^0$ and, therefore, for convenience μ_N^0 is set equal to infinity for $1 < N < m$. Substitution of this μ_N^0 in eqn (2.2)–(2.3) leads to

$$\frac{X_N}{N} = e^{-\alpha}\, Y^N, \qquad S = X_1 + e^{-\alpha} \sum_{N=m}^{\infty} N Y^N, \qquad\qquad (2.11)$$

where

$$Y = X_1 \exp\,(\mu_1^0 - \mu_\infty^0)/kT. \qquad\qquad (2.12)$$

The summation in eqn (2.11) can be carried out to give

$$S = X_1 + \frac{m Y^m\, e^{-\alpha}}{1-Y}\left(1 + \frac{Y}{m(1-Y)}\right), \qquad\qquad (2.13)$$

whence from eqn (2.5)

$$\overline{N} = m + \frac{Y}{1-Y}\left(1 + \frac{1}{m(1-Y)+Y}\right). \qquad\qquad (2.14)$$

For small Y, in fact for $m(1-Y) \gg 1$, $\overline{N} \simeq m$ and as Y increases to 1, $\overline{N} \to \infty$. But if $m(1-Y) \ll 1$ so that $Y \approx 1$, we find

$$(S - X_1)\, e^{\alpha} \simeq \frac{1}{(1-Y)^2} \gg m^2. \qquad\qquad (2.15)$$

Hence $Y \approx 1$ corresponds to large and physically unreasonable micelle concentrations $(S - X_1)$ unless α is large. That is, very large micelles can only occur at reasonable concentrations (typically in the range 10^{-6}–10^{-2} mol dm^{-3}) if α is large. In this case

$$\overline{N} \simeq \frac{2}{1-Y} \simeq 2\sqrt{S\, e^{\alpha}}. \qquad\qquad (2.16)$$

This means that $\overline{N}$ is large and varies very rapidly with concentration S once the value of α exceeds about 25. In the other limit $m(1-Y) \gg 1$, we find

$$(S - X_1)\, e^{\alpha} \simeq \frac{m Y^m}{(1-Y)}. \qquad\qquad (2.17)$$

This limit, therefore, corresponds to values of $(S - X_1)\, e^{\alpha}$ small compared with m^2. If we take $m(1-Y) = x$ (with m large and $x \sim 1$) we have

$$\overline{N} \simeq \frac{m(x^2 + 2x + 2)}{x(x+1)}$$

$$(S - X_1)\, e^{\alpha} \simeq \frac{m^2\, e^{-x}}{x} \qquad\qquad (2.18)$$

so that if x is chosen to make $(S - X_1)\, e^{\alpha} \simeq 1$, we have $\overline{N} \simeq m$. Therefore, with the form of μ_N^0 given by eqn (2.10), when α is small only small micelles ($\overline{N} \simeq m$) can occur at reasonable micelle concentrations $(S - X_1) \ll 1$. These micelles will be narrowly dispersed. When α is large, micelles will be large and broadly dispersed.

1530   ASSEMBLIES OF HYDROCARBON AMPHIPHILES

Another form which will be of interest in connection with the formation of bilayers is

$$\mu_N^0 = \mu_\infty^0 + \alpha \frac{kT}{N^{\frac{1}{4}}} \qquad N \geqslant m$$

$$= \infty \qquad\qquad 1 < N < m \qquad\qquad (2.19)$$

which leads to the expression for the concentration

$$S = X_1 + \sum_{N=m}^{\infty} N Y^N \exp\left(-\alpha N^{\frac{1}{4}}\right), \qquad\qquad (2.20)$$

where Y is given by eqn (2.12). This series has the property of converging on its circle of convergence ($|Y| = 1$), which leads to an interesting consequence. For a sufficiently large α, it will converge to a value of $S \ll 1$ for $Y = 1$. Once this concentration is exceeded the amphiphiles will assemble spontaneously into infinite bilayers. ($Y = 1$ or $\mu_1^0 + kT \ln X_1 = \mu_\infty^0$ is equivalent to the condition that the chemical potential of amphiphiles in solution equals that of amphiphiles incorporated into infinite bilayers.) It is possible that the system will have a cmc before it reaches this phase transition.

Consider as yet another possible example

$$\mu_N^0 = \mu_\infty^0 + \alpha kT/N^2; \qquad N \geqslant m \qquad\qquad (2.21)$$

which we shall use to illustrate the case when $\mu_N^0 - \mu_\infty^0$ vanishes faster than $1/N$. We find now

$$S - X_1 = \sum_{N=m}^{\infty} N Y^N e^{-\alpha/N}. \qquad\qquad (2.22)$$

Since the result does not depend significantly on the lower limit of the sum when $\bar{N} > m$ we shall replace m by 1 for convenience.

Consider the sum

$$\sum = \sum_{1}^{\infty} Y^N e^{-\alpha/N} = \tfrac{1}{2} \sum_{-\infty}^{\infty} e^{-\beta|N| - \alpha/|N|}, \qquad\qquad (2.23)$$

where $\beta = -\ln Y$. Applying the Poisson summation formula [5] to this, we find

$$\sum = \tfrac{1}{2} \sum_{N=-\infty}^{\infty} \int_{-\infty}^{\infty} e^{2\pi i N x} e^{-\beta|x| - \alpha/|x|} \, dx$$

$$= \int_{0}^{\infty} e^{-\beta x - \alpha/x} \, dx + 2 \sum_{1}^{\infty} \int_{0}^{\infty} \cos\left(2\pi N x\right) e^{-\beta x - \alpha/x} \, dx. \qquad\qquad (2.24)$$

By the substitution $x = y\sqrt{\alpha/\beta}$, this equation may be written

$$\sum = \left(\frac{\alpha}{\beta}\right)^{\frac{1}{4}} \int_{0}^{\infty} \exp\left\{-(\alpha\beta)^{\frac{1}{4}}(y + 1/y)\right\} \, dy +$$

$$2 \sum_{1}^{\infty} \left(\frac{\alpha}{\beta}\right)^{\frac{1}{4}} \int_{0}^{\infty} \cos\left[2\pi N \left(\frac{\alpha}{\beta}\right)^{\frac{1}{4}} y\right] \exp\left\{-(\alpha\beta)^{\frac{1}{4}}(y + 1/y)\right\} \, dy. \qquad (2.25)$$

Consequently if $\beta = -\ln Y$ is small and α large, which is the limit of most interest to us, the period of the cosine in the integrand of eqn (2.25) is much smaller than the width of the exponential term (the precise condition for this is $\alpha^{\frac{1}{4}}/\beta^{\frac{1}{4}} \gg 1$) and consequently the sum of eqn (2.25) may be neglected leaving

$$\sum \approx \left(\frac{\alpha}{\beta}\right)^{\frac{1}{4}} \int_{0}^{\infty} \exp\left\{-(\alpha\beta)^{\frac{1}{4}}(y + 1/y)\right\} \, dy. \qquad\qquad (2.26)$$

J. N. ISRAELACHVILI, D. J. MITCHELL AND B. W. NINHAM 1531

Let us further restrict ourselves to the case $(\alpha\beta)^{\frac{1}{2}} \gg 1$. The integral may then be evaluated by the method of steepest descents,[6] yielding

$$\sum \approx \frac{\pi^{\frac{1}{2}}\alpha^{\frac{1}{4}}}{\beta^{\frac{3}{4}}} e^{-2\sqrt{\alpha\beta}}. \tag{2.27}$$

Therefore

$$S - X_1 = -\frac{\partial}{\partial\beta}\sum \simeq \frac{\pi^{\frac{1}{2}}\alpha^{\frac{3}{4}}}{\beta^{\frac{1}{4}}} e^{-2\sqrt{\alpha\beta}} \tag{2.28}$$

and

$$\bar{N} = -\frac{\partial}{\partial\beta}\ln(S - X_1) \simeq \sqrt{\alpha/\beta}. \tag{2.29}$$

This allows us to rewrite (2.28) as

$$S - X_1 = \frac{\pi^{\frac{1}{2}}}{\alpha^{\frac{1}{2}}}\bar{N}^{\frac{3}{2}} e^{-2\alpha/\bar{N}}, \tag{2.30}$$

giving us an implicit expression for $\bar{N}$ as a function of $S - X_1$. We have made the approximations $(\alpha\beta)^{\frac{1}{2}} \simeq \alpha/\bar{N} \gg 1$ and $\alpha^{\frac{3}{4}}/\beta^{\frac{1}{4}} = \bar{N}^{\frac{3}{2}}/\alpha \gg 1$, i.e., $\bar{N} \ll \alpha \ll \bar{N}^{\frac{3}{2}}$. One sees immediately from eqn (2.30) that under these conditions a large change in lipid concentration $S - X_1$ is required to produce a significant change in $\bar{N}$, i.e., $\bar{N}$ is a slowly varying function of $S - X_1$. In order that N may be a rapidly varying function of $S - X_1$, it would be necessary that $\alpha \sim \bar{N}$, whereupon $S - X_1$ would be unphysically large. We conclude, therefore, that the form (2.21) would lead to lipid aggregates with little polydispersity. More generally, it can be shown that for a system with a chemical potential of the form $\mu_N^0 = \mu_\infty^0 + \alpha kT/N^P$, a phase transition to macroscopic aggregates will occur only if $P < 1$. For $P > 1$, the system will be narrowly dispersed about a finite aggregation number. When $P = 1$, the system will be broadly dispersed with a large aggregation number which is very sensitive to total amphiphile concentration.

3. FORMULATION OF A MODEL FOR THE FREE ENERGIES

No further progress can be made without some explicit model which gives the free energies μ_N^0, μ_1^0 per amphiphile in both the aggregated and dispersed states. First consider the aggregated state. Contributions to the free energy μ_N^0 fall into two classes, bulk and surface terms. It is well established that for most amphiphiles their hydrocarbon interiors in micelles and bilayers are in a liquid-like state above $0°C$.[4, 7-11] Hence the bulk free energy per amphiphile, g, will be a function only of temperature T and the number of carbon atoms n. Surface contributions are of two kinds: (1) those arising from the attractive hydrophobic or surface tension forces, and (2) from opposing repulsive forces, in main of electrostatic origin. Since the hydrocarbon interior exists in a liquid-like state, we expect that the attractive force contribution can be well represented by an interfacial free energy per unit area of aggregate γ, where γ is close to 50 erg cm^{-2} characteristic of the liquid hydrocarbon–water interface. This value has been shown to be essentially the surface tension of water minus the dispersion energy contributions at the water–hydrocarbon interface.[12, 13] As the surface tension of water varies by $< 1\%$ over the range $0–0.5$ mol dm^{-3} NaCl,[14] we can expect the interfacial tension to be likewise insensitive to ionic strength. The surface area a per amphiphile is measured at the hydrocarbon–water interface, and the choice of this interface is a contentious issue. Herman[15] and Tanford et al.[4, 16] choose this interface at a distance equal to the radius of one water molecule beyond the van der Waals boundary of the outermost methylene groups. (Since dispersion contributions to γ are negligible, this is equivalent to choosing the Gibbs dividing surface as the interface.) There appears to be no substantial reason for this choice. Indeed, thermodynamic arguments,[17] together with an analysis of some solubility data,[18] show that for various molecules the interface should be taken at, or even inside, the van der Waals boundary. We take the van der Waals boundary to define the interface. It will be seen subsequently that this choice goes some way towards removing a puzzling anomaly in our understanding of hydrophobic interactions. We remark on one further problem in defining surface tension contributions: The question as to whether we should take γa or $\gamma(a-a')$, where a' is the area per amphiphile covered by the head group and therefore not in direct contact with water. Since $\gamma a'$ is constant, it can be absorbed into the bulk term g, to which it makes a small contribution.

The repulsive surface terms are much more difficult to handle. The shape, size, and orientation of charged head groups, surface charge density, specific ionic adsorption, unknown dielectric constant of the surface region, certain occurrence of Stern layers and associated discreteness of charge effects all conspire to inhibit any rigorous analysis. For example, in recent work on ionic micelles, Stigter[11] showed that use of a Chapman–Gouy model for the double layer based on the non-linear Poisson–Boltzmann equation gives an outward electrostatic pressure which varies by only 20 % over an electrolyte concentration of 0–0.4 mol dm^{-3} NaCl. On the other hand, the cruder Debye–Hückel approximation taken with an arbitrary factor of the order of $\frac{1}{2}$ gives a reasonable description of the critical micelle concentration and other properties of ionic lipid micelles.[4, 19, 20] As regards the Stern layer, Stigter[11, 21] has carried out an analysis for a variety of ionic micelles and concludes that the counterions are effectively in a layer of thickness 4–5 Å. Despite the apparent intractability of the problem, the various approaches suggest that all of these complications can be subsumed by simple phenomenological forms. Various authors since Debye[22] have attempted such an analysis.[11, 19-26] Thus, a repulsive energy contribution which varies as a constant/a has been shown by Tanford[19] to give a realistic description of micelle size and cmc. A repulsive energy of this form would arise from a double layer of charge as in a capacitor, with charge e/a per unit area, separation D of the capacitor planes, and dielectric constant ε. The magnitude of the constant is then $2\pi e^2 D/\varepsilon$. We adopt this form.

We stress that this choice for the repulsive energy is not critical. All the subsequent analysis can be carried out with other repulsive forms, e.g., constant/a^2. The advantage of the capacitance model is that it gives about the simplest expression for the energy with any semblance of physical reality which is amenable to exact analysis. Such an assumption clearly disguises a multitude of sins, and can only be justified *a posteriori*.

One further advantage of this form is that for zwitterionic amphiphiles where the head groups are normal to the surface (e.g., as is believed to be the case for lecithin[27, 28]), the value of D can be readily identified with zwitterionic charge group separation. The expression we have chosen is to be expected here (see Appendix A). Additional support for the universal phenomenological form constant/a is furnished by the observation[29] that both ionic and zwitterionic amphiphiles have very similar micellar properties. If these assumptions are granted, the free energy μ_N^0 per amphiphile in the aggregate is

$$\mu_N^0 = \gamma a + \frac{2\pi e^2 D}{\varepsilon a} + g. \tag{3.1}$$

J. N. ISRAELACHVILI, D. J. MITCHELL AND B. W. NINHAM   1533

(a) CURVATURE CORRECTIONS

The form eqn (3.1) is to hold strictly for planar surfaces. We have already alluded to curvature corrections to the effective surface tension contribution. If the micellar surface is spherical, the effective surface tension γ for both a drop and a hole in the liquid is less than that for a planar interface by a factor $(1-\delta/R)$, where δ is of the order of one or two molecular radii and R is the position of the Gibbs dividing surface.[17, 18] Corresponding corrections to the electrostatic contributions are expected to be of much more importance, and can be handled within the framework of a capacitance description. Thus for a spherical capacitance the energy per unit area is from electrostatics

$$E_s = \frac{2\pi e^2 D}{\varepsilon a(1+D/R)}. \tag{3.2}$$

Except for two instances in the subsequent development, curvature corrections can be ignored.

(b) MONOMER FREE ENERGY

The monomer free energy μ_1^0 comprises two main contributions, a hydrophobic term g' due to the self-energy of the hydrocarbon chain in water, and an electrostatic self-energy associated with the (charged) head group. The term g' is undefined by itself, but the important quantity $g'-g$ has a meaning as the usual hydrophobic energy [4] required to take a hydrocarbon chain from water to bulk hydrocarbon. For this quantity we use the measured value so that a precise model for this contribution is unnecessary. Characteristically [4] $g'-g \simeq 825$ cal mol^{-1} (CH$_2$ group)$^{-1}$ and 2100 cal mol^{-1} (CH$_3$ group)$^{-1}$. (We note in passing that in analysing the measured hydrophobic free energy changes for n-alkanes, various authors [15, 16] assign a value to the hydrophobic interfacial energy of $\gamma_h = 20$–33 cal mol^{-1} Å^{-2}, equivalent to 14–23 erg cm^{-2}. The range of values reflects the particular choice of " cratic " contributions.[4] The assignment of such values depends critically on choice of the surface of tension, normally taken at the Gibbs dividing surface. As already remarked, this is probably incorrect, and if the van der Waals boundary of the hydrocarbon is chosen, a higher value close to the bulk hydrocarbon–water interfacial energy obtains, *i.e.*, 50 erg cm^{-2}.)

Similarly, the change in electrostatic self free energy of the head group between aggregated and dispersed state is the relevant entity required. As shown in Appendix A, the capacitance form for the repulsive surface term includes only head group–head group, head group–counterion and counterion–counterion interactions; *i.e.*, we have chosen a reference state in which the head group self-energy is omitted. Although the head group in the aggregated state may have a different self-energy from that in the dispersed state due to its proximity to a region of low dielectric constant, both free energies, and therefore their difference, can be expected to be much smaller than the hydrophobic free energy change. (A very crude estimate of the electrostatic self-energy is provided by the Born expression $e^2/2\varepsilon r$. For $\varepsilon \approx 80$, $r \approx 2$ Å, we have $e^2/2\varepsilon r \approx 1.7\,kT$ while typically hydrophobic contributions to $\mu_N^0 - \mu_1^0$ amount to $\gtrsim 15\,kT$.) Hence we ignore changes in electrostatic self-energies. The interaction free energy of a head group in the dispersed state with counterions and other head groups is certainly negligible, as is evident from measured activity coefficients of salt solutions.[30]

For zwitterionic amphiphiles there will be a small additional electrostatic contribution to $\mu_N^0 - \mu_1^0$ of the form $e^2/(\varepsilon|r_+ - r_-|)$.

This completes the formulation of our model for micellar free energies, except for one caveat: to examine now the consequences of the model, the thermodynamic formulation of section 2 requires that the relation between free energy (a function of a) and the aggregation number N be spelt out. The connection between a and N is determined by the shape and size of the aggregate. For a given a, fixed hydrocarbon volume per amphiphile and aggregation number, there are geometric or packing constraints which specify the allowed shape and size. An examination of these packing constraints and their consequences will be deferred to sections 5 and 6. For the moment we shall be concerned with testing the model for the simpler problem of spherical micelles, and comparing with experiment, in order to develop confidence that the assumptions embodied in eqn (3.1) provide a satisfactory basis for prediction.

4. PRELIMINARY APPLICATION TO SPHERICAL MICELLES

To illustrate and test the viability of the model, we now consider its application to the formation of spherical micelles. In this application we ignore the important geometric constraints dealt with in the following section. These complications will be built into the theory in section 6. We consider spheres for the following reasons. If we ignore packing constraints, of all the possible shapes for aggregates, the sphere is the smallest shape with a given surface area per molecule and therefore has the lowest aggregation number. Consequently a sphere will be the thermodynamically favoured state, even in the absence of curvature corrections to electrostatic free energies, or of end-effects which will occur for rod-like micelles or planar bilayers. If curvature and end-effects are included, non-spherical shapes, *e.g.*, cylinders and bilayers, turn out to be even less favoured. Although spherical micelles are unlikely to occur in general due to packing constraints,[29, 31] restriction of the discussion to these aggregates gives us an idea of the behaviour to be expected of globular aggregates as opposed to large cylinders or bilayers. Distortions from spherical shape change the relation between area per amphiphile a and aggregation number N and this modification can be incorporated subsequently.

The thermodynamic equations of section 2 are expressed in terms of N, whereas $\mu_N^0 - \mu_1^0$ given in section 3 involves a. We need a relation between a and N. The geometry of a spherical micelle is illustrated in fig. 1 and if v denotes the volume of the hydrocarbon tail of an amphiphile we have the relations

$$N = \frac{4\pi}{3}\frac{R^3}{v} \tag{4.1}$$

and

$$a = \frac{4\pi R^2}{N} = (3v)^{\frac{2}{3}}\frac{(4\pi)^{\frac{1}{3}}}{N^{\frac{1}{3}}} = \frac{3v}{R}. \tag{4.2}$$

The first relation assumes that the density of hydrocarbon is constant in the interior of the micelle.[11] If we ignore curvature corrections the free energy per amphiphile in a micelle is from eqn (3.1) and (4.2)

$$\mu_N^0 = \gamma\left(a + \frac{a_0^2}{a}\right) + g; \qquad a_0 \equiv \sqrt{\frac{2\pi e^2 D}{\varepsilon\gamma}}. \tag{4.3}$$

Henceforth a_0 will be referred to as the optimal surface area per amphiphile, being that area at which the free energy per amphiphile in a micelle is a minimum. From eqn (2.8) the distribution of aggregates is given by

$$X_N = N\left[\frac{X_M}{M}\exp\left(-\frac{M}{kT}(\mu_N^0 - \mu_M^0)\right)\right]^{N/M}. \tag{4.4}$$

J. N. ISRAELACHVILI, D. J. MITCHELL AND B. W. NINHAM   1535

Take M to be the aggregation number at which μ_N^0 takes its optimal value, *i.e.*,

$$\mu_M^0 = 2\gamma a_0 + g, \qquad M = \frac{4\pi(3v)^2}{a_0^3}, \qquad R_M = \frac{3v}{a_0} \tag{4.5}$$

so that eqn (4.4) takes the form

$$X_N = N\left[\frac{X_M}{M}\exp\left\{-\frac{\gamma a_0 M}{kT}\left[\left(\frac{M}{N}\right)^{\frac{1}{3}}-\left(\frac{N}{M}\right)^{\frac{1}{6}}\right]^2\right\}\right]^{N/M}. \tag{4.6}$$

Note that since $a_0 = (4\pi)^{\frac{1}{3}}(3v)^{\frac{2}{3}}/M^{\frac{1}{3}}$, and the tail volume of an amphiphile is known, the only parameters which occur in the distribution eqn (4.6) are X_M, M and γ, the interfacial tension of the hydrocarbon–water interface.

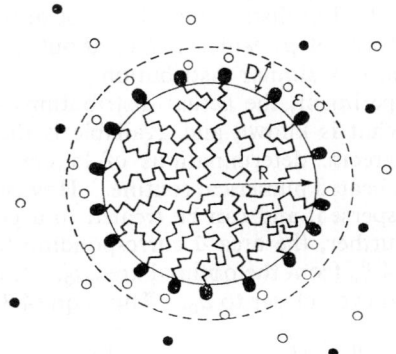

FIG. 1.—Spherical micelle.

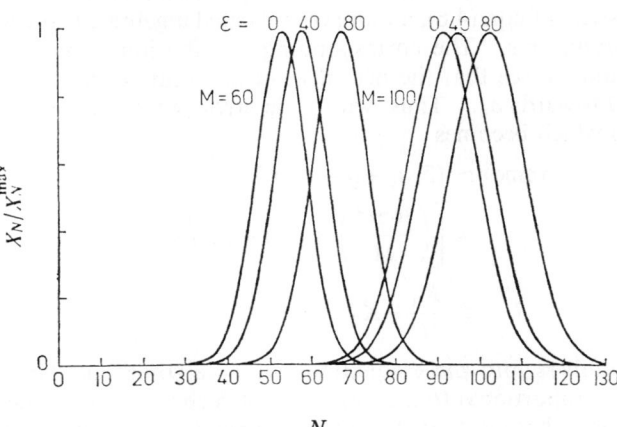

FIG. 2.—Concentration of amphiphiles X_N in spherical micelles of aggregation number N. The curves are plotted for three assumed values of dielectric constant ε. The curve for $\varepsilon = 0$ is equivalent to ignoring curvature corrections to the electrostatic energy. The curves are normalised to the same maximum value. X_M is taken as 10^{-4} mole fraction (5 mmol dm^{-3}).

Fig. 2 shows a plot of X_N against N for $M = 60$ and $M = 100$. For these plots v was taken to be 350 Å^3 corresponding to a hydrocarbon chain of 12 carbon atoms,[4, 11] and $\gamma = 50$ erg cm^{-2}. Several features deserve comment. The maximum of the distribution is insensitive to X_M, and therefore to total lipid concentration. The curves are essentially Gaussians distributed about the maximum so that the mean aggregation number $\bar{N}$ and the maximum are identical. $\bar{N}$ is reduced from the optimal value M by about 10 %, so that in view of eqn (4.2) and (4.5), the area $\bar{a}$ at the maximum is only about 4 % greater than a_0, or less for higher values of M. The standard deviations of the curves are given by

$$\sigma \simeq \left(\sqrt{\frac{9kT}{2\gamma(4\pi)^{\frac{1}{3}}(3v)^{\frac{2}{3}}}} \right) M^{\frac{5}{6}} \simeq 0.4\, M^{\frac{5}{6}} \qquad (4.7)$$

and the spread about the maximum is therefore a narrow one. Also, as expected from eqn (2.7) a change in lipid concentration X_M by an order of magnitude changes the mean aggregation number by only 1 or 2. These distributions are similar to those derived by Tanford.[20] The distribution does not depend on the specific form for μ_N. Any admissible form of μ_N which comes about due to a balance between opposing forces would lead to a similar distribution.

In comparing with experiment, the actual distributions are not known and are difficult to determine. What is known and available is the cmc under a range of conditions. In general, precise determinations of ln (cmc) for a given choice of parameters involve a complicated numerical routine. However, as already remarked, the system is narrowly dispersed, and can be treated to a good approximation as a monodisperse system. Further, the area $\bar{a}$ corresponding to the mean aggregation number $\bar{N}$ differs by only 4 % from the optimal area a_0. If curvature corrections to μ_N^0 are included $\bar{a}$ is shifted even closer to a_0. Thus eqn (4.3) becomes

$$\mu_N^0 = \gamma\left(a + \frac{a_0^2}{a[1 + D/R]} \right) + g. \qquad (4.8)$$

Since $D/R \equiv 6v\varepsilon\gamma/Me^2(M/N)^{\frac{1}{3}}$, after a little algebra, repetition of our previous analysis gives, instead of eqn (4.6), a more complicated algebraic expression involving an additional parameter ε. The corresponding distribution is plotted in fig. 2 for $\varepsilon = 80$ and 40, and we see that the net effect of curvature corrections is to shift $\bar{N}$ towards M and $\bar{a}$ towards a_0. Thus, with little error, we can take $\bar{a} = a_0$, $\bar{N} = M$, and use eqn (2.9) which becomes

$$\ln (\text{cmc}) \simeq (2\gamma a_0 + g - g')/kT$$

$$= \left[\left(\frac{36\pi v^2}{M} \right)^{\frac{1}{3}} 2\gamma + g - g' \right] \bigg/ kT$$

$$\equiv \frac{K_1}{M^{\frac{1}{3}}} - K_2. \qquad (4.9)$$

Before we put numbers into these equations we note some qualitative predictions. Since v is roughly proportional to the number n of carbon atoms in the hydrocarbon tail, eqn (4.5) shows that $(\bar{R} \simeq R_M)$ is proportional to n as observed.[32] Further, a_0 is independent of n. This is consistent with experiments on monolayers at the oil–water interface where the pressure-area isotherms are found to be insensitive to chain length above the transition area.[33] As $g' - g$ is a linear function of n, ln (cmc) is also linear in n, again as observed.[4]

The much studied sodium dodecyl sulphate (SDS) micelle provides a more substantial test. Huisman[34] measured the variation of cmc and $\bar{N}$ for these micelles as a function of NaCl concentration from 0 to 0.3 mol dm^{-3} (see also Emerson and Holtzer[35]). In eqn (4.9) all parameters except M are weakly dependent on salt concentration. Fig. 3 shows the plot of ln (cmc) against $M^{-\frac{1}{3}}$. The best fit straight

J. N. ISRAELACHVILI, D. J. MITCHELL AND B. W. NINHAM   1537

line gives $K_1 = 44$, $K_2 = 20$. With $v_{SDS} = 350 \text{ Å}^3$,[4] $K_1 = 44$ implies that $\gamma \simeq 37$ erg cm^{-2}. This value, while reasonable, is somewhat less than the value 50 dyn cm^{-2} for the oil–water interface. This empirical value for γ allows an estimate to be made of the capacitance distance D. Taking $\varepsilon = 80$, D varies in the range 8 to 5 Å as salt concentration varies from 0 to 0.3 mol dm^{-3}. For smaller values of ε the values of D would be proportionately smaller. We do not place much emphasis on the magnitude of this parameter, but note that it is of the order of magnitude one would expect for the size of the Stern layer.[11] Next consider the value of $(g'-g) = K_2 kT = 20 \, kT \equiv 12\,000$ cal mol^{-1}. This compares with the expected hydrophobic free energy for transport of 1 CH$_3$ + 11 CH$_2$ groups from water to bulk hydrocarbon of $\sim 11\,000$–$12\,000$ cal mol^{-1}.[4] If we make allowance for the area a' covered by the head group, the best fit hydrophobic energy would be reduced to $\sim 10\,500$ cal mol^{-1}. If in addition we admit curvature corrections this value is reduced further to $\sim 10\,000$ cal mol^{-1}.

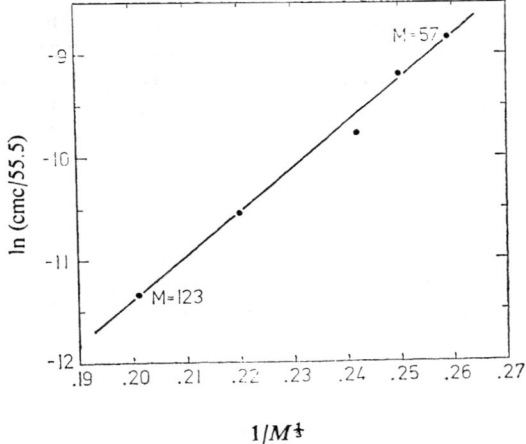

FIG. 3.—Variation of ln (cmc/55.5) with $M^{-\frac{1}{3}}$ for SDS in NaCl solutions as given by eqn (4.9). Points: experimental results. Theoretical curve is based on the assumption that the micelles are spherical. cmc/55.5 is in mole fraction units.

We could go further at this stage and check the predicted n dependence, effect of temperature on cmc, and analyse changes with salt in more detail. All of these are in good semi-quantitative agreement with experiment. To sum up: the model as developed so far gives good agreement with observation, although it might be considered that the surface free energy appears 30 % too low. However, as we cannot yet explain the existence of rod-like micelles or bilayers which suggests that geometric constraints cannot be ignored. We postpone a more precise analysis until the effects of these constraints have been investigated.

5. PACKING CONSTRAINTS AND NON-SPHERICAL MICELLES

We have already alluded to geometric limitations which place restrictions on the allowed shapes of micelles, and it is clear that packing constraints must be invoked for a proper treatment of self-assembly, for, in the absence of any such restrictions, spherical micelles will always be thermodynamically favoured over other shapes like cylindrical micelles or bilayers. There must then be some overriding factor that forces some amphiphiles to assemble into these larger structures which appear to be thermodynamically disfavoured.

In this section therefore, we study the packing properties of amphiphiles which come about due to geometry. The most general form of the problem is exceptionally tedious because thermodynamics and packing are inextricably linked and cannot be considered in isolation. Further, the general problem of relating aggregation number to shape involves a difficult problem in the calculus of variations. Nonetheless we can make considerable progress in elucidating the role of packing provided we assume that the mean area per amphiphile will always be close to the optimum area a_0. As shown in section 4, this assumption is fair for systems which exhibit little polydispersity, and will allow us to disentangle energetic and geometric factors from the thermodynamics. This leads us to consider the shape and size of the smallest micelle consistent with $a = a_0$. This we call the critical packing shape. We shall find that beyond some characteristic critical aggregation number N_c, both spherical and ellipsoidal micelles are prohibited. Geometric factors lead us to consider different shapes once these are excluded.

Consider first a spherical micelle (fig. 1). Here the radius R, hydrocarbon core volume v and surface area a per amphiphile at the hydrocarbon–water interface are related by

$$\frac{3v}{a} = R. \tag{5.1}$$

Then, since the radius of a spherical micelle cannot exceed a certain critical length, l_c, roughly equal to but less than the fully extended length of the hydrocarbon chain, it is apparent from eqn (5.1) that once $v/a_0 l_c > \frac{1}{3}$, spherical micelles will not be able to form unless the surface area $a > a_0$. This gives a critical condition for the formation of spheres:

$$\frac{v}{a_0 l_c} = \frac{1}{3}. \tag{5.2}$$

On the assumption that the surface area per amphiphile is everywhere equal to or close to the optimum area a_0, we must therefore look for alternative non-spherical shapes once $v/a_0 l_c > \frac{1}{3}$. (The way v and l_c are actually determined from the nature of the hydrocarbon chains is model-dependent and not of immediate concern here.) Similarly, the critical condition for cylindrical micelles and planar bilayers is readily found to be respectively

$$\frac{v}{a_0 l_c} = \frac{1}{2}, \qquad \frac{v}{a_0 l_c} = 1. \tag{5.3}$$

The transition shapes of amphiphilic structures, as they go from spheres to cylinders [36, 37] have not been studied; this will be our next concern.

Regardless of shape, any aggregated structure must satisfy the following two criteria: (1) no point within the structure can be farther from the hydrocarbon–water surface than l_c; (2) the total hydrocarbon core volume of the structure V and the total surface area A must satisfy $V/v = A/a_0 = N$. This criterion is only approximate since it assumes that the mean surface area per amphiphile is equal to a_0.

(a) ELLIPSOIDAL MICELLES

Tarter [31] and more recently Tanford [29] have examined the way amphiphiles can pack into prolate and oblate spheroids once they can no longer pack into spheres.

J. N. ISRAELACHVILI, D. J. MITCHELL AND B. W. NINHAM   1539

The shorter ellipsoid semi-axis was put equal to l_c and the longer axis was then determined by applying criterion (2). Both oblate and prolate ellipsoids are found to satisfy criterion (2), though for a given v, l_c and a_0 oblate spheroids have a slightly lower aggregation number N. Clearly, criteria (1) and (2) can be satisfied by many different shapes. Closer scrutiny of spheroids shows that these shapes are in fact untenable and leads us to a stronger packing condition. This can be seen as follows.

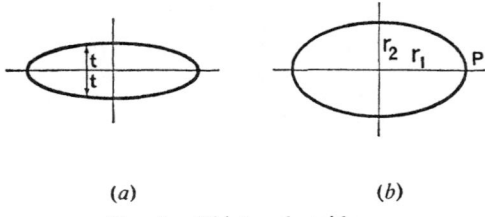

(*a*)   (*b*)

FIG. 4.—Oblate spheroids.

Consider an oblate spheroid of large eccentricity [fig. 4(*a*)] satisfying criterion (2) and with minor semi-axis equal to l_c. Any local region of the oblate spheroid looks very much like a bilayer. However, for the amphiphiles to be so packed we require that $at = v$ (a constant), where $2t$ is the (varying) hydrocarbon thickness. Consequently the area per amphiphile satisfies $a \neq a_0$ almost everywhere so that the amphiphiles of the spheroid must also be energetically unfavourable almost everywhere. [Any form for the opposing surface forces will give the surface energy per unit area as constant $\times a_0$ + constant $\times (a-a_0)^2$.] Criterion (2) disguises this energetically unfavourable situation since it contains the volume V and area A of the whole spheroid, found by averaging the amphiphile volumes and surface areas over the whole structure. By averaging the volumes and areas in this way the discrepancy in the local surface area and energy per amphiphile is averaged to zero, even though the local packing is almost everywhere energetically unfavourable. A similar situation holds for a prolate spheroid of large eccentricity. Thus criteria (1) and (2) are necessary but insufficient conditions for viable micellar structures. Another criterion is required which would indicate whether amphiphiles are packing in an energetically favourable way at all parts of an aggregate and not only on the average.

(*b*) LOCAL PACKING CRITERION

Consider an amphiphile (fig. 5) of surface area a and liquid hydrocarbon core volume v in a micelle (or bilayer vesicle) where the local radii of curvature are R_1 and R_2. Referring to fig. 5 we obtain by elementary geometry a "packing equation"

$$v/a = l\left[1 - \frac{l}{2}\left(\frac{1}{R_1} + \frac{1}{R_2}\right) + \frac{l^2}{3R_1 R_2}\right]; \qquad (5.4)$$

where l is the length of the hydrocarbon region of the amphiphile. This equation is exact for spherical surfaces ($R_1 = R_2$), cylindrical surfaces ($R_2 = \infty$) and planar surfaces ($R_1 = R_2 = \infty$), and holds to a high degree of approximation, with an error of not more than 1 %, for surfaces of arbitrary curvature.

The weaker criteria (1) and (2) are embodied in eqn (5.4) if we stipulate that $l \leqslant l_c$. Application of eqn (5.4) to any part of a micelle or bilayer provides a test for whether or not amphiphiles there can pack in an energetically favourable way, *i.e.*, with $a = a_0$. There are no "new concepts" here; the packing equation is simply a necessary geometric expression that relates v, a, l, R_1 and R_2.

To illustrate the application of the packing equation we now show that spheroids cannot accommodate amphiphiles in their regions of greatest curvature once these amphiphiles can no longer pack into spheres, *i.e.*, once $v/a_0 l_c$ exceeds $\frac{1}{3}$. Consider the oblate spheroid of fig. 4(*b*). Since by assumption spheres are prohibited, we have $r_1 > l_c$, and by criterion (2), $r_2 \leqslant l_c$. In the region of greatest curvature P,

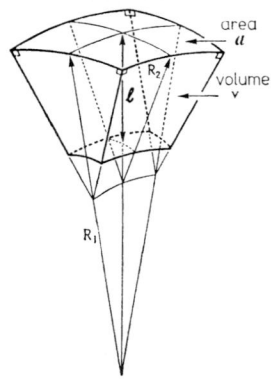

FIG. 5.—Geometric packing of a hydrocarbon region of volume v and surface area a at a surface with two radii of curvature R_1 and R_2. Eqn (5.4) gives the relation between v, a, R_1, R_2 and the length of the hydrocarbon l. For both $R_1, R_2 > l$, a void region is formed behind the hydrocarbon region.

the principal radii of curvature are $R_1 = r_1$, and $R_2 = r_2^2/r_1$, so that $R_2 < l_c$. Since $v/a_0 > l_c/3$, it is sufficient to show that v/a_0 as given by the packing equation is less than $l_c/3$. We have from eqn (5.4)

$$\frac{v}{a} = l\left[1 - \frac{l}{2}\left(\frac{1}{R_1} + \frac{1}{R_2}\right) + \frac{l^2}{3R_1 R_2}\right] \leqslant \frac{R_2}{2}\left(1 - \frac{R_2}{3R_1}\right) \tag{5.5}$$

since l must satisfy $l \leqslant R_2$. Successive application of the inequalities $R_1 > R_2$, $R_2 < l_c$ gives further $v/a < R_2/3 < l_c/3$ which establishes the result. Thus the oblate spheroid, rather than being a viable alternative shape when amphiphiles can no longer pack into spheres, is seen to be energetically unfavourable the moment the spherical shape itself becomes unfavourable. Indeed, even when amphiphiles can pack comfortably into spheres, eqn (5.4) shows that only spheroids of low eccentricities can accommodate the amphiphiles as well. Likewise for prolate spheroids.

(*c*) PROPERTIES OF ALLOWED SHAPES

Strictly, the packing equation should be all that is needed for a complete solution of the problem of finding the shape (or shapes) that can pack amphiphiles of a given v, a_0 and l_c. However, as already remarked, the exact solution to the problem is both difficult and uncalled for. On reflection, it is apparent that the simple shape which satisfies all the packing conditions is not unlike a distorted oblate spheroid. Recall that the unacceptability of an oblate spheroid comes about because the peripheral regions had too great a curvature while the central regions were too thick. If, however, the central region is flattened so that locally it approaches a bilayer and simultaneously the curvature of the peripheral regions is reduced, the packing criteria can be satisfied. Of a number of shapes that have the general features of such a distorted oblate spheroid, a simple analytically tractable shape which satisfies all our criteria is illustrated in fig. 7. This shape may be termed an ellipse of revolution. Qualitatively, it has the following properties: when amphiphiles can still pack into spheres the ellipse is a circle centred at the origin and the shape is therefore a sphere; as v/a_0 increases above $l_c/3$ the circle distorts into an ellipse of small eccentricity

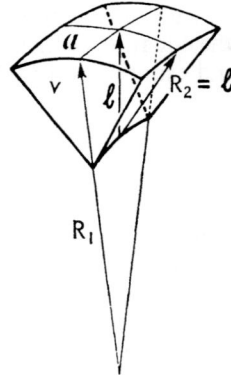

FIG. 6.—Wedge-shaped volume take up by hydrocarbon when $R_2 = l$. Note that the void region existing when $R_1 > l$ and $R_2 > l$ has disappeared (*cf.* fig. 5).

whose centre moves progressively outwards from the symmetry axis, and the micelle becomes globular. If this process were continued indefinitely, at some value of v/a_0 the shape would become toroidal-like. Ultimately, as v/a_0 approached $l_c/2$, the critical condition for cylinders, the ellipse would have become a circle once again giving an infinite cylinder.

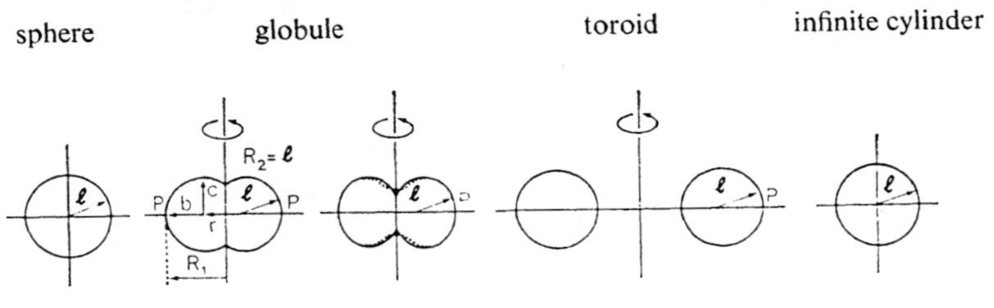

	sphere	globule	toroid	infinite cylinder
$v/a_0 l$	0.33	0.36	0.38	0.44	0.50
r/l	0	0.47	0.68	1.93	∞
b/l	1	0.72	0.70	0.85	1
c/l	1	0.85	0.84	0.92	1
$M/(4\pi l^3/3v)$	1	1.4	1.9	7.1	∞

FIG. 7.—Approximate transition shapes from spheres to cylinders, based on purely geometric considerations. For any given shape the area per amphiphile a_0 is uniform. The minimum aggregation number M is achieved at $l = l_c$ as explained in the text. More realistic transition shapes may be different due to thermodynamic requirements, discussed in section 6.

In the globular regime, *i.e.*, when micelles are still near spherical, we expect, as shown in section 4, that the distribution of sizes will be narrow, so that the approximation $a \simeq a_0$ is good; any shape which satisfies the packing criteria will be adequate here. When we come to larger aggregation numbers, we are on much more tentative ground, because the assumption of monodispersity is no longer valid, and the average micelles can be far from optimal. Indeed, when thermodynamic considerations are involved, the initial growth phase discussed in section 6 turns out to be by a process which is in fact energetically unfavourable, through rods with spherical-like ends. Nonetheless the conclusion that large tori are energetically and geometrically favoured is a useful one, as the optimum micelle has a finite aggregation number. This is important. Otherwise, as we shall show, cylinders would continue to grow without bound, a prediction often in conflict with observation.

For completeness, we now give the equations which describe the transition micellar shapes from spheres to cylinders, and examine some of their properties.

(d) THE SPHERE-CYLINDER TRANSITION SHAPES

The simplest transition shape consistent with all three packing criteria are ellipses of revolution, depicted in fig. 7. The ellipse is centred at a distance r from the origin and has major and minor axes equal to c and b. At the point P, where the curvature will be a maximum, the principal radii of curvature are

$$R_1 = r+b, \qquad R_2 = c^2/b. \tag{5.6}$$

For this shape the following conditions must apply:

$$N = V/v = A/a_0 \qquad \text{[criterion (2)]} \tag{5.7}$$

$$l = R_2 = c^2/b \leqslant l_c, \tag{5.8}$$

where l the length of the hydrocarbon tail of an amphiphile at P. The assumption $l = R_2$ is not obvious. Intuitively, the reason for this choice can be seen in fig. 5 and 6. A truncated pyramid (or cone) would lead to a micelle with a region behind the hydrocarbon tail end of the amphiphile which cannot be filled up, *i.e.*, there results a micelle with a hole in it. Setting $R_2 = l$ gives instead a wedge-shaped region (fig. 6) which avoids this difficulty. The packing equation [eqn (5.4)] determines another relation of the form

$$R_1 = (r+b) = \frac{l}{3[1-2v/a_0l]}$$

or

$$\frac{r}{b} = \frac{(c/b)^2}{3[1-2v/a_0l]} - 1. \tag{5.9}$$

The transition shapes may be divided into a globular regime and a toroidal regime.

(e) GLOBULAR SHAPES $(r < b)$

For this shape we find the following expressions for the total volume V and surface area A of the hydrocarbon core:

$$V = 2\pi c[r^2 \sin \theta_0 + rb(\theta_0 + \tfrac{1}{2} \sin 2\theta_0) + b^2(\sin \theta_0 - \tfrac{1}{3} \sin^3 \theta_0)] \tag{5.10}$$

$$A = 4\pi c\left[r\theta_0 + b \sin \theta_0 - \frac{[1-(b/c)^2]}{2}\left(\frac{r\theta_0}{2} - \frac{r}{4} \sin 2\theta_0 + \frac{b}{3} \sin^3 \theta_0\right)\right], \tag{5.11}$$

where

$$\cos \theta_0 = -(r/b). \tag{5.12}$$

The expression for the volume V is exact, while that for the area A is approximate to first order in $[1-(b/c)^2]$, which is entirely adequate for our problem since (b/c) is

J. N. ISRAELACHVILI, D. J. MITCHELL AND B. W. NINHAM   1543

always close to 1.   Applying the condition given by eqn (5.6), (5.10) and (5.11), we obtain the following equation :

$$\left(\frac{2v}{la_0}\right)\left(\frac{c}{b}\right)^2 \frac{\left[\left(\frac{[1-(b/c)^2]}{4}-1\right)\cos\theta_0 \cdot \theta_0+\left(1-\frac{[1-(b/c)^2]}{12}(2+\cos^2\theta_0)\right)\sin\theta_0\right]}{[\frac{1}{3}\sin\theta_0(2+\cos^2\theta_0)-\theta_0 \cdot \cos\theta_0]}=1.$$
(5.13)

Thus for a given value of v/a_0l, eqn (5.9), (5.12) and (5.13) may be solved to find the unique value of (b/c) from which all the other parameters such as b, c, r and N may be readily calculated.   It can be seen that $N/(4\pi l^3/3v)$ is a function of v/a_0l.

(f) TOROIDAL SHAPES ($r > b$)

For this shape the total volume V and area A of the hydrocarbon core is given by :

$$V = 2\pi^2 bcr$$
(5.14)

$$A = 4\pi^2 cr[1-\frac{1}{4}[1-(b/c)^2]],$$
(5.15)

where again the expression for the area A is to first order in $[1-(b/c)^2]$.   For this shape (b/c) is readily obtained and is given by

$$(b/c)^2 = \frac{3v/a_0l}{2-v/a_0l}$$
(5.16)

from which the values of b, c, r and N may be readily calculated.

For both the globular and toroidal shapes the values of N have been calculated numerically.   For fixed v and a_0, N is a decreasing function of l, so that its minimum value M occurs when $l = l_c$, i.e., the maximum value of l.   Since we require the

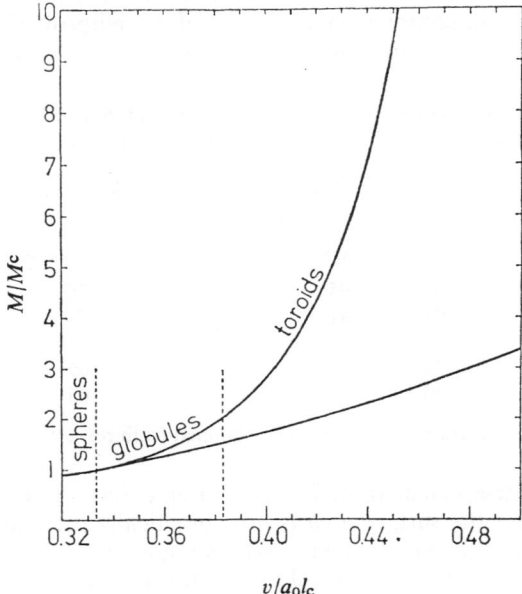

FIG. 8.—Plot of optimal aggregation number M against (v/a_0l_c) for transition shapes of the ellipsoid of revolution model. $M^c = 4\pi l_c^3/3v$ is the largest possible aggregation number for spheres consistent with packing.   Lower curve corresponds to spheres with no packing constraints.

shape with the minimum aggregation number we take $l = l_c$ hereafter.   The variation of $M/(4\pi l_c^3/3v)$ with $v/a_0 l_c$ is plotted in fig. 8.   The range of values $v/a_0 l_c$ from $\tfrac{1}{3}$ to $\tfrac{1}{2}$ correspond to the transition from the purely spherical to the purely cylindrical shape. The results show that as $v/a_0 l_c$ increases above $\tfrac{1}{3}$ the spherical shape becomes progressively more globular (see fig. 7), the two radii of the ellipse become progressively smaller than l_c while the ratio b/c decreases below 1.   The transition from the globular to the toroidal shape occurs when $v/a_0 l_c$ has reached 0.383 (or about 15 % higher than its value of $\tfrac{1}{3}$ when the spherical shape is no longer viable).   At this point b/c has reached its minimum value of 0.74, and the micelle contains about twice the number of amphiphiles in a spherical micelles of the same value of v and l_c. As $v/a_0 l_c$ further decreases the toroid grows outward ; both b and c increase towards l_c and the ellipse approaches a circle once more.   At $v/a_0 l_c = \tfrac{1}{2}$, the critical condition for cylinders is reached ; the ellipse is now a circle and both b and c equal l_c.   For each of these shapes it may be verified that no point within the micelle is farther from the surface than l_c, and that the total curvature at the outermost regions is greater than at any other part, so that the packing equation is everywhere satisfied.

As mentioned earlier, the above model does not depend on the way v and l_c are calculated for an amphiphile.   This is a separate matter, and has been analysed by Tanford,[1, 29] who used the X-ray data of Reiss–Husson and Luzzati [38] to obtain the following relations

$$v = (27.4+26.9\,n)\,\text{Å}^3 \text{ per hydrocarbon chain} \Big\}$$
$$l_c = (1.5+1.265\,n)\,\text{Å per hydrocarbon chain} \Big\} \,, \qquad (5.17)$$

where n is some value close to but smaller than the number of carbon atoms per chain.   Tanford [29] used the above relations to interpret some experimental results of Swarbrick and Daruwala [39] who measured the aggregation numbers of n-alkyl betaines in water as a function of alkyl chain length from C_8 to C_{15}.   He calculated the surface area per amphiphile on the basis of an ellipsoidal model and found, somewhat unexpectedly, that this area fell considerably with increasing chain length.

TABLE 1.—AREA PER HEAD GROUP CALCULATED FOR n-ALKYL BETAINES USING PRESENT MODEL, FOR VARIOUS METHODS OF ESTIMATING v AND l_c

alkyl chain n	N	$v = 27.4+26.9(n-2)$ $l_c = 1.5+1.265(n-2)$			$v = 27.4+26.9n$ $l_c = 0.8(1.5+1.265n)$	$v = 27.4+26.9n$ $l_c = 1.5+1.265n$
		Tanford [29]		present model	present model	present model
		prolate	oblate			
C_8	24	97	95	58	70	65
C_{10}	34	92	91	59	70	66
C_{11}	58	79	77	56	67	60
C_{13}	87	74	71	56	66	59
C_{15}	130	70	66	55	65	57

Experimental values of $\overline{N}$ obtained from ref. (38).

We have repeated these calculations based on our ellipse of revolution model (see table 1), and find that the surface area is now practically invariant, *i.e.*, it is almost insensitive to chain length, as expected.   We may note, too, that in Tanford's calculations, the surface areas were calculated at a distance of about 3 Å from the surface of the hydrophobic core, whereas in our calculations the surface areas were calculated at the surface of the hydrophobic core.   This is consistent with our earlier conclusion (section 3) that it is at this interface that the surface area per amphiphile a_0 should be

J. N. ISRAELACHVILI, D. J. MITCHELL AND B. W. NINHAM 1545

determined, and that it is this area that should remain fairly constant with changes in curvatures and/or alkyl chain length. Our results for the n-alkyl betaines bear out this conclusion and also lend support to our micelle shape.

6. NON-SPHERICAL MICELLES: THERMODYNAMIC CONSEQUENCES OF PACKING

The thermodynamics of spherical micelle formation as discussed in section 4 provides an encouraging preliminary test of our model. We now wish to carry out a more refined comparison between experiment and theory, modified to include packing constraints. The analysis of this section falls into two parts. (1) A discussion of globular micelles. Here the model gives even better quantitative agreement with observation. (2) A discussion of the behaviour of very large aggregates. Again we compare with experiment.

(a) GLOBULAR MICELLES

As discussed in section 5, once aggregation numbers exceed a certain limit, it is no longer possible to pack amphiphiles into spheres. If the aggregation number is not too large, we can expect micelles to take a globular, near spherical, shape and now consider corrections to the theory which arise from this non-sphericity.

(i) CHANGES IN SALT CONCENTRATION

We proceed as in section 4, where it was shown that because the mean aggregation number $\bar{N}$ is close to the (energetically) " optimal " aggregation number M, the area per lipid is also close to a_0; the consequent error in assuming $a = a_0$ is not large.

TABLE 2

experimental results [34]			$a_0(36\pi v^2)^{\frac{1}{3}}$		
			spherical approx.		
cmc/mol dm^{-3}	ln (cmc/55.5)	M	$1/M^{\frac{1}{3}}$	$M^c = 64$	$M^c = 57$
8.14	−8.83	57.3	0.259	0.259	0.259
5.6	−9.20	64.2	0.250	0.250	0.251
3.13	−9.78	70.8	0.240	0.243	0.245
1.47	−10.54	93.4	0.220	0.229	0.233
0.66	−11.34	123.0	0.201	0.218	0.223
$K_1 = 2\gamma(36\pi v^2)^{\frac{1}{3}}/kT$			43.5	61.5	69
γ/erg cm^2			37	52	58
K_2			20	24.5	26.5
K_2kT/kcal ($a' = 0$)			12	14.5	16
K_2kT/kcal ($a' = 20$ Å^2)			11	13	14.5

For globular micelles it would be extremely difficult to attempt a complete thermodynamic analysis. Therefore, we shall assume that the micelles are narrowly dispersed and take on the optimal size and shape discussed in section 5, *i.e.*, the shape and size of the smallest micelle which has a head group area a_0. The only consequence then of packing constraints is that the relation between a_0 and M is modified. This relation is exhibited in fig. 8. Clearly, the value one obtains for a_0 depends on what one assumes to be the largest aggregation number possible for a spherical micelle. For SDS we have taken this cut-off size (M^c) to be either one of the two smaller values 57 or 64. Were we to take $M^c > 123$ the graph of fig. 3 [in which $a_0/(4\pi)^{\frac{1}{3}}(3v)^{\frac{2}{3}}$ is taken to be precisely $M^{-\frac{1}{3}}$] would be reproduced. The results for $M^c = 57, 64$ are tabulated in table 2 together with values of K_1, K_2 and γ which best fit the experimental data exhibited in fig. 9. K_1, K_2 are defined by

$$\ln (cmc) = [2\gamma(a_0 - a') + g - g']/kT$$
$$= K_1[a_0/(4\pi)^{\frac{1}{3}}(3v)^{\frac{2}{3}}] - K_2. \qquad (6.1)$$

The best fit for $M^c = 64$ corresponds to an interfacial energy at the micelle–water interface of $\gamma = 52$ erg cm^{-2} in agreement with the measured value of γ, at the bulk oil–water interface. The interfacial energies of liquid hydrocarbon–water interfaces [13] vary from about 50 to 54 erg cm^{-2} (from section 4 with the spherical approximation, the best fit γ was 37 erg cm^{-2}). With $a' \simeq 20$ Å, the value of K_2 corresponds to a hydrophobic energy of 13 300 cal mol^{-1}. Curvature corrections (see below) reduce this to about 12 000 cal mol^{-1} close to the expected value deduced in section 4.

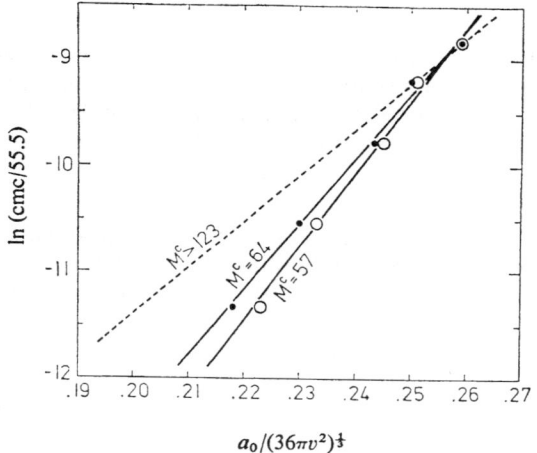

FIG. 9.—Variation of ln (cmc/55.5) with $a_0/(36\pi v^2)^{\frac{1}{3}}$ for SDS in NaCl. Data listed in table 2. O, $M^c = 57$; ●, $M^c = 64$; - - -, $M^c > 123$ (this line ignores packing and is equivalent to fig. 3).

(ii) EFFECT OF CHANGES IN CHAIN LENGTH

Our model can be subjected to a further test. From eqn (6.1) for ln (cmc) we find that the only term which depends on n is the hydrophobic term $g - g'$. It then follows that ln (cmc) must be a linear function of n. This is found to be so experimentally. The experimental results for ln (cmc) for non-ionic surfactants or ionic surfactants at 0.5 mol dm^{-3} salt, correspond to a hydrophobic free energy [4] of about 720 cal mol^{-1} (CH$_2$ group)$^{-1}$. The expected hydrophobic energy as deduced from solubility data for hydrocarbons in both water and in hydrocarbon is about 820 cal mol^{-1} (CH$_2$ group)$^{-1}$.[4] This discrepancy can be resolved if we specialise our model to include the curvature effects discussed in section 3. Unfortunately there is some uncertainty as to the correct value of D which makes it difficult to give a proper quantitative account of curvature corrections, but we can attempt a rough accounting as follows. First, let us determine a range of values for D. We have from eqn (4.3) $a_0 = \sqrt{2\pi e^2 D/\varepsilon\gamma}$ and from table 2 we find

$$0.22 < \frac{a_0}{(4\pi)^{\frac{1}{3}}(3v)^{\frac{2}{3}}} < 0.26, \qquad (6.2)$$

J. N. ISRAELACHVILI, D. J. MITCHELL AND B. W. NINHAM 1547

where for SDS [4, 11] $v = 350 \text{ Å}^3$. Whence D is in the range

$$10^{-9} < \frac{D}{\varepsilon} < 1.4 \times 10^{-9} \text{ cm} \tag{6.3}$$

over the range of salt concentrations 0 to 0.3 mol dm^{-3}. Taking $\varepsilon = 40$ we have $4 < D < 5.6 \text{ Å}$, and for $\varepsilon = 80$, $8 < D < 11 \text{ Å}$. The first choice seems more reasonable as the choice $\varepsilon = 40$ corresponds more nearly with expectations on the dielectric constant near the oil-water interface. Now, assuming an approximately near spherical shape, the curvature correction to be added to eqn (6.1) is

$$\gamma a_0 \left(\frac{1}{1+D/R} - 1 \right) \Big/ kT = - \frac{\gamma a_0 D}{kT(R+D)}, \tag{6.4}$$

where $R \simeq 3v/a_0$ is the mean radius of the globule. This expression assumes that $a \simeq a_0$. The parameters γ, a_0, D are independent of n, and R is a linear function of n. As a typical example, take $a_0 = 60 \text{ Å}^2$ corresponding to $M^c = 64$, $\gamma = 50 \text{ erg cm}^{-2}$, and $D \simeq 5 \text{ Å}$. Then we obtain for various values of n using $v = 27.4 + 26.9\, n \text{ Å}^3$ [eqn (5.19)], the results in table 3. It can be seen from the table that curvature corrections are in the right direction in that they reduce the hydrophobic energy and are qualitatively correct whatever choice is made for D.

TABLE 3

n	$R/\text{Å}$	$\gamma a_0 D/kT(D+R)$		difference (CH$_2$ group)$^{-1}$		difference/ cal mol^{-1} (CH$_2$ group)$^{-1}$	
		$D = 5 \text{ Å}$	$D = 10 \text{ Å}$	$D = 5 \text{ Å}$	$D = 10 \text{ Å}$	$D = 5 \text{ Å}$	$D = 10 \text{ Å}$
8	12.1	2.12	3.27				
				0.15	0.17	90	102
9	13.4	1.97	3.10				
				0.14	0.18	84	108
10	14.8	1.82	2.92				
				0.11	0.14	66	84
12	16.1	1.72	2.78				
				0.11	0.15	66	90
13	17.5	1.61	2.63				
				0.89	0.12	54	72

For alkyl sulphates in the absence of salt both D and a_0 will be larger, leading to larger curvature effects. The observed discrepancy here however is about 400 cal mol^{-1} (CH$_2$ group)$^{-1}$,[4] and this is too large a discrepancy to be accounted for by the simple capacitance model. At very low salt concentrations with ionic surfactants we expect highly non-linear behaviour for the repulsive electrostatic terms, so that this simple model can be expected to break down.

(iii) TEMPERATURE EFFECTS

We turn now to the temperature dependence of the cmc. Eqn (6.1) gives little indication of how the cmc should vary with temperature. The observed cmcs all have a minimum between 20-30°C, for sodium n-dodecyl, n-decyl and 2-decyl sulphates, and for n-dodecyl trimethylammonium bromide.[40-42] While the temperature dependence of the term $2\gamma a_0/kT$ could be estimated from data on the surface tension of water and decreases with temperature, we have no information on the temperature dependence of the hydrophobic term $g-g'$. However, evidence [4] on the solubilities and enthalpy of solution of the sequence ethane, propane, butane, together with corresponding data for the saturated alcohols from ethanol to n-pentanol, indicates that $\partial/\partial T[(g-g')/kT]$ increases as the number n of CH_2 groups increases.   Since $2\gamma a_0$ is independent of n, eqn (6.1) predicts that the slope of the cmc against T curve should increase as n increases.   This qualitative prediction is in accord with Flockhart's results [40] on sodium n-alkyl sulphate micelles for $n = 10$, 12, 14.

One further prediction from eqn (6.1) is that the slope of the cmc against T curve should increase as the ionic strength increases.   This follows because an increase in ionic strength causes a decrease in a_0 while other parameters in eqn (6.1) remain constant. Such an effect has been observed by Matijević and Pethica [43] with SDS micelles in 0.01-0.20 mol dm^{-3} NaCl solutions.

(b) ROD-SHAPED MICELLES

When a_0 is small so that the optimal size is much larger than the largest possible sphere, we have argued in section 5 that the amphiphiles would pack into large toroids if the optimal size were indeed attained.   However, as can be seen from fig. 8, aggregates much smaller than the optimal may have an area per molecule, a, only marginally larger than a_0 and, consequently, self-energies μ_N^0 only marginally higher.   It follows from the arguments of section 2 that in this case the system will be polydisperse and we may expect to find aggregates somewhat smaller than the optimal size.   This is in contrast to the situation with globular micelles and with lipid vesicles to be discussed in the next section.   The problem of determining the shape of the micelles is much more difficult.   For micelles of a given aggregation number N, and, therefore, of a given volume Nv, and a given total area A, the minimum μ_N^0 can be shown to correspond to uniform area $a = A/N$, but only provided this shape is allowed by packing.   The minimum μ_N^0 for the given aggregation number N would be determined by the smallest value of a.   The functional relationship between this minimum area a and the given aggregation number N is the same as that between a_0 and the minimum aggregation number N illustrated in fig. 8.   Use of this form leads to results which do not agree with the observed behaviour of large aggregates.   However, for some values of N packing restrictions may prohibit the formation of a micelle with uniform area a, and the micelle with minimum μ_N^0 may have a non-uniform area per molecule.   A possible candidate is a cylindrical micelle with $a = a_0$ having globular ends with $a > a_0$.   Such a shape would have an average area per molecule very close to a_0.   Experimental evidence seems to indicate that micelles of large aggregation number formed from one-tail-amphiphiles are rod-shaped.[36, 37]   Such a micelle would have a μ_N^0 of the form

$$\mu_N^0 = \mu_\infty^0 + \frac{\alpha kT}{N}, \tag{6.5}$$

where μ_∞^0 is the self-energy per molecule of an infinite cylinder $= 2\gamma a_0$.   The term $\alpha kT/N$ arises because the molecules in the globular ends have a larger area and the number of these is independent of N.   As was shown in section 2, this form for μ_N^0 leads to very polydispersed systems whose mean aggregation number increases rapidly with amphiphile concentration, provided α is large.   It is experimentally observed that the mean aggregation number of these (apparently rod-shaped) micelles does change rapidly with concentration, eventually levelling out to a fixed number.[4]

A model which might be expected to give a rough estimate of α is shown in fig. 10. Consider a cylinder of radius $r = 2v/a_0$ with spherical ends of radii l_c.   Then the molecules in the hemispheres on the ends have area $a = 3v/l_c$ and all other molecules

J. N. ISRAELACHVILI, D. J. MITCHELL AND B. W. NINHAM 1549

in our model have $a = a_0$. Then if $\eta = 4\pi l_c^2/a$ denotes the number of molecules in a spherical micelle of radius l_c we find that μ_N^0 is given by

$$\mu_N^0 = 2\gamma a_0 + \eta\gamma a_0(\sqrt{a/a_0} - \sqrt{a_0/a})^2/N \tag{6.6}$$

with

$$a = 3v/l_c. \tag{6.7}$$

This gives

$$\alpha = \frac{4\pi x l_c^2 \gamma}{kT}(\sqrt{x} - 1/\sqrt{x})^2 \tag{6.8}$$

with

$$x = a_0/a = a_0 l_c/3v \tag{6.9}$$

so that x lies between 1 (corresponding to the smallest a_0 attainable by spheres) and $\frac{2}{3}$ (corresponding to the smallest a_0 attainable by infinite cylinders).

In table 4 we tabulate values of α for various values of x, assuming different values for l_c with $\gamma = 50$ erg cm^{-2}.

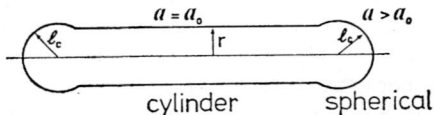

cylinder   spherical

FIG. 10.—Cylindrical micelle with globular ends. Aggregation number can be less than that of the optimal torus.

TABLE 4

x	$l_c = 15$ Å α	$l_c = 25$ Å α
1	0	0
0.95	0.9	2.4
0.90	3.5	10.0
0.85	7.9	21.8
0.80	14.0	38.8
0.75	21.8	61
0.70	31.4	—
0.60	38.8	—

TABLE 5

α	$S = 10^{-10}$	$N = 2\sqrt{Se^x}$ $S = 10^{-8}$	$S = 10^{-6}$
22	—	—	120
24	—	—	326
26	—	—	885
28	—	241	2410
30	—	654	6540
32	180	1800	1.8×10^4
34	480	4.8×10^3	4.8×10^4
36	1300	1.3×10^4	1.3×10^5
38	3000	3.6×10^4	3.6×10^5
40	10^4	10^5	10^6

Now it has already been shown in section 2 [*cf.* eqn (2.16)] that for $S \geqslant$ cmc and $\alpha \geqslant 1$, $\bar{N} = 2\sqrt{S e^{\alpha}}$ (as was obtained by Mukerjee,[44] by a different approach). In table 5 we tabulate $\bar{N}$ for various values of α at different concentrations in mole fraction units.

Some interesting features of this model deserve comment. The values of α which best fit available experimental data were found by Mukerjee[44] to lie in the range 25-35. For the amphiphiles concerned $l_c \sim 20$-25 Å. It can be seen from tables 4 and 5 that the model used here does in fact produce values of α which straddle this range. An increase in salt concentration decreases a_0 and therefore x, so that $\bar{N}$ increases rapidly with increase in salt concentration. This is in keeping with experimental observation.[36] Also an increase in chain length increases l_c and consequently increases α. Therefore amphiphiles with larger chain lengths will tend to have micelles with larger aggregation numbers which increase faster with concentration.[4]

There is a further interesting consequence. As the lipid concentration is increased and the aggregation number is, as a consequence, increased, the " optimal size " will be approached. After this value μ_N^0 will remain constant and the micelles will cease to grow rapidly. This seems to be in keeping with experimental observation although there is some disagreement about interpretation of the results.[4, 44]

(c) BILAYERS

It is of interest to note here that the behaviour of lipid aggregates for which the optimal shape is a spherical vesicle or an infinite bilayer will be quite different from that of rod-shaped micelles. One might anticipate by analogy with rods that finite disc-like bilayers could form with cylindrical rims. For such a system the free energy would take the form $\mu_N^0 = \mu_\infty^0 + \alpha k T / N^{\frac{1}{4}}$. From section 2, for large α, the system will assemble spontaneously into infinite bilayers. Repetition of an analysis similar to that for rods leads to values of α sufficiently large so that this process always occurs.

7. THE STRUCTURE AND STABILITY OF ONE-COMPONENT BILAYERS

The preceding section has established that the predictions of our model accord both qualitatively and quantitatively with available observations on globular and rod-like micelles. We can, therefore, now turn with some degree of confidence to an area as yet imperfectly explored, but whose importance is not in dispute. We shall apply packing criteria to a consideration of those amphiphiles which coalesce naturally into vesicles or planar bilayers. Attention will be confined to biological phospholipids : these have anionic or zwitterionic head groups ; they commonly contain two hydrocarbon chains with 16-18 carbons per chain, and display some unsaturation. It is this unsaturation that ensures that the hydrocarbon chains are above their " melting temperature ", *i.e.*, that they are in a fluid-like state, above 0°C.[4, 7, 45] Because these molecules have two hydrocarbon chains, they tend to have double the volume of the single chain molecules previously considered, for which $v/a_0 l_c \gtrsim \frac{1}{3}$; thus for phospholipid molecules $v/a_0 l_c > \frac{2}{3}$, and globular and cylindrical micelles are prohibited.

We first analyse the free energy of a one-component spherical vesicle bilayer (fig. 11), and investigate its stability in the absence of packing restrictions. It should be emphasised that without packing no sensible results emerge unless we include curvature corrections. Subsequently, when packing is built into the model, curvature corrections become of secondary importance, so that the principal results do not depend on a detailed model for the repulsive stabilising forces. In accordance with the terminology of the preceding sections, the free energy per amphiphile in a spherical vesicle of outer and inner radii R_1 and R_2 is given by

$$\mu_N^0 = \left\{ 4\pi(R_1^2+R_2^2)\gamma + \frac{e^2 D}{2\varepsilon}\left[\frac{n_1^2}{R_1(R_1+D)} + \frac{n_2^2}{R_2(R_2-D)} \right] \right\} \Big/ N \tag{7.1a}$$

$$= 4\pi\gamma \left\{ (R_1^2+R_2^2) + a_0^2\left[\frac{n_1^2}{R_1(R_1+D)} + \frac{n_2^2}{R_2(R_2-D)} \right] \right\} \Big/ N, \tag{7.1b}$$

where n_1 and n_2 are the number of amphiphiles in the outer and inner layers of the bilayer, e is the charge per polar head group, ε is the dielectric constant of the head group region, and D is the effective capacitor thickness of the head group. The interfacial energy of the hydrocarbon–water interface is given by γ, and the outer and inner radii, R_1 and R_2, are referred to these two hydrocarbon–water interfaces. Eqn (7.1a) is written in the more convenient form eqn (7.1b) in which the elusive parameters e, D and ε are replaced by γ and the optimum surface area per amphiphile in the bilayer a_0, as given by eqn (4.3).

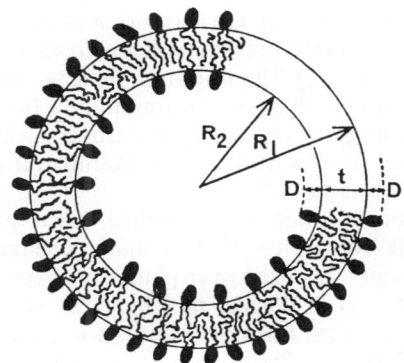

FIG. 11.—Spherical bilayer vesicle of hydrocarbon thickness t.

The aggregation number N and hydrocarbon thickness t of the vesicles are given by

$$N = n_1 + n_2 = 4\pi(R_1^3 - R_2^3)/v \tag{7.2}$$

and

$$t = R_1 - R_2, \tag{7.3}$$

where v is the hydrocarbon volume per amphiphile, it being assumed that, to a first approximation, v is constant and independent of vesicle size. If a_1 and a_2 are the surface areas per amphiphile on the outer and inner hydrocarbon–water interfaces, we may write

$$a_1 = 4\pi R_1^2/n_1 \tag{7.4}$$

and

$$a_2 = 4\pi R_2^2/n_2. \tag{7.5}$$

Eqn (7.1)-(7.5) allow us to determine the minimum free energy configuration for a particular value of N or R_1. This is done in Appendix B where, in the absence of any packing restrictions, we obtain the following results:

$$\mu_N^0(\min) \approx 2a_0\gamma\left[1 - \frac{2\pi Dt}{Na_0}\right] \tag{7.6}$$

$$a_1 \approx a_0\left[1 - \frac{(R_1 + 3R_2)D}{4R_1R_2}\right] \tag{7.7}$$

$$a_2 \approx a_0\left[1 + \frac{(3R_1 + R_2)D}{4R_1R_2}\right] \tag{7.8}$$

$$t \approx \frac{2v}{a_0}\left[1 + \frac{t^2}{6R_1^2}\right], \tag{7.9}$$

where a_0 is the surface area per amphiphile in the planar bilayer (*i.e.*, when $N = \infty$ or $R_1 = \infty$). From the above results we may conclude the following: (1) the minimum free energy per amphiphile is slightly lower than that in a bilayer, for which $\mu_N^0(\min) = 2a_0\gamma$. Since a spherical vesicle must have a much lower aggregation number N than a planar bilayer we conclude that spherical vesicles are thermodynamically favoured over planar bilayers and that, in the absence of any packing constraints, the vesicles will have very small radii. Indeed, if there were no packing constraints the vesicles would shrink to such a small size that they would actually form into micelles. As will be shown, packing constraints prevent vesicles from shrinking much below a certain critical packing radius. (2) The optimum area per amphiphile on the outer surface a_1 is found to be smaller than the optimum area for a bilayer a_0, while the inner area a_2 is found to be larger. (3) The hydrocarbon thickness t is slightly greater than that of a bilayer, given by $t = 2v/a_0$. For example, if $t = 30\,\text{Å}$, $R_1 = 100\,\text{Å}$, the value of t is increased by about 1.5 %.

While some of the above results are modified when packing constraints are introduced, the important conclusion that remains unchanged is that spherical vesicles are thermodynamically favoured over planar bilayers. The effect of packing is only to place limits on their size. We may note here that while sonication is generally employed in the laboratory for their formation this is not essential: Batzri and Korn [46] have reported the formation of vesicles without sonication. In addition, vesicles, once formed, appear to be stable for many days.[46-50] We return to this matter later.

(*a*) THE EFFECT OF PACKING CONSTRAINTS ON BILAYER STRUCTURES

In section 5, a geometric "packing equation" was derived that relates the amphiphile surface area a, hydrocarbon volume v, and hydrocarbon thickness l to the curvature at the amphiphile surface. For a spherical vesicle of outer radius R_1 the packing equation, eqn (5.4), for the outer layer amphiphiles is

$$v/a_1 = l\left[1 - \frac{l}{R_1} + \frac{l^2}{3R_1^2}\right] \tag{7.10}$$

so that

$$R_1 = \frac{l[3 + \sqrt{3(4v/a_1l - 1)}]}{6(1 - v/a_1l)}. \tag{7.11}$$

It will be shown subsequently that the thermodynamically favoured vesicles have their hydrocarbon chains maximally extended. Then $l = l_c$, and R_1 takes the minimum possible value for a given a_1. This value will be called the critical packing radius R_1^c. To a first approximation the outer amphiphile surface area a_1 may be put equal to a_0, and the value of R_1^c may then be readily obtained from eqn (7.11). As an example of

J. N. ISRAELACHVILI, D. J. MITCHELL AND B. W. NINHAM 1553

eqn (7.11) we shall use some accurately determined data [51, 52] for egg phosphatidyl-choline (egg PC) for which $v = 1063$ Å^3, $a_0 = 71.7$ Å^2, and we obtain the following critical vesicle radii as a function of l_c. $R_1^c(l_c = 19$ Å$) \approx 80$ Å, $R_1^c(l_c = 17.5$ Å$) \approx 108$ Å, $R_1^c(l_c = 16$ Å$) \approx 212$ Å.

Before we proceed with a thermodynamic analysis it is worth considering the qualitative consequence of geometric packing in order to understand why double chained amphiphiles aggregate so differently from single chained amphiphiles. In section 5 we found, as a rule of thumb, that those amphiphiles for which $v/a_0 l_c$ lies between $\frac{1}{3}$ and $\frac{1}{2}$ form into micelles whose shapes vary gradually from spherical to cylindrical. For values of $v/a_0 l_c$ between $\frac{1}{2}$ and 1 we might expect a gradual variation from a cylindrical micelle to a planar bilayer. However, there is no way that a cylinder may be gradually distorted into a planar bilayer consistent with the geometric packing criteria. On the other hand, once the critical condition for cylinders is reached, *i.e.*, once $v/a_0 l_c = \frac{1}{2}$, it may be readily verified that a very small spherical vesicle may already be packed with an outer radius R_1 equal to $(1 + 1/\sqrt{3})l_c \approx 1.6 \, l_c$. Thus we do not need to look for a gradual transition from cylindrical micelles to bilayers. At some value of $v/a_0 l_c$ close to $\frac{1}{2}$ the spherical vesicle will become viable as regards packing, and since its aggregation number is bound to be much smaller than that of a long cylinder it will take over as the most favoured structure. As $v/a_0 l_c$ increases above $\frac{1}{2}$ the vesicle will grow, becoming a planar bilayer when $v/a_0 l_c$ reaches 1.

Tanford [4, 29] has concluded that almost all of the commonly studied amphiphiles with one hydrocarbon chain cannot pack into spherical micelles. Such amphiphiles invariably form into small globular micelles or long rod-like micelles. Thus their values of $v/a_0 l_c$ lie in the range $\frac{1}{3}$ to $\frac{1}{2}$. If the only effect of doubling the number of hydrocarbon chains is to double $v/a_0 l_c$, we may conclude that the values of $v/a_0 l_c$ for diacyl chained phospholipids should lie in the range $\frac{2}{3}$ to 1.* By this criterion, using eqn (7.11), such phospholipids should form either planar bilayers or spherical vesicles of radii $R_1 > \sim 3l_c$.

(b) THERMODYNAMIC CONSEQUENCES OF PACKING CONSTRAINTS

For vesicles that are larger than the critical radius R_1, there are no packing restrictions on any of the amphiphiles and the minimum free energy configuration is given by eqn (7.5)-(7.9). Such vesicles, however, are not thermodynamically favoured over smaller vesicles which have lower aggregation numbers N. The vesicles will therefore tend to shrink; but once their outer radii become lower than the critical packing radius R_1^c, the surface areas of the outer amphiphiles a_1 must now increase above the optimal area [given by eqn (7.7)] and this increase in a_1 leads to an increased free energy per amphiphile as N or R_1 falls. We must, therefore, analyse how the free energy per amphiphile varies with N when packing constraints are included. To obtain exact solutions to the initial eqn. (7.1)-(7.4) is difficult when packing constraints, given by eqn (7.10), are introduced. But since we are concerned with vesicles where the effective capacitor length D is much smaller than the outer and inner radii R_1 and R_2 approximate solutions may be obtained by ignoring curvature effects, which are now small. This is done in Appendix B where it is first shown that the optimum surface area per amphiphile on

---

* As already discussed, the assumption that the surface area a_0 is not much affected by doubling the number of chains is supported by the observation that these areas are independent of chain length, and have much the same values for both single chained and double chained amphiphiles. This is also expected on theoretical grounds for hydrocarbon chains in the fluid state.

the inner surface a_2 is unaffected by the packing constraints, which, as is to be intuitively expected, only affects the outer layer amphiphiles. The inner layer amphiphiles are always in a favourable packing state; their hydrocarbon chains simply fill up the inner layer volume with their inner surface areas a_2 remaining at their optimal values, close to a_0. In Appendix B it is further shown that ignoring curvature effects the mean free energy per amphiphile is given by

$$\mu_N^0 \approx 2a_0\gamma + \frac{4\pi l_c^2\gamma}{N}(1 - R_1/R_1^c)^2 \tag{7.12}$$

when R_1 is smaller than the critical packing radius R_1^c given by eqn (7.11). For $R_1 > R_1^c$ we have

$$\mu_N^0 \approx 2a_0\gamma. \tag{7.13}$$

To obtain the vesicle size distribution we substitute the above expressions into eqn (2.8). For $R_1 < R_1^c$ we obtain the following distribution law for the concentration of vesicles X_{R_1} of radius R_1:

$$X_{R_1} = N\left(\frac{X_{R^c}}{N^c}\right)^{N/N^c} \exp\left\{-4\pi l_c^2\gamma(1 - R_1/R_1^c)^2/kT\right\}, \tag{7.14}$$

where N/N^c is the ratio of the aggregation numbers of vesicles with radii R_1 and R_1^c, and is given approximately by

$$N/N^c = \{R_1^2 + (R_1 - t)^2\}/\{(R_1^c)^2 + (R_1^c - t)^2\}. \tag{7.15}$$

When $R_1 > R_1^c$ the distribution function reduces to

$$X_{R_1} = N\left(\frac{X_{R_1^c}}{N^c}\right)^{N/N^c}. \tag{7.16}$$

Eqn (7.14)-(7.16) give the concentration of vesicles as a function of R_1. As an application of this distribution we shall use the values previously obtained for egg phosphatidylcholine. Fig. 12 shows how the egg PC vesicle concentration varies with outer radius R_1 for two values of l_c and their corresponding values of R_1^c determined by eqn (7.11), i.e., $l_c = 17.5\,\text{Å}$, $R_1^c = 108.3\,\text{Å}$, and $l_c = 16.0\,\text{Å}$, $R_1^c = 212.5\,\text{Å}$. We have assumed that $\gamma = 50\,\text{erg cm}^{-2}$, $t = 29.6\,\text{Å}$,[52] $T = 20°\text{C}$ and have plotted the distributions for $(X_{R_1^c}/N^c) = 10^{-8}$ and $(X_{R_1^c}/N^c) = 10^{-14}$. The value of $(X_{R_1^c}/N^c) = 10^{-8}$ corresponds roughly to a concentration of 1 mg phospholipid per cm^3 water.

A number of general conclusions for egg PC may be drawn from the distribution.

(1) The concentration of vesicles has a near-Gaussian profile which peaks at an outer radius R_1 close to, but slightly smaller than, the critical packing radius R_1^c. At the peak, the surface area per amphiphile is slightly greater (by $\sim 1\%$) than that of the bilayer a_0, though the difference is diminished as the total amphiphile concentration increases.

(2) The profile is given approximately by $\exp\{-(R_1 - R_1^c)^2/2\sigma^2\}$, where the standard deviation σ is given by

$$\sigma = \frac{R_1^c}{l_c}\sqrt{\frac{kT}{8\pi\gamma}}. \tag{7.17}$$

Thus the standard deviation σ is found to be proportional to the vesicle size R_1^c, and is approximately equal to 3% of R_1^c. Further, the theoretical value of σ is fairly insensitive to the value chosen for the hydrocarbon–water interfacial energy γ.

J. N. ISRAELACHVILI, D. J. MITCHELL AND B. W. NINHAM 1555

(3) The mean outer radius decreases as the lipid concentration is lowered, but the magnitude of the shift is very small over a wide range of concentrations. A million-fold change in concentration shifts the peak by only a few Ångstrom units.

The implications so far are that egg PC vesicles should be fairly homogeneous in size, and that this size is effectively independent of the total amphiphile concentration even down to extremely low concentrations. In general, the factors that are expected to affect vesicle size are changes in the optimal surface area a_0 and the ratio v/l_c. For anionic amphiphiles the surface area a_0 may be expected to be changed by variations in pH, ionic strength, $[Ca^{2+}]$, etc.,[53] while zwitterions may be expected to be less sensitive to such changes. On the other hand, v/l_c may be expected to vary with the degree and type (*cis*, *trans*) of unsaturation.

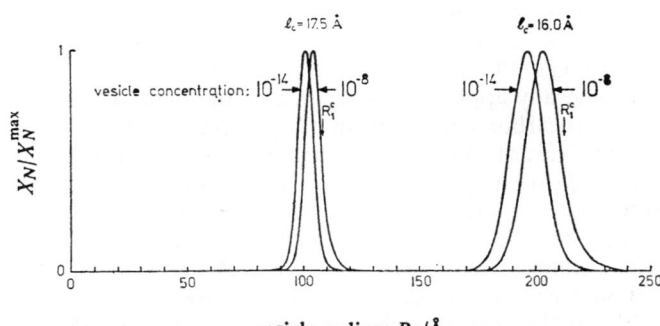

FIG. 12.—Concentration of phospholipids X_N in spherical vesicles of outer radii R_1. The curves are based on eqn (7.14)-(7.16) and are plotted for $\gamma = 50$ erg cm^{-2}, $t = 29.6$ Å (corresponding to egg PC), for two values of l_c. The vesicle concentration $(X_{R_1^q}/N^c)$ is in mole fraction units (55.5 mol dm^{-3}).

We now return to fig. 12 in order to test its predictions for egg PC vesicles. These vesicles have been found experimentally [54] to have a total outer radius of 105 ± 4 Å. (A knowledge of the exact outer radius is not essential in the foregoing analysis.) Since the outer radius R_1 is measured at the hydrocarbon–water interface, a total radius of 105 Å corresponds to a value of R_1 closer to 100 Å or even less (depending on the length of the polar head group). Fig. 12 shows that for the vesicle size distribution to peak at $R_1 \approx 100$ Å requires the maximally extended hydrocarbon length to be $l_c \approx 17.5$ Å. The theoretical results also predict that the vesicle diameters will have a standard deviation of $\simeq 7$ Å, which may be compared with the experimental value of ± 8 Å,[54] and that the aggregation number (at $R_1 = 100$ Å) is 2600 as also observed.[54] Further, the ratio of the number of egg PC molecules on the outer and inner vesicle layers should be given roughly by $[R_1/(R_1-t)]^2$, where t is the hydrocarbon thickness and not that of the whole vesicle bilayer. Putting $t = 29.6$ Å yields an outer PC/inner PC ratio of $(100/70.4)^2 = 2.0$. If instead of 29.6 Å a value of 40 Å were chosen for t (as is commonly done) we should have obtained a ratio of $(105/65)^2 = 2.6$, while for $t = 45$ Å the ratio rises to 3.1. We may note that experimental determinations of this ratio vary greatly and depend on the method of measurement. The reported values to date are: 1.6,[48] 1.7,[55] 1.85-1.95,[56-59] 2.0,[60] 2.2,[61] 2.3,[49-50] 2.6[9] (the average value is about 2.0).

As regards the outer and inner surface areas of the amphiphiles, the results indicate that these should be about the same, and close (within $\sim 1\%$) to the area in lamellar bilayers, as observed.[9, 62]

The value obtained for the maximally extended hydrocarbon length of 17.5 Å is more difficult to compare with experimental values. In section 5 we found that 80-100 % of the fully extended hydrocarbon chain appears to be a reasonable estimate of l_c of saturated chains (see also Tanford [4, 29]). Since egg PC is mainly 16 : 0, 18 : 0 and 18 : 1,[63] we obtain a theoretical estimate for l_c of $(0.8-1.0) \times [1.5 + (1.265 \times 16)] \approx 17.4\text{-}19.4$ Å. Alternatively, the fully extended length may be obtained from the limiting surface area at which monolayers collapse. For egg PC this occurs at a limiting area [64, 65] of 55-62 Å^2 corresponding to a fully extended hydrocarbon region (assuming $v = 1063$ Å^3) of $l_c = 17\text{-}19$ Å. (We note that the value of l_c at collapse may be somewhat greater than in the vesicle bilayer since the collapse surface area is appreciably less than the vesicle surface area. The difference is about 10 Å^2 and implies that at collapse there will be some additional hydrocarbon chain available to the interior region.)

The above conclusions appear to account fairly well for the experimentally observed behaviour of egg PC vesicles. The qualitative features of the results, however, should be generally applicable to all one-component vesicles composed of amphiphiles whose hydrocarbon chains are in a liquid-like state.

A more detailed analysis of the problem, that includes the electrostatic effects of curvature, is difficult and probably unnecessary. In section 4 the effects of curvature on micelle size were analysed and it was found that, depending on the value of the dielectric constant ε, the micelle distribution was shifted to slightly larger sizes (fig. 2). In the case of vesicles a similar result is expected. The electrostatic effect of curvature will lower the optimum area per outer amphiphile below a_0 as R_1 decreases [see eqn (7.7)]. This will shift the critical packing radius R_1^c to larger values and thereby shift the distribution curve to larger radii. On the other hand, the effect of curvature should lower the minimum energy per amphiphile as R_1 falls [see eqn (7.6)], and this will shift the distribution curve to smaller radii. These two effects will partly cancel out and the net shift could be in either direction. If, as expected, the net shift is to larger radii the implication is that l_c is slightly greater than 17.5 Å (from fig. 12 we see that a small change in l_c causes a large shift in the distribution curve).

Finally, the above analysis allows us to calculate a value for the Young's modulus of a planar bilayer under compression or tension. Rewriting the free energy per lipid, in terms of the volume v and hydrocarbon thickness t ($a = v/t$), the energy per two lipids on either side of a bilayer is $\mu_N^0 = 2\gamma(a + a_0^2/a) = 2\gamma(v/t + ta_0^2/v)$. Thus the force F on compressing the bilayer from its initial equilibrium thickness t_0 to $(t_0 - \Delta t)$ is

$$F = -\frac{\partial(2\mu_N^0)}{\partial t} = -2\gamma\left(-\frac{v}{t^2} + \frac{a_0^2}{v}\right)$$

$$= \frac{2\gamma a_0}{t_0}\left(\frac{t_0^2}{t^2} - 1\right)$$

$$\approx 4\gamma a_0 \frac{\Delta t}{t_0^2}. \tag{7.18}$$

The Young's modulus of the bilayer is therefore

$$\left. \begin{aligned} Y &= \frac{F/a_0}{\Delta t/t_0} \approx \frac{4\gamma}{t_0} \\[2mm] i.e., \qquad Yt_0 &\approx 4\gamma \approx 200 \text{ dyn cm}^{-1} \end{aligned} \right\}. \tag{7.19}$$

J. N. ISRAELACHVILI, D. J. MITCHELL AND B. W. NINHAM 1557

For a bilayer of total thickness 50 Å the effective Young's modulus is of order 4×10^8 dyn cm^{-2} when an interfacial energy of $\gamma \approx 50$ erg cm^{-2} is taken. Experimentally, it is only possible to measure Yt_0. Though experimental determinations of Y vary greatly, Requena *et al.*[66] have recently argued that the black lipid membranes have a value of Y close to 2×10^8 dyn cm^{-2}. The theoretical value of $Yt_0 \approx 200$ dyn cm^{-1} may also be compared with measured values in the range 51-357 dyn cm^{-1} for red cell membranes.[67]

We note that the elasticity discussed above is only for planar bilayers under compression or tension and does not extend to the bending of bilayers. On the contrary, our analysis has shown that the bending of a bilayer is favoured down to the critical packing radius (assuming that the lipids can freely rearrange by lateral movement and/or flip-flop), and that " bending elasticity " sets in only for radii smaller than this critical value. The elasticity of a fluid bilayer is therefore seen to be profoundly different from that of a classical elastic plate or shell.

(c) FURTHER ASPECTS OF BILAYER STRUCTURES

In general we find that, when amphiphiles aggregate, entropy favours the smallest structures, but that packing restricts the sizes and shapes of these structures. Single chained amphiphiles form small globular micelles or long rod-like micelles, whereas double chained amphiphiles form bilayers. But there are other important differences between single and double chained amphiphiles that are of some biological significance. The first is to do with the large dispersity of micellar size and aggregation number, and the great sensitivity of micellar size to changes in the total amphiphile concentration. By contrast, double chained amphiphiles form very homogeneous structures that are unaffected by the total amphiphile concentration. Further, while single chained amphiphiles have relatively high critical micelle concentrations, of order 10^{-3} mol dm^{-3}, double chained amphiphiles such as biological phospholipids, have very low critical micelle concentrations, of order 10^{-10} mol dm^{-3}.[4] It is these properties that make phospholipids such excellent building blocks for the formation of homogeneous and stable bilayer membranes, unaffected by drastic changes in the surrounding amphiphile concentration. However, there is always a continuous exchange of lipids with other membranes that is largely energy independent,[68, 69] and which may be important in membrane–membrane interactions.[70]

In the other extreme of very high amphiphile concentrations a very different picture emerges as bilayers (or micelles) now begin to interact with each other. We stress that our treatment applies only to dilute systems, where there are no interactions between the aggregated structures. Thermodynamically, this was expressed by putting the activity coefficient equal to unity in eqn (2.1). If such interactions are present the expressions for the free energy in both the aggregated and dispersed states will be modified by additional terms due to long range van der Waals and electrostatic interactions. Such interactions become important at high concentrations and lead to the formation of ordered mesomorphic phases.[7, 51, 52, 71-73] Our present treatment clearly does not apply to these structures, though some of the observation on mesomorphic phases have a bearing on bilayer vesicles. One of these concerns the observation that phospholipids do not always readily form into vesicles, but rather into multilamellar bilayers or myelin figures. Sonication is often, but not always, necessary to break up and transform these bilayers into vesicles. But once formed, these vesicles are homogeneous and stable, and unaffected by the length of time and intensity of sonication.[54] It seems that as soon as bulk phospholipid is placed in aqueous solution the lipids initially form into the " neat mesomorphic phase " (*i.e.*, bilayers);[7, 51, 52, 71, 72] since as the lipids diffuse out of the bulk into the water the local lipid concentration is initially very high. Once formed, the bilayers cannot reform into vesicles without first having to break up. For this to occur, a certain activation energy must be overcome, and it is this energy that is furnished by the sonication process. But even then the total lipid concentration must be sufficiently low for the vesicles to be dilute enough not to interact significantly with each other.

Thus whereas we conclude that spherical vesicles, if allowed by packing, are thermodynamically favoured over extended bilayers, we add the reservation that they need not necessarily form spontaneously from bilayers.

It is worth examining how a vesicle might form spontaneously from a bilayer; this is shown in fig. 13. Our analysis suggests that the initial stages of the formation of a spherical bulge is energetically favourable [*cf.* eqn (7.6)], but the final stage of separation involves adhesion and fusion at the neck as the vesicle comes away from the bilayer. This often requires a certain amount of energy and brings us to the mechanism of membrane fusion which is beyond the scope of this paper.

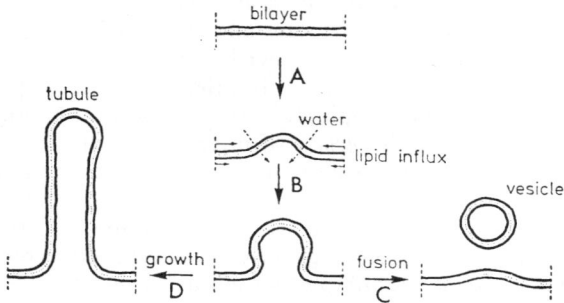

FIG. 13.—Stages of vesicle and tubule formation from a one-component bilayer. Stages A and B are energetically favourable if allowed by packing, but require free lateral diffusion of lipids in the bilayer and water flow across the bilayer. Stage C involves fusion. Stage D is energetically unfavourable for a bilayer in the fluid-state. A mechanism of vesicle and alveoli formation from a membrane similar to that shown here has been found to occur in a variety of cell types and in micropino-cytosis.[74, 75]

So far we have only looked at spherical or planar bilayers. More generally, the packing equation [eqn (5.4)] shows that for a given value of a_0, v and l_c, the critical packing curvature, $\frac{1}{2}(1/R_1^c + 1/R_2^c)$, is approximately constant (where R_1^c and R_2^c are the two outer radii of a curved bilayer at critical packing). Thus for a tubular vesicle, putting $R_2^c = \infty$, its critical radius R_1^c would be about half that of a spherical vesicle. Tubular vesicles, however, have a larger aggregation number than spherical vesicles so that they should not form and they have not been reported. In addition, we show in Appendix B that the free energy per amphiphile in a tubular bilayer is higher than that in a spherical or planar bilayer. Thus we conclude that tubular protrusions should not grow spontaneously out of bilayers (fig. 13), and that bilayers in a fluid-like state should not transform into tubules. For tubular membranes to exist they must either have a resilient cytoplasmic skeleton that gives them their structure, or the membranes must be rigid. Though this conclusion may not necessarily extend to multicomponent lipid bilayers or biological membranes, we have not found any evidence for unsupported tubular membranes that are in the fluid-like state. Thus tubular cell surface projections such as filopodia, flagella, cilia and microvilli all

J. N. ISRAELACHVILI, D. J. MITCHELL AND B. W. NINHAM 1559

appear to be supported by a cytoskeleton of rigid microfilaments or microtubules.[75-82] Those tubular structures that do not have a supporting skeleton appear to have rigid (non-fluid) membraneous walls, for example, microtubules,[82] invertebrate photo-receptor membranes,[83, 84] tubular mitochondrial cristae.[75, 85, 86]

We now turn to the factors that affect vesicle properties. For anionic phospho-lipids, the optimum surface area a_0 may be expected to be larger at higher pH or lower ionic strength.[53] This will result in the formation of smaller vesicles, as observed for phosphatidylserine at different ionic strengths.[87]

A very different matter is to change the properties of a solvent that already contains stable vesicles in suspension, for now any changes in pH, ionic strength, [Ca²⁺], temperature, other lipid levels, *etc.*, will all induce strains on the already existent vesicles. These strains will be asymmetrical about the outer and inner faces of the vesicle bilayer,[88] and the way the vesicles will respond to these changes will depend on such factors as the flip-flop rate, the permeability of the bilayer to water and ions, and the strength and energy of rupture of the vesicle. For example, on addition of Ca²⁺ anionic vesicles are bound ultimately to rupture.[89, 90] The ruptured bilayers may then reform into larger vesicles or extended bilayers,[90] and it is probably this two-step process that the " fusion " of vesicles under the influence of Ca²⁺ occurs.[91] On the other hand the extraction of Ca²⁺ (by EDTA, for example) may cause existing vesicles to shrivel up. We have not attempted to analyse all the possibilities, though in many respects our conclusions are in agreement with our earlier model for the packing of lipids in bilayers,[92] and with the " bilayer-couple " hypothesis recently proposed by Sheetz and Singer.[93]

8. CONCLUSION

The main aim of this paper has been geared towards the building and testing of a simple model for self-assembly of ionic and zwitterionic amphiphiles. This model exists on two levels. At a non-specific level, as underlined in earlier work by Tanford,[4] the principle of opposing forces, notions concerning geometric packing properly delineated, and thermodynamics, form a framework for a model of such fundamental and general application that its predictions cannot be seriously in dispute. It is reassuring that this indeed turns out to be so, both qualitatively and quantitatively. At the level of the simple model, in dealing with comparisons of theory and experiment on micellisation, the major new point of departure in our work has been the recognition of and an attempt to quantify the role of packing. When packing constraints are recognised a number of apparently disconnected facts begin to fall into place. Thus, for globular micelles, for the effects of salt changes on cmc, the best fit occurs for an interfacial tension at the oil–water interface close to 50 dyn cm⁻¹, the known value. The rapid growth of rod-shaped micelles from globular micelles with changes of amphiphile concentration again emerges in a very natural and inevitable way once packing constraints are allowed. Further, not only do predicted parameters for the growth process agree with observation, but the packing model also predicts the observed levelling of mean aggregation number with increasing number per micelle, and the dramatic change in $\bar{N}$ with increasing chain length.

At another and more restricted level, we have imposed much more stringent requirements, as curvature effects stem from a highly specific form for the repulsive electrostatic forces. It is again reassuring, although not a necessary requirement, that such a decoration removes at least in part two sources of difficulty. These are the annoying discrepancy of 100 cal mol⁻¹ (CH₂ group)⁻¹ between calculated and observed hydrophobic energies of alkane chains, and the apparent non-existence of tubular bilayer vesicles.

Especially pleasing is the circumstance that our model gives a unified framework which includes bilayers with no additional assumptions. Various physical properties of bilayer vesicles are readily accounted for. Even the Young's modulus of bilayers appears to be given correctly from elementary considerations.

As far as simple modelling of self-assembly is concerned, the treatment of single component lipid molecules given here has probably been pushed as far as it can. The refinement of our theory of self-assembly requires a proper examination of Stern layers, consequences of deviations from liquid-like properties of hydrocarbon chains,[94] head group steric effects, specific ion adsorption and other effects. While such a more rigorous analysis would undoubtedly provide specific insights into the properties of particular molecules, it is doubtful if a more refined theory will provide a better overview.

The simple model does allow an entry point into the study of self-assembly of multicomponent lipid systems, lateral phase separation (clustering), membrane asymmetry, and in particular how these relate to curvature through packing. These form a central class of problems in membrane biology.

This work arose from some earlier unpublished ideas of Dr. L. R. White and one of us (J. N. I.), which were presented at the 49th National Colloid Symposium, Potsdam, N.Y., in 1975.

APPENDIX A

ELECTROSTATIC ENERGY OF ZWITTERIONIC ARRAYS

We give here an example which illustrates how the capacitance form for the electrostatic interaction energy comes about for the special case of zwitterionic head groups.

Consider a planar lattice of parallel " dipoles " oriented normal to the plane. The positive and negative charges of each " dipole " representing a zwitterionic head group are separated by a distance D, and the array is immersed in a medium of dielectric constant ε. Let a denote the area of an elementary cell of the lattice, *i.e.*, the area per dipole. The electrostatic interaction energy per dipole is then

$$u_{el} = -\frac{e^2}{\varepsilon}\left[\frac{1}{D} + \sum_{m,n=-\infty}^{\infty}{}' \left(\frac{1}{(r^2+D^2)^{\frac{1}{2}}} - \frac{1}{r}\right)\right], \tag{A1}$$

where r denotes the position vector of a dipole relative to a fixed dipole and the sum ranges over the whole lattice excluding the reference dipole ($m = n = 0$). For a square lattice $r = |r| = \sqrt{(m^2+n^2)}a$. We see immediately that $u_{el} = e^2/\varepsilon D \times$ a function of $D/\sqrt{a}$. In the limit $D \to 0$, u_{el} reduces to the energy of a lattice of point dipoles of dipole moment eD. We are concerned with the opposite limiting case $D/\sqrt{a} \to \infty$, and first obtain the leading asymptotic term, which follows quite easily. Taking Mellin transforms [95] with respect to D, and inverting, we can estimate the integral representation

$$\frac{1}{(r^2+D^2)^{\frac{1}{2}}} - \frac{1}{r} = \frac{1}{4\pi i}\int_\gamma \left(\frac{r}{D}\right)^s \Gamma(\tfrac{s}{2})\frac{\Gamma(\tfrac{1}{2}-s/2)}{\Gamma(\tfrac{1}{2})}\,ds, \tag{A2}$$

where the contour of integration is a line parallel to the imaginary axis which satisfies $-2 < \operatorname{Re} s < 0$. This integral representation can be substituted into eqn (A1) and the summation and integration interchanged provided we now choose γ such that

J. N. ISRAELACHVILI, D. J. MITCHELL AND B. W. NINHAM   1561

$-2 < \mathrm{Re}\, s < -1$, since the sum $\Sigma\, r^{s-1}$ converges when $\mathrm{Re}\, s < -1$. Thus we obtain

$$u_{el} = -\frac{e^2}{\varepsilon}\left[\frac{1}{D} + \frac{1}{4\pi i}\int_\gamma D^{-s}\,\frac{\Gamma()\Gamma\frac{s}{2}(\frac{1}{2}-\frac{s}{2})}{\Gamma(\frac{1}{2})}\sum r^{s-1}\,\mathrm{d}s\right]. \tag{A3}$$

The asymptotic form of u_{el} as $D \to \infty$ is now given by the residue at the pole $s = -1$. To evaluate the residue we observe that the sum $\Sigma\, r^{s-1}$ has the same singularity as the integral

$$\frac{1}{a}\int |r|^{s-1}\,\mathrm{d}^2 r \sim \frac{2\pi/a}{(s+1)}.$$

Evaluating the residue of the integrand of eqn (A3) at the pole $s = -1$, we have

$$u_{el} \underset{(D\to\infty)}{\sim} \frac{2\pi e^2 D}{\varepsilon a}. \tag{A4}$$

The above argument does not depend on the assumption that the dipoles lie on a lattice since only the residue of the sum $\Sigma\, r^{s-1}$ at its pole at $s = -1$ comes into the final answer, and this will be the same even if the dipoles are irregularly distributed so long as they have an average density $1/a$. Indeed, if we consider the correlation energy of a statistical distribution of dipoles we have

$$u_{el} = -\frac{e^2}{\varepsilon}\left[\frac{1}{D} + \frac{1}{a}\int g(r)\,\mathrm{d}^2 r\left(\frac{1}{(r^2+D^2)^{\frac{1}{2}}} - \frac{1}{r}\right)\right], \tag{A5}$$

where $g(r)$ is the radial distribution function (which tends to unity as $r \to \infty$). This expression leads to the same asymptotic form eqn (A4), and since the asymptotic result is linear in the coupling constant e^2, so, therefore, will be the free energy.

To establish the range of validity of eqn (A4) we need further terms in a complete asymptotic expansion. This is a more difficult problem and to derive such expansions we return to eqn (A3). The sum can be carried out exactly [96] in terms of known transcendental functions to give

$$\sum r^{s-1} = \sum_{m,n=-\infty}^{\infty}\!{}'\,\frac{(\sqrt{a})^{s-1}}{(m^2+n^2)^{(1-s)/2}} = 4(\sqrt{a})^{s-1}\,\zeta\!\left(\frac{1}{2}-\frac{s}{2}\right)\beta\!\left(\frac{1}{2}-\frac{s}{2}\right), \tag{A6}$$

where $\zeta(z)$ is the Riemann zeta function, and $\beta(z)$ is a related entire function [96] defined for $\mathrm{Re}\, z > 1$ as

$$\beta(z) = \sum_{n=0}^{\infty}\frac{(-1)^n}{(2n+1)^z}. \tag{A7}$$

The properties of $\beta(z)$ are enumerated in the important paper of Glasser.[96] Thus

$$u_{el} = -\frac{e^2}{\varepsilon}\left[\frac{1}{D} + \frac{2}{2\pi i\sqrt{a}}\int_\gamma\left(\frac{D}{\sqrt{a}}\right)^{-s}\frac{\Gamma(\frac{s}{2})\Gamma(\frac{1}{2}-s/2)}{\Gamma(\frac{1}{2})}\zeta\!\left(\frac{1}{2}-\frac{s}{2}\right)\beta\!\left(\frac{1}{2}-\frac{s}{2}\right)\right] \tag{A8}$$

with $-2 < \mathrm{Re}\, s < -1$. Translating the contour to the right we have poles at $s = -1$ with residue (-2) from the zeta function, a pole of $\Gamma(s/2)$ at $s = 0$ with residue $(+2)$ and a pole at $s = 1$ with residue (-2) due to $\Gamma(\frac{1}{2}-s/2)$, and find

$$u_{el} = \frac{e^2}{\varepsilon}\left[\frac{2\pi D}{a} + \frac{4}{\sqrt{a}}\,\zeta(\tfrac{1}{2})\,\beta(\tfrac{1}{2}) - \frac{4}{2\pi i\sqrt{a}}\int_{c-i\infty}^{c+i\infty}\left(\frac{2D\pi}{\sqrt{a}}\right)^{-s}\Gamma(s)\,\zeta\!\left(\frac{1}{2}+\frac{s}{2}\right)\beta\!\left(\frac{1}{2}+\frac{s}{2}\right)\mathrm{d}s\right];$$
$$1 < c \tag{A9}$$

To obtain this expression, note that the contribution from the pole at $s = 1$ cancels with the term in $1/D$ in eqn (A8). We have also used the Riemann relation [97] linking $\zeta(s)$ with $\zeta(1-s)$, a corresponding relation [96] for $\beta(s)$ and the duplication formula for the gamma function.[97] If we now use eqn (A6) again the last integral can be easily re-expressed as a sum, and since $\zeta(\frac{1}{2}) = -1.460$, $\beta(\frac{1}{2}) = 0.6677$,[96] we have

$$u_{el} = \frac{2\pi D e^2}{\varepsilon a}\left[1 - \frac{0.6206\sqrt{a}}{D} - \frac{\sqrt{a}}{2\pi D}\sum_{m,n=-\infty}^{\infty}{}' \frac{1}{(m^2+n^2)^{\frac{1}{2}}}\exp\left\{-\frac{2\pi D}{\sqrt{a}}\sqrt{m^2+n^2}\right\}\right]. \quad (A10)$$

Note that if $D \gtrsim \sqrt{a}$, the sum converges with extreme rapidity and can be ignored, so that to sufficient approximation

$$u_{el} \underset{(D/\sqrt{a}\gtrsim 1)}{\simeq} \frac{2\pi D e^2}{\varepsilon a}\left[1 - \frac{0.62\sqrt{a}}{D}\right]. \quad (A11)$$

For small $D/\sqrt{a}$, the corresponding expansion follows directly from eqn (A8) and gives

$$u_{el} \simeq -\frac{e^2}{\varepsilon}\left[\frac{1}{D} - \frac{2D^2\,\zeta(\frac{3}{2})\,\beta(\frac{3}{2})}{a^{\frac{3}{2}}} + O\left(\frac{D^3}{a^2}\right)\right], \quad (A12)$$

where $\zeta(\frac{3}{2}) = 2.612$, $\beta(\frac{3}{2}) = 0.864$.

These techniques can also be developed further to deal with Stern layer problems. Note that for $D \simeq \sqrt{a}$, the "effective" capacitance distance D is reduced by a factor $(1 - 0.62\sqrt{a}/D)$. If image effects are included, it can be shown that (when the zwitterions are immersed in a high dielectric medium $\varepsilon = 80$, adjacent to a hydrocarbon surface $\varepsilon = 2$), image effects can virtually double the electrostatic energy so that the two effects may partially cancel out. Nonetheless the observation that u_{el} is reduced from the intuitive capacitance form is of some interest, as it is known [4] that for ionic micelles use of Debye–Hückel theory (at low salt equivalent to the capacitance form) gives results for the repulsive free energy too large by a factor of 2.

APPENDIX B

(1) MINIMUM FREE ENERGY FOR VESICLES

We outline here the steps necessary to derive the results quoted in section 7.

(a) SPHERICAL VESICLES: NO PACKING CONSTRAINTS

Consider first a spherical vesicle comprising N amphiphiles. The free energy of the vesicle is

$$G \equiv N\mu_N^0 = 4\pi\gamma(R_1^2+R_2^2) + \frac{1}{2}\frac{n_1^2 e^2 D}{\varepsilon R_1(R_1+D)} + \frac{1}{2}\frac{n_2^2 e^2 D}{\varepsilon R_2(R_2-D)}, \quad (B1)$$

where R_1, n_1; R_2, n_2 refer to outer and inner surfaces. When packing is ignored, the only constraints are those due to conservation of volume and number, *viz*

$$Nv = \frac{4\pi}{3}(R_1^3-R_2^3) \quad (B2$$

$$N = n_1+n_2. \quad (B3)$$

Since N is fixed, there are only two variables which can be taken as n_1, R_1. Then

J. N. ISRAELACHVILI, D. J. MITCHELL AND B. W. NINHAM 1563

n_2 and R_2 are determined. We first minimise G with respect to n_1, taking R_1, R_2 constant. This gives

$$\frac{\partial G}{\partial n_1} = \frac{e^2 D}{\varepsilon}\left\{\frac{n_1}{R_1(R_1+D)} - \frac{n_2}{R_2(R_2-D)}\right\} = 0. \tag{B4}$$

Physically this condition is equivalent to the statement that there is no potential difference across the vesicle in a situation of minimum free energy. Using eqn (B3) and (B4) we have

$$\min_{n_1} G = 4\pi\gamma(R_1^2+R_2^2) + \frac{1}{2\varepsilon}\frac{N^2 e^2 D}{R_1^2+R_2^2+(R_1-R_2)D}. \tag{B5}$$

We now minimise with respect to the remaining variable R_1, using $\partial R_2/\partial R_1 = R_1^2/R_2^2$, which follows from eqn (B2). The condition for a minimum is

$$\frac{\partial}{\partial R_1}(\min_{n_1} G) = 0 \tag{B6}$$

or, after a little algebra

$$\left(\frac{4\pi}{N}(R_1^2+R_2^2)\right)^2 = \frac{2\pi e^2 D}{\varepsilon\gamma}\frac{\left[1-(R_1-R_2)\dfrac{D}{2R_1 R_2}\right]}{\left[1+(R_1-R_2)\dfrac{D}{(R_1^2+R_2^2)}\right]}. \tag{B7}$$

This equation can be solved by iteration as follows: clearly, for $R_1, R_2 \gg D$ there is only one solution. The first approximation can be found by dropping terms in D inside the square brackets. This gives

$$\bar{a}^2 \equiv \left(\frac{4\pi(R_1^2+R_2^2)}{N}\right)^2 = \frac{2\pi e^2 D}{\gamma} \tag{B8}$$

or $a_1 = a_2 = a_0$. To find the second approximation to the solution of eqn (B7), we find solutions R_1, R_2 of the equations

$$4\pi(R_1^2+R_2^2) = Na_0$$

$$\frac{4\pi}{3}(R_1^3-R_2^3) = Nv. \tag{B9}$$

These approximate values can then be substituted inside the square brackets of eqn (B7) to yield a better value for $R_1^2+R_2^2$ on the left side. Writing $t = R_1 - R_2$, the solution of eqn (B9) is

$$t = \frac{2v}{a_0}\left(1+\frac{t}{6R_1^2}+O(t^3)\right) \simeq \frac{2v}{a_0}\left(1+\frac{16\pi v^2}{3Na_0^3}\right) \tag{B10}$$

$$R_1 = \left\{t+\sqrt{\frac{Na_0}{2\pi}-t^2}\right\}\Big/2, \qquad R_2 = R_1 - t. \tag{B11}$$

Substitution of these values into eqn (B7) then gives the average area $\bar{a}$ to second approximation

$$\bar{a}^2 = a_0^2\left\{1-\frac{tD}{\left(\dfrac{Na_0}{4\pi}-t^2\right)}\right\}\Big/\left(1+\frac{4\pi tD}{Na_0}\right). \tag{B12}$$

Optimal areas a_1, a_2 are now given as

$$a_1 = \frac{4\pi R_1^2}{n_1} = a_0 \left\{ 1 - \frac{D(R_1 + 3R_2)}{4R_1 R_2} \right\}$$

$$a_2 = \frac{4\pi R_2^2}{n_2} = a_0 \left\{ 1 + \frac{D(3R_1 + R_2)}{4R_1 R_2} \right\}. \tag{B13}$$

The minimum free energy is now from eqn (B5)

$$G_{\min} = N\gamma \left[\bar{a} + \frac{a_0^2}{\bar{a}\left(1 + \dfrac{4\pi D(R_1 - R_2)}{N\bar{a}}\right)} \right]$$

$$= 2N\gamma a_0 - 4\pi\gamma Dt, \tag{B14}$$

where t is given by eqn (B10).

(b) CRITICAL RADIUS

Let l_c be the maximum extended length of a lipid molecule. As N decreases we eventually reach a critical radius below which packing constraints become important. This radius R_1^c is determined by

$$n_1 v = \frac{4\pi R_1^2}{a_1} v = \frac{4\pi}{3}[R_1^3 - (R_1 - l_c)^3] \tag{B15}$$

with a_1 given by (B13). To a sufficient approximation $a_1 \approx a_0$ (B15) becomes

$$l_c = \frac{v}{a_0} \left[1 + \frac{v}{a_0 R_1^c} + \frac{5}{3}\left(\frac{v}{a_0 R_1^c}\right)^2 + \cdots \right]. \tag{B16}$$

Eqn (B15) determines R_1^c and is a special case of eqn (5.4).

(c) PACKING CONSTRAINTS

Now suppose that $N < N^c$. We wish to find the configuration of such a vesicle with minimum free energy. The additional constraint is now eqn (B15), and for fixed N, l_c the free energy involves only a single variable. Note now that we must have a potential across the membrane. It is convenient here to rewrite eqn (B1) in the form

$$G = n_1 \gamma \left(a_1 + \frac{a_0^2}{a_1} \right) + n_2 \gamma \left(a_2 + \frac{a_0^2}{a_2} \right), \tag{B17}$$

where we have ignored curvature corrections to the electrostatic contribution to a first approximation. A complete solution to the problem of minimising this function is very difficult and we proceed as follows: we first show that $\partial a_2 / \partial a_1 \gg 1$. This follows from the constraints

$$\frac{4\pi}{3}(R_1^3 - R_2^3) = Nv \Rightarrow \frac{\partial R_2}{\partial R_1} = \frac{R_1^2}{R_2^2}$$

$$\frac{4\pi}{3}[R_1^3 - (R_1 - l_c)^3] = n_1 v \Rightarrow \frac{\partial n_1}{\partial R_1} = \frac{4\pi l_c}{v}(2R_1 - l_c)$$

$$n_1 + n_2 = N \Rightarrow \frac{\partial n_2}{\partial R_1} = -\frac{4\pi l_c}{v}(2R_1 - l_c)$$

$$\frac{4\pi R_1^2}{n_1} = a_1 \Rightarrow \frac{\partial a_1}{\partial R_1} = 4\pi\left\{\frac{2R_1}{n_1} - \frac{R_1^2}{n_1^2}\frac{\partial n_1}{\partial R_1}\right\}$$

$$= \frac{8\pi R_2}{n_1}\left\{1 - \frac{2\pi R_1 l_c}{n_1 v}(2R_1 - l_c)\right\}$$

$$\frac{4\pi R_2^2}{n_2} = a_2 \Rightarrow \frac{\partial a_2}{\partial R_1} = 4\pi\left\{\frac{2R_1^2}{n_2 R_2} + \frac{4\pi R_2^2 l_c}{n_2^2 v}(2R_1 - l_c)\right\}$$

$$= \frac{8\pi R_2}{n_2}\left\{\frac{R_1^2}{R_2^2} + \frac{2\pi R_2 l_c}{n_2 v}(2R_1 - l_c)\right\}.$$

Hence

$$\frac{\partial a_2}{\partial a_1} = \frac{\partial a_2}{\partial R_1}\bigg/\frac{\partial a_1}{\partial R_1} = \frac{n_1^2 R_2^2}{n_2^2 R_1^2}\frac{\left\{\dfrac{n_2 R_1^3}{n_1 R_2^3} + \dfrac{2\pi R_1 l_c}{n_1 v}(2R_1 - l_c)\right\}}{\left\{1 - \dfrac{2\pi R_1 l_c}{n_1 v}(2R_1 - l_c)\right\}}. \tag{B18}$$

Now from eqn (B15)

$$\frac{2\pi R_1 l_c}{n_1 v}(2R_1 - l_c) = \frac{4\pi R_1^2 l_c}{n_1 v}\left(1 - \frac{l_c}{2R_1}\right) = \frac{a_1 l_c}{v}\left(1 - \frac{l_c}{2R_1}\right) \simeq \left(1 + \frac{l_c}{2R_1}\right). \tag{B19}$$

Hence (B18) becomes

$$\left|\frac{\partial a_2}{\partial a_1}\right| \gtrsim \frac{n_1 R_1}{n_2 R_2}\left(\frac{2R_1}{l_c}\right) \simeq \frac{2R_1^4}{R_2^3 l_c} \gg 1. \tag{B20}$$

Typically for egg PC, $R_1 \sim 100\,\text{Å}$, $R_2 \sim 70\,\text{Å}$, $l_c \sim 17\,\text{Å}$, $|\partial a_2/\partial a_1| > 30$. Consequently [cf. eqn (B17)] a small change in a_1 produces a dramatic increase in a_2 and therefore in G. Thus we can conclude that when G is near its minimum value $a_2 \simeq a_0$, and after rewriting (B17) in the form

$$G = 2N\gamma a_0 + \frac{n_1\gamma(a_1 - a_0)^2}{a_1} + \frac{n_2\gamma(a_2 - a_0)^2}{a_2} \tag{B21}$$

we may ignore the term in n_2. Thus $G = 2N\gamma a_0 + n_1\gamma a_1(1 - a_0/a_1)^2$. From (B15), (B16)

$$\frac{a_0}{a_1} \approx \frac{1 - l_c/R_1}{1 - l_c/R_1^c} \approx 1 - \frac{l_c}{R_1}\left(1 - \frac{R_1}{R_1^c}\right).$$

Hence

$$G \approx 2N\gamma a_0 + \frac{n_1\gamma a_1 l_c^2}{R_1^2}\left(1 - \frac{R_1}{R_1^c}\right)^2$$

$$\approx 2N\gamma a_0 + 4\pi\gamma l_c^2\left(1 - \frac{R_1}{R_1^c}\right)^2. \tag{B22}$$

The corrections to this result due to curvature can, to the lowest approximation, be computed ignoring packing, and added to (B22).

(2) TUBULAR VESICLES

Consider now a tubular vesicle. We shall show that the free energy per lipid of such an assembly is greater than that for a bilayer, so that tubular vesicles will not form, being disfavoured both energetically and entropically. To do this, it is not necessary to deal with the more difficult problem which includes complications due to packing. In the capacitance model for the repulsive forces, the free energy per unit length of a tubular vesicle is

$$\frac{G}{L} = 2\pi\gamma(R_1 + R_2) + \frac{n_1^2 e^2}{\varepsilon} \ln\left(\frac{R_1 + D}{R_1}\right) - \frac{n_2^2 e^2}{\varepsilon} \ln\left(\frac{R_2 - D}{R_2}\right) \tag{B23}$$

where R_1, R_2 are outer and inner radii, n_1, n_2 denote number of head groups per unit area. For a given aggregation number N per unit length, the constraint equations corresponding to eqn (B2), (B3) are

$$N = n_1 + n_2, \qquad \pi(R_1^2 - R_2^2) = Nv. \tag{B24}$$

Again we take n_1, R_1 as the two independent variables, and first minimise with respect to n_1 and fixed R_1 remembering that $n_1 + n_2 = N$. This gives

$$\min_{n_1}\left(\frac{G}{L}\right) = 2\pi\gamma(R_1 + R_2) + \frac{Nn_1 e^2}{\varepsilon} \ln\left(\frac{R_1 + t}{R_1}\right). \tag{B25}$$

Now, minimising with respect to R_1 we find

$$\frac{G}{L}\min = \min_{R_1}\left\{\min_{n_1}\left(\frac{G}{L}\right)\right\} = 2\pi\gamma(R_1 + R_2) +$$

$$\frac{N^2 e^2}{\varepsilon}\left\{\frac{\ln\left(\frac{R_1 + t}{R_1}\right)\ln\left(\frac{R_2}{R_2 - t}\right)}{\ln\left(\frac{R_1 + t}{R_1}\right) + \ln\left(\frac{R_2}{R_2 - t}\right)}\right\}. \tag{B26}$$

It is now straightforward to demonstrate that the optimal free energy per lipid G/NL (min) satisfies

$$\frac{G}{NL}(\min) > \gamma\left[a + \frac{a_0^2}{a}\left(1 + \frac{D^2}{12R_1 R_2}\right)\right] > 2\gamma a_0. \tag{B27}$$

If the packing constraints are included the free energy per lipid would be even larger than $2\gamma a_0$, the optimal energy for a bilayer. Hence we conclude that tubular vesicles can not exist, *i.e.*, that planar or spherical bilayers are the preferred assembly.

[1] V. A. Parsegian, *Ann. Rev. Biophys. Bioeng.*, 1973, **2**, 221.
[2] D. G. Hall and B. A. Pethica, in *Nonionic Surfactants*, ed. M. J. Schick (Marcel Dekker, New York, 1967).
[3] T. L. Hill, *Thermodynamics of Small Systems* (W. A. Benjamin, New York, 1964), vol 2.
[4] C. Tanford, *The Hydrophobic Effect* (John Wiley & Sons, New York, 1973).
[5] M. J. Lighthill, *Fourier Analysis and Generalized Functions* (Cambridge University Press, London, 1970).
[6] G. N. Watson, *Treatise on the Theory of Bessel Functions* (Cambridge University Press, London, 1966).
[7] R. M. Williams and D. Chapman, *Progr. Chem. Fats and Other Lipids*, 1970, **11**, 3.
[8] Y. K. Levine and M. H. F. Wilkins, *Nature NB*, 1971, **230**, 69.
[9] E. G. Finer, A. G. Flook and H. Hauser, *Biochim. Biophys. Acta*, 1972, **260**, 49.
[10] A. Wishnia, *J. Phys. Chem.*, 1963, **67**, 2079.
[11] D. Stigter, *J. Colloid Interface Sci.*, 1967, **23**, 379.
[12] F. M. Fowkes, *J. Phys. Chem.*, 1972, **66**, 382.
[13] J. F. Padday and N. D. Uffindell, *J. Phys. Chem.*, 1968, **72**, 1407.
[14] *Handbook of Chemistry and Physics*, ed. R. C. Weast (Chemical Rubber Co., Ohio, 1971).
[15] R. B. Hermann, *J. Phys. Chem.*, 1972, **76**, 2754.

J. N. ISRAELACHVILI, D. J. MITCHELL AND B. W. NINHAM   1567

[16] J. A. Reynolds, D. B. Gilbert and C. Tanford, *Proc. Nat. Acad. Sci. U.S.A.*, 1974, **71**, 2925.
[17] R. C. Tolman, *J. Chem. Phys.*, 1949, **17**, 333.
[18] D. S. Choi, M. S. Jhon and H. Eyring, *J. Phys. Chem.*, 1970, **53**, 2608.
[19] C. Tanford, *J. Phys. Chem.*, 1974, **78**, 2469.
[20] C. Tanford, *Proc. Nat. Acad. Sci.*, 1974, **71**, 1811.
[21] D. Stigter, *J. Phys. Chem.*, 1975, **79**, 1008, 1015.
[22] P. Debye, *J. Phys. Colloid Chem.*, 1949, **53**, 1.
[23] P. Mukerjee, *J. Phys. Chem.*, 1969, **73**, 2054.
[24] Y. Ooshika, *J. Colloid Sci.*, 1954, **9**, 254.
[25] A. Veis and C. W. Hoerr, *J. Colloid Sci.*, 1960, **15**, 427.
[26] D. C. Poland and H. A. Scheraga, *J. Colloid Interface Sci.*, 1966, **21**, 272.
[27] M. C. Phillips, E. G. Finer and H. Hauser, *Biochim. Biophys. Acta*, 1972, **290**, 397; H. Hauser, personal communication, and presented at the 49th National Colloid Symposium (Potsdam, N.Y., 1975).
[28] D. A. Cadenhead, R. J. Demchak and M. C. Phillips, *Kolloid-Z.*, 1967, **220**, 59.
[29] C. Tanford, *J. Phys. Chem.*, 1972, **76**, 3020.
[30] R. A. Robinson and R. H. Stokes, *Electrolyte Solutions* (Butterworth, London, 2nd edn., 1959).
[31] H. V. Tartar, *J. Phys. Chem.*, 1955, **59**, 1195.
[32] A. E. Alexander and P. Johnson, *Colloid Science* (Clarendon Press, Oxford, 1950), chap. 34.
[33] J. Mingins and J. A. G. Taylor, *Proc. Roy. Soc. Med.*, 1973, **66**, 383; J. Mingins, personal communication.
[34] J. F. Huisman, *Thesis* (Utrecht, 1964). Data also given in M. J. Sparnaay, *The Electrical Double Layer* (Pergamon, New York, 1972), p. 232.
[35] M. F. Emerson and A. Holtzer, *J. Phys. Chem.*, 1967, **71**, 1898.
[36] K. Kalyanasundaram, M. Grätzel and J. K. Thomas, *J. Amer. Chem. Soc.*, 1975, **97**, 3915.
[37] P. Debye and W. Anacker, *J. Phys. Colloid Chem.*, 1951, **55**, 644.
[38] F. Reiss–Husson and V. Luzzati, *J. Phys. Chem.*, 1967, **71**, 957.
[39] J. Swarbrick and J. Daruwala, *J. Phys. Chem.*, 1970, **74**, 1293.
[40] B. D. Flockhart, *J. Colloid Sci.*, 1961, **16**, 484.
[41] G. Pilcher, M. N. Jones, L. Espada and H. A. Skinner, *J. Chem. Thermodynamics*, 1969, **1**, 381.
[42] L. Espada, M. N. Jones and G. Pilcher, *J. Chem. Thermodynamics*, 1970, **2**, 1.
[43] E. Matijević and B. A. Pethica, *Trans. Faraday Soc.*, 1958, **54**, 587.
[44] P. Mukerjee, *J. Phys. Chem.*, 1972, **76**, 565.
[45] R. J. Cherry, *FEBS Letters*, 1975, **55**, 1.
[46] S. Batzri and E. D. Korn, *Biochim. Biophys. Acta*, 1973, **298**, 1015.
[47] C. Huang and J. P. Charlton, *Biochim. Biophys. Acta*, 1972, **46**, 1660; M. Roseman, B. J. Litman and T. E. Thompson, *Biochemistry*, 1975, **14**, 4826.
[48] L. W. Johnson, M. E. Hughes and D. B. Zilversmit, *Biochim. Biophys. Acta*, 1975, **375**, 176.
[49] J. E. Rothman and E. A. Dawidowicz, *Biochemistry*, 1975, **14**, 2809.
[50] H. Hauser and M. D. Barratt, *Biochim. Biophys. Res. Comm.*, 1973, **53**, 399.
[51] D. M. Small, *J. Lipid Res.*, 1967, **8**, 551.
[52] F. Reiss–Husson, *J. Mol. Biol.*, 1967, **25**, 363.
[53] A. J. Verkleij, B. de Kruyff, P. H. J. Th. Vervegaert, J. F. Tocanne and L. L. M. van Deenen, *Biochim. Biophys. Acta*, 1974, **339**, 432.
[54] T. E. Thompson, C. Huang and B. J. Litman, in *The Cell Surface in Development*, ed. A. A. Moscona (John Wiley, 1974), chap. 1.
[55] D. M. Michaelson, A. F. Horwitz and M. P. Klein, *Biochemistry*, 1973, **12**, 2637.
[56] R. D. Kornberg and H. M. McConnell, *Biochem.*, 1971, **10**, 1111.
[57] M. P. Sheetz and S. I. Chan, *Biochem.*, 1972, **11**, 4573.
[58] J. A. Berden, R. W. Barker and G. K. Radda, *Biochim. Biophys. Acta*, 1975, **375**, 186.
[59] V. F. Bystrov, Y. E. Shapiro, A. V. Viktorov, L. I. Barsukov and L. D. Bergelson, *FEBS Letters*, 1972, **25**, 337.
[60] B. de Kruijff, P. R. Cullis and G. K. Radda, *Biochim. Biophys. Acta*, 1975, **406**, 6.
[61] C. H. Huang, J. P. Snipe, S. T. Chow and R. Bruce–Martin, *Proc. Nat. Acad. Sci. U.S.A.*, 1974, **71**, 359.
[62] S. M. Johnson, *Biochim. Biophys. Acta*, 1973, **307**, 27.
[63] D. Papahadjopoulos and N. Miller, *Biochim. Biophys. Acta*, 1967, **135**, 624.
[64] L. de Bernard, *Bull. Soc. Chim. Biol.*, 1958, **40**, 11.
[65] D. A. Shah and J. H. Schulman, *J. Lipid Res.*, 1967, **8**, 227.
[66] J. Requena, D. A. Haydon and S. B. Hladky, *Biophys. J.*, 1975, **15**, 77.
[67] R. P. Rand, *Biophys. J.*, 1964, **4**, 303.
[68] L. Huang and R. E. Pagano, *J. Cell Biol.*, 1975, **67**, 38.
[69] D. Papahadjopoulos and G. Poste, *Biophys. J.*, 1975, **15**, 945.

[70] R. E. Pagano and L. Huang, *J. Cell Biol.*, 1975, **67**, 49.
[71] R. P. Brand, D. O. Tinker and P. G. Fast, *Chem. Phys. Lipids*, 1971, **6**, 333.
[72] V. Luzzati and F. Husson, *J. Cell Biol.*, 1962, **12**, 207.
[73] V. A. Parsegian, *Trans. Faraday Soc.*, 1966, **62**, 848.
[74] M. Locke and P. Huie, *Science*, 1975, **188**, 1219.
[75] D. W. Fawcett, *An Atlas of Fine Structure: The Cell, Its Organelles and Inclusions* (W. B. Saunders, Philadelphia & London, 1966).
[76] K. R. Porter and M. A. Bonneville, *Fine Structure of Cells and Tissues* (Lea and Febiger, Philadelphia, 4th edn, 1973).
[77] K. M. Yamada, B. S. Spooner and N. K. Wessels, *Proc. Nat. Acad. Sci. U.S.A.*, 1970, **66**, 1206.
[78] R. E. Fine and D. Bray, *Nature N.B.*, 1971, **234**, 115.
[79] M. Daniels, *Ann. New York Acad. Sci.*, 1975, **253**, 535.
[80] K. T. Edds, *J. Cell Biol.*, 1975, **67**, 103a.
[81] B. S. Eckert and R. H. Warren, *J. Cell Biol.*, 1975, **67**, 103a.
[82] see *The Biology of Cytoplasmic Microtubules* (*Ann. New York Acad. Sci.*), 1975, **253**.
[83] J. N. Israelachvili and M. Wilson, *Biological Cybernetics*, 1976, **21**, 9.
[84] J. N. Israelachvili, R. A. Sammut and A. W. Snyder, *Vision Res.*, 1975, **16**, 44.
[85] B. Tandler and C. L. Hoppel, *Mitochondria* (Academic Press, N.Y. and London, 1972), pp. 1-10.
[86] A. L. Lehninger, *The Mitochondrion* (W. A. Benjamin, N.Y. and Amsterdam, 1964), pp. 23-29.
[87] D. Atkinson, H. Hauser, G. G. Shipley and J. M. Stubbs, *Biochim. Biophys. Acta*, 1974, **339**, 10.
[88] D. Papahadjopoulos and S. Ohki, *Science*, 1969, **164**, 1075.
[89] D. Papahadjopoulos, G. Poste, B. E. Schaeffer and W. J. Vail, *Biochim. Biophys. Acta*, 1974, **352**, 10.
[90] D. Papahadjopoulos, W. J. Vail, K. Jacobson and G. Poste, *Biochim. Biophys. Acta*, 1975, **394**, 483.
[91] J. Lansman and D. H. Haynes, *Biochim. Biophys. Acta*, 1975, **394**, 335.
[92] J. N. Israelachvili and D. J. Mitchell, *Biochim. Biophys. Acta*, 1975, **389**, 13.
[93] M. P. Sheetz and S. J. Singer, *Proc. Nat. Acad. Sci. U.S.A.*, 1974, **71**, 4457.
[94] S. Marčelja, *Biochim. Biophys. Acta*, 1974, **367**, 165.
[95] A. Erdélyi, W. Magnus and F. G. Tricomi, *Tables of Integral Transforms* (McGraw-Hill, New York, 1954), vol. 1.
[96] M. L. Glasser, *J. Math. Phys.*, 1973, **14**, 701.
[97] A. Erdélyi, W. Magnus, F. Oberhettinger and F. G. Tricomi, *Higher Transcendental Functions* (McGraw-Hill, New York, 1954), vol. 1.

(PAPER 5/2173)

Liquid–Liquid Interfaces

Jeremy Frey

School of Chemistry, University of Southampton, UK SO17 1BJ

Commentary on: **Ionic equilibria and phase-boundary potentials in oil–water systems**,
F. M. Karpfen and J. E. B. Randles, *Trans. Faraday Soc.*, 1953, **49**, 823.

It was with mixed feelings that I agreed to choose and comment on a significant paper in the area of liquid–liquid interfacial chemistry as this is a field with which I have only recently been involved. However, when I started to be interested in the chemistry of liquid surfaces and the liquid–liquid boundary I had many useful conversations with Roger Parsons. I therefore took the opportunity of a departmental celebration to mark his award of the Royal Society Davy Medal to ask him for a key reference published in a Faraday journal that covered some of the ideas we had talked about when I was exploring the analysis of our second harmonic generation experiments at the oil–water interface. Roger immediately suggested a paper that was conveniently published exactly 50 years ago (almost to the month).

The study of liquid–liquid interfaces is currently frequently justified in terms of its importance in areas ranging from the role played in biological systems to the role in industrial liquid-liquid extraction processes at the other extreme of applied chemistry. The selected paper highlights the need to understand the origin of bio-electric phenomena, for which we now have considerable insight in terms of ion pumps, polarisable membranes and the whole panoply of the cell's molecular machinery. Even 50 years on there are still relatively few techniques for directly probing the actual liquid–liquid interface. Second harmonic generation has provided a more molecular probe to supplement the thermodynamic approach of measuring interfacial tension, but as with many interfacial phenomena the analysis of the electrochemical behaviour extends the possibilities in the investigation of the liquid-liquid interface. This is as true now as it was 50 years ago.

With considerable insight Karpfen and Randles chose a system to investigate that would simplify the study of the interfacial boundary rather than attempt to study a more obviously relevant biological system. The careful choice, combined with the careful investigation of all the background issues, was the origin of their success in providing convincing experimental proof of the way in which the phase-boundary potential is related to the distribution coefficients for ionic species between the oil and water phases. It is striking how almost all of the background experiments needed to justify the choice of the system, as well as the choice of the salt bridge materials, had to be determined by the authors, and relatively little information could be obtained from the existing literature. This paper is a classic in its achievement of its object by a series of elegant experiments, which were done in the classical physical chemical way with apparatus entirely constructed by the authors. As well as demonstrating the elegance of traditional classical physical chemistry undertaken by a couple of people, it contrasts with how we now have to study many aspects of modern physical chemistry.

The understated conclusion that the results are consistent with the formulae derived from a direct consideration of an electrochemical equilibrium coupled with the need for electroneutrality belies the fundamental confirmation that liquid junction potential at the oil/water boundary are small; a fact only possible to derive because of the very careful minimisation of all other effects in the work of Karpfen and Randles.

IONIC EQUILIBRIA AND PHASE-BOUNDARY
POTENTIALS IN OIL-WATER SYSTEMS

By F. M. Karpfen and J. E. B. Randles

Dept. of Chemistry, University of Birmingham

Received 2nd February, 1953

The influence of the distribution of electrolytes between water and oils on the electrical potential change at the interface has been investigated. Methods for comparing these potentials were developed and applied to a number of simple systems.

The electrical potential at an oil + water interface has been the subject of many investigations aimed at discovering the part it plays in bio-electric phenomena. These investigations tried to relate the changes of this potential to the nature of the ions in the aqueous solution. The observed results have been attributed to adsorption potentials,[1] diffusion potentials [2] and thermodynamic phase-boundary potentials.[3] It has been shown that the first of these suggestions is definitely false [4] and it seems likely that diffusion potentials and phase-boundary potentials have both made a contribution in the systems investigated hitherto. The attempts at quantitative correlation [2] can hardly be considered successful.

The thermodynamic basis of ionic equilibria in oil + water systems may be summarized as follows. The electrochemical potential of a univalent positive ion M^+ in a phase whose potential is ψ may be expressed

$$\bar{\mu}_{M_+} = \mu^{\circ}_{M^+} + RT \ln a_{M^+} + \psi F, \tag{1}$$

where $\bar{\mu}$ is the electrochemical potential, μ° is the standard chemical potential, a is the activity of the ion, and R, T and F have their usual significance.

The similar expression for a univalent negative ion X^- is

$$\bar{\mu}_{X^-} = \mu^{\circ}_{X^-} + RT \ln a_{X^-} - \psi F. \tag{2}$$

When a salt MX is distributed between oil and water, the condition of equilibrium is that the electrochemical potential of each ion must be the same in both phases. Also each bulk phase must be electrically neutral. Using these conditions it may be shown from (1) and (2) that

$$\psi_{MX} = \frac{RT}{2F} \ln \left(\frac{B}{B_X} \right) \frac{f_w^+ f_0^-}{f_w^- f_0^+}, \tag{3}$$

824   OIL-WATER SYSTEMS

where ψ_{MX} is the potential of the oil phase relative to the aqueous phase,

$$\ln B_M = (\mu_{M_w^+}^\circ - \mu_{M_0^+}^\circ)/RT,$$

$$\ln B_X = (\mu_{X_w^-}^\circ - \mu_{X_0^-}^\circ)/RT,$$

f is an activity coefficient, and superscripts and subscripts indicate the ion and the phase.

By rearranging an expression derived by Shedlovsky,[5] it can be shown that

$$S = (S^\circ)^2 C_w (f_w^\pm)^2/K + S_0 f_w^\pm/f_0^\pm, \tag{4}$$

where S is the distribution coefficient of the salt MX between oil and water when its concentration in the aqueous phase is C_w, S° is the distribution coefficient at infinite dilution, and is equal to $(B_M B_X)^{\frac{1}{2}}$, K is the dissociation constant of the salt in the oil phase (dissociation in the aqueous phase is assumed to be complete) and $f_0^\pm$ and $f_w^\pm$ are mean ionic activity coefficients. Thus the phase-boundary potentials and the distribution coefficient can be related to the same constants B_M and B_X, and hence to each other.

The object of the present investigation was to test this relationship by measuring distribution coefficients and phase-boundary potentials for several different salts. The work fell into three parts:

(a) the measurement of distribution coefficients,
(b) the construction of a non-aqueous electrolyte bridge to minimize diffusion potentials,
(c) the measurement of differences in phase-boundary potentials.

THE MEASUREMENT OF DISTRIBUTION COEFFICIENTS

After extensive preliminary experiments it was decided to use di-*iso*propyl ketone as oil phase and picrates as electrolytes. Such systems bear no chemical resemblance whatever to living-cell membranes, but they are free from interferences such as hydrolysis of the oil phase, and the concentration of the electrolyte may easily be determined colorimetrically. For the initial working out of a satisfactory method for the measurement of phase-boundary potential, such simplifications are essential.

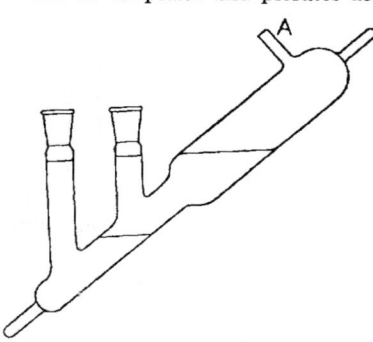

FIG. 1.—Equilibrating vessel, shown in position for withdrawal of solutions, or insertion of electrodes.

EXPERIMENTAL

APPARATUS.—The vessel used for bringing di-*iso*propyl ketone into equilibrium with an aqueous solution of the picrate is shown in fig. 1. The vessel was rocked in a thermostat, during which time the side-arms were closed by ground-glass stoppers and the opening A by a rubber "policeman". The water level in the thermostat was always below the tops of the side-arms, so that no water could leak into the vessel.

MATERIALS.—Commercial di-*iso*propyl ketone was washed with acid, alkali and water in that order. It was then dried over anhydrous sodium sulphate and distilled, using a 3-ft. fractionating column. The fraction boiling between 122·5 and 124·0° C was used for the experiments. This fraction was given a final washing with water and used moist unless the contrary is specifically stated.

Sodium and potassium picrate were prepared from picric acid and the alkali carbonate. They were purified by recrystallizing from water. Tetraethyl ammonium picrate was prepared by mixing equivalent quantities of tetraethyl ammonium iodide and sodium picrate solution and recrystallizing from water.

RESULTS

20 ml of a suitable aqueous picrate solution were brought to equilibrium with 40 ml of pure ketone in the apparatus described above. Trial experiments showed that equilibrium was usually reached in 4 h; in the recorded experiments the vessels were rocked over-night. The two layers were made to rise up their respective side-arms and allowed to settle in the thermostat for 20 min. Suitable volumes of each layer were then removed by a pipette, diluted to 50 ml and the light-absorption of each solution was measured on a Spekker photo-electric absorptiometer. The equilibrium concentrations of the picrate in the two phases (C_0 and C_w) were then read off from calibration graphs. The distribution coefficients ($S = C_0/C_w$) were plotted against $C_w(f_w^{\pm})^2$ in accordance with eqn. (4), and the graphs obtained are shown on fig. 2. (Activity coefficient data for picrates were not available. However, up to the maximum concentration used, the activity coefficients of most uni-univalent electrolytes in water are almost identical; the values for potassium chloride at equivalent concentrations were therefore employed instead.)

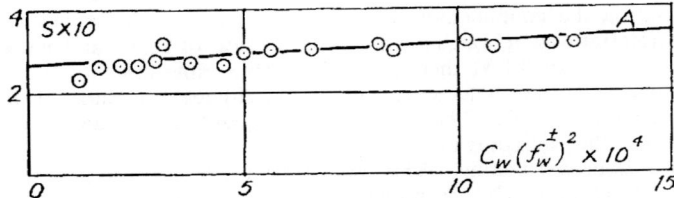

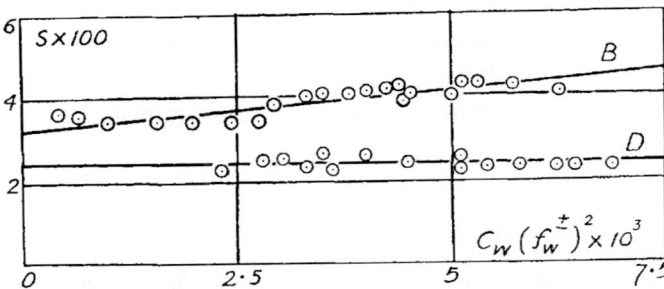

FIG. 2.—Distribution coefficients of picrates between water and di-*iso*propyl ketone
C_w is in moles/l.
A, (Et)$_4$NPi; B, KPi; D, NaPi.

The graphs showed a linear dependence of S on $C_w(f_w^{\pm})^2$ over the concentration range investigated, indicating the constancy of the ratio $f_w^{\pm}/f_0^{\pm}$. The best straight line through each set of points was found by the method of least squares. The distribution coefficients S° at infinite dilutions and dissociation constants K in the ketone were determined from the intercepts and slopes of these lines. The former are probably accurate to $\pm 5\%$ but the slopes of the lines are so small that the probable accuracy of the dissociation constants is much lower. This is unimportant since they are not used for any subsequent calculations but served in practice to indicate the extent of the dissociation of the salts in the solutions used. Sodium picrate appeared to be fully dissociated over the range investigated.

TABLE 1

salt	S°	K mole/l.
tetraethylammonium picrate	$2 \cdot 6 \times 10^{-1}$	$1 \cdot 5 \times 10^{-3}$
potassium picrate	$3 \cdot 3 \times 10^{-2}$	$0 \cdot 6 \times 10^{-3}$
sodium picrate	$2 \cdot 4 \times 10^{-2}$	—

THE PREPARATION OF A NON-AQUEOUS SALT BRIDGE

In order to measure changes in phase-boundary potential an electrode must be connected to each phase by way of a salt bridge. It is essential that the liquid-junction potential between each salt bridge and the phase to which it connects shall be, so far as possible, independent of the ionic content of the latter. The electrolyte in the salt bridge must therefore satisfy two conditions: (i) the two ions must have almost equal mobilities, (ii) the ionic concentration in the bridge solution must be large compared with that in the solution to which it connects. If the solvent is the same on both sides of the junction, these conditions will ensure that the junction potential itself is small and changes in it will be small or negligible. If the solvent in the salt bridge is different from, but miscible with, that in the phase to which it connects, the junction potential may not be small but it may reasonably be expected to be almost independent of the nature and concentration of ions in the latter.

By measuring the conductance of several picrates in di-*iso*propyl ketone at different concentrations, it was shown by the method of Fuoss and Kraus [6] that up to concentration of 0·1 M there is no detectable triple ion formation. Thus concentrations high enough to satisfy condition (ii) are attainable without the formation of multiple ions. The results of semi-quantitative preliminary experiments indicated that tetraethylammonium and picrate ions had nearly the same mobility in di-*iso*propyl ketone. This was confirmed by measuring the transport number of the picrate ion by the moving-boundary method. The conditions for the successful use of the moving-boundary method have been fully examined by Longsworth and MacInnes.[7] A simplified apparatus was used and is shown in fig. 3; camphor-sulphonate was found to be a suitable indicator ion.

The electrolyte system can be represented as

$$+ \text{ Hg } \text{HgPi} \mid (\text{Et})_4\text{NPi} \mid (\text{Et})_4 \text{ N-camphor-sulphonate} \mid \text{Pt}^-,$$
$$\mid \text{ soln.} \qquad \mid \qquad \text{soln.}$$

where Pi represents the picrate ion and the other symbols have their usual significance.

EXPERIMENTAL

MATERIALS. An aqueous solution of tetraethylammonium camphor-sulphonate was prepared by neutralizing camphor-sulphonic acid with the hydroxide. This solution was evaporated to dryness under reduced pressure, and the salt was purified by recrystallizing from benzene.

Mercurous picrate was prepared by allowing a solution of pure mercurous nitrate to run dropwise into a hot solution of sodium picrate. The precipitate was filtered off and washed with water till no nitrate ions could be detected in the washings. It was then dried in a desiccator, and before use it was ground into a paste with mercury and successive portions of the picrate solution.

The solutions of the two salts were prepared directly by dissolving weighed amounts in 100 ml of di-*iso*propyl ketone.

RESULTS

The whole apparatus except the cathode compartment was rinsed out with the prepared picrate solution. The tube containing the mercury and mercurous picrate paste was introduced into the anode compartment, which was then filled with picrate solution from the reservoir. When the anode compartment and U-tube were completely full the anode connector was put in position and the apparatus was clamped in a thermostat, so that the tap was just above the surface of the liquid. After the filled part of the apparatus had reached the temperature of the thermostat, the reservoir was emptied via the cathode compartment; both these were then washed with alcohol and dried by suction. The cathode, which consisted of a piece of platinum gauze surrounded by solid tetraethyl-ammonium camphor-sulphonate, and the inner tube B were put in position and the cathode

F. M. KARPFEN AND J. E. B. RANDLES 827

compartment was filled with camphor-sulphonate solution via the reservoir. The cell was then connected to a 500 V d.c. power-pack in series with a galvanometer and 2 variable resistances. When the cathode compartment had reached thermal equilibrium, it was connected to the U-tube by turning the tap, and the current was switched on.

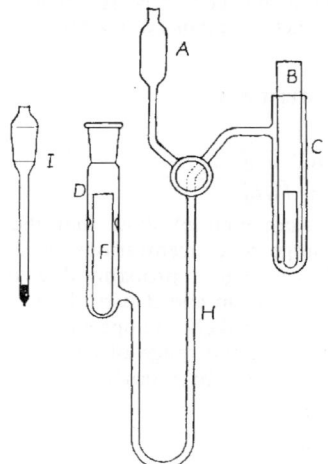

FIG. 3.—Transport number apparatus.

A, reservoir; C, cathode compartment; D, anode compartment; F, tube containing mercury-mercurous picrate anode; H, graduated tube; I, anode connector.

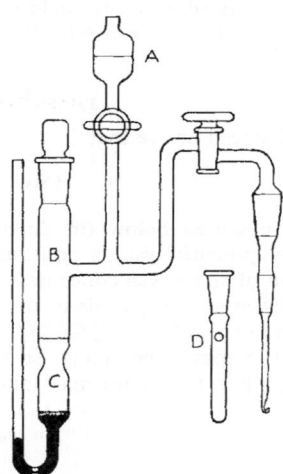

FIG. 4.—Picrate electrode.

A, reservoir for bridge solution; B, $(Et)_4NPi$ in di-*iso*propyl ketone; C, $(Et)_4NPi$ in water; D, guard tube.

As the boundary moved down the U-tube, the resistance of the solution increased; the galvanometer deflection was kept constant by decreasing the variable resistances. The movement of the boundary was followed with a travelling microscope, and the time taken to reach successive graduations was recorded on a stop-watch. Immediately after the experiment the galvanometer was calibrated to get an accurate value of the current used. The measurements and results are tabulated below.

TABLE 2.—TRANSPORT NUMBER t^- OF THE PICRATE ION IN A SOLUTION OF TETRAETHYL-AMMONIUM PICRATE IN MOIST DI-*iso*PROPYL KETONE

temp. = $25 \pm 0.1°$ C; concentration of indicator solution = 1.90×10^{-3} M

concentration of picrate solution C (moles/cm³)	time T (sec)	current i (A)	quantity of electricity iT (coulombs)	volume swept out by boundary in time T V (cm³)	$\dfrac{V}{iT}$	$t^- = \dfrac{VCF}{iT}$
1.660×10^{-6}	1932	4.560×10^{-5}	8.810×10^{-2}	2.823×10^{-1}	3.204	0.513
	965	4.560	4.400	1.373	3.120	0.500
	1033	4.560	4.710	1.445	3.068	0.492
	480	4.560	—	—	—	—
	611	3.790	4.505	1.420	3.152	0.505
	1135	3.790	4.302	1.352	3.143	0.504
2.289×10^{-6}	1326	4.560	6.047	1.364	2.256	0.498
	1434	4.560	6.539	1.459	2.231	0.493
	1353	4.560	6.170	1.373	2.225	0.492
	1452	4.560	6.621	1.445	2.182	0.482

average = 0.498 ± 0.003

When the experiments were repeated with dry di-*iso*propyl ketone as the solvent, the results showed a similar degree of agreement among themselves, and the average value for t^- was 0·543. The experiments showed that sharp boundaries can be obtained without the elaborate apparatus usually used for this purpose. With a better optical arrangement a higher degree of accuracy could easily have been obtained. The results were good enough to justify the use of tetraethyl ammonium picrate in moist di-*iso*propyl ketone as a salt-bridge.

PHASE BOUNDARY POTENTIALS

We deduced above that

$$\psi_{MX} = \frac{RT}{2F} \ln \left(\frac{B_M}{B_X}\right) \frac{f_w^+ f_0^-}{f_w^- f_0^+} \tag{3}$$

Results presented below (in detail for potassium picrate) show that the phase boundary potential ψ_{MX} is constant over a range of concentrations of the salt. Deviation at the lower concentration end of this range is probably due to interference by other ions, while at the higher concentration end it is probably caused by change of the factor $f_w^+ f_0^-/f_w^- f_0^+$. This factor is unlikely to approach constancy except as it approaches unity, so that in and below the range of concentrations, where ψ_{MX} is constant we may take it as unity and (3) becomes

$$\psi_{MX} = \frac{RT}{2F} \ln \frac{B_M}{B_X}.$$

Further, $(S_{MX}^\circ)^2 = B_M B_X,$

hence the difference between the phase-boundary potentials for two different salts, in dilute solution, is given by

$$\Delta\psi_{(MA/NX)} = \psi_{MA} - \psi_{NX} = \frac{RT}{2F} \ln \frac{B_M B_X}{B_N B_A} \tag{5}$$

$$= \frac{RT}{2F} \ln \left(\frac{S_{MX}^\circ}{S_{NA}^\circ}\right)^2. \tag{6}$$

If the two salts have one ion in common, $\Delta\psi$ should be independent of the nature of the common ion. Further, eqn. (6) shows the relationship between $\Delta\psi$ and the infinite dilution distribution coefficients.

If the system contains two salts with a common ion it may be shown that the phase-boundary potential at equilibrium is given by

$$\psi_{(MA + MX)} = \frac{RT}{2F} \ln \left\{\frac{B_M}{B_A(1 - x) + xB_X}\right\}.$$

$$\therefore \; \psi_{(MA + MX)} - \psi_{MA} = \frac{RT}{2F} \ln \left\{\frac{1}{(1 - x) + xB_X/B_A}\right\}, \tag{7}$$

where x is the fraction of the total anion concentration, in the aqueous phase at equilibrium, provided by X^-. The work described below was designed to test the experimental measurements in the light of eqn. (5), (6) and (7).

EXPERIMENTAL

APPARATUS AND MATERIALS.—The electrode for making connection to the non-aqueous phase can be represented diagrammatically as

Hg HgPi | (Et)₄NPi (in water) | (Et)₄NPi (in ketone).
C_w C_0

The symbols have the same significance as before. The two solutions were prepared by shaking 20 ml of 0·1 M tetraethylammonium picrate solution with 20 ml of pure ketone till equilibrium was reached.

The design of the electrode vessel was derived from Lewis, Brighton and Sebastian ; [8] a guard (shown in fig. 4) was introduced to impede the diffusion of bridge electrolyte into the bulk of the oil phase. The preparation of the materials for the picrate electrode has been described in previous sections. The aqueous electrode, made up in a similar vessel, may be represented diagrammatically as

$$\text{Hg, Hg}_2\text{Cl}_2 \mid \text{KCl aq. 3 M.}$$

Calomel was prepared by electrolyzing hydrochloric acid, using a mercury anode.[9] Both electrodes were stored in a thermostat.

For measurements on mixed salts, tetraethylammonium chloride solution was required. It was prepared by grinding the quaternary base iodide with an excess of silver chloride and some water, filtering, diluting and standardizing by silver nitrate titration. The design of the potentiometer circuit was derived from Penick,[10] whose method of adjustment was also used. This circuit was very sensitive to changes of external electric fields, and the thermostat, which contained the cell, had to be stirred by water circulated by a rotary pump, placed at some distance from the thermostat.

METHOD OF MEASUREMENT.—The complete cell used in the comparison of phase-boundary potentials may be represented as follows :

Hg, Hg$_2$Cl$_2$	KCl	M$^+$X$^-$	M$^+$X$^-$	(Et)$_4$NPi	(Et)$_4$NPi	HgPi, Hg
	3M	c	c_0	C_0	C_w	
	ψ_1	ψ_2	ψ_3	ψ_4		

calomel electrode   picrate electrode,

where MX is the salt whose phase-boundary potential ψ_2 is under investigation, c_w is the concentration of the salt in water, c_0 is the concentration of the salt in the oil phase, ψ_1 and ψ_3 are diffusion potentials, and ψ_4 is the potential at the tetraethylammonium picrate phase boundary. Since ψ_1 and ψ_3 are presumed to be small and ψ_4 is constant, any large variations in the e.m.f. of such a cell must be due to changes in ψ_2.

The solutions of MX were prepared in the vessel shown in fig. 1 as described above. When equilibrium had been reached, the two layers were separated and the vessel was clamped in the thermostat so that the side arms were vertical. After running a little solution out of the tips of the electrode vessels, and removing excess with filter paper, clean dry guards were slipped into position. When the electrodes were lowered into their respective side-arms, the guards filled with MX solution. By making the liquid-liquid junctions in the guards, the diffusion of the bridge electrolyte into the bulk of the MX solution was delayed, and the e.m.f. was fairly steady. Contamination of the bulk-phase by minute amounts of bridge electrolyte led to a rapid drift in the e.m.f. of the cell. The calomel electrode was connected to the potentiometer, the picrate electrode to grid of the electrometer valve, and the e.m.f. of the cell was recorded at regular intervals. The potentiometer was checked before each measurement. The equilibrium concentrations of the salt MX were determined colorimetrically after the e.m.f. measurements had been completed. A typical set of readings is given below :

salt in the equilibrium vessel (MX) = potassium picrate (KPi) ;
equilibrium concentration in the aqueous phase (c_w) = 2.24×10^{-3} M.

e.m.f. of cell (V)	0·3442	0·3428	0·3417	0·3410	0·3408	0·3408
time after making the junction (min)	3	6	10	15	20	30

Final average value = 0·341 V ; e.m.f. of reference cell = 0·391 V.

$$\therefore (\Delta\psi(\text{Et})_4\text{N}^+/\text{K}^+) \ 25° \text{ C} = 0.050 \text{ V}.$$

To provide a check on the behaviour of the external electrodes, each set of measurements began and ended with a measurement of the e.m.f. of a standard reference cell. The solutions in the equilibrium vessel for this cell were identical with those in the picrate electrode, and the complete cell may be represented :

Hg Hg$_2$Cl$_2$	KCl	(Et)$_4$NPi	(Et)$_4$NPi	(Et)$_4$NPi	(Et)$_4$NPi	HgPi, Hg
	3M	C_w	C_0	C_0	C_w	

calomel electrode   picrate electrode

It was found that, provided that the same picrate electrode was always used, the e.m.f. of this cell was reproducible to within 1 mV.

830   OIL-WATER SYSTEMS

Experiments on mixed electrolytes were carried out using chloride + picrate mixtures. 20 ml of a solution containing tetraethylammonium chloride and picrate was shaken with 20 ml of pure di-*iso*propyl ketone, and the e.m.f. of the corresponding cell and the equilibrium concentration of the picrate were measured as described above. As very little of the chloride passed into the non-aqueous phase, the initial value of the aqueous concentration was used to calculate the chloride fraction x.

RESULTS

(i) *The dependence of* $\Delta\psi$ *on concentration*

salt MX	concentration in aqueous phase	e.m.f. of cell at 25° C (V) (picrate electrode is always positive)	$\Delta\psi Et_4N . Pi/MX$ (mV)	
Et$_4$N . Pi	0·08 M	0·391	0	—
	0·008 M	0·391	0	—
	0·001 M	0·394	3	—
KPi	0·45 × 10^{-3} M	0·350	41	mean of 3 exp.
	0·55 ,,	0·349	42	—
	0·72 ,,	0·347	44	—
	0·89 ,,	0·345	46	—
	1·08 ,,	0·343	48	—
	1·44 ,,	0·340	51	—
	1·80 ,,	0·341	50	—
	2·24 ,,	0·341	50	mean of 3 expt.
	2·41 ,,	0·341	50	—
	3·00 ,,	0·341	50	—
	3·60 ,,	0·340	51	—
	4·80 ,,	0·339	52	mean of 3 expt.
	10·3 ,,	0·338	53	,,
NaPi	0·80 ,,	0·343	48	—
	4·00 ,,	0·327	64	mean of 2 expt.
	5·00 ,,	0·327	64	mean of 3 expt.

Calculation of $\Delta\psi$ from distribution data :

$$\psi Et_4NPi - \psi KPi = \frac{RT}{2F} \ln \left(\frac{0.260}{0.033}\right)^2 = 52 \text{ mV at } 25° \text{ C,}$$

$$\psi Et_4NPi - \psi NaPi = \frac{RT}{2F} \ln \left(\frac{0.260}{0.024}\right)^2 = 61 \text{ mV at } 25° \text{ C.}$$

(ii) *The dependence of* $\Delta\psi$ *on the nature of the common ion and the distribution coefficient of 2 chlorides.*

salt in equilibrium vessel (MX)	e.m.f. of the complete cell (V)	$\Delta\psi(Cl^-/Pi^-)$ (mV)	$S°$
Et$_4$NCl	0·615	} 224	4·1 × 10^{-5}
Et$_4$NPi	0·391		
KCl	0·572	} 231	5·2 × 10^{-6}
KPi	0·341		

The values of $S°$ for Et$_4$NCl and KCl in the last column are calculated from the experimental values for the picrates and the values of $\Delta\psi$ in column 3, using eqn. (6).

(iii) $\Delta\psi$ *for mixed electrolytes.*

composition of the aqueous phase at equilibrium		$\frac{1-x}{(x = [Cl^-]/}$ $([Cl^-] + [Pi^-]))$	$\psi(Et)_4NPi + (Et)_4NCl - \psi(Et)_4NPi$	
(Et)$_4$N Pi	(Et)$_4$N Cl		calc. from eqn. (7)	measured
			mV	mV
1·31 × 10^{-3} M	4·96 × 10^{-3} M	2·09 × 10^{-1}	20·1	23
0·97 × 10^{-3} M	1·24 × 10^{-2} M	7·27 × 10^{-2}	33·6	34
0·493 × 10^{-3} M	2·48 × 10^{-2} M	1·95 × 10^{-2}	50·5	52
0·349 × 10^{-3} M	4·96 × 10^{-2} M	6·99 × 10^{-3}	63·6	66

(iv) *Measurements in nitrobenzene.*

Davies [11] has given values for the distribution coefficients of several simple salts between water and nitrobenzene. These figures were substituted in eqn. (6) and compared with the e.m.f. measurements obtained from the cells of the following type:

$$Hg, Hg_2Cl_2 \left| \begin{array}{c} KCl \\ 3M \\ water \end{array} \right. \vdots \left. \begin{array}{c} M^+X^- \\ \\ water \end{array} \right| \left. \begin{array}{c} M^+X^- \\ \\ nitro\text{-} \\ benzene \end{array} \right. \vdots \left. \begin{array}{c} (Et)_4NPi \\ \\ di\text{-}iso propyl \\ ketone \end{array} \right| \left. \begin{array}{c} (Et)_4NPi \\ \\ water \end{array} \right| HgPi, Hg.$$

$$\psi_1 \qquad\qquad \psi_2 \qquad\qquad \psi_3 \qquad\qquad \psi_4$$

As in the previous cases changes in the e.m.f. of such a cell when MX is changed should be due to the changes in ψ_2. The experimental procedure was the same as was used for measurements in di-*iso*propyl ketone, and the results are tabulated below.

calculations from distribution experiments			e.m.f. measurements		
salt used by Davies [11]	distribution coefficient $1/S$	$\frac{RT}{2F}\ln\left(\frac{S_1}{S_2}\right)^2$; * (mV)	salt used (MX)	e.m.f. of the cell (mV)	$\Delta\psi$ (mV)
NaI	$3{\cdot}9 \times 10^4$	$-\ 50$	NaPi	171	$-\ 53$
KI	$5{\cdot}5 \times 10^3$		KPi	224	
(Et)$_4$NI	$4{\cdot}5 \times 10^1$	$-\ 124$	(Et)$_4$NPi	350	$-\ 126$
KCl	$2{\cdot}3 \times 10^5$	$+\ 95$	(Et)$_4$NCl	593	$+\ 102$
KI	$5{\cdot}5 \times 10^3$		(Et)$_4$NI	491	

* S_1 and S_2 refer to the indicated pair of values from the second column.

Reviewing all the results, we may say that there was reasonable agreement between theory and experiment except when the electrolyte concentration in the oil phase was very low. It was concluded that the variations in liquid junction potential had been made negligible, and when the same oil phase was used throughout the cell, the liquid-junction potentials themselves were probably very small.

One of us (F. M. K.), is indebted to the University Authorities for a maintenance grant.

[1] Baur *et al.*, *Z. Elektrochem.*, 1913, **19**, 590; *Z. physik. Chem.*, 1916, **92**, 81 Ehrensvärd and Sillén, *Z. Elektrochem.*, 1939, **45**, 440.

[2] Osterhout, *J. Gen. Physiol.*, 1943, **26**, 293 ;, 1943, **27**, 91.

[3] Beutner, Leob, *et al.*, *Z. Elektrochem.*, 1913, **19**, 319, 467; *Z. physik. Chem.*, 1914, **87**, 385; *Biochem. Z.*, 1912, **41**, 1; 1912, **44**, 303; 1913, **51**, 288; 1914, **59**, 195.

[4] Craxford, Gatty and Rothschild, *Nature*, 1938, **141**, 1098. Dean, *Trans. Faraday Soc.*, 1940, **36**, 166.

[5] Shedlovsky, *Cold Spring Harb. Symp. Quant. Biol.*, 1936, **4**, 27.

[6] Fuoss and Kraus, *J. Amer. Chem. Soc.*, 1933, **55**, 21.

[7] Longsworth and MacInnes, *Chem. Rev.*, 1932, **11**, 171.

[8] Lewis, Brighton and Sebastian, *J. Amer. Chem. Soc.*, 1917, **39**, 2245, fig. 4.

[9] Ellis, *J. Amer. Chem. Soc.*, 1916, **38**, 740.

[10] Penick, *Rev. Sci. Instr.*, 1935, **6**, 115.

[11] Davies, *J. Physic. Chem.*, 1950, **54**, 185.

Electrochemistry

P. N. Bartlett

Department of Chemistry, University of Southampton, UK SO17 1BJ

Commentary on: **Kinetics of Rapid Electrode Reactions**, J. E. B. Randles, *Discuss. Faraday Soc.*, 1947, **1**, 11.

Faraday Discussion meetings are one of the distinctive activities of the Faraday Division of the RSC. The meetings trace their origins back to the original meetings of the Faraday Society and they are distinctive in that all of the contributions are printed and circulated to participants well before the meeting. Then, at the meeting itself, the authors are given just five minutes to highlight the important points of their paper and the bulk of the time is given over to questions from the audience and discussion of the work. This discussion is recorded and appears in print in the Discussion volume along with the paper. This unique format means that the meetings provide a vivid record of the state of development of the particular topic of the discussion and highlight areas of disagreement or uncertainty. The very first discussion meeting of the Faraday Society in 1907 concentrated on the topic of "Osmotic Pressure" and the record of the discussion appeared in volume 3 of *Faraday Transactions*, the Faraday Society's journal. In 1947 the papers for the discussion meetings appeared for the first time in a separate journal, *Discussions of the Faraday Society*. This continues to the present day.

The topic of the first Faraday Discussion to appear in the new journal was "Electrode Processes"—a very appropriate choice given that Faraday himself set out many of the basic principles of electrochemistry and gave us much of the basic terminology, coining the terms "electrode" and "electrolysis", "anion" and "anode", "cation" and "cathode".[1] The paper that I have selected from this first volume of *Discussions of the Faraday Society* is a landmark contribution to the field by J. E. B. Randles, at that time working at the University of Birmingham. It is a paper which continues to be widely cited more than 50 years after its publication. In 1947 the field of electrode kinetics was in its infancy. This was a period when attention was shifting from thermodynamic studies in electrochemistry to studies of the dynamics of electrode processes and new techniques and instrumentation was being developed.

One of the central features of electrochemical reactions is that they are heterogeneous processes - the electron transfer reaction occurs at the electrode surface. Therefore to understand and measure the kinetics of electrode reactions it is necessary to know the concentration of the reactive species at the electrode surface. This concentration will not be the same as the bulk value because the reactant is being consumed by the electrode reaction. Thus there are contributions to the overall process from both the interfacial electrode kinetics and from mass transport of species to and from the electrode surface. To study the kinetics of electron transfer reactions it is essential to separate these contributions. This is the topic dealt with by Randles in his paper and also by other contributors to the Discussion volume, notably Levich[2] and Agar.[3] The central problem is very clearly laid out by Randles in the introduction to his paper. Whereas Levich and Agar approach the problem by using forced convection to control the rate of mass transport, an approach which was to lead to the development of the rotating disc electrode and the well known Levich equation,[4] Randles adopts a different approach and applies a small amplitude sinusoidal modulation about the rest potential of the electrode. In his paper he lays out the theory for the resulting sinusoidal modulation in the current taking account of both the interfacial electrode kinetics and the mass transport of species to the electrode by diffusion. Randles realised that this system could be treated in terms of an equivalent circuit made up of a collection of resistors and capacitors (see Fig. 3 of his paper). If the values of these resistors and capacitors were measured then the electrochemical rate constant, k, could be calculated.

This approach, and the "Randles equivalent circuit", have been very widely used in the years since 1947. The impact of Randles' paper can be gauged by the fact that it continues to be cited, typically attracting 10 citations a year in the period since 1981, and by the breadth of research fields on which it has impacted. Thus in the last few years alone Randles' paper has been cited in papers on biosensors,[5] corrosion,[6] physiology,[7] displays,[8] and materials research[9] as well, of course, as in electrochemistry.

References

1 K. J. Laidler, *The World of Physical Chemistry*, OUP, Oxford, 1993.

2 B. Levich. *Faraday Discuss.*, 1947, **1**, 37.

3 J. N. Agar, *Faraday Discuss.*, 1947, **1**, 26.

4 V. G. Levich, *Physicochemical Hydrodynamics*, Prentice Hall, Englewood Cliffs, NJ, 1962.

5 M. Zayats, E. Katz, I. Willner, *J. Am. Chem. Soc.*, 2002, **124**, 14724.

6 M. Hosseini, S. F. L. Mertens, and M. R. Arshadi, *Corrosion Science*, 2003, **45**, 1473.

7 I. C. Stefan, Y. V. Tolmachev, Z. Nagy, M. Minkoff, D. R. Merrill, J. T. Mortimer, D. A. Scherson, *J. Electrochem. Soc.*, 2001, **148**, E73.

8 A. E. Aliev and H. W. Shin, *Displays*, 2002, **23**, 239.

9 J. Jamnik, *Solid State Ionics*, 2003, **157**, 19.

KINETICS OF RAPID ELECTRODE REACTIONS.

By J. E. B. Randles.

Received 5th March, 1947.

There are three factors [1] which may operate in controlling the speed of an electrode reaction, (1) the rate of the electrode process itself, (2) the rates of diffusion of the reactant and product, and (3) the ohmic resistance of the electrolyte. Reactions which have received the most study are those for which (1) is very slow, so that rate control due to (2) and (3) has been negligible or has been easily rendered so. A few investigations [2] of moderately rapid electrode processes have been made in which diffusion control has been either disregarded, or minimised. However, most of

[1] Bowden and Agar, *Proc. Roy. Soc. A*, 1938, **169,** 206 ; *Ann. Reports*, 1938, 90.
[2] Erdey Gruz and Volmer, *Z. physik. Chem.*, 1931, **157,** 165. Roiter, Poluyan and Juza, *Acta Physicochim.*, 1939, **10,** 389; 845. Essin, *ibid.*, 1942, **16,** 102. Bonnemay, *J. Chim. Physique*, 1944, **41,** 218.

12   KINETICS OF RAPID ELECTRODE REACTIONS

the work on rapid electrode reactions has been done from the point of view of diffusion control alone, the electrode processes themselves being loosely classified as " reversible," i.e. so rapid that their influence on the overall reaction rate can be neglected. Any approach to the measurement of the rates of such electrode processes necessitates as precise as possible a knowledge of rate control due to the combined influence of the electrode process and diffusion. Calculation of this is impeded by the difficulty of solving the dynamic diffusion equation in most circumstances. In the experimental method to be described the working conditions are such as to make possible the solution of this equation.

Theory of the Method.

The basis of the method is the application of an alternating potential of small amplitude to a micro-electrode at which the relevant electrode reaction is proceeding, or is in equilibrium. Let us consider the reaction

$$M^{n+} + ne = M \text{ (Hg)} \qquad . \qquad . \qquad . \qquad (1)$$

occurring between an aqueous solution containing the metal ion M^{n+} at a low concentration C (moles. per cc.), and a solution of the metal in mercury also at a concentration C. (Equal concentrations of M and M^{n+} give the most favourable working conditions and will, for simplicity, be assumed from the start. The effect of inequality of concentrations is discussed briefly in the appendix to this paper.) In order that the diffusion of M^{n+} in the aqueous solution shall not be complicated by migration, a large excess of an indifferent electrolyte such as KNO_3 (concentration molar while C is of the order of 10^{-3} M.) is also present.

Let the mean potential of the electrode be such that reaction (1) is in equilibrium, and let us take this potential arbitrarily as zero. An alternating potential $v = V \cos \omega t$ is now applied between electrode and aqueous solution and in consequence an alternating current will flow which we shall denote as

$$i = I \cos (\omega t + \phi) \qquad . \qquad . \qquad . \qquad (2)$$

where I and ϕ are to be determined. It may be assumed that as a result of this current there will be a harmonic variation in the concentrations of M and M^{n+} close to the interface, which will spread away into the nearby solution. For the concentration C_1 of M in the mercury we may write

$$C_1 = C + \delta C_1$$

and, close to the interface

$$C_1^\circ = C + \delta C_1^\circ \qquad . \qquad . \qquad . \qquad (3)$$

where

$$\delta C_1^\circ = \Delta C_1^\circ \cos (\omega t + \theta) \qquad . \qquad . \qquad (4)$$

where ΔC_1° is the amplitude of the variation of C_1 close to the interface, and θ the phase angle (relative to the applied alternating potential). A general expression for $\delta C_{1x, t}$ can be obtained by solving the equation for linear diffusion, $\dfrac{\partial C_1}{\partial t} = D_1 \dfrac{\partial^2 C_1}{\partial x^2}$, for the boundary condition (4) at $x = 0$, x being the distance from the electrode surface (regarded as planar) and D_1 the diffusion coefficient of M in mercury. This gives *

$$\delta C_{1x, t} = \Delta C_1^\circ e^{-\sqrt{\frac{\omega}{2D_1}} \cdot x} \cos \left(\omega t - \sqrt{\frac{\omega}{2D_1}} \cdot x + \theta\right) \qquad . \qquad (5)$$

which represents a diffusion " wave " of length $2\pi\sqrt{\dfrac{2D_1}{\omega}}$ and whose amplitude declines exponentially with distance from the interface.

* Compare the solution of the corresponding problem in thermal conduction : Carslaw, *Conduction of Heat in Solids* (Macmillan, 1921), p. 47.

J. E. B. RANDLES 13

The current flowing (taking a positive value of i corresponding to the left to right direction of reaction (1)) is given by

$$i = - nFAD_1\left(\frac{\partial C_1}{\partial x}\right)_{x=0} = - nFAD_1\left(\frac{\partial(\delta C_1)}{\partial x}\right)_{x=0}$$

where F is the Faraday and A the area of the electrode. Using (5) we obtain

$$i = nFAD_1\Delta C_1^{\circ}\sqrt{\frac{\omega}{2D_1}}\ [\cos(\omega t + \theta) - \sin(\omega t + \theta)] \qquad . \qquad (6)$$

$$= nFA\Delta C_1^{\circ}\sqrt{\frac{\omega D_1}{2}}\cos\left(\omega t + \theta + \frac{\pi}{4}\right). \qquad . \qquad . \qquad . \qquad (7)$$

Hence $\qquad I = nFA\Delta C_1^{\circ}\frac{\sqrt{\omega D_1}}{2}; \qquad \phi = \theta + \frac{\pi}{4}. \qquad . \qquad . \qquad . \qquad (8)$

Comparison of (7) with the corresponding expression based on the diffusion of M^{n+} in the aqueous solution, shows that

$$-\frac{\Delta C_1^{\circ}}{\Delta C_2^{\circ}} = \sqrt{\frac{D_2}{D_1}}$$

where subscript 2 indicates quantities appertaining to M^{n+}. Since[3] D_2 is generally approximately equal to D_1 we shall assume that

$$\Delta C_2^{\circ} = - \Delta C_1^{\circ} \qquad . \qquad . \qquad . \qquad . \qquad . \qquad (9)$$

the minus sign showing that they are 180° out of phase. Having assumed $D_1 \approx D_2$ we shall from now on omit the subscript on D.

The current may also be expressed in terms of the rates of the two opposing processes of equation (1), as

$$i = nFA[k_2C_2^{\circ}e^{\alpha vnF/RT} - k_1C_1^{\circ}e^{-(1-\alpha)vnF/RT}] \qquad . \qquad . \qquad (10)$$

where C_1° and C_2° are the respective concentrations of M and M^{n+}, close to the interface, k_1 and k_2 are the rate constants for the two reactions, v is the potential of the aqueous solution relative to the mercury and α is the fraction of the potential difference operative on the aqueous solution side of the energy barrier.[4] Since we are taking v as zero for the equilibrium potential of the electrode when $C_1^{\circ} = C_2^{\circ} = C$, then $k_1 = k_2 = k$ where k defines the rates of the opposing reactions at the equilibrium potential as kC moles per sec. per unit area of electrode surface. Substituting for C_1° and C_2° by means of (3) and (9), (10) becomes

$$i = nFAk[(C - \delta C_1^{\circ})e^{\alpha vnF/RT} - (C + \delta C_1^{\circ})e^{-(1-\alpha)vnF/RT}] \qquad . \qquad (11)$$

We now use (6) and (11) to obtain ΔC_1° and θ. It is convenient first to differentiate both with respect to time. Provided that nvF/RT is small (< 0.2), then to a close approximation, and independent of α (but the approximation is closest if $\alpha \approx \frac{1}{2}$), equation (11) leads to

$$\frac{di}{dt} = 2nFAkC\left(\frac{nF}{2RT}\frac{dv}{dt} - \frac{1}{C}\frac{d(\delta C_1^{\circ})}{dt}\right)$$

$$= 2nFAkC\left[-\frac{nF}{2RT}\omega V\sin\omega t + \frac{1}{C}\Delta C_1^{\circ}\omega\sin(\omega t + \theta)\right] \qquad . \qquad (12)$$

From (6) we have

$$\frac{di}{dt} = - nFA\Delta C_1^{\circ}\sqrt{\frac{\omega D}{2}}[\omega\sin(\omega t + \theta) + \omega\cos(\omega t + \theta)] \qquad . \qquad (13)$$

[3] Kolthoff and Lingane, *Polarography* (Interscience, 1941), p. 151 ; *Landolt Bornstein Tabellen*, 5th Ed., I, 249.
[4] Glasstone, Laidler and Eyring, *Theory of Rate Processes* (McGraw-Hill, 1941), p. 575, etc.

14   KINETICS OF RAPID ELECTRODE REACTIONS

Expanding (12) and (13) and equating coefficients of cos ωt gives

$$- \cot \theta = 1 + 2k\sqrt{\frac{2}{\omega D}}$$

whence

$$\cot \phi = 1 + \frac{1}{k}\sqrt{\frac{\omega D}{2}} \qquad . \qquad . \qquad . \qquad . \qquad (14)$$

Thus ϕ may vary from zero to $\dfrac{\pi}{4}$ according as $\dfrac{1}{k}\sqrt{\dfrac{\omega D}{2}}$ varies from infinity to zero.

Equating coefficients of sin ωt in (12) and (13) leads to

$$\Delta C_1{}^\circ = - \frac{kCnFV}{RT}\sqrt{\frac{2}{\omega D}}\sin \theta = \frac{nFCV}{\sqrt{2}RT}\sin \phi$$

whence

$$\frac{I}{V} = \frac{n^2F^2AC\sqrt{\omega D/2}}{RT}\sin \phi \qquad . \qquad . \qquad . \qquad (15)$$

For a rapid reaction (k large), (15) becomes $n^2F^2AC\sqrt{\omega D}/2RT$ depending on D but not on k, while for k small it becomes n^2F^2ACk/RT.

Since the current vector I leads the voltage vector V by an angle ϕ it is convenient to compare the system with its electrical equivalent, a condenser and resistance in series. Denoting these by C_r and R_r we have

$$\frac{I}{V} = \frac{1}{R_r}\cos \phi \; ; \qquad \omega C_r R_r = \cot \phi.$$

By means of (14) and (15) the following relations are derived

$$R_r = \frac{RT}{n^2F^2AC}\left(\sqrt{\frac{2}{\omega D}} + \frac{1}{k}\right) \qquad . \qquad . \qquad . \qquad (16)$$

$$C_r = \frac{n^2F^2AC}{RT}\sqrt{\frac{D}{2\omega}} \qquad . \qquad . \qquad . \qquad . \qquad (17)$$

and

$$R_r - \frac{1}{\omega C_r} = \frac{RT}{n^2F^2AC}\cdot\frac{1}{k} \qquad . \qquad . \qquad . \qquad (18)$$

Measurement of R_r and C_r will therefore enable the rate constant k to be calculated.

Experimental.

Apparatus.—The micro-electrode used was a capillary dropping electrode similar to a polarographic electrode except that a dilute amalgam replaced the mercury. Such an electrode has the advantages that its surface is continually renewed, and that the surface area is easily measured by weighing the drops. The amalgams were prepared by electro-deposition of the required amount of metal from solution, on to a measured volume of mercury acting as cathode, in the apparatus shown in Fig. 1. The electrolyte was generally NH_4Cl solution with the appropriate amount of metal salt added. A mercury anode was used, contained in a small glass cup filled up with KNO_3 solution and with a loose cotton-wool plug inserted. During the electrolysis the counter migration of mercurous ions, and Cl′ ions from the main electrolyte, precipitated Hg_2Cl_2 in the cotton wool thus preventing the entry of mercury ions into the main electrolyte but avoiding the contamination of the mercury surface. Air was displaced from the electrolysis vessel by nitrogen. The amalgam was used direct from the vessel by means of the long capillary syphon shown in the diagram, the negative potential of the amalgam relative to the electrolyte being maintained in order to prevent re-solution of the dissolved metal.

Electrical measurements were carried out with the capillary electrode immersed in a suitable solution contained in a cell of the type used in polarography. The other electrode was a mercury pool, and an auxiliary electrode of platinum gauze was fitted over the lower end of the capillary, flush with the end as shown in Fig. 1 (a). The solutions contained indifferent electrolyte (KNO_3, KCl, etc.) at molar concentration and a low concentration (of the order of 10^{-3} M.) of the ion under investigation. Atmospheric oxygen was expelled by a stream of nitrogen. The cell was placed in a thermostat at $25° \pm 0.1°$ C.

The electrical circuit is shown in Fig. 2; D.E., G.E., and M.P. are the dropping electrode, gauze electrode and mercury pool electrode respectively. A valve oscillator connected to points A and B supplied alternating current of adjustable frequency. The current passed by the dropping electrode flowed also through the variable resistance R and the variable capacity C (a set of standardised condensers). The potential difference across R and C was amplified by a D.C. push-pull amplifier connected to XX, the output of which was applied to one pair of deflector plates of a cathode ray tube. The potential difference between the dropping electrode and the gauze electrode was similarly amplified and applied to the other pair of deflector plates. A negligible current flowed through the gauze electrode therefore its potential followed exactly that of the surrounding solution. By making the distance between it and the dropping electrode as small as possible the minimum of potential drop due to electrolyte resistance was included in the measured potential difference. The mercury pool was connected to R and C and the dropping electrode to earth in order to avoid the difficulty of electrical screening of the latter.

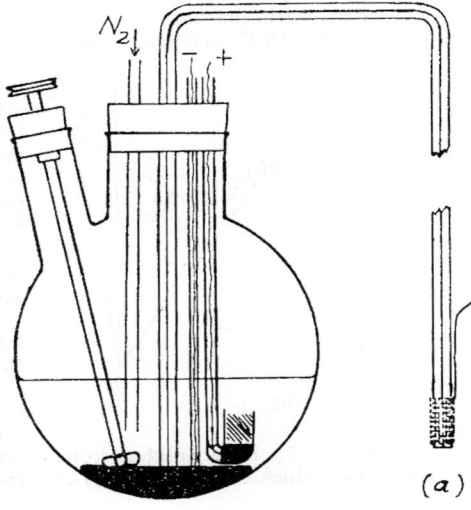

FIG. 1.

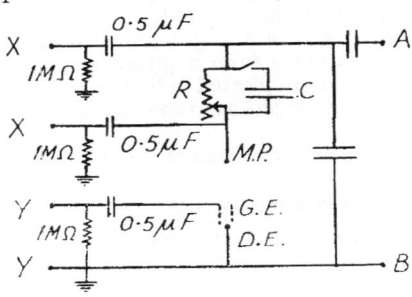

FIG. 2.

Method of measurement and calculation of results.—With capacity C zero and resistance R in the circuit the cathode ray trace is an ellipse. By switching out one amplifier at a time, the horizontal and vertical amplitudes for maximum size of the amalgam drop were measured. When corrected for overall sensitivity for the two directions the ratio of these amplitudes gives RI/V where R is known. The total voltage swing, $2V$, never exceeded about 7 mv. as required in the derivation of equation (12). The voltage-current phase difference, ϕ', for the dropping electrode-gauze electrode was obtained by adjusting C and R until the trace became a straight line at maximum drop size. The phase angle

16   KINETICS OF RAPID ELECTRODE REACTIONS

for the C-R combination given by $\tan \phi' = \omega CR$ is then equal to that for the electrode system.

The whole electrode system is electrically equivalent to the circuit shown in Fig. 3 (a), where C_r and R_r are the capacity and resistance equivalent to the electrode reaction, C_l the ordinary double layer capacity of the electrode surface and R_e the electrolyte resistance between the electrode and the platinum gauze. R_e was obtained by measuring the total impedance, with the usual indifferent electrolyte, but cutting down the impedance due to C_r and R_r to about 3 ohms by using a high (M./20) concentration of lead or cadmium in the amalgam and solution. Thirty ohms is a typical value for R_e. C_l was measured in a series of experiments using pure mercury and pure indifferent electrolyte for various potentials of the electrode ; values of C_l per unit area agreed well with accepted values.

A vector diagram, Fig. 3 (b), was used for calculating C_r and R_r. Measured values of V and I are first drawn at angle ϕ'. A vector IR_e, parallel to I is subtracted from V giving V_r.

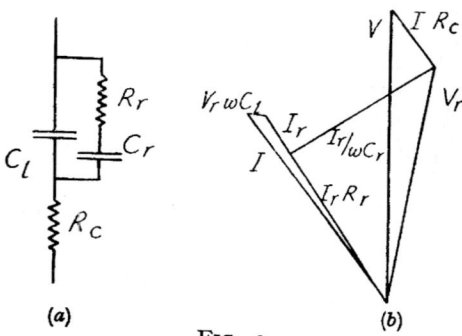

(a)   **(b)**

FIG. 3.

$V_r \omega C_l$, normal to V_r, is subtracted from I giving I_r. A normal from V_r to I_r gives the values of $I_r R_r$ and $I_r/\omega C_r$, from which R_r and $1/\omega C_r$ are calculated.

Results.

The results presented in this section are exploratory in character and much more extensive work is intended. Only a few metal-metal ion reactions have been investigated. However, the data obtained is sufficient to verify the theoretical equations and to illustrate the type of result obtainable.

Expressions (16) and (17) show that R_r and C_r plotted against $1/\sqrt{\omega}$ should give straight lines. Instead of C_r it is more profitable to plot the capacitative impedance $1/\omega C_r$, which should give a straight line through the origin parallel to the R_r line, and at a distance from it given by (18). Results for copper, zinc, cadmium and thallium are plotted in Figs. 4 and 5. The dotted lines in the cadmium and thallium graphs are those calculated

for $\dfrac{1}{\omega C_r} = \dfrac{RT}{n^2 F^2 A C} \sqrt{\dfrac{2}{D}} \cdot \dfrac{1}{\sqrt{\omega}}$, using for D the geometric mean of the diffusion

coefficients of the ion in water and the metal in mercury, i.e.

$$\sqrt{0.65 \times 1.5} \times 10^{-5} \text{ and } \sqrt{2.0 \times 1.0} \times 10^{-5}, \text{ cm.}^2 \text{ sec}^{-1}.,$$

for cadmium and thallium respectively.

Inspection of the graphs shows that constancy of $\left(R_r - \dfrac{1}{\omega C_r} \right)$ is

satisfactory for the reactions of moderate rapidity but it is not so good for the very rapid reactions (cadmium and thallium). This is only to be expected as in the latter case it represents a small difference between two comparatively large quantities. There appears in these cases to be a tendency for it to decrease as the frequency increases. It has not yet been discovered whether this is due to some effect inherent in the detailed mechanism of diffusion plus electrode process, or whether it is an artefact of some sort. In any case since rate of diffusion must tend to a limiting

J. E. B. RANDLES

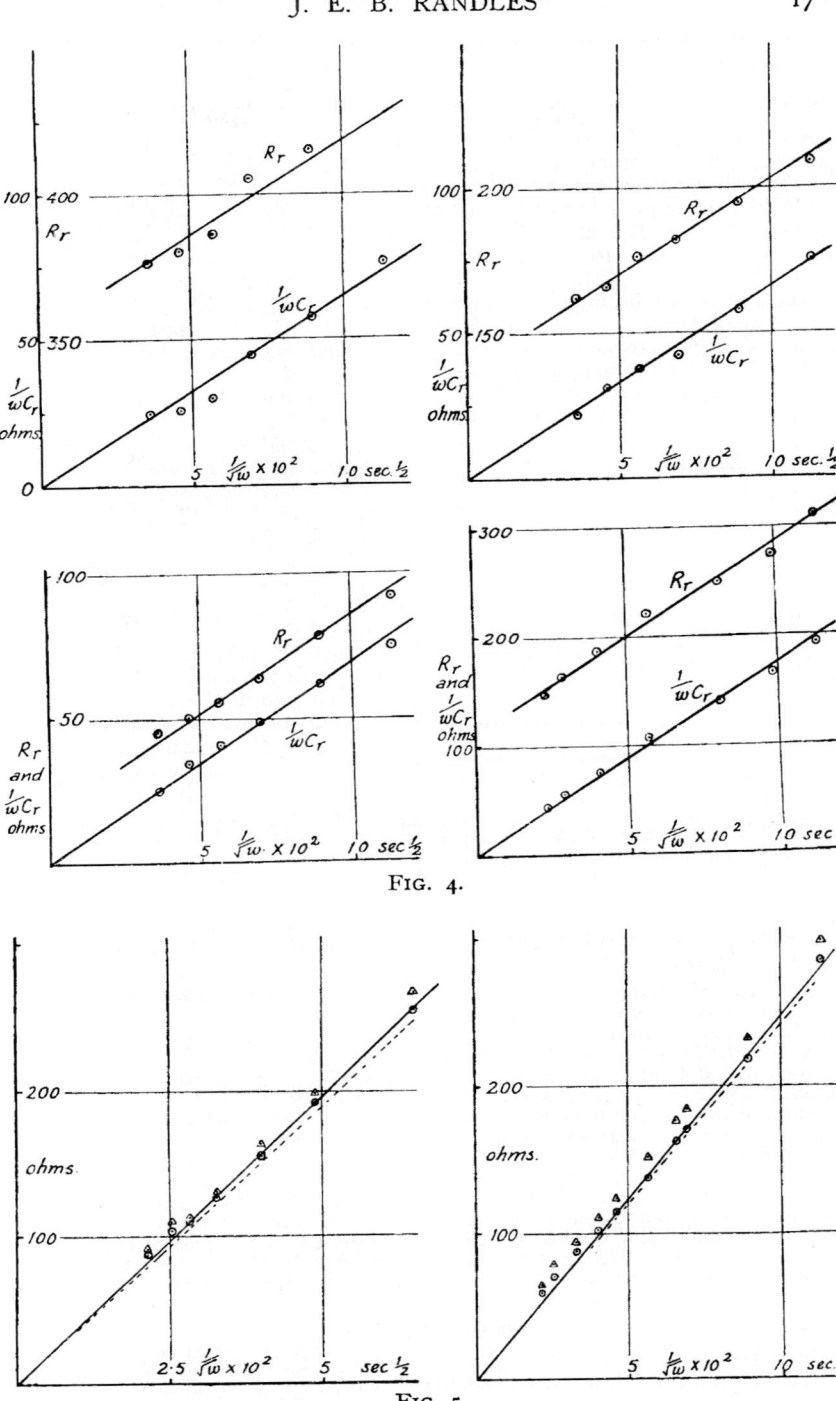

FIG. 4.

FIG. 5.

18   KINETICS OF RAPID ELECTRODE REACTIONS

value for very high concentration gradients such as exist close to the electrode surface when the electrode process is very rapid, it is not certain how much of the rate control represented by $R_r - \dfrac{1}{\omega C_r}$ may be due to this cause. Thus for such cases as thallium, cadmium, lead (similar to cadmium) the constant can at present only be regarded as a lower limit for the true rate constant of the electrode process.

Rate constants for zinc as the hydrated ion and in complexes with bromide and iodide ions are given in Fig. (4). Those for zinc in M. KCl and KCNS are 6×10^{-3} and $1 \cdot 7 \times 10^{-2}$. Thus k increases in the order $k(NO_3^-) < k(Cl^-) < k(Br^-) < k(CNS^-) < k(I^-)$. Preliminary results for nickel and cobalt indicate a similar order for $k(Br^-)$, $k(CNS^-)$ and $k(I^-)$; these reactions are a great deal slower than for zinc. It appears possible that increasing covalency of bonding between the metal ion and its addenda, lowers the activation energy for discharge.

The addition of small amounts of surface active substances, such as gelatin or methyl red, to the solution has an interesting effect on the rates of rapid reactions. A very small amount has no effect, but as the amount is increased there is a sudden increase in R_r for the reaction, but little or no effect on $1/\omega C_r$. This implies a slowing down of some process at the electrode surface or very close to it, but not of the general rate of diffusion. With further increase in the concentration of the colloid, there is no further change in the reaction rate until there occurs another sudden decrease and the reaction becomes too slow for accurate measurement. These results are indicated in the graph of log k for M./2000 Cd++ in M. KNO$_3$ solution, against concentration of gelatin, Fig. 6. Further investigation of this effect is in progress.

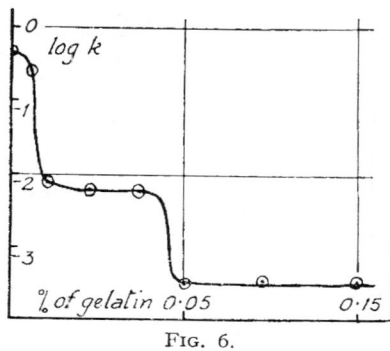

FIG. 6.

Appendix.

A.C. Conductance of a Dropping Mercury Electrode, at which a Reduction Process is Occurring.

A note on this subject is worth while in view of a recent publication [5] on the topic. At a mercury electrode at which a reduction process is occurring, both the reactant and product are present at the interface. When the current passing is equal to half the limiting diffusion current, the concentrations of the two are nearly equal if their diffusion coefficients are similar. The electrode therefore behaves similarly to the amalgam electrode described, except that its mean potential is not the equilibrium potential, but this difference is small for a rapid reaction. Starting from expression (10) it can be shown that the A.C. conductance of the electrode is a maximum when $C_1 = C_2$, i.e. for the dropping mercury electrode, when the D.C. component is half the limiting diffusion current. This is the result obtained by the authors of the paper cited, but their theoretical deduction of an expression in agreement with this, from the equation for the ordinary polarographic current-voltage curve, appears to the present author to be quite erroneous. That equation refers to what is, in effect, a steady state of the diffusion layer, and is not applicable to the alternating current process.

Summary.

A theoretical investigation has been made of the current passed by an electrode at which an electrochemical reaction is in equilibrium, when

[5] Breyer and Gutmann, *Trans. Faraday Soc.*, 1946, **42**, 645.

J. E. B. RANDLES 19

it is subjected to a small alternating potential relative to the solution. Consideration of the electrode process, and diffusion, shows that the reaction is equivalent, electrically, to a capacity and resistance in series. Measurement of these quantities enables the rate constant for the electrode reaction to be calculated. Experiments have been carried out using a dropping amalgam electrode and results are presented in support of the theoretical equations. Rate constants for the discharge of a few metal ions are compared. Evidence has been obtained of a marked slowing down of rapid electrode reactions by the addition of surface active colloids to the aqueous solution.

Résumé.

On a fait l'étude du courant qui passe par une électrode, siège d'une réaction électrochimique en équilibre, lorsque la solution est soumise à de faibles potentiels alternatifs. Diverses considérations sur le processus à l'électrode et la diffusion montrent que la réaction est électriquement équivallente à une capacité et une résistance en série. La mesure de ces quantités permet de calculer la constante de vitesse. Des expériences, qui employaient une électrode à goutte d'amalgame, ont apporté des résultats, qui viennent appuyer les équations théoriques. Les constantes de vitesse pour la décharge de quelques ions métalliques sont comparées. On a mis en évidence un ralentissement marqué des réactions rapides à l'électrode par addition, à la solution aqueuse, de colloïdes actifs sur les surfaces.

Zusammenfassung.

Es wurde eine theoretische Untersuchung des Stromdurchgangs durch eine Elektrode, an der ein elektrochemischer Vorgang im Gleichgewichtszustand ist, wenn dieser im Verhältnis zur Lösung eine kleine Wechselspannung versetzt wird, durchgeführt. Eine Erwägung des Elektrodenprozesses und der Diffusion ergibt, dass—in elektrischer Beziehung—die Reaktion einer Kapazität und einem Widerstand in Hintereinanderschaltung äquivalent ist. Durch Messung dieser Grössen kann die Geschwindigkeitskonstante der Reaktion an der Elektrode berechnet werden. Es werden mit Benützung einer Amalgamtropfelektrode erhaltene Versuchsergebnisse angeführt, welche die theoretischen Gleichungen unterstützen. Die Geschwindigkeitskonstanten für die Abscheidung von einigen Metallionen werden verglichen. Beobachtungen weisen darauf hin, dass die Beifügung von oberflächenaktiven Kolloiden zu einer wässerigen Lösung rasche Elektrodenreaktionen merklich verzögert.

Department of Chemistry,
University of Birmingham.

Gas–Solid Surface Science

M. W. Roberts

School of Chemistry, Cardiff University, PO Box 912, Cardiff, UK CF10 3TB

Commentary on: **Catalysis: Retrospect and prospect**, H. S. Taylor, *Discuss. Faraday Soc.*, 1950, **8**, 9.

In his lecture Taylor emphasised both the phenomenological similarities and differences that existed between adsorption processes on technical catalysts and those observed with idealised catalysts produced by the Beeck evaporated film method to form oriented or non-oriented metal surfaces. These were inherent to the two schools of surface chemistry that had emerged, there were those who studied technical catalysts (Eucken, Emmett, Brunauer) and those who favoured the "clean surface" approach,[1] where particular attention was given to vacuum conditions and surface preparation and exemplified by Beeck in Emeryville and Roberts in Cambridge. Kinetic studies were dominant with Trapnell, Tompkins and Kemball following the evaporated film approach in the 1950s and sixties.[2–4]

In 1952 under the supervision of Keble Sykes (one of Hinshelhood's students) I began my doctoral studies; the aim was to unravel the role of sulphur as a catalyst in the formation of nickel carbonyl at a nickel surface. Two aspects were to be addressed (a) the kinetics of carbonyl formation and (b) characterising the nickel surface (formed by the reduction of NiO) by adsorption studies. I decided that it would be logical to characterise first the surface through chemisorption (hydrogen at −183 °C) and physical adsorption (krypton at −183 °C) and compare the data with those of Otto Beeck. This led me directly into the controversy between those who advocated the "technical catalyst" approach and those who preferred the more "physics approach" of the "clean surface" school the latter recognising the need to develop ultra high vacuum techniques. Could nickel powder surfaces be prepared so as to have characteristics similar to evaporated (clean) nickel films and if not, why not?[5] That was my problem, and a pre-requisite for any meaningful studies of the kinetics of carbonyl formation. The Taylor lecture addressed these two approaches and provided the philosophical arguments and the experimental pre-requisites for unravelling the mechanism of a surface catalysed reaction.

Taylor drew attention to surface poisons (particularly the role of oxygen) and promoters and whether one should approach oxide catalysis via solid state physics invoking band theory and concepts from semiconductor science, an aspect that the Bristol School (Garner, Gray and Stone) had favoured. The significance of variable oxidation states (for which direct surface evidence was not available) and their control through doping, and Mott's views of the role of impurity centres as "great swollen atoms extending over 10–20 lattice parameters" were challenging aspects requiring both new theoretical and experimental approaches. As indeed were the views being expressed[6] by Boudart, Dowden and Eley correlating catalytic activity with the d-band character of the metal and whether Schwab's studies of metal alloy systems could be interpreted in terms of the extent of completion of Brillouin zones. Surface composition (at the atomic level), and appreciation of the density of electron states and the orbitals involved in surface bonding were unhelpful at that time. The challenge to both experimentalists and theoreticians was clear.

Heats of adsorption were scarce, limiting any thermodynamic analysis of catalytic reactions, while the significance of surface structure had not been addressed. Where heats of adsorption were available there was a conflict between data obtained by those advocating the clean surface (Beeck) and technical catalyst (Eucken) approaches. Taylor drew attention to Muller's (then recent) field emission microscope and the potential of deuterium in isotopic exchange studies an aspect that Kemball, who had worked in Princeton used to much advantage in unravelling the mechanism of hydrocarbon reactions at metal surfaces. Volkenstein's views based on the role of lattice defects were analysed in detail but there was a lack of definitive data. Nevertheless it was an aspect central to catalysis as illustrated much later by the electron microscopic studies of Thomas.[7]

Taylor recognised the advantage of running in parallel studies of metal and oxide systems, a view central to my postgraduate work and encapsulated in his statement: *"The catalyst itself, rather than the reactions which occur on it, seems to be the principal objective for future research in the coming years."* I wonder whether this is what convinced me that I should concentrate first on surface characterisation rather than the kinetics of carbonyl formation? There were, however, some fundamental difficulties with this approach in that there was a lack of experimental methods that were surface sensitive. Nevertheless adsorption studies coupled with developing ultra-high-vacuum techniques was, in my view, a way forward in 1952. The downside was that no progress was made with the kinetics of carbonyl formation!

In the next Faraday Discussion devoted to heterogeneous catalysis in 1966 we began to see some of the experimental problems recognised in the Taylor lecture being addressed. Emphasis was given to chemisorption studies at single crystal metal surfaces, field emission, infrared and electron spin resonance spectroscopy. Characterisation of the catalyst and surface species were now being given priority. Nevertheless in 1966 only adsorbed carbon monoxide, due to its high extinction coefficient, could be studied by infrared spectroscopy and only when the metal was dispersed in a high area support such as alumina. There was clearly an incentive to develop surface sensitive spectroscopies of wider applicability and which could also be applied to small area single crystal surfaces.

My inclination in 1960 was to seek more direct experimental evidence and to address some of the issues raised by Taylor and also my postgraduate and postdoctoral work (with Tompkins). Chemisorbed oxygen at metal surfaces had been a recurrent theme in my work whether as a surface poison (Taylor's view) or whether its transformation to an "oxide" state was energetically facile and catalytically significant. I chose changes in work function,[8] making use of the capacitor method described by Mignolet in the 1950 Faraday Discussion, and later monitored the kinetic energy of photoelectrons[9] as a means of following (at low temperatures) the transformation of chemisorbed oxygen to 'oxide'. This was the essence of our paper[10] in 1966 and which led, in 1970, to the development of an ultra-high-vacuum photoelectron spectrometer with both X-ray and UV sources of radiation.[11] The chemistry of defective oxygen states became (and still is!) a major interest of mine[12,13] with the atom-resolved capability of STM and the surface sensitivity of XPS of crucial significance.[14]

The emergence in the 1960s of ultra-high-vacuum techniques, structural studies through LEED and surface sensitive spectroscopies became recognised as "the surface science" approach to catalysis, answering many of the issues raised in Taylor's paper. It was, however, the last paragraph of his paper that made a particular impression on me:

".....we may anticipate a reconciliation of the several attitudes that sometimes have appeared to divide us, but are in reality a spur to further and continued effort towards the mastery of our science in an era which is of deep significance in all human affairs. In prospect, therefore, the future of our science is both challenging and bright".

References

1 A. Wheeler, in *Structure and Properties of Solid Surfaces*, ed. R. Gomer and C. S. Smith, University of Chicago Press, 1953, ch. 13.

2 M. A. H. Lanyon and B.M.W Trapnell, *Proc. R. Soc London, Ser. A*, 1955, **227**, 387.

3 M. W. Roberts and F.C Tompkins, *Proc. R. Soc. London, Ser. A*, 1959, **251**, 369.

4 C. Kemball, *Proc. R. Soc. London, Ser. A*, 1954, **223**, 377.

5 M. W Roberts and K. W Sykes, *Proc. R. Soc. London, Ser. A*, 1957, **242**, 534; M. W Roberts and K. W. Sykes, *Trans. Faraday Soc.*, 1958, **54**, 548.

6 *Chemisorption*, ed. W. E. Garner, Butterworths Scientific Publications, 1957.

7 J. M. Thomas, E. L. Evans and J. O. Williams, *Proc. R. Soc. London, Ser. A*, 1972, **331**, 417.

8 C. M. Quinn and M. W Roberts, *Trans. Faraday Soc.*, 1964, **60**, 899.

9 C. M. Quinn and M. W Roberts, *Trans. Faraday Soc.*, 1965, **61**, 1775.

10 M. W. Roberts and B. R. Wells, *Discuss. Faraday Soc.*, 1966, **41**, 162.

11 C. R. Brundle and M. W. Roberts, *Proc. R. Soc. London, Ser. A*, 1972, **331**, 383.

12 C. S. McKee, L. V. Renny and M. W. Roberts, *Surf. Sci.* 1978, **75**, 92.
13 A. F. Carley, P. R. Chalker and M. W. Roberts, *Proc. R. Soc. London, Ser. A*, 1985, **399**, 167.
14 A. F. Carley, P. R. Davies and M. W. Roberts, *Catal. Lett.*, 2002, **80**, 25.

5th SPIERS MEMORIAL LECTURE

CATALYSIS : RETROSPECT AND PROSPECT

By Hugh S. Taylor

The finest memorial to the first secretary of the Faraday Society is the maintenance of the General Discussions at the highest level of excellence. Wherever physical chemistry is prosecuted, wherever students are acquiring physicochemical science, there the record of the discussions which Secretary Spiers initiated and organized for so many years will be found indispensable.

There is a curious discrepancy in the archives of the Society concerning the General Discussions. In the index of Volume III for 1907 two general discussions are recorded, one on " Osmotic Pressure," one on " Hydrates in Solution ". Sir Oliver Lodge was then President of the Society. In the index to the first twenty volumes of the *Transactions* the earliest recorded General Discussion is in Volume IV, on " The Constitution of Water ". I am at a loss to account for such a discrepancy unless it is a simple oversight. The last General Discussion organized by Secretary Spiers was in October 1925 on " Photochemical Reactions in Liquids and Gases " and was held in Oxford. I was present on that occasion, which marked, as the General Discussions so frequently do, a milestone in the development of the particular phase of physical chemistry then under consideration. For such reasons, and with a peculiar sense of privilege and honour, I am happy on this occasion to introduce a general discussion on " Heterogeneous Catalysis " as the Spiers Memorial Lecturer, in my own University, where, as a student in the departmental library, I first learned to appreciate how greatly these General Discussions can contribute to the definition of the present state of science and in what direction progress may develop.

The present Discussion is the third in a sequence which includes two famous predecessors. In 1922 the Discussion on Catalysis brought forward two basic concepts, that of Lindemann on the nature of unimolecular kinetics and that of Langmuir on reaction at surfaces between adjacently adsorbed reactants, with adsorption restricted to monolayers. The discussion on " Adsorption " in 1932 was concerned, in the main, with slow processes of sorption and evoked, from our President, a definition of van der Waals' and of chemisorption in terms of the potential energies between an impinging molecule and a surface. The diagram brought out clearly the manner in which activation energy of adsorption might be involved, and sharply differentiated physical and chemical processes of adsorption, the differences between which had been indistinct up to that time. It emerged that physical adsorption had little relevance to catalysis at surfaces ; the chemisorbed monolayers of Langmuir were the loci of such reactions.

It is pertinent here to emphasize that the Langmuir formulation of surface kinetics was restricted to those surface reactions in which the velocity of interaction on the surface was the rate-determining process. This condition was indeed fulfilled in the classic researches of Langmuir and in further developments by Hinshelwood, Rideal, Schwab and others.

A * 9

It is useful, however, to recall an example which does not conform to this condition, the decomposition of ammonia on doubly-promoted iron synthetic ammonia catalysts as studied by Love and Emmett.[1] They found a kinetic equation,

$$ - \frac{d[NH_3]}{dt} = k \frac{[NH_3]^{0.6}}{[H_2]^{0.9}}, $$

which, on the Langmuir basis, would suggest a strong adsorption of hydrogen on the iron surface and a moderate ammonia adsorption on the surface bare of hydrogen, with nitrogen an inert constituent. We now know that such a view is erroneous ; the slow step is the desorption of nitrogen from the surface, the slow sorption being rate-determining in synthesis. This example alone would have justified the emphasis on slow sorption in the monolayer, first discussed by the Society in 1932. The availability of isotopic forms of a given molecular species has revealed the variation in temperatures at which these species will interact on surfaces which still further emphasizes the necessity for the concept of activation energy accompanying their chemisorption on catalyst surfaces. Some examples are cited in Table I.

TABLE I.—INTERACTION OF ISOTOPIC MOLECULES ON SURFACES

Catalyst	Isotope Reaction	Temp. ° C. for Measurable Rates	Reference
Fe synthetic ammonia catalyst, doubly promoted .	$H_2 + D_2$	-195	1
,, ,, ,,	$NH_3 + ND_3$	$+ 25$	2
,, ,, ,,	$N_2^{14} + N_2^{15}$	$+450$	1, 3
Rhenium	$H_2 + D_2$	$+ 25$	3
,, ...	$NH_3 + ND_3$	$+100$	3
,, ...	$N_2^{14} + N_2^{15}$	$+500$	3
Nickel	$H_2 + D_2$	-195	4
,, ...	$CH_4 + CH_4$	$+150$	5
Osmium	$N_2^{14} + N_2^{15}$	$+250$	6

[1] Kummer and Emmett, *Brookhaven Conference Report* (Dec. 1948), p. 1.
[2] Taylor and Jungers, *J. Amer. Chem. Soc.*, 1935, **57**, 660.
[3] McGeer, *Thesis* (Princeton, 1949).
[4] Sadek and Taylor, *J. Amer. Chem. Soc.*, 1950, **72**, 1168.
[5] Wright and Taylor, *Can. J. Res. B*, 1949, **27**, 303.
[6] Guyer, Joris and Taylor, *J. Chem. Physics*, 1941, **9**, 287.

It is important for the further argument that these variations found with technical catalysts are also to be found with the idealized catalysts produced by the Beeck method of evaporating metals to form oriented or non-oriented films. In these cases, also, where purity and cleanliness of surface have been raised to the highest attainable standards, similar variations in the temperature at which chemisorption occurs are found, although there will be differences between films and technical catalysts in the actual temperature range involved. Thus, on nickel films, Beeck has shown chemisorption of hydrogen even at 15° K, but examines interaction between ethane and nickel films only in the temperature range 200-250° C. In contrast to Beeck's measurements on nickel films are those of Eucken and Hunsmann[2] with reduced nickel from the oxide where the adsorption at 20° K shows a heat of van der Waals' adsorption. Chemisorption with 5 kcal. of heat of adsorption is found only at 50° K. While ;iron films chemisorb nitrogen (as molecules ?) at liquid-air temperatures

[1] Love and Emmett, *J. Amer. Chem. Soc.*, 1941, **63**, 3297.
[2] Eucken and Hunsmann, *Z. physik. Chem. B*, 1939, **44**, 163.

HUGH S. TAYLOR II

there is a slow activated adsorption of nitrogen, with a much higher heat of adsorption, around room temperatures.   This latter fact is to be contrasted with the studies of Emmett and Brunauer with iron synthetic ammonia catalysts [3] where the slow sorption of nitrogen was measured in the temperature interval from 224 to 449° C.

This example takes us at once to the heart of a problem which it ought to be the objective of this Discussion finally to resolve.   In 1925, a concept of the catalytic surface was formulated which emphasized heterogeneity or, as it came to be expressed, the concept of " active centres ".   A variety of evidence on the properties of technical catalysts, which were the only catalysts then extensively studied, contributed to this concept of active centres.   This evidence included observations on adsorption by catalysts both active and inactivated by heat treatment.   It attempted to account for the great influence of poisons and promoters, present in

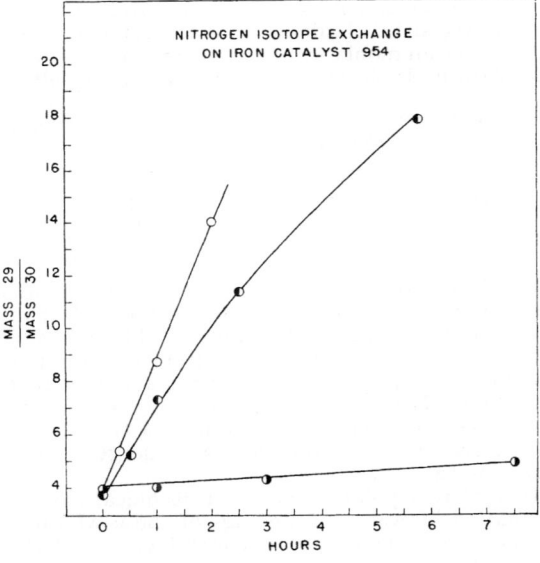

Fig. 1

minimal amounts, and invoked the existence of centres of high activity, very sensitive to poisoning.   The quantitative measurements of poisoning by Pease in the hydrogenation of ethylene and by Almquist and Black on the poisoning action of oxygen on water vapour in ammonia synthesis on iron catalysts were conspicuous examples of such studies. Over the 25-year period, and largely due to the work of Balandin in Russia, of Roberts on the properties of a clean tungsten wire surface and of Beeck and his co-workers on evaporated films, a contrary view has emerged which has sought the interpretation of catalysis solely in terms of the properties of plane faces of crystalline materials, which, by specialized techniques, could be studied in a " clean " condition.   It is my purpose to attempt a reconciliation of these two points of view in a more generalized and unifying concept.

A recent research by McGeer [4] on the exchange reaction between light and heavy nitrogen on iron synthetic ammonia catalysis and on rhenium surfaces at 450° C provides a clue to such an attempt.   The velocity of

[3] Emmett and Brunauer, *J. Amer. Chem. Soc.*, 1940, **62**, 1732.
[4] McGeer, *Thesis* (Princeton, 1949).

12   RETROSPECT AND PROSPECT

reaction was shown to be very sensitive to the reduction process to which the catalyst was submitted prior to the velocity measurements. In Fig. 1 the slowest rate is that of an iron-alumina catalyst (No 954, 1·5 % Al_2O_3) which had been prepared by reduction at 450° C with a fast stream of tank hydrogen containing 0·15 % oxygen. The intermediate rate shows the rate of exchange when the same catalyst was subjected to reduction in a stream of hydrogen from which oxygen and water vapour had been removed by the normal techniques of oxygen removal and drying of the gas stream. The residual water vapour was probably well below 0·01 %. The fastest rate of exchange was obtained when excessive precautions were taken to ensure dry, oxygen-free hydrogen for reduction. Similar findings resulted with the rhenium catalyst. The data suffice to show how very sensitive the exchange reaction is to the residual traces of oxygen that are left on an iron surface even with good reduction techniques. In the terminology of 20 years ago this is a typical example of " active centres " on a technical catalyst. It parallels entirely, as it should, the quantitative data obtained by Almquist and Black [5] on an iron-alumina catalyst in ammonia synthesis in presence of water vapour as a poison. The data in Table II recall some of these results.

TABLE II.—AMMONIA YIELD AT A SPACE VELOCITY OF 25,000, 444° C AND
1 ATM. PRESSURE

Mg : O_2 retained by catalyst . . .	0	5·5	13
% NH_3 produced . . .	0·22	0·06	0·03

The very marked effect of the first 5 mg. of oxygen retained is the more striking since, from B.E.T. measurements of the surface area of this catalyst, only 10 to 15 % of the total surface would be covered by this oxygen. Because the ammonia synthesis reaction is determined in rate by the slow step of chemisorption of nitrogen it is apparent that we must seek the interpretation of the " active centres " or quasi-heterogeneity in the effect of adsorbed oxygen on the surfaces of the iron crystallites. The presence of oxygen results in an increased energy of activation of nitrogen adsorption on the iron surface.

A generalization of this point of view for technical catalysts makes possible a reconciliation between the findings of those who have worked with clean tungsten wires and evaporated metal films and those whose attention has centred on the properties of technical catalysts. In each case one is concerned with the properties of one or more crystal faces of a particular catalytic species. In the case of technical catalysts, however, these properties may be profoundly modified by the presence of poisons as adsorbed oxygen or added ingredients such as alumina in the case just considered.

Our knowledge in this area is being rapidly increased by reason of the studies of electron emission from hot wires and by studies of the properties of semi-conductors. One need only cite the beautiful studies of Müller [6] and of Jenkins [7] relative to the emission of electrons from fine tungsten points as revealed by the field-emission projection electron microscope, and the influence of adsorbed oxygen, barium, thorium and sodium on the emission process. The varying activity of different crystal faces, the preferential adsorption of the poison on particular faces and the migration of poisons at definite temperatures can be visually demonstrated. Especially, however, the data accumulating on semi-conductors demonstrate how profoundly the conducting properties of a pure substance

[5] Almquist and Black, *J. Amer. Chem. Soc.*, 1926, **48**, 2814.
[6] Müller, *Z. physik.*, 1949, **126**, 642.
[7] Jenkins, *Reports Prog. Physics*, 1943, **9**, 177.

HUGH S. TAYLOR   13

can be modified by traces of a prescribed impurity. One may cite in this respect the recent data of Pearson and Bardeen [8] on the activation of silicon as a semi-conductor by additions of boron. The addition of 0·0013 atomic per cent of boron significantly changes the energy required to release an electron, whilst a concentration of 0·013 atomic per cent lowers this energy to zero. As Mott pointed out recently in the Kelvin Lecture [9] impurity centres may be imagined as " great swollen ' atoms ' extending over 10-20 lattice parameters ", a concept pertinent to the whole problem of interaction between adsorbed species on a surface. We must assume, in the case of the iron-synthetic ammonia catalyst already discussed, that the presence of impurity centres raises the activation energy of adsorption of nitrogen as oxygen would raise the activation energy of conduction in zinc oxide semi-conductors. In this manner the findings of Beeck on evaporated iron films can be correlated with those of Emmett and Brunauer and of McGeer with the technical iron catalysts. It is, on this basis, the impurity centre randomly distributed over the plane face of a crystal which would confer on such a crystal face quasi-heterogeneity.

We can draw from the data on semi-conductors further analogies to problems of surface catalysis. Verwey,[10] cited by Mott, has prepared semi-conducting NiO by dissolving in the lattice Li_2O, the radii of the Li^+ and Ni^{++} being practically identical. On heating in air, Li^+ replaces Ni^{++} in the lattice and for each Li^+ introduced a Ni^{+++} ion is formed. The latter are carriers of current by electron displacement. The magnetic measurements carried out by Selwood and his co-workers on valence induction in nickel and other oxides are illustrative of the same effect.[11] Magnetic measurements, on nickel oxide impregnated on γ-alumina supports, indicate that the nickel is present in the trivalent form. On magnesia, the isomorphous divalent oxide is present. On the rutile structure of titania the nickel assumes a valency of four, the oxide being bright yellow in colour. Manganese and iron oxides behave similarly. The relation or valence induction to semi-conductors and to the whole problem of promoter action in catalysis is illuminated by such studies. Let us recall that the activation energy of adsorption of hydrogen by zinc oxide is markedly diminished by incorporating chromium oxide in the preparation.

On the viewpoint here presented poisons and promoters become impurity centres in the normal lattice of the catalyst, the former tending to raise the activation energy of adsorption, the latter to lower it.

A further factor which can be influenced by such impurity centres in a catalyst surface is the heat of adsorption. The data are scanty but what data are available tend to indicate that the measured heats of chemisorption are markedly less on technical catalysts than on the corresponding evaporated films. The data of Beeck recorded for this Discussion on a variety of metal films for hydrogen, ethylene and nitrogen are conspicuously higher heats of adsorption than have been recorded by Beebe, Eucken and others with technical catalysts. Indeed, the high heats of adsorption on evaporated films constitute a grave disadvantage of these films regarded as catalysts. They are, indeed, clean, but they " die " after brief experimentation because they are self-poisoned owing to the high heats of adsorption of one or more of the reactants. The technical catalysts, though " dirty ", at least " live ", largely because of the lower heats of binding to surfaces having " impurity centres ". From such circumstances it can result that a technical catalyst may have a higher activity than an evaporated film. Wright [12] has made one such

[8] Pearson and Bardeen, *Physic. Rev.*, 1949, **75,** 865.
[9] Mott, *Proc. Inst. Electr. Eng.*, 1949, **96,** 253.
[10] Verwey, Haayman and Romeyn, *Chem. Weekblad*, 1948, **44,** 705.
[11] Selwood, *Bull. Soc. chim. France D*, 1949, 489.
[12] Wright, *Thesis* (Princeton, 1949).

concrete comparison. With 27 mg. of a nickel-chromia (80 % Ni) catalyst in a reaction volume of 350 cm.^{3}, i.e. 0·077 mg. catalyst per cm.^{3} gas, containing an equimolar hydrogen-ethylene mixture Wright found a half-life of $t_{1/2} = 4$ min. at $-78°$ C. From Beeck's data and his temperature coefficient one computes a half-life of 45 min. at $-78°$ C with 30 mg. nickel film in a reaction volume of 400 cm. or 0·075 mg. catalyst per cm.^{3} of reacting gas. On this basis, the technical catalyst is 11 times as efficient as the evaporated film. The cause is obvious from an examination of Beeck's paper to this Discussion on the reactions of hydrocarbons. Most of his surface is covered with " acetylenic residues " by reason of the high heat of chemisorption of ethylene, thereby becoming ineffective for the catalytic interaction. Beeck's data on the heat of adsorption of hydrogen on clean tantalum, 45 kcal., and on a nitrided tantalum film, 27 kcal. are evidence for a lower heat of adsorption on metal surfaces covered in part with other constituents. Another method of formulating the same idea is that, with technical catalysts, the areas responsible for the high initial heats of adsorption found with metal films are already occupied by impurity centres.

Beeck's data for heats of adsorption of hydrogen on his metal films vary directly with the change in d-band character of the metals. Boudart has recently [13] shown that Beeck's measurements of activities in the hydrogenation of ethylene on metal films increase with increasing d-band character of the metallic bond as given by Pauling, with rhodium of maximum activity. This suggests that " the lattice parameter is not to be considered solely as a cause but as an effect. The primary cause has to be sought in the electronic structure of the metal and a deeper insight into the latter may be obtained by means of Pauling's theory." From this point of view there is a notable discrepancy between the findings with films and technical catalysts in the case of copper. Beeck reports no hydrogen adsorption on films of copper whose d-band is filled. Technical copper catalysts have revealed chemisorption of hydrogen from the earliest studies of adsorption on catalysts, with a heat of adsorption of some 10 kcal. It is pertinent to ask whether this discrepancy is to be associated with the intrinsic nature of technical copper catalysts, to what extent it may be a function of a mixed $Cu^+—Cu^{++}$ structure deriving from impurity centres of a promoting character.

Similar considerations, involving activation energy or heat of adsorption or both, animate the research work on catalysis by metal alloy systems. Schwab associates the catalysis in the decomposition of formic acid, on alloys of silver and of gold,[14] with a variety of metals giving both homogeneous and heterogeneous phases, with an entry of protons into the interlattice planes and electrons dissolving in the electron gas of the metal, the required activation energy depending on the degree of completion of Brillouin zones which define the allowed electron energies in the metallic structure. Similarly Couper and Eley [15] have associated the increase in activation energy of the hydrogen-deuterium exchange with the filling of the partly empty d-band of palladium by alloying with gold. Dowden [16] has treated the problem in detail theoretically and Reynolds [17] has applied the treatment to experimental findings on homogeneous solid solution binary alloys of Group VIII metals with copper. In all such cases both activity and activation energy should be studied ; it is preferable also to study exchange reactions between isotopic molecules in order to minimize the influence of other factors such as displacement effects when two different molecular species are competing for a given surface. In this area there is a wealth of experimental opportunity.

[13] Boudart, *J. Amer. Chem. Soc.*, 1950, **72,** 1040.
[14] Schwab, *Trans. Faraday Soc.*, 1946, **42,** 689.
[15] Couper and Eley, *Nature*, 1949, **164,** 578.
[16] Dowden, *Chem. and Ind.*, 1949, 320 ; *J. Chem. Soc.*, 1950, 242.
[17] Reynolds, *Chem. and Ind.*, 1949, 320 ; *J. Chem. Soc.*, 1950, 265.

HUGH S. TAYLOR 15

Parallel studies with oxide systems will be equally interesting. In these, the properties of the catalytic oxide viewed as a semi-conductor will be useful aids to understanding. One example is copper oxide, recently examined by Garner, Gray and Stone,[18] in terms of the electrical conductivities of thin films during formation, on reduction and after adsorption of gases. Oxygen enhances, hydrogen and carbon monoxide depress the conductivity of the Cu_2O—CuO surface. The conductivity is interpreted in terms of a movement of electrons across an array of Cu^+ and Cu^{++} ions. In contrast to cuprous oxide which is an oxidation semi-conductor, zinc oxide is a reduction semi-conductor losing oxygen in a vacuum, the conductivity increasing and the excess zinc atoms entering interstitially into the lattice as ions, the electron being held in the field of the positive charge. In this way the oxide acquires the characteristics of a metal, hydrogenating in character. Foreign ions can enter the lattice substitutionally and interstitially, electrons being trapped in their fields, the radius of the orbit of the trapped electron extending over several interatomic distances. The conductivity of the semi-conductors varies with the temperature and there is evidence that a whole spectrum of energies may be involved.

Our knowledge of the properties of solid catalysts conferred by impurity centres or admixtures of two or more constituents is still very much in the qualitative stage. It is known that the admixture of aluminium oxide in magnetite increases considerably the difficulty of reduction of the iron oxide. On the other hand the admixture of alumina has little influence on the dissociation pressure of copper oxide, but increases considerably the oxygen dissociation pressure with cerium oxide, CeO_2. Chromium oxide admixture strongly increases the dissociation pressure of copper oxide. These findings on dissociation pressure are paralleled by the results of Rienäcker [19] on the catalytic activity of such mixed oxides in the oxidation of carbon monoxide. The added oxide influences both activity and activation energy of the basic catalyst, with marked lowering of the activation energy in those cases where admixture of the second oxide increases the oxygen dissociation pressure, and with increase of activation when the added oxide decreases the reducibility of the basic catalyst.

In the older literature there are many such examples. The old data of Kendall and Fuchs [20] on the influence of added oxides on oxygen release from barium peroxide, silver and mercuric oxides need to be recalled. The influence of added oxides on the slow sorption of hydrogen as illustrated by the accelerating influence of chromium oxide and the retarding effect of molybdenum oxide on the adsorption of hydrogen by zinc oxide is an example in another area. Such material should now be re-examined in the light of data revealed by Selwood on valence induction and with the newer techniques stemming from the study of semi-conductors. The catalyst itself, rather than the reactions which occur on it, seems to be the principal objective for future research in the coming years.

This is not to deny the importance of a study of such reactions, since they, also, can reveal the nature and action of the catalyst surface. An excellent example in this respect is the recent work on the nature and functions of cracking catalysts, as exemplified by the recent conclusions of Greensfelder, Voge and Good [21] based on earlier work and newer investigations of the cracking of cetane, cetene and other hydrocarbons thermally and over catalysts such as activated carbon, alumina, silica and alumina-silica commercial cracking catalysts. Two fundamental types of cracking emerge, characteristic both as to the type of

[18] Garner, Gray and Stone, *Proc. Roy. Soc. A*, 1949, **197,** 294.
[19] Rienäcker, *Z. anorg. Chem.*, 1949, **258,** 280.
[20] Kendall and Fuchs, *J. Amer. Chem. Soc.*, 1921, **43,** 2017.
[21] Greensfelder, Voge and Good, *Ind. Eng. Chem.*, 1949, **41,** 2573.

16   RETROSPECT AND PROSPECT

primary and secondary reactions.   Two types of mechanism are involved.
The one involves free radical fragments and the pattern of cracking is
described by the Rice-Kossiakoff theory of cracking.   This mechanism
is characteristic of thermal cracking.   The other mechanism is ionic in
nature conforming closely to acid-activated, carbonium ion mechanism.
Commercial acid-treated clays and synthetic silica-alumina catalysts are
of the latter type.   Activated carbon, an active, non-acidic catalyst
gives a product distribution which can be interpreted as a free-radical
type of cracking, quenched, as compared with thermal cracking, by
reaction of free radicals with chemisorbed hydrogen on the catalyst to
yield the corresponding normal paraffin.   With silica-alumina, acidic-
type catalysts, isomerization accompanies cracking to yield its particular
spectrum of products.   Such acidic surfaces are poisoned specifically
by organic bases such as quinoline and by potash, ammonia and other
alkalis.   Oblad and his co-workers indicate that some 4 % of the total
surface is involved in such catalyses.

Volkenstein [22] has recently analyzed the concepts of adsorption and of
adsorption kinetics from the standpoint of the solid catalyst viewed as a
structure containing a certain concentration of lattice defects, two-fold
in nature.   On the one hand, there are macro-defects, such as cracks,
whose perturbations exceed those of the crystallographic unit.   On the
other hand, there are micro-defects which exercise perturbations of the
order of the unit cell, the periodic structure being re-established at a
distance of a few lattice parameters.   In the terminology of semi-
conductors, these defects include holes or vacant sites, interstitial atoms
or ions, ions in a heteropolar lattice with normal position but anomalous
charge, or foreign atoms in substitutional or interstitial positions.   Such
defects deform a region of the lattice and it is the region which should
be regarded as the defect.   It is with the properties of such regions that
catalysis may well be associated.   Volkenstein introduces into his considerations the concept of mobility of micro-defects with an associated
activation energy determined by the nature of the defect and the lattice
and by the direction of migration.   These micro-defects may react, attract
or repel each other, dependent on the charges involved.   Two defects
may interact to produce a new defect with different properties.

Volkenstein envisages the " disorder " in terms of the total number
of defects which is small compared with the total number of unit cells.
Disorder may be " biographical ", that which is present at $0°$ K and arising
from the circumstances of the catalyst preparation or it may be " thermal "
or that which results from the effect of temperature.   The biographical
disorder x, at $0°$ K will increase progressively with temperature to a
maximum y at $T = \infty$, the thermal disorder at any temperature varying
from o at $T = 0°$ K to $y-x$ at $T = \infty$.   The relative importance of
biographical and thermal disorder depends both on the temperature and
the " biography ".

In the older theories of adsorption one assumes (a) constancy of adsorption centres, (b) immobility of adsorption sites, and (c) invariance
of sites with coverage.   None of these assumptions enter necessarily into
Volkenstein's treatment.   Volkenstein assumes that adsorption occurs in
the defect region.   With one kind of defect the surface is energetically
homogeneous.   With differing defects there will be heterogeneity.   Adsorption creates new centres.   Increase of temperature increases the
number of adsorption centres up to but not beyond the maximum, y.
Volkenstein shows that the Langmuir isotherm results when the whole
disorder is biographical.   When, however, the defect is thermal involving
an absorption of energy u, then, depending on the nature of the process
producing the new defect whether unimolecular or by bimolecular interaction of two defects, the mathematics shows that, with no heterogeneity

[22] Volkenstein, *Zhurr. Fiz. Khim.*, 1949, **23,** 917.

of sites and without interaction between adsorbed atoms, the adsorption isotherm may conform to the Freundlich equation, $v_{ads} = kp^{1/n}$, and the differential heat of adsorption can fall from an initial value q to a minimum $q - u$. The curve obtained is reverse sigmoid falling at first slowly, then rapidly and finally asymptotically to $q - u$ at complete coverage. The adsorption centres on the crystal surface operate as a kind of plane gas the concentration of which increases on filling up and the change of energy of which is measured by the differential heat of adsorption.

For the kinetics of adsorption the bimolecular production of sites can be shown to yield either a rapid Langmuir kinetics determined by the rate at which molecules strike the surface or an adsorption proportional to the square root of the time. For a unimolecular production of sites the kinetic expression for adsorption is that typical of the so-called activated adsorption, $h = h_0 \exp(-E_{act}/RT)$. In the theory of activated adsorption a potential barrier between adsorption sites and impinging molecule is assumed. In Volkenstein's treatment it is the number of sites which increases with temperature and proportionally to $\exp(-u/RT)$. No potential barrier is assumed. An alternative method of stating the same is to say that the number of sites remains constant but the number of excited centres which are able to absorb increases as observed.

There are several observations from the older literature on activated adsorption which can readily be interpreted on the bases of the Volkenstein concepts. In the General Discussion for 1932 it was pointed out that the velocity of activated adsorption of hydrogen as measured on zinc oxide was very much smaller than the number of molecules striking the surface with the necessary activation energy. Our subsequent knowledge, from B.E.T. measurements, of surface area indicates that the discrepancy between calculation and observation was of the order of 10^8. A similar discrepancy was noted by Emmett and Brunauer [3] in their careful measurements of the velocity and activation energy of adsorption of nitrogen on synthetic ammonia catalysts. In this case the discrepancy was at least of the order of 10^6. From Volkenstein's point of view these data would involve both the concentration of defect centres and the activation energy of adsorption of the molecule. In the case of doubly-promoted ammonia synthesis catalysts this might well be associated with the iron-potassium-oxide-aluminium oxide centres where the varying valency of iron, the univalent potassium and the trivalent aluminium ions suggest at once the possibility of lattice defects, as was pointed out to me by Boudart. Boudart has also emphasized in this regard the observations of Pace and myself [23] on the identity in the velocities of adsorption of hydrogen and deuterium on oxide catalysts which led to the conclusion that " the activation energy necessary is required by the solid adsorbent ". Other data on the influence of pressure on the velocity of chemisorption of hydrogen on oxides such as chromium oxide gel are also in agreement with the view that velocity is not determined by the number of molecules striking the surface. To account for the slow sorptions observed it was earlier suggested that interaction must occur between the adsorbent and van der Waals' adsorbed gas. The Volkenstein concept represents a mechanism by which such would be achieved.

The apparent saturation capacity of an oxide surface for hydrogen adsorption at a given temperature and the large change to a new apparent saturation at another temperature, facts familiar to all who have studied the slow sorption processes on oxides, should be re-studied in reference to Volkenstein's assumption that the sites available for adsorption, the " thermal " sites, vary with temperature. On this view the measurements of Shou-Chu Liang [24] and the writer would gain new significance. In brief, measurements of slow sorption on oxide surfaces need to be

[23] Pace and Taylor, *J. Chem. Physics*, 1934, **2**, 578.
[24] Taylor and Liang, *J. Amer. Chem. Soc.*, 1947, **69**, 1306.

18   MOLECULE NEAR METAL SURFACE

re-examined from the standpoint of the entropy as well as the energy of activation.

In retrospect, three decades of scientific effort devoted to the science of catalytic phenomena have revealed a wealth of detail and understanding not available to the technologist in the empirical developments of the nineteenth and early twentieth centuries. The background of scientific theory and data that has been accumulated since Mr. Spiers first organized a General Discussion on Catalysis has led to a surer and swifter attack on any new catalytic problem that emerges. Out of the discussions which will ensue, here in Liverpool, we may anticipate a reconciliation of the several attitudes that sometimes have appeared to divide us, but are, in reality, a spur to further and continued effort towards the mastery of our science in an era which is of deep significance in all human affairs. In prospect, therefore, the future of our science is both challenging and bright.

Chemistry Department,
Princeton University,
Princeton, N.J.

Biophysical Chemistry

R. H. Templer

Department of Chemistry, Imperial College London, UK SW7 2AZ

Commentary on: **Energy landscapes of biomolecular adhesion and receptor anchoring at interfaces explored with dynamics force spectroscopy**, E. Evans, *Faraday Discuss.*, 1998, **111**, 1.

Much of the chemical dynamics within cellular systems is characterised by the making and breaking of non-covalent interactions. Folding of proteins, growth and consumption of actin filaments, formation of membrane pores, aggregation of protein signalling complexes; the list of important cellular phenomena that fall into this category is virtually without end. It is therefore not too surprising that many of the most important and interesting conceptual and technical developments in Biophysical Chemistry over the past ten years have attempted to advance our understanding of such interactions.

One of the most promising developments has been that of dynamic force spectroscopy, a technique whose development began less than ten years ago. A founding figure of the field, Professor Evan Evans of the Universities of British Columbia and Boston, gave the first comprehensive account of the technique in a Faraday Discussion organised by the Biophysical Chemistry Group of the Faraday Division in 1998.[1–4]

In dynamic force spectroscopy the piezoelectric displacement techniques of atomic force microscopy are used to pull on a molecule held between a spring and substrate. Knowing the device's spring constant and measuring spring deflection enables force to be calculated. However, as Evans points out, the thermally agitated buffeting that any molecule in a cell will experience means that macroscopic pictures of force must be abandoned. The fluctuations in thermal energy mean that over some long time any bond can be broken without the application of any external force, *i.e.* over infinite time bond strength is zero. If however we pull on a bond faster than the average thermal impulse time then the bond will have its maximum strength. This makes it clear that the strength of molecular interactions depend on the rate at which load is applied.[5] Evans' paper elegantly demonstrates how making measurements over a range of loading rates on a single molecule at a time, can reveal an entire hierarchy of molecular interactions, which would otherwise remain hidden from view.

Evans' approach has now become the common currency of dynamic force spectroscopy, a field which has grown at a staggering rate since 1998.[6] However, it is surprising to note that his use of aspirated white blood cells as his spring is still the technique that can scan the greatest range of loading rates. His technique's great advantage lies in the fact that one is able to modify the device's spring constant by tensioning the cell membrane, something that can be controlled very precisely during the experiment.

The clarity and depth of thought in the article, the palpable sense of excitement it conveys and the simple provocation of the ideas it contains mark it out.

References

1 E. Evans, and K. Ritchie, *Biophys. J.*, 1997, **72**, 1541
2 E. Evans, *Faraday Discuss.*, 1999, **111**, 1
3 R. Merkel, P. Nassoy, A. Leung, K. Ritchie, and E. Evans,, *Nature*, 1999, **397**, 50
4 E. Evans, *Annu. Rev. Biophys. Biomol. Struct.*, 2001, **30**, 105
5 H. A. Kramers, *Physica (Amsterdam)*, 1940, **7**, 284
6 For example: T. Strunz, *et al.*, *Proc. Natl. Acad. Sci. USA*, 1999, **96**, 11277; C. Friedsam, *et al.*, *J. Phys.: Condens. Matter*, 2003, **15**, S1709; U. Seifert, *Phys. Rev. Lett.*, 2000, **84**, 2750; F. Kienberger, *et al.*, *Ultramicroscopy*, 2003, **97**, 29

Introductory Lecture

Energy landscapes of biomolecular adhesion and receptor anchoring at interfaces explored with dynamic force spectroscopy

Evan Evans

Departments of Physics and Pathology, University of British Columbia, Vancouver, BC, Canada V6T 1Z1 and Department of Biomedical Engineering, Boston University, Boston, Massachusetts, USA 02215

Received 21st December 1998

Beyond covalent connections within protein and lipid molecules, weak noncovalent interactions between large molecules govern properties of cellular structure and interfacial adhesion in biology. These bonds and structures have limited lifetimes and so will fail under any level of force if pulled on for the right length of time. As such, the strength of interaction is the level of force most likely to disrupt a bond on a particular time scale. For instance, strength is zero on time scales longer than the natural lifetime for spontaneous dissociation. On the other hand, if driven to unbind or change structure on time scales shorter than needed for diffusive relaxation, strength will reach an adiabatic limit set by the maximum gradient in a potential of mean force. Over the enormous span of time scales between spontaneous dissociation and adiabatic detachment, theory predicts that bond breakage under steadily rising force occurs most frequently at a force determined by the rate of loading. Moreover, the continuous plot (spectrum) of strength expressed on a scale of $\log_e$(loading rate) provides a map of the prominent barriers traversed in the energy landscape along the force-driven pathway and reveals the differences in energy between barriers. Illustrated with results from recent laboratory measurements, dynamic strength spectra provide a new view into the inner complexity of receptor–ligand interactions and receptor lipid anchoring.

Introduction

Well-recognized in biology, ligand–receptor interactions are the fundament of nanoscale chemistry in recognition, signalling, activation, regulation, and other processes from outside to inside cells. Thus, following the advent of atomic force microscopy (AFM) a decade ago,[1] it was no surprise that researchers quickly seized the opportunity to test strengths of receptor–ligand bonds. Since then, AFM and other sensitive force probes have been used to pull on a variety of molecules embedded in—or adhesively bonded to—surfaces. Applying these techniques, experimentalists often imagine that probe force establishes a well-defined property of an interaction between molecules. Such expectations originate from the age-old creed of physics, which states that strength is the maximum gradient $-(\partial E/\partial x)_{max}$ of an interaction potential or energy contour $E(x)$ defined along the direction (x) of separation. Hence, it is anticipated that detachment forces for different types of molecular interactions will follow a scale set by the ratio of bond energy to the effective range of the interaction (bond length). This seems consistent with the standard model of biochemistry where the scale for bond strength is the free energy $\Delta G°$ reduction when molecules combine in solution as found from the equilibrium ratio $k_{eq}[\sim \exp(\Delta G°/k_B T)]$ of bound to free constituents. As such, the criterion for a *strong* bond should be simply a binding energy much larger than the thermal energy per molecule $k_B T$. In marked contrast to these two paradigms, we will see that even bonds with binding energies $>40\ k_B T$ can fail under minuscule forces—more than 100-fold lower than the maximum energy gradient implied by energy/distance. Indeed, we will find that measurement of molecular detachment force—no matter how precise the technique or how carefully performed—is not in itself a fundamental property of a molecular interaction. So what is the appropriate framework for describing *strength* of molecular bonds and how can we relate measurements of these forces to nanoscale chemistry?

When we test strength of molecular cohesion or adhesion at surfaces, we determine the maximum level of force that a molecular attachment can support at the instant of failure. Unlike intimate covalent connections within protein and lipid molecules, biomembrane structure and interfacial adhesion bonds involve noncovalent interactions between large macromolecules, which have limited lifetimes and thus will fail under any level of force if pulled on for the right length of time. In other words, when we speak of strength, we should think of the force that is most likely to disrupt an adhesive bond or structural linkage on a particular time scale. At equilibrium for example, bonds dissociate and reform under zero force. Thus, an isolated bond has no strength on time scales longer than its natural lifetime $t_0 = 1/k_{off}^0$ for spontaneous (entropy-driven) dissociation. On the other hand, if detached within the time needed for diffusive relaxation over the range of molecular interaction (*e.g.* $x_\beta \lesssim 1$ nm $\rightarrow x_\beta^2/D < 10^{-9}$ s in water), the strength of a bond will reach and even exceed the adiabatic limit $f_\infty \approx |\Delta E|/x_\beta$ set by the maximum gradient in a potential of mean force. This is the situation in molecular dynamics (MD) simulations.[2,3]

From the slow limit set by spontaneous transition (from µs to months) to the ultrafast limit set by diffusive relaxation ($<$ns), strength is governed by thermally activated kinetics under external force and thus depends on how the force is applied over time. Since application of force always requires a finite interval of time, the simplest way to parameterize the history of loading is to treat force as a ramp in time set by a constant loading rate $r_f = \Delta f/\Delta t$. In fact, a ramp of force is what single molecular attachments experience when a force probe and test surface are separated at constant speed (*i.e.* loading rate = probe stiffness × speed). Using this parameterization and some nearly sixty year old physics[4] for Brownian dynamics of chemical reactions in liquids, we have shown that bond dissociation under steadily rising force occurs most frequently at a time determined by the rate of loading.[5] Since loading rate is constant, the time of dissociation specifies the most likely rupture force—strength—which has the same dependence on loading rate. Of particular significance, the continuous plot of strength expressed on a scale of $\log_e$(loading rate) maps the most prominent barriers traversed in the energy landscape to distances along the force-driven pathway and reveals the splitting in energy between barriers. Thus, strength *vs.* $\log_e$(loading rate) establishes the basis for a dynamic force spectroscopy (DFS) to probe the inner world of molecular-scale chemistry. Testing bond strength or structural transitions at different loading rates effectively probes the lifetime of a molecular complex under different levels of force. The experimental challenge is to measure forces over many orders of magnitude in loading rate. This dynamic requirement is dictated by the exponential of the energy difference ΔE_b between the highest and lowest barriers divided by thermal energy, *i.e.* $\exp(\Delta E_b/k_B T)$, which can be enormous!

The strength spectra to be presented here will show that we can now cover six orders of magnitude in loading rate from <0.1 pN s^{-1} to $\sim 10^5$ pN s^{-1} with a rather simple dynamic force probe, which could be extended to $\sim 10^7$ pN s^{-1} with complementary measurements using other probes. But more important than demonstrations of technique, these spectra provide a new level of insight into the complexity of macromolecular interactions and structural linkages. First, results from biotin–(strept)avidin[6] and carbohydrate–L-selectin[7] bond tests will show that a cascade of sharp energy barriers exists in receptor–ligand bonds where each barrier governs strength on a different time scale. We see then that these bonds, cannot be simply idealized by a sole energy barrier, and we cannot rely on the classical intuition about kinetics implicit in the detailed balance $k_{eq} = k_{on}/k_{off}$ where k_{on} and k_{off} are constants. Second, tests of lipid extraction[8] from membranes will show that anchoring of receptors to surface structure plays an important role in adhesion strength and can introduce unexpected transitions in the strengths of receptor–ligand attachments. In other words, we should not assume that a structural linkage of several molecules will fail at a specific weak connection nor that we can uniquely attach *strong* or *weak* labels to bonds in a linkage. Simultaneous kinetics over different energy landscapes in serial molecular linkages can lead to strength or weakness on different time scales. Dynamic crossovers in strength switch the site of failure from one location to another. Taken together, these insights show that mechanical force can tune and switch time scales for kinetics in biomolecular reactions governed by complex energy landscapes, which exposes a potentially new dimension in biochemical regulation and control.

Theory of molecular kinetics under force in liquids

We begin with an abstract of the physics that underlies the kinetics of bond dissociation and structural transitions in a liquid environment. Developed from Einstein's theory of Brownian motion, these well-known concepts take advantage of the huge gap in time scale that separates rapid thermal impulses in liquids ($< 10^{-12}$ s) from slow processes in laboratory measurements (*e.g.* from 10^{-4} s to min in the case of force probe tests). Three equivalent formulations describe molecular kinetics in an overdamped liquid environment. The first is a microscopic perspective where molecules behave as particles with instantaneous positions or states $x(t)$ governed by an overdamped Langevin equation of motion,

$$dx/dt = D/k_BT[f - \nabla E + \delta f]$$

The rate of change in state equals the instantaneous force scaled by the mobility of states or inverse of the damping coefficent $\gamma (= k_BT/D)$. The deterministic force ($-\nabla E + f$) includes both the local gradient in molecular interaction potential $E(x)$ and the external force f. An uncorrelated random force δf from thermal impulses modulates the deterministic force and obeys the *fluctuation–dissipation theorem* where the integrated square fluctuation in a window of time can be modeled as a Gaussian distribution $\sim \exp[-\int \delta f^2 dt/(4k_BT)]$ with variance set by temperature and viscous damping.[9] The microscopic physics also defines a stochastic process that has become the foundation of an important computational technique—*Brownian dynamics* or *smart* Monte Carlo (SMC)[10] simulations. In this description, the likelihood $P(x + \Delta x, t + \Delta t \,|\, x, t)$ that a state $x(t)$ will evolve to a new state $x + \Delta x$ over a time increment Δt is specified by the product of the equilibrium (long-time) Boltzmann weight for the step and a Gaussian weight for dynamics,

$$P(x + \Delta x, t + \Delta t \,|\, x, t) \sim \exp\{-(\Delta E - f \cdot \Delta x)/k_BT\}\exp\{-|\Delta x - (D/k_BT)f\Delta t|^2/(4D\,\Delta t)\}/(D\,\Delta t)^{1/2}$$

Finally, on time scales that include many thermal impulses, the overdamped dynamics can be cast in a continuum representation where the density of states $\rho(x, t)$ at location x and time t obeys Smoluchowski transport,[9]

$$d\rho/dt = -\nabla \cdot J$$

where the *flux* of states $J = D[f - \nabla E)\rho/k_BT - \nabla\rho]$ reflects both convection by force and spread by diffusion. Although each description of the ultrafast kinetics brings to light important features, Kramers[4] demonstrated that Smoluchowski transport readily predicts the rate of escape from a deeply bound state when a large number of thermally activated steps are needed to pass a barrier in a dissipative environment.

Escape from a bound state confined by a single barrier

Starting far from equilibrium with all states confined inside the barrier, the kinetics of escape are idealized as a stationary flux of probability density along a preferential path from the deep energy minimum outward past the barrier *via* a saddle point in the energy surface. In real molecular interactions, there can be many such paths and the paths can map out complex trajectories in configuration space. However, application of an external pulling force acts to select the reaction path, which we express by a scalar coordinate x. Assumed to be bounded by steeply rising energy in other directions, the energy landscape $E(x)$ along this coordinate is illustrated schematically in Fig. 1(a). Governed by orientation θ relative to the microscopic reaction coordinate, external force adds a mechanical potential $-fx(\cos \theta)$ that tilts the energy landscape and diminishes the energy barrier E_b at the transition state ($x = x_{ts}$). When the tilted landscape is introduced into the Smoluchowski equation, the stationary solution ($J = \text{constant}$ in 1-D or $= \text{constant}/x^{d-1}$ in d-

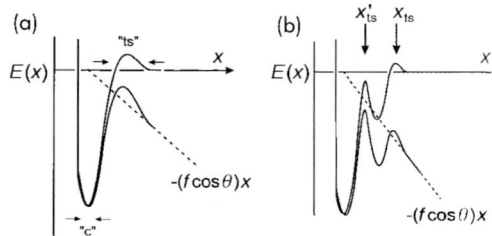

Fig. 1 Conceptual energy landscapes for bound states "c" confined by sharp activation barriers. Oriented at an angle θ to the molecular coordinate x, external force f adds a mechanical potential $-(f \cos \theta)x$ that tilts the landscape and lowers the barrier. For sharp barriers, the energy contours local to barriers—transition states "s"—are highly curved and change little in shape or location under force. (a) A single barrier under force. (b) A cascade of barriers under force. The inner barrier emerges to dominate kinetics when the outer barrier is driven below it by $\geqslant k_B T$.

dimensions) yields a generic expression for rate of escape from bound to unpopulated free states under force,[5]

$$k_{off} \approx (D/l_c l_{ts})\exp[-E_b(f)/k_B T]$$

The diffusive nature of kinetics in liquids is embodied in the attempt frequency, $D/l_c l_{ts}$, which is the reciprocal of a characteristic time $t_D = l_c l_{ts}(\gamma/k_B T)$ set by damping and two length scales. The first length l_c represents confinement in the bound state and defines the entropy gradient ($\partial \rho/\partial x \approx 1/l_c$), which drives escape. In a harmonic approximation, l_c is derived from curvature $\kappa_c = (\partial^2 E/\partial x^2)_c$ of the energy landscape local to the minimum, *i.e.* $l_c = (2\pi k_B T/\kappa_c)^{1/2}$. The second length l_{ts} is the energy-weighted width of the barrier $l_{ts} = \int dx \exp[\Delta E(x)_{ts}/k_B T]$ local to the transition state $x = x_{ts}$, also determined by curvature $\kappa_{ts} = (\partial^2 E/\partial x^2)_{ts}$ of the energy landscape, *i.e.* $l_{ts} = (2\pi k_B T/\kappa_{ts})^{1/2}$. Although force can displace and deform the width of the barrier [*i.e.* $(\kappa_{ts}/2\pi k_B T)^{1/2} \approx g(f)$], the major impact of force arises in the thermal likelihood of reaching the top of the energy barrier, $\exp[-E_b(f)/k_B T]$. For a sharp energy barrier, the shape and location of the transition state are insensitive to force but force lowers the barrier in proportion to the thermally averaged projection $x_\beta = <x_{ts} \cos \theta>$, *i.e.* $E_b(f) = E_b - fx_\beta$. As such, thermal activation introduces the characteristic scale for force through the ratio of thermal energy to the distance x_β, *i.e.* $f_\beta = k_B T/x_\beta$, which can be surprisingly small since $k_B T \approx 4.1$ pN nm at room temperature and $x_\beta \approx 0.1$–1 nm. On this scale, the rate of escape increases exponentially with force, $k_{off} \approx (1/t_0) \exp(f/f_\beta)$, as first postulated by Bell[11] twenty years ago. But in contrast to the resonant frequency of bond excitations described in Bell's model, Kramers showed that the relevant attempt frequency is $1/t_D = (\kappa_c \kappa_{ts})^{1/2}/2\pi\gamma$ for overdamped transitions in liquids, which is at least 1000-fold slower. With the Arrhenius dependence on initial barrier height and the attempt frequency, Kramers classic result for spontaneous escape in the overdamped limit sets the scale for the transition rate, *i.e.* $1/t_0 = (1/t_D)\exp(-E_b/k_B T)$.

Escape from a bound state confined by several barriers

Although a naive model of chemical binding, the single-sharp barrier model already captures the profound impact of force on thermally activated kinetics: *i.e.* exponential amplification of the forward rate for dissociation (and suppression of backward rate for reassociation) characterized by a small force scale $k_B T/x_\beta$ well below the adiabatic limit $>E_b/x_\beta$! However, the energy landscapes of biomolecular bonds are expected to be much more complex because there are many sites of interaction involving large numbers of small molecules. This should produce a *rough* topography of barriers in an energy landscape and many possible pathways for unbinding. If again conceptualized as precipitous (*sharp*) energy maxima along a single pathway, these prominent barriers are predicted to emerge under increasing force and dominate kinetics in succession as demonstrated by the sketch in Fig. 1(b).[5] An inner barrier is exposed when the force exceeds a crossover force $f_\otimes \approx \Delta E_b/\Delta x_\beta$ set by the splitting ΔE_b between barrier heights and separation in projected positions Δx_β. Depending on the difference in barrier energies, the crossovers occur at forces much larger than the local thermal forces given by $k_B T/x_\beta$. Thus, marked by these crossovers, the kinetic rate constant is predicted to rise in a staircase of force-dependent exponentials that amplify the rate of transition less and less with each increase in thermal force scale. The transition rate for escape past a cascade of n sharp barriers is easily predicted with Kramers–Smoluchowski theory,

$$k_{off}(f) \approx (1/t_0)\exp(f/f^0_\beta)/\{1 + \sum_{i \to n} l_i \exp[f\Delta x_\beta - \Delta E_b]\}$$

which at low force begins with the steepest exponential dominated by the outermost barrier. At larger forces, the rate crosses over to more shallow exponentials. The transition from one exponential regime to the next depends on the ratio of widths $l_i \approx l_{ts}/l^0_{ts}$ $[=(\kappa^0_{ts}/\kappa_{ts})^{1/2}]$ plus differences in location $\Delta x_\beta = x^0_\beta - x_\beta$ and energy $\Delta E_b = E^0_b - E_b$ of inner barriers relative to the outermost barrier as defined by E^0_b, l^0_{ts}, and x^0_β. We see then that a major consequence of structured energy landscapes is to make molecular interactions more durable (survive longer) at higher forces.

Theory of force distributions in probe experiments

Even with ultrasensitive probes and high resolution detection, tests of molecular detachment yield a spread in force values. To understand the origin of the intrinsic uncertainty in force, we have to examine the generic process of bond dissociation in laboratory experiments. Typically, a probe decorated with a small amount of ligand—and a substrate studded with specific molecular receptors—are repeatedly touched together through steady precision movement to/from contact. If the surfaces are prepared with a sufficiently low density of reactive sites, point contacts between the probe tip and the test surface will occasionally result in attachments (*e.g.* one attachment for every 5–10 touches). Under controlled touch, infrequent bonding ensures a high probability of forming single molecular bonds ($\sim 95\%$ confidence for 1 attachment out of 10 touches). An attachment is exposed when the force transducer exhibits an extension or deflection Δx_t during surface separation. Identified by rapid recoil at breakage, the rupture force is given by the maximum transducer extension Δx_t, *i.e.* $f = k_f \cdot \Delta x_t$ where k_f is the spring constant of the transducer. Following many measurements, detachment forces are then cumulated into a histogram. The peak in the distribution is the most likely the rupture force, which is labelled bond strength. This approach has been reported many times in the literature over the past decade including studies of bond strength using AFM[12-16] and other techniques.[17,18] The exception to the generic description is that the frequency of attachment in most tests has been one for every touch, which represents many molecular bonds and yields broad force distributions.

Given that only a single molecular attachment forms on contact, the crucial feature of the generic method is that the force experienced by the attachment prior to rupture is not constant but increases in time. This is shown clearly by two traces of attachment force *vs.* time in Fig. 2 taken from our experiments[6] on single receptor–ligand bonds using a biomembrane force probe (BFP).[19] In probe tests like these, the linear rise of force with time is set by the product of separation speed v_t and transducer spring constant k_f, which is called the loading rate $r_f = k_f v_t$. (Note: if soft structures like long polymers link the bond to a stiff probe, the loading history can be nonlinear in time.[20]) Very different levels of force and time frame characterize the two detachment processes in Fig. 2. Comparing these, we see then that bond survival and breakage force depend on the rate of loading in reciprocal ways: *i.e.* high speed loading → short lifetime but large detachment force whereas low speed loading → long lifetime but small detachment force, which is the direct consequence of thermally activated kinetics.

Statistics of transitions under increasing force

To analyse bond breakage under steady loading, we take advantage of the enormous gap in time scale between the ultrafast Brownian diffusion ($t_D \approx 10^{-10} - 10^{-9}$ s) and the time frame of laboratory experiments ($\sim 10^{-4}$ s to min). This means that the slowly increasing force in laboratory experiments is essentially stationary on the scale of the ultrafast kinetics. Thus, dissociation rate merely becomes a function of the instantaneous force and the distribution of rupture times can be described in the limit of large statistics by a first-order (Markov) process with time-dependent rate constants. As force rises above the thermal force scale, *i.e.* $r_f t > k_B T/x_\beta$, the forward transition

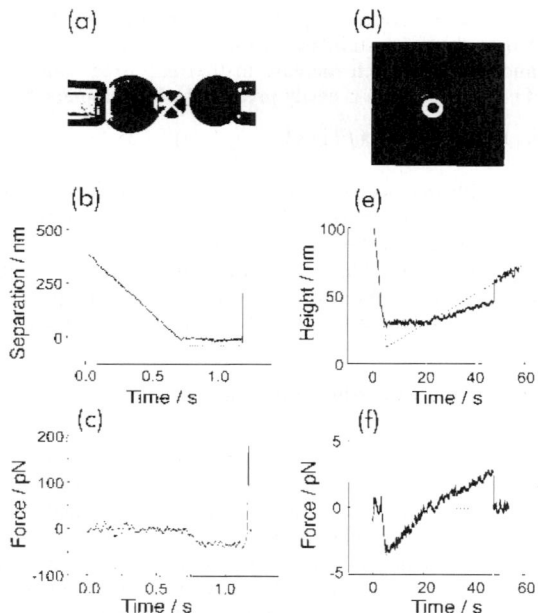

Fig. 2 Testing strength of single molecular attachments with the biomembrane force probe (BFP). The spring component of the BFP is a pressurized membrane capsule.[19] Membrane tension sets the force constant k_f (force/capsule extension), which is controlled by micropipet suction P and radius R_p, $k_f \approx P \times R_p$. Using a red blood cell as the transducer, the BFP stiffness can be selected between 0.1 and 3 pN nm^{-1} to measure forces from 0.5 to 1000 pN. At the BFP tip, a glass microbead of 1–2 μm diameter is glued to the membrane. The probe tip and red cell surfaces are bound covalently with heterobifunctional polyethylene oxide PEG polymers that carry glue components and test ligands.[6] The BFP is operated in two orientations (modes) on the stages of inverted microscopes as illustrated by the following examples of fast and slow bond detachment: (a) first, the BFP (on the left) is kept stationary in the horizontal mode and the microbead test surface (on the right) is translated to/from contact with the BFP tip by precision piezo control. Video image processing is used to track the bead as shown by the simulated cursor; a single high speed (~ 1000 frames s^{-1}) scan through the center of the bead is used to track deflection of the transducer (force) on a fast time scale at a resolution of 8–10 nm. Parts (b) and (c) show the BFP tip–substrate separation and force *vs.* time for rapid bond detachment in the horizontal mode. (b) The test microbead was moved towards the probe tip at a speed of ~ 500 nm s^{-1}. Stopped for ~ 0.5 s after sensing contact at a preset impingement force of ~ -30 pN, the test surface was then retracted at speed of $\sim 30\,000$ nm s^{-1}. (c) Loaded at extremely fast rate, the bond held the tip to the surface for ~ 0.003 s (spike in force) and broke at ~ 180 pN as the piezo continued to retract the test surface. The force fluctuations were due to position uncertainties $\times$ BFP stiffness. (d) In the vertical mode, reflection interference contrast is used to image the BFP tip as it is translated by piezo control along the optical axis to/from contact with a coverglass test surface. Standard video (30 frames s^{-1}) processing of the circular interference pattern reveals elevation of the tip at a resolution of 2–5 nm. Transducer deflection (force) is obtained from the difference between piezo translation and bead displacement. Parts (e) and (f) show the BFP tip–substrate separation and force *vs.* time for a slow bond detachment in the vertical mode. (e) The probe was moved towards the coverglass test surface at a speed of ~ 20 nm s^{-1}. After sensing contact at a preset impingement force of ~ -3 pN, the probe was retracted at slow speed of ~ 1 nm s^{-1}. (f) Loaded at extremely slow rate, the bond held the tip to the surface for ~ 24 s and broke at ~ 3 pN as the piezo continued to retract the probe (dashed trajectory). The fluctuations in tip position were due to thermal excitations of the BFP with mean square displacement $\sim k_B T/k_f$. Stretch of the PEG polymers that linked the bond to the glass surfaces is shown by the slight upward movement of the tip (~ 15 nm) under force prior to detachment. Due to polymer compliance, the true loading rate felt by a bond at nominal rates ($k_f v_t$) below 10 pN s^{-1} had to be obtained from the actual force *vs.* time.

(escape) rate increases extremely rapidly. Moreover, the molecules drift apart faster than diffusion can recombine them from positions beyond the confining barrier so the backward rate for re-association quickly vanishes ($k_{on} \rightarrow 0$). Thus, the likelihood $S(t)$ of remaining in the bound state is dominated by the forward process, *i.e.* $dS(t)/dt \approx -k_{off}(t)S(t)$ or equivalently $S(t) = \exp[-\int_{0 \rightarrow t} k_{off}(t')dt']$. The probability density $p(t) = k_{off}(t)S(t)$ for detachment between times t and $t + \Delta t$ describes the distribution of lifetimes. Since instantaneous force is the product of time and loading rate ($f = r_f t$), the probability density $p(f)$ for detachment between forces f and $f + \Delta f$ is given by the distribution of lifetimes $p(t)$,[5]

$$p(f) = (1/r_f)k_{off}(f)\exp[-(1/r_f)\int_{0 \rightarrow f} k_{off}(f')df']$$

noting the statistical identity $p(t)dt = p(f)df$. The peak in the distribution of forces defines the force f^* for most frequent transition, which is strength. Analytically, the location of a distribution peak is found from $\partial p(f)/\partial f = 0$ which establishes a transcendental equation that relates the strength f^* to loading rate r_f,

$$[k_{off}]_{f=f*} = r_f[\partial \log_e(k_{off})/\partial f]_{f=f*}$$

Although somewhat forbidding, this expression yields a simple result for strength as a function of loading rate in the case of a single sharp energy barrier,

$$f^* = f_\beta \log_e(r_f/r_f^0)$$

recalling that the rate is modelled by an exponential in force, $k_{off} \approx (1/t_0)\exp(f/f_\beta)$. Governed by a thermal scale for loading rate $r_f^0 = f_\beta/t_0$, the most likely force—strength—simply shifts upward linearly with the logarithm of loading rate multiplied by the thermal force f_β. Similarly, the curvature of the distribution local to the peak, $1/\Delta_f^2 = -[1/p(f)][\partial^2 p(f)/\partial f^2]_{f=f*}$, can be used to estimate a Gaussian width for uncertainty in the force distribution,

$$\Delta_f^2 = 1/\{[\partial \log_e(k_{off})/\partial f]^2 - [\partial^2 \log_e(k_{off})/\partial f^2]\}_{f=f*}$$

For a sharp energy barrier, this again yields a simple result, $\Delta_f = f_\beta$. Hence, even without experimental uncertainty, the distribution of forces is broadened by thermal activation (kinetics)!

Dynamic force spectroscopy

In the context of experiments, the signature of a major sharp barrier is predicted to be a straight line in a plot of most frequent probe force f^* *vs.* log(loading rate) as illustrated in Fig. 3(a). This linear regime can span orders of magnitude in rate as determined by the ratio of barrier energy E_b

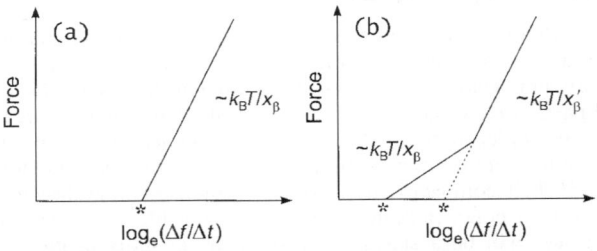

Fig. 3 Dynamic strength spectra defined by most likely bond detachment force f^* *vs.* $\log_e$(loading rate = r_f/r_f^0), where the loading rate scale $r_f^0 = (f_\beta/t_D)\exp(-E_b/k_B T)$ is set by thermal force $f_\beta = k_B T/x_\beta$, diffusive attempt frequency $1/t_D$, and height E_b of the activation barrier. (a) Linear spectrum predicted for a single sharp energy barrier. The logarithmic intercept at zero force (represented by *) is determined by the barrier height and the microscopic diffusion time, $\log_e(r_f^0) = -E_b/k_B T + \log_e(f_\beta/t_D)$. (b) Piece-wise linear spectrum for a cascade of two sharp energy barriers. The abrupt increase in slope from one thermal force scale to the next shows that the outer barrier has been suppressed and that the inner barrier has become the dominant kinetic impedance to detachment [*cf.* Fig. 1(b)]. The difference between logarithmic intercepts (represented by *) is governed by the splitting in barrier energies and the ratio of thermal force scales, $\log_e(\Delta r_f^0) \approx -\Delta E_b/k_B T + \Delta\log_e(f_\beta)$.

to thermal energy $k_B T$. The slope f_β of this line maps the thermally averaged projection of the microscopic transition state to a distance $x_\beta = \langle x_{ts} \cos \theta \rangle$ along the direction of force. Moreover, the logarithmic intercept at zero force reflects the magnitude of barrier energy as given by $\log_e(r_f^0) = -E_b/k_B T + \log_e(f_\beta/t_D)$. Setting the scale for loading rate, the ratio f_β/t_D involves the microscopic attempt frequency $1/t_D$. Assuming that attempt frequency is weakly affected by point mutations, the simple linear-log behavior exposes a unique opportunity to quantitate the resulting chemical modifications in energy and/or location of barriers. Such changes in microscopic properties can be derived from the shift in the logarithmic intercept and/or change in slope, $\Delta E_b/k_B T \approx -\Delta\log_e(r_f^0) + \Delta\log_e(f_\beta)$. Taken together, these features demonstrate that the plot of most frequent probe force *vs.* log(probe loading rate) represents a dynamic spectral image of an activation barrier. [Although unknown, attempt frequency can be estimated from the damping factor indicated by MD simulations. Values for damping factor seem to be typically on the order of $\gamma \approx 10^{-8}$ pN s nm^{-1} (equivalent to Stokes drag on a 1 nm size sphere in water), *e.g.* $\gamma \approx 2 \times 10^{-8}$ pN s nm^{-1} in simulations of biotin–streptavidin separation[2] and $\gamma \approx 5 \times 10^{-8}$ pN s nm^{-1} in simulations of lipid extraction from a bilayer.[3] Since the product of molecular lengths $l_c l_{ts}$ should lie in the range ~ 0.01–0.1 nm^2, the attempt frequency is expected to be in the range $1/t_D \approx 10^9$–10^{10} s^{-1} and the microscopic scale for loading rate in the range $f_\beta/t_D \approx 10^{10}$–$10^{11}$ pN s^{-1}. The effective loading rate in the slowest MD simulations[2,3] is even higher $\geqslant 10^{12}$ pN s^{-1}.]

As described earlier, the most idealized view of a complex molecular energy landscape is a cascade of sharp activation barriers, which leads to a staircase of exponential increases in the rate constant under force. Using this prescription, the most likely force *vs.* log(loading rate) is predicted to follow a simple spectrum of piece-wise continuous linear regimes with ascending slopes as shown in Fig. 3(b). The abrupt increase in slope from one regime to the next signifies that an outer barrier has been suppressed by force and that an inner barrier has become the dominant kinetic impedance to escape as sketched in Fig. 1(b). These dynamic crossovers occur at somewhat higher forces than the stationary crossovers in rate constant as shown by the analytical approximation,

$$f_\otimes^{dyn} \approx \Delta E_b/\Delta x_\beta + k_B T[\log_e(x_\beta'/x_\beta)]/\Delta x_\beta$$

where $\Delta x_\beta = x_\beta' - x_\beta$ and $\Delta E_b = E_b' - E_b$ represent adjacent prominent barriers. In contrast to the idealized theory, the shape of a strength spectrum could be nonlinear and a challenge to interpret because force can distort physical potentials and molecular structure. Surprisingly, the results from recent probe experiments to be shown next yield linear plots for strength *vs.* log(loading rate) with one or more well-defined regimes, which allows the spectra to be interpreted in terms of sharp activation barriers.

Energy landscapes of receptor–ligand bonds

Not well-appreciated in biology is that energy landscapes of receptor–ligand bonds can be *rugged* terrains with more than one prominent activation barrier. The inner barriers are undetectable in test-tube assays but are important since they establish different time scales for kinetics under force. With two unrelated pairs of molecules, we will demonstrate that dynamic force measurements can be used to reveal these hidden barriers. The first pair of molecules will be the ligand biotin (a vitamin) and the protein receptor streptavidin (from bacteria) or avidin (a closely similar protein from hen egg white).[21] This complex is used widely in biotechnology because it has one of the highest affinity noncovalent bonds in biology with a force-free lifetime on the order of days.[22] The second pair of molecules will be a sialylated (carbohydrate) short peptide ligand† and the L-selectin receptor resident in the outer membrane of blood leukocytes. Although weaker in affinity with a lifetime of ~ 1 s or less, the carbohydrate–L-selectin bond plays a crucial role in the initial capture of leukocytes from blood circulation at sites of injury or infection.[23] In preparation for both experiments, the ligand was covalently anchored to a glass microbead along with a chemical glue for attachment of the bead to the BFP transducer [as noted in Fig. 2(a)]. A similarly pre-

---

† Note: the actual ligand used in the tests was a short peptide chimera of the biological molecule called P-selectin glycoprotein ligand (PSGL1), which was constructed by Genetics Institute and obtained through collaboration with Scott Simon at Baylor College of Medicine. The generic label *carbohydrate* will be used for convenience.

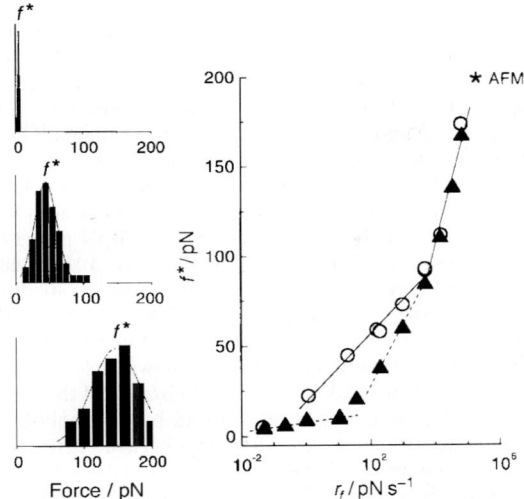

Fig. 4 On the left are examples of force histograms taken from tests of single biotin–streptavidin bonds, which demonstrate the shift in peak location and increase in width with increase in loading rate (top histogram, 0.05 pN s^{-1}, middle histogram, 20 pN s^{-1}, bottom histogram, 60000 pN s^{-1}). Superposed on the histograms are Gaussian fits used to determine the most frequent rupture force–bond strength. Governed ideally by the thermal force f_β, standard deviations σ_f of the distributions also reflected uncertainties in position Δx and video sampling time Δt_v, *i.e.* $\sigma_f \approx [f_\beta^2 + (k_f \Delta x)^2 + (r_f \Delta t_v)^2]^{1/2}$. As σ_f increased from ± 1 pN at the slowest rate to ± 60 pN at the fastest rate, the standard error in mean force—the uncertainty in strength—ranged from ± 0.3 to ± 5 pN. On the right are complete dynamic strength spectra for both biotin–streptavidin (open circles) and biotin–avidin (closed triangles) bonds.[6] Defined as thermal energy $k_B T$/distance x_β, the slopes of the linear regimes seen in the spectra map activation barriers at positions along the direction of force. The common high strength regime in the biotin–streptavidin and biotin–avidin spectra place the innermost barrier at $x_\beta \approx 0.12$ nm. Separate intermediate strength regimes place the next barrier at $x_\beta \approx 0.5$ nm for biotin–streptavidin and $x_\beta \approx 0.3$ nm for biotin–avidin (with a slight reduction in slope below 38 pN suggesting that the biotin–avidin barrier extends to ~ 0.5 nm). Only well-defined in the biotin–avidin spectrum, a low strength regime implies a distal barrier at $x_\beta \approx 3$ nm. Also marked ($*_{\text{AFM}}$) is the biotin–streptavidin strength measured recently by AFM at $\sim 10^5$ pN s^{-1} using a carbon nanotube as the tip.[14] This and the earlier measurements of biotin–avidin bond strength[13] at loading rates of $\sim 6 \times 10^4$ pN s^{-1} also correlate with the high strength regime shown here.

pared microbead was used as the test surface for probing biotin–(strept)avidin bonds whereas a white blood cell (granulocyte) taken from a small blood sample was used as the test surface for probing carbohydrate–L-selectin bonds.

[Methods: In testing molecular bonds, the density of reactive sites must be reduced significantly as mentioned earlier so that only 1 out of 7–10 touches results in a molecular attachment. Assumed to be governed by Poisson statistics, this ensures that 90–95% of the attachments are single bonds. To obtain strength spectra with the BFP technique, detachment forces are measured over a six order of magnitude range in loading rate from 0.05 pN s^{-1} to 100000 pN s^{-1}. The loading rate $k_f v_t$ is preselected by setting the transducer force constant k_f in the range 0.1–3 pN nm^{-1} and the piezo retraction speed v_t in the range 1–30000 nm s^{-1} as described in Fig. 2. From thousands of repeated touches at fixed loading rate, histograms of detachment forces are compiled at many rates and Gaussian fits are used to locate the peak in each histogram. These most probable values of force are then plotted as a function of log$_e$(loading rate), which yields the dynamic strength spectrum.]

Biotin (strept)avidin bonds

Because of high affinity, the first ligand–receptor pair chosen by researchers for testing with AFM was biotin and streptavidin; which was soon followed by biotin and avidin.[12–14] Deduced from a broad distribution of AFM forces, it was concluded that the strength of a biotin–streptavidin bond lies in a range of ~ 200–300 pN and somewhat lower for biotin–avidin ~ 160 pN. However, the examples of force histograms and the strength spectra[6] in Fig. 4 show that biotin–streptavidin (and biotin–avidin) bond strengths fall continuously from ~ 200 pN to $\sim$pN with each decade increase in time scale for rupture from 10^{-3} to 10^2 s, which clearly demonstrates the thermally activated nature of bond breakage. Moreover, distinct linear regimes with abrupt changes in slope imply sharp barriers, which can be analysed using the idealized theory.[6] First, above 85 pN, there is a common high strength regime for both biotin–streptavidin and biotin–avidin with a slope of $f_\beta \approx 34$ pN. This locates a barrier deep in the binding pocket at $x_\beta \approx 0.12$ nm. Below 85 pN, the $f_\beta \approx 8$ pN slope in the biotin–streptavidin spectrum maps the next activation barrier at $x_\beta \approx 0.5$ nm whereas the steeper slope $f_\beta \approx 13$–14 pN between 38 and 85 pN in the biotin–avidin spectrum indicates that its next barrier maps to $x_\beta \approx 0.3$ nm. Interestingly, a slight curvature and reduction in slope between 38 and 11 pN suggests that the barrier in biotin–avidin extends to ~ 0.5 nm. Below 11 pN, the biotin–avidin spectrum exhibits a very low strength regime (dashed line) with a slope of $f_\beta \approx 1.4$ pN that maps to $x_\beta \approx 3$ nm. A similar low strength regime is indicated by results from the slowest test of biotin–streptavidin bonds; but it was not possible to perform tests at loading rates below 0.05 pN s^{-1} as needed to verify the existence of this regime. In addition to the map of barrier locations, the logarithmic intercepts found by extrapolation of each linear regime to zero force also yield estimates of the energy differences between activation barriers within each landscape as well as energy differences between related barriers of biotin–avidin and biotin–streptavidin landscapes. However, instead of discussing barrier heights, it is more illuminating to examine how the 1-D map of barrier locations compares with detailed molecular simulations of biotin–(strept)avidin interactions.

In separate MD simulations,[2] biotin was extracted from a binding pocket of streptavidin and avidin by pulling on the outer end with a pseudo-mechanical spring. Consistent with the numerous bonds to small molecules in the binding pocket, simulations yield a fluctuating superposition of many attractions—buffeted by steric collisions—along the unbinding trajectories. This is shown by a profile of instantaneous energy calculated over a *slow* ~ 500 ps extraction of biotin from avidin [Fig. 5(a), kindly provided by Professor K. Schulten and coworkers, Beckman Institute, University of Illinois]. Even with the enormous and fast changes in energy, simple qualitative features appear in the profile that provide important clues to the thermally averaged free energy landscape relevant on laboratory time scales. In particular, transition states are readily identified by regions with rarified statistics where biotin passes quickly. Taking a simple coarse-grained average over ~ 20 ps windows [Fig. 5(b)] smooths over the strong rapid fluctuations and exposes locations of activation barriers. First, within an initial displacement of 0.1–0.2 nm, the spring force in the simulations revealed abrupt detachment of biotin from a nest of hydrogen bonds, water bridges, and nonpolar interactions deep in the binding pocket. Next, forces reached maximal

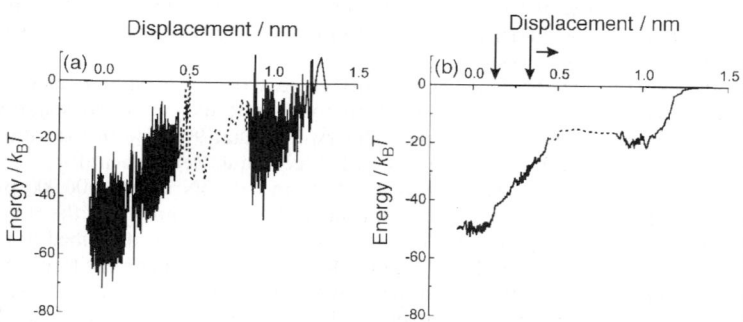

Fig. 5 (a) Profile of instantaneous energy computed for interaction between biotin and avidin over a half-nanosecond extraction from the binding pocket in the simulations of Israilev *et al.*[2] (kindly provided by Professor K. Schulten and coworkers, University of Illinois). Separating regions of rapid intense fluctuations, locations of rarified statistics coincide with maximal forces in the simulations, which signify the presence of transition states. (b) Coarse-grained average over the fast degrees of freedom which yields an approximate potential of mean force.[5] Arrows mark barrier locations derived from the intermediate and high strength regimes of the spectrum for biotin–avidin in Fig. 4.

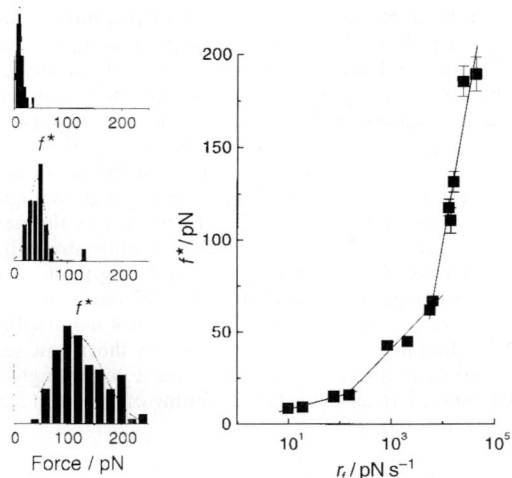

Fig. 6 On the left are examples of force histograms taken from tests of single carbohydrate (sialylated PSGL1 short peptide chimera)–L-selectin bonds, which demonstrate shift in peak location and increase in width with increase in loading rate (top histogram, 10 pN s^{-1}, middle histogram, 850 pN s^{-1}, bottom histogram, 13000 pN s^{-1}). Superposed on the histograms are Gaussian fits used to determine the most frequent rupture force— bond strength. On the right is the complete dynamic spectrum of strength *vs.* log(loading rate). Defined as thermal energy $k_B T$/distance x_β, the slope of the high strength regime places the innermost barrier at $x_\beta \approx 0.06$ nm. The intermediate strength regime places the next barrier at $x_\beta \approx 0.3$ nm. The low strength regime implies a barrier further out at $x_\beta \approx 1.2$ nm.

values followed by sudden displacements of biotin at a distance of ∼0.4 nm (and ∼0.5 nm in the biotin–streptavidin simulation). Finally, biotin was observed to still cling to peripheral polar groups at ∼1.4 nm in the avidin simulation. As labelled in Fig. 5(b), the locations of activation barriers derived from the high and intermediate strength regimes of the laboratory spectra in Fig. 4 correlate well with regions of rarified statistics and the qualitative appearance of the energy landscape. The conclusion is that these transition states inferred from the simulations persist on long time scales. However, the outer barrier indicated by the low strength regime is 2–3-fold more distant than the last transition state seen in the MD simulation; this is perhaps due to interactions with the peripheral flexible loops[24–27] which border the channel that leads to the binding pocket. More puzzling, however, extrapolation of the lowest strength regime to zero force implies that bond strength vanishes below a threshold loading rate of ∼0.0006 pN s^{-1} for biotin–avidin. In other words, the spontaneous off rate would be ∼1 per hour. This is 50-fold faster than the rate of ∼1 per 55 hours that we measured for spontaneous dissociation of PEG–biotin from probe tips in free solution and found previously for biotin by others.[22] Hence, some nontrivial effect remains that accounts for the significant increase in rate of dissociation under extremely small forces below 5 pN.

Carbohydrate–L-selectin bonds

In contrast to the high affinity biotin–(strept)avidin bonds, carbohydrate–L-selectin bonds with modest affinity stop white cells at vessel walls in the circulation.[23] Numerous bonds to other surface (*integrin*) receptors then form between the white cell and vessel wall to sustain adhesion and enable subsequent movement into the surrounding tissue. On its initial arrest from the blood flow, the white cell can be subjected to forces of ∼100 pN in a time frame of milliseconds, which implies loading rates of 10^4–10^5 pN s^{-1}. With this functional requirement in mind, we now examine recent tests of carbohydrate–L-selectin bonds under dynamic loading in probe tests.

From the results[7] presented in Fig. 6, we again see a sequence of high, intermediate, and low strength regimes for carbohydrate–L-selectin bonds where strength also falls continuously from

~ 200 pN to $\sim$ pN but over fewer decades in time scale for detachment from 10^{-3} to 1 s. The high strength regime has a very steep slope of $f_\beta \approx 70$ pN that maps an inner barrier to a small distance $x_\beta \approx 0.06$ nm along the direction of force. Although we lack detailed molecular information about L-selectin binding, the small value of x_β seems to imply that the microscopic reaction coordinate deviates significantly from the macroscopic orientation of force. For example, if the ligand was bound to the side of the receptor, pulling parallel to the axis of the receptor along the surface normal could result in a large orientation angle θ relative to the microscopic pathway and weak coupling of force to the energy landscape. Departing from the high strength regime below 70 pN, the intermediate strength regime with a slope of $f_\beta \approx 13$ pN places the next activation barrier at $x_\beta \approx 0.3$ nm. Finally, below ~ 20 pN, the spectrum exhibits a low strength regime with a slope of $f_\beta \approx 3.4$ pN that sets the outermost barrier at $x_\beta \approx 1.2$ nm. Using the logarithmic intercepts found by extrapolation of each linear regime to zero force, the differences in energy between the inner activation barriers are calculated to be only 2–3 $k_B T$. As for biotin–(strept)avidin bonds, the inner-most barrier deep in the binding pocket provides strength on short time scales (<0.03 s), which is sufficient to meet the functional requirements noted earlier. Even though only 4–6 $k_B T$ higher in energy, the outermost activation barrier extends the lifetime of the bond almost 100-fold (to ~ 1 s) beyond that set by the innermost barrier.

Energy landscapes for anchoring lipids in membranes

Lipids and acylated proteins are anchored in bilayers by hydrophobic interactions. The handbook[28] correlation for free energy of transfer from aggregates (*e.g.* micelles or bilayers) to water is a linear proportionality of ~ 1 $k_B T$ per aliphatic carbon for lysophosphatidylcholines (PCs) and not quite double ~ 1.7 $k_B T$ per carbon for diacylPCs, although little evidence exists for diacyl lipids with chain lengths longer than 10–12 carbons. This reinforces the established view that anchoring potential increases with hydrophobic surface area embedded in the bilayer.[29] Partition and solubility provide important static-equilibrium assays but represent energetic measures of strong *vs.* weak anchoring—not strength—which is the force needed to extract a molecule. Based on the hydrophobic energy scale for exposure to water, the energy landscape for hydrophobic anchoring in bilayers should simply rise linearly with displacement along the bilayer normal. Treating the embedded molecule as a cylinder with radius r_m, the surface energy per unit area for creating a water/nonpolar interface and the circumference of the cylinder (*i.e.* energy/length ≈ 30 mJ m^{-2} $\times 2\pi r_m$ or 7 $k_B T$ nm^{-2} $\times 2\pi r_m$) suggest naively that the molecular extraction force should be a constant set by molecular size, $f \approx 180$ pN $\times r_m$(nm). Taking a radius ~ 0.5 nm for a lipid, the anchoring force would be ~ 100 pN. On the other hand, we will see next that lipids can be extracted from membrane bilayers with forces as small as ~ 1 pN if performed over seconds! Over the range of anchoring strengths between 0 and 100 pN, the missing ingredient is thermally activated kinetics. By comparison, lipid pull-out forces in MD simulations[3] were >200 pN even under the slowest extraction of $\sim 10^{-8}$ s and increased with speed apparently due to viscous damping.

Strength of hydrophobic anchoring in fluid membranes was tested by extraction of single receptor lipids from giant bilayer vesicles prepared with two lipid compositions: pure stearoyloleoyl-phosphatidylcholine (SOPC) (C18:0/1) and a 1 : 1 mixture of SOPC plus cholesterol (CHOL)—somewhat similar to membranes that encapsulate cells. Doped in the vesicle bilayers at extremely low concentration ($<0.0001\%$), the receptor lipids were a special lipid construct of biotin–PEG–distearoylphosphatidylethanolamine (DSPE) (diC18:0) kindly provided by INEX Pharmaceuticals, Burnaby, B.C., Canada. Plotted in Fig. 7, we see little structure in the spectra for receptor lipid anchoring and much lower forces compared to the spectra in Figs. 5 and 6 for receptor–ligand bonds.[8] Over nearly four orders of magnitude in loading rate, only a single linear strength regime is found for extraction of the receptor lipids from SOPC : CHOL bilayers. The low slope of $f_\beta \approx 2.4$ pN places a barrier at a distance $x_\beta \approx 1.7$ nm along the direction of force. Two linear regimes are found for lipid extraction from pure SOPC bilayers with a modest difference in slopes. The initial slope of $f_\beta \approx 3.4$ pN locates an outer barrier at $x_\beta \approx 1.2$ nm and the second slope of $f_\beta \approx 6.1$ pN implies an inner barrier at $x_\beta \approx 0.7$ nm. Consistent with the simple concept of hydrophobic interaction, the locations of the outermost barriers for both types of bilayers are comparable to (but slightly less than) the hydrophobic half thickness of the bilayer,

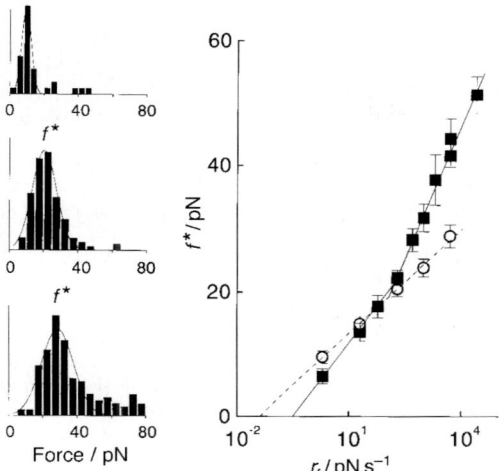

Fig. 7 On the left are examples of force histograms taken from tests of receptor lipid (biotin–PEG–DSPE) extraction from mixed SOPC : CHOL vesicle bilayers, which demonstrate shift in peak location and increase in width with increase in loading rate (top histogram, 2 pN s^{-1}, middle histogram, 200 pN s^{-1}, bottom histogram, 5000 pN s^{-1}). Superposed on the histograms are Gaussian fits used to determine the most frequent extraction force—anchoring strength. On the right are the complete dynamic spectra of strength *vs.* log(loading rate) for extraction of receptor lipids from SOPC (closed boxes) and mixed SOPC : CHOL bilayers (open circles). Defined as thermal energy $k_B T$/distance x_β, the slopes of the initial linear regimes map activation barriers at $x_\beta \approx 1.2$ nm for extraction from SOPC and $x_\beta \approx 1.7$ nm for extraction from SOPC : CHOL along the direction normal to the bilayer. Not seen in the SOPC : CHOL spectrum, the break in slope for the SOPC spectrum places a weak inner barrier at $x_\beta \approx 0.7$ nm.

which is increased by cholesterol. Addition of cholesterol to SOPC bilayers increases the outer activation barrier by $\sim 2\, k_B T$ as shown by the shift between logarithmic intercepts of the initial regimes for SOPC : CHOL and SOPC. Quite unexpected, the break in slope in the spectrum for SOPC reveals an inner transition state near the middle of the hydrophobic monolayer, which appears to be $\sim 2\, k_B T$ below the outer barrier. Perhaps coincidental, the location of this transition state derived from the thermal force scale correlates with the position of the unsaturated bond in the oleoyl chain of SOPC. Completely speculative, the split in activation barriers could reflect an entropic bottle neck as chains transiently pass the average position of the unsaturated group. Very puzzling, the bilayer residence time of ~ 0.01 s derived from the logarithmic intercept of these spectra at zero force is much shorter than implied by the lack of perceptible dissociation from an isolated vesicle over the the time scale of 1 h. Without an explanation at present, we see again (as for biotin–avidin bonds) that very small forces must strongly affect the shape of the soft outer transition state. In any case, anchoring of lipids and acylated proteins in bilayers will always be weak unless the molecules are extracted very rapidly from the bilayer.

Strong *vs.* weak bonds in serial linkages

For a serial linkage of n identical bonds, the rate of breakage under force is simply increased by a factor n, $k_{off} \approx (n/t_0)\exp(f/f_\beta)$, if the bond kinetics are uncorrelated. This increases the thermal scale for loading rate, $r_f^0 = n f_\beta/t_0$, and shifts the strength spectrum along the $\log_e$(loading rate) axis by a factor of $\log_e(n)$, which reduces strength at a given loading rate by $-f_\beta \log_e(n)$. In contrast to a simple shift along the log(loading rate) axis, we expect the strength of a multiple linkage of dissimilar bonds to be limited by the weakest bond and naively that strong *vs.* weak should be defined by the energy barriers sustaining the bonds. However, theory shows that this anticipated hierarchy is only correct for some sets of bonds; other sets will exhibit unexpected switching from strong to weak and *vice versa* as loading rate increases. In the determination of strong *vs.* weak at a particular retraction rate, the important parameters are both the spontaneous rates of dissociation set by barrier energies and the thermal force scales that characterize *e*-fold changes in the dissociation rates of bonds under force. Again invoking the simple sharp barrier model, we can easily establish a phase diagram [*cf.* Fig. 8(a)] of the most likely site for breakage in a two-bond linkage. Assuming that both bonds are characterized by the same diffusive time scale t_D for simplicity, the rate of uncorrelated breakage is the combined rates for each bond,

$$k_{\text{off}} \approx (1/t_0)\exp(f/f_\beta)\{1 + \exp[-\Delta E_b/k_B T + f\Delta(1/f_\beta)]\}$$

where $1/t_0$ and f_β specify the spontaneous rate and thermal force scale of the fast bond (smallest barrier energy) as the reference; ΔE_b and $\Delta(1/f_\beta)$ represent the differences in barrier energy and reciprocal thermal force scale for the slow bond (*i.e.* $\Delta E_b > 0$) relative to the fast bond. The combined rate and the predicted strength spectrum predict that the fast bond will remain the expected weak bond so long as the following inequality holds: $\Delta E_b/k_B T > f\Delta(1/f_\beta)$. This will always be the case when the thermal force scale for the fast bond is less than the thermal force scale for the slow bond [*i.e.* $\Delta(1/f_\beta) < 0$ or equivalently $\Delta(x_\beta) > 0$]. On the other hand, if the thermal force scale for the fast bond is larger, then there will be a crossover force where $\Delta E_b/k_B T \leqslant f\,\Delta(1/f_\beta)$; this strong $\rightleftharpoons$ weak bond phase boundary is sketched in Fig. 8(a). Now, the fast bond will be the strong bond and the slow bond will be the most likely site of failure, which is not anticipated in the traditional view.

To demonstrate the importance of this concept, imagine that the selectin receptor was linked to a vesicle bilayer by a lipid anchor and then the strengths of carbohydrate ligand bonds to the selectin were probed as in the leukocyte tests. Purely hypothetical, Fig. 8(b) shows that at slow loading rates, the carbohydrate–selectin bond would most likely detach because lipid anchoring is

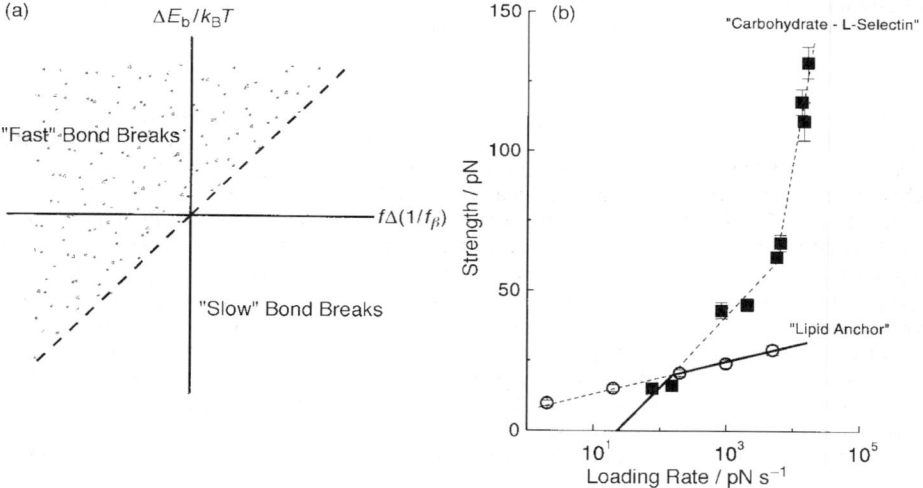

Fig. 8 (a) Phase diagram for definition of strong $\rightleftharpoons$ weak bonds in a serial linkage of two bonds sustained by single sharp energy barriers. The vertical axis is the difference in barrier heights ΔE_b for the two bonds ($\Delta E_b > 0$ characterizes a slow bond relative to a fast bond as defined by spontaneous off rates); the horizontal axis is the product $f\Delta(1/f_\beta)$ of applied force f and the difference $\Delta(1/f_\beta)$ in reciprocal thermal force scales. In the traditional view, strong : weak equates to slow : fast. But for bonds in series, this diagram shows that there can be unexpected switching of these attributes under force. (b) Hypothetical strength of carbohydrate bonds to selectins if linked by lipid anchors to membranes. Simultaneous kinetics over different energy landscapes for the carbohydrate–selectin bond and lipid anchoring predicts a dynamic crossover in site for detachment when pulled on by a probe decorated with the carbohydrate ligand. At slow rates of loading, the lipid anchor is stronger than the carbohydrate–selectin bond, which is the most likely site of detachment. On the other hand, at fast rates of loading, the carbohydrate–selectin bond is stronger than lipid anchoring strength, which then becomes the most likely site of detachment.

stronger. On the other hand, under fast loading, the carbohydrate–selectin bond becomes strong and the lipid anchor weak by comparison. Hence, the lipid-anchored selectin would most likely be pulled out of the membrane by the carbohydrate ligand attached to the probe. In contrast to the image of inner activation barriers in a complex bond, we see that the signature of a strong to weak bond metamorphosis in a serial linkage of bonds is an abrupt reduction in slope from one linear strength regime to the next with increase in loading rate.

Summary

Recent laboratory probe experiments confirm that bond breakage and molecular detachment occur at forces determined by the loading rate. Measured under steadily rising force over an enormous span of loading rates, the spectra of strength *vs.* log$_e$(loading rate) yield images of the prominent barriers traversed in the energy landscapes along force-driven pathways in unbinding. Simple analysis of the spectra provides a view into the inner complexity of biomolecular interactions and structural cohesion as noted in the following list of highlights.

(1) Examining two unrelated receptor–ligand bonds, we find a similar sequence of linear strength regimes *vs.* log(loading rate). These regimes reveal a cascade of three activation barriers for both receptor–ligand interactions, although quite different in energy scale. The innermost barrier deep in the binding domain is responsible for the high strength perceived on short time scales and the major portion of total activation energy. The more distal barriers lead to weakness on long time scales but significantly extend bond lifetime in the absence of force. The intriguing question is why did nature structure energy landscapes in receptor–ligand bonds and create a sequence of time scales for amplification of kinetics under force? Answering this question is likely to introduce a new perspective of biological chemistry.

(2) No surprise, anchoring strengths of lipids in bilayers are consistent with nearly structureless hydrophobic potentials, although small inner barriers do appear in some cases. Most significant, the small thermal force scale set by acyl chain length results in very weak anchoring strength unless the molecules are extracted extremely rapidly from the bilayer. Still to be confirmed, integral membrane proteins should be much more strongly anchored to membranes since hydrophilic groups at the interfaces will contribute major activation barriers with large thermal force scales.

(3) Dissimilar bonds in a serial linkage can unexpectedly switch from strong to weak and shift the most likely site for failure between bonds as loading rate increases. Such behaviour is not only a major factor in cohesive and adhesive strength but is likely to be important in signalling and regulation of biochemical pathways inside cells.

Acknowledgement

It is important to credit the individuals who carried out the experiments and developed the instrumentation described in this paper since many of the results are yet to be published. Tests of biotin–(strept)avidin bonds were performed by Andrew Leung (University of British Columbia) and Pierre Nassoy (now at l'Institut Curie in Paris). Computer control of the biomembrane force probe assembly and video image processing was developed by Ken Ritchie (now at Nagoya University in Japan). Tests of carbohydrate–L-selectin bonds were also performed by Andrew Leung in collaboration with Scott Simon (from Baylor College of Medicine in Houston) and Dan Hammer (from University of Pennsylvania in Philadelphia). Tests of lipid anchoring in bilayer membranes were performed by Florian Ludwig (University of British Columbia).

The work was supported by grants HL54700 and HL 31579 from the US National Institutes of Health, grant MT7477 from the Medical Research Council of Canada, and the Canadian Institute for Advanced Research Program in Science of Soft Surfaces and Interfaces.

References

1   G. Binnig, C. F. Quate and C. H. Gerber, *Phys. Rev. Lett.*, 1986, **56**, 930; B. Drake, C. B. Prater, A. L. Weisenhorn, S. A. C. Gould, T. R. Albrecht, C. F. Quate, D. S. Cannell, H. G. Hansma and P. K. Hansma, *Science*, 1989, **243**, 1586.
2   H. Grubmuller, B. Heymann and P. Tavan, *Science*, 1996, **271**, 997; S. Izrailev, S. Stepaniants, M. Balsera, Y. Oono and K. Schulten, *Biophys. J.*, 1997, **72**, 1568.

3   S-J. Marrink, O. Berger, P. Tieleman and F. Jahnig, *Biophys. J.*, 1998, **74**, 931.
4   H. A. Kramers, *Physica (Amsterdam)*, 1940, **7**, 284; P. Hanggi, P. Talkner and M. Borkovec, *Rev. Mod. Phys.*, 1990, **62**, 251.
5   E. Evans and K. Ritchie, *Biophys. J.*, 1997, **72**, 1541.
6   R. Merkel, P. Nassoy, A. Leung, K. Ritchie and E. Evans, *Nature (London)*, 1999, **397**, 50.
7   S. Simon, A. Leung, D. Hammer and E. Evans, to be submitted.
8   F. Ludwig and E. Evans, to be submitted.
9   M. Doi and S. F. Edwards, *The Theory of Polymer Dynamics*, Clarendon Press, Oxford, 1986; N. G. van Kampen, *Stochastic Processes in Physics and Chemistry*, North-Holland, Amsterdam, 1981.
10 P. J. Rossky, J. D. Doll and H. L. Friedman, *J. Chem. Phys.*, 1978, **69**, 4628.
11 G. I. Bell, *Science*, 1978, **200**, 618.
12 G. U. Lee, D. A. Kidwell and R. J. Colton, *Langmuir*, 1994, **10**, 354.
13 E-L. Florin, V. T. Moy and H. E. Gaub, *Science*, 1994, **264**, 415.
14 V. T. Moy, E-L. Florin and H. E. Gaub, *Science*, 1994, **264**, 257.
15 P. Hinterdorfer, W. Baumgartner, H. J. Gruber, K. Schilcher and H. Schindler, *Proc. Natl. Acad. Sci. USA*, 1996, **93**, 3477.
16 S. S. Wong, E. Joselevich, A. T. Woolley, C. L. Cheung and C. M. Lieber, *Nature (London)*, 1998, **394**, 52.
17 S. P. Tha, J. Shuster and H. L. Goldsmith, *Biophys. J.*, 1986, **50**, 117.
18 E. Evans, D. Berk and A. Leung, *Biophys. J.*, 1991, **59**, 838.
19 E. Evans, K. Ritchie and R. Merkel, *Biophys. J.*, 1995, **68**, 2580.
20 E. Evans and K. Ritchie, *Biophys. J.*, 1999, in press.
21 N. M. Green, *Adv. Protein Chem.*, 1975, **29**, 85.
22 A. Chilkoti and P. S. Stayton, *J. Am. Chem. Soc.*, 1995, **117**, 10622.
23 R. Alon, D. A. Hammer and T. A. Springer, *Nature (London)*, 1995, **374**, 539; K. D. Puri, S. Chen and T. A. Springer, *Nature (London)*, 1998, **392**, 930.
24 P. C. Weber, D. H. Ohlendorf, J. J. Wendoloski and F. R. Salemme, *Science*, 1989, **243**, 85.
25 O. Livnah, E. A. Bayer, M. Wilchek and J. L. Sussman, *Proc. Natl. Acad. Sci. USA*, 1993, **90**, 5076.
26 S. Freitag, I. Le Trong, L. Klumb, P. S. Stayton and R. E. Stenkamp, *Protein Sci.*, 1997, **6**, 1157.
27 V. Chu, S. Freitag, I. Le Trong, R. E. Stenkamp and P. S. Stayton, *Protein Sci.*, 1998, **7**, 848.
28 D. Marsh, *Handbook of Lipid Bilayers*, CRC Press, Boca Raton, FL., 1990, p. 275–280.
29 C. Tanford, *The Hydrophobic Effect: Formation of Micelles and Biological Membranes*, John Wiley and Sons, New York, NY, 1973.

Paper 8/09884K

Solid State Chemistry

C. R. A. Catlow

Royal Institution, 21 Albemarle Street, London, UK W1S 4BS

Commentary on: **Intracrystalline channels in levynite and some related zeolites**,
R. M. Barrer and I. S. Kerr, *Trans. Faraday Soc.*, 1959, **55**, 1915–1923.

The Faraday Society journals have made major contributions to the science of solid state chemistry over the last hundred years, and many of the major developments in the subject have been first published in the Society's journals. Here we highlight a fascinating study[1] reported in the *Transactions of the Faraday Society* in 1959 by that versatile and prolific physical and solid state chemist, Richard Barrer, which in a number of ways foreshadows modern developments of the subject.

Contemporary solid state chemistry is increasingly concerned with the *synthesis and characterisation of complex functional materials*, where perhaps microporous materials provide the most elegant example. This extraordinary class of materials, comprising framework-structured silicas, aluminosilicates and aluminophosphates, with major applications in catalysis, gas separation and ion exchange, has developed over the last hundred years from a minor branch of mineralogy into a major theme of current solid state and materials chemistry. The work of Barrer was central to these developments. As well as pioneering new synthetic strategies, which have been extraordinarily influential, Barrer contributed profoundly to the fundamental physical chemistry of these materials with the first detailed studies of sorbate diffusion (see for example refs. 2 and 3): a key process influencing both catalytic and molecular sieving properties. Barrer and Kerr's paper[1] is remarkable in that it links structural properties of four related zeolites, levynite (whose framework structure is shown in Fig. 1), chabazite, erionite and gmelinite, with the diffusional properties of sorbed molecules, a major theme in the subsequent development of zeolite science. The paper discusses in detail the free diameters, stereochemistry and interrelation between the ring and channel patterns for the four zeolites, which are correlated with the magnitude and anisotropy of the diffusion and molecular sieving behaviour within the solids.

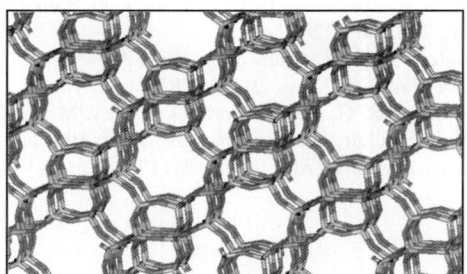

Fig. 1 Framework structure of levynite.

This and other papers of Barrer illustrate the role of crystallographical techniques in advancing knowledge of the physical properties of these materials. The more general role of advanced characterisation techniques in the science of complex materials, including microporous materials, is well illustrated by pioneering science reported in the Society's journals. Magic angle spinning NMR (MASNMR) has, for example, revolutionised structural solid state chemistry. An elegant example is the application to Si/Al ordering in zeolites by Thomas *et al.*[4] which, building on earlier work, revealed new Si/Al ordering schemes in zeolite Y. HREM, EXAFS and high resolution powder diffraction methods have also played key roles in the growth and development of the field, as again highlighted by articles published in Faraday Society journals (for example refs. 5 and 6).

Zeolite science also has profited enormously from another major development in solid state chemistry, the growth of computer modelling techniques. A seminal contribution here was made well before the computer age by a remarkable paper published in *Transactions of the Faraday Society* by Mott and Littleton,[7] which established the basic methodology required to calculate the energies of defects in solids. It needed the advent of high speed computers before this methodology could be widely exploited. The work of Norgett and colleagues at the Harwell laboratories in the 1970s developed general purpose computer codes based on the Mott–Littleton approach, the success and influence of which are evident from the volume published in *Faraday Transactions* in 1988 celebrating the 50th anniversary of the original paper.[8]

The role of computer modelling in the science of complex solids including microporous materials was surveyed in Faraday Discussion 106 held in 1997. These techniques have now an increasingly predictive role.[9] They can, for example, predict new microporous structures, design templates for their synthesis and model the static and dynamical behaviour of sorbed molecules within their pores,[10] a topic of enduring importance and one of particular interest to Barrer. Computer modelling methods are, of course, most effective when used in a complementary manner with other physical techniques. Ref. 6 nicely illustrates this theme. Here EXAFS and quantum mechanical methods are used in a concerted manner to elucidate the structure of the active site in microporous titanosilicate catalysts. Articles in *Faraday Discussions*, vol. 106 again illustrate the complementarity of computational and experimental techniques.

The growth of solid state chemistry can therefore be charted in Faraday Society Discussions and Transactions. The paper of Barrer and Kerr is one of many that we could have highlighted. Its theme, that of linking detailed microscopic structures with understanding and prediction of the physical properties of the material, is, however, one that is central to the contemporary subject.

References

1 R. M. Barrer and I. S. Kerr, *Trans. Faraday Soc.*, 1959, **55**, 1915.
2 R. M. Barrer, *Trans. Faraday Soc.*, 1944, **40**, 555.
3 R. M. Barrer, *Discuss. Faraday Soc.*, 1949, **7**, 135.
4 J. Klinowski, S. Ramdas, J. M. Thomas, C. A. Fyfe and J. S. Hartman,
    J. Chem. Soc., Faraday Trans. 2, 1982, **78**, 1025.
5 J. M. Thomas, G. R. Millward, S. Ramdas, L. A. Borsill and M. Audier,
    Discuss. Faraday Soc., 1981, **73**, 346.
6 C. M. Barker, D. Gleeson, N. Kaltsoyannis, C. R. A. Catlow, G. Sankar and J. M. Thomas,
    Phys. Chem. Chem. Phys., 2002, **4**, 1228.
7 N. F. Mott and M. J. Littleton, *Trans. Faraday Soc.*, 1938, **34**, 485.
8 See articles in *J. Chem. Soc. Faraday Trans. 2*, 1989, **85**(5).
9 C. R. A. Catlow, L. Ackermann, R. G. Bell, F. Corà, D. H. Gay, M. A. Nygren, J. C. Pereira,
    G. Sastre, B. Slater and P. E. Sinclair, *Faraday Discuss.*, 1997, **106**, 1.
10 See articles in *J. Chem. Soc., Faraday Trans.*, 1991, **87**(13).

INTRACRYSTALLINE CHANNELS IN LEVYNITE AND SOME RELATED ZEOLITES

By R. M. Barrer and I. S. Kerr

Physical Chemistry Laboratories, Chemistry Dept.,
Imperial College, London, S.W.7

Received 27th April, 1959

Chabazite, gmelinite, levynite and erionite form a group of porous crystals which exhibit certain similarities. Structures have previously been proposed for the anionic frameworks of chabazite,[5] gmelinite [5] and erionite.[8] A structure is now proposed for levynite. On the basis of the crystallographic data, diffusion anisotropy and molecular sieve behaviour of the four types of framework have been examined and the channel systems described and compared.

Increasing attention has been given to the structures of zeolitic crystals,[1-6] not only on account of their practical value as selective sorbents, but also because of the remarkable pore systems which have been revealed. As a result considerable new information exists about the anionic frameworks, although the disposition of the relatively mobile intracrystalline water and cations is intrinsically more difficult to determine. Four structures which have certain related features, and which are of interest as molecular sieves, are those of chabazite, gmelinite, levynite and erionite, for which hexagonal unit cells [7, 8] may be given as follows:

	a	c
chabazite, $(Ca, Na_2) O . Al_2O_3 . 4SiO_2 . 6H_2O$	$13 \cdot 7_4, 13 \cdot 7_8$	$14 \cdot 8_3, 15 \cdot 0_6$
gmelinite, $(Na_2, Ca) O . Al_2O_3 . 4SiO_2 . 6H_2O$	$13 \cdot 7_2, 13 \cdot 7_6$	$10 \cdot 0_2, 10 \cdot 0_6$
levynite, $CaO . Al_2O_3 . 4SiO_2 . 6H_2O$	$13 \cdot 3_2$	$22 \cdot 5_1$
erionite, $(Ca, Mg, Na_2, K_2) O . Al_2O_3 . 6SiO_2 . 6H_2O$	$13 \cdot 2_6$	$15 \cdot 1_2$

Anionic frameworks have been proposed for chabazite,[5] gmelinite [5] and erionite.[8] In this paper we suggest an anionic framework for levynite, and compare, on the basis of the proposed frameworks, the ease and degree of anisotropy of molecule diffusion, and the possible molecular sieve behaviour, for the four zeolites. It has already been shown that diverse intracrystalline channel systems can arise in structures such as analcite, nosean-sodalite minerals, cancrinite, faujasite, and Linde Sieve A.[9, 10]

THE ANIONIC FRAMEWORK OF LEVYNITE

Strunz [11] has drawn attention to a relation between the hexagonal unit cells of levynite and chabazite, which have a-dimensions roughly equal, and c-axes standing in the ratio 3:2. A similar relationship also exists between these cells for levynite and erionite; in fact this second relation is the closer of the two, because the respective a-axes are very nearly equal. Similarities in certain single crystal X-ray diffraction patterns of levynite and chabazite, together with the axial relationships, have led us to a structure for levynite based upon a rhombohedral cell of space group $R\bar{3}m$. This unit cell, which is merely an alternative to the hexagonal cell, has $a = 10 \cdot 75$ Å and $\alpha = 76° 25'$. The composition of levynite approaches $Ca(Al_2Si_4O_{12})6H_2O$, and there are then three such units in each rhombohedral cell, for which the framework atoms lie approximately in the following positions :

	x	y	z	no. of positions
(Si, Al) (1)	$\cdot31_5$	$\cdot82_5$	$\cdot07$	12
(Si, Al) (2)	$\cdot25$	$-\cdot25$	$\cdot50$	6
O (1)	$\cdot28$	$-\cdot28$	0	6
O (2)	$\cdot45_5$	$\cdot76$	$\cdot11$	12
O (3)	$\cdot03$	$\cdot03$	$\cdot68$	6
O (4)	$\cdot20$	$\cdot20$	$\cdot85$	6
O (5)	$\cdot37$	$\cdot37$	$\cdot72$	6

The chabazite framework can be built by stacking layers of hexagonal prisms, made of two parallel rings of six (Al, Si)O_4 tetrahedra linked by six oxygen bridges to form the prism. Every fourth layer is superposable upon the first. Gmelinite can be similarly regarded as having an aluminosilicate framework made from layers of the same hexagonal prisms, every third layer being superposable upon the first.[5] The levynite framework now proposed can, on the other hand, be

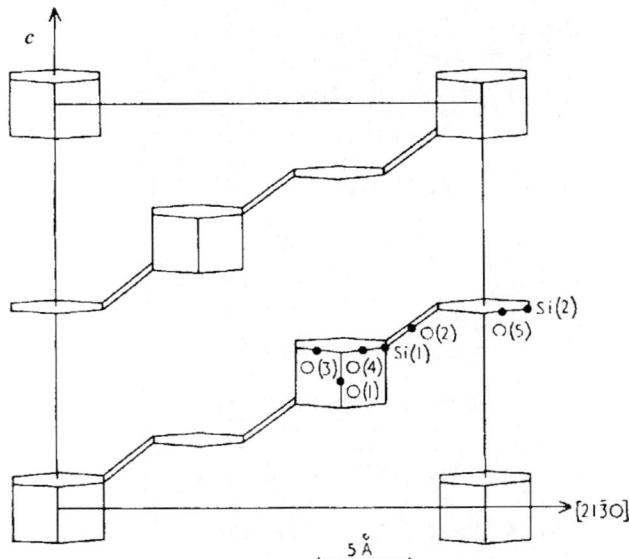

FIG. 1.—The sequence of layers in the proposed structure for levynite, showing alternation between hexagonal prisms and hexagonal rings.

considered to consist of alternate layers of the hexagonal prisms and of single six-membered rings, alternating according to the formal arrangement shown in fig. 1, every seventh layer being superposable upon the first. The whole structure in three dimensions forms a comparatively rigid network having notable porosity, with channels and cavities permeating the structure. Fig. 2 shows the nature of the frameworks produced in the three crystals. In these models, Al or Si atoms are at the corners of the hexagons. Framework oxygen atoms are not shown, but their centres are near, although not on, the mid-points of the edges joining (Al, Si) positions.

X-RAY EVIDENCE

The X-ray examination was carried out on a sample of levynite from Breiddalur, Iceland, and involved the taking of Weissenberg photographs for which a small crystal was rotated respectively about the c and the b hexagonal axes. The intensities of reflection within the c-zone were then found to be similar to those of

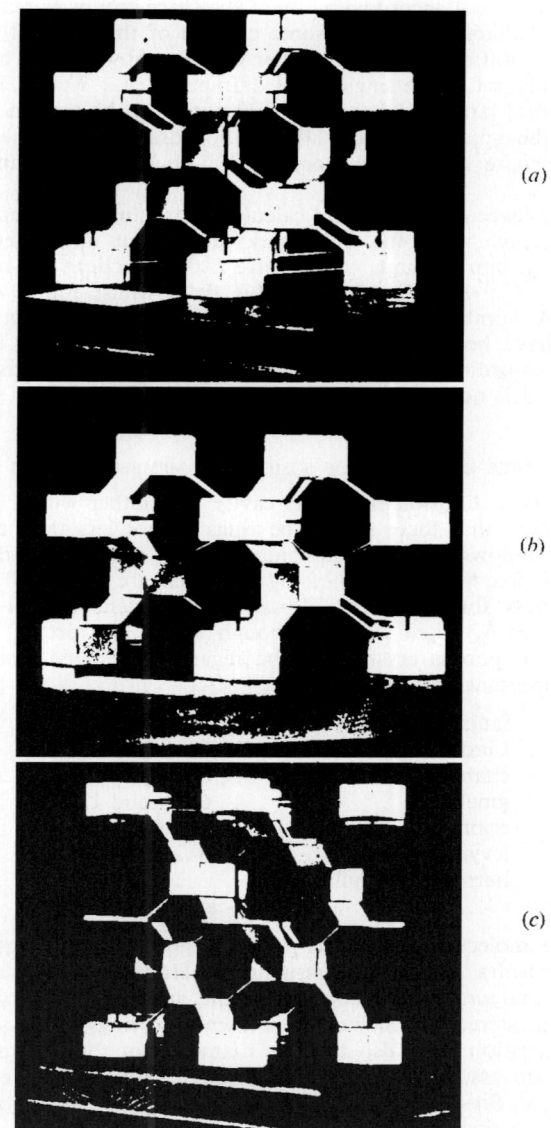

FIG. 2.—A comparison of the framework structures of (*a*) chabazite, (*b*) gmelinite, and (*c*) levynite. Al and Si atoms occupy every apex. Oxygen atoms are not shown.

chabazite, only a few of the diffractions showing minor differences.   This similarity is required by the proposed levynite structure, since the projections of the two frameworks on the *ab*-plane are practically identical.

The Weissenberg photographs obtained by rotation around the *b*-axis showed diffractions falling into two sets only, for which either $h + l$ or $h - l$ equals $3n$. This behaviour does not accord with any of the space groups but can be explained by analogy with chabazite. Thus, some crystals of this mineral were found to give both sets of diffractions, whilst other crystals show one to be the brighter, or indeed the only, set.   The single set of diffractions, $h + l = 3n$, is an indication of a rhombohedral lattice, and the mixed patterns can be explained as the result of twinning on the *c*-plane (i.e. the (111) rhombohedral plane).   These observations suggest that levynite is likewise rhombohedral, but that our sample is heavily twinned.

Agreement between observed and calculated structure amplitudes is moderate at the present stage when the framework atoms, but not water molecules or cations, are taken into account.   For twelve $hk0$ diffractions the reliability factor $R_{hk0} = \Sigma \mid F_0 - F_c \mid \div \Sigma \mid F_0 \mid$ has a value 0.40, whilst for 30 $h0l$ diffractions $R_{h0l} = 0.43$.   A number of configurations of the water molecules within the large cavities have been found to bring R_{hk0} to values below 0.2.   However, the possible Z co-ordinates and their effect on the $h0l$ diffractions are still being investigated, so it is not yet possible to give the arrangement for the best overall agreement.

FREE DIMENSIONS OF SOME EIGHT-MEMBERED RINGS

Ready molecule diffusion from one cavity to another within porous crystals appears to require that there should be rings of not less than eight (Al, Si)O_4 tetrahedra as windows leading from one cavity to others.   Six-membered rings have in general free * diameters (2.2-2.7 Å) which are too small to allow most molecules to pass through them.   Possibly, hydrogen, with a cross-sectional diameter of ~ 2.4 Å, might do so, although this is not proved with certainty. The structures of porous crystals can be regarded as made from linked rings, examples of important rings being:

faujasite	4-, 6- and 12-rings
Linde Sieve A	4-, 6- and 8-rings
chabazite	4-, 6- and 8-rings
gmelinite	4-, 6-, 8- and 12-rings
erionite	4-, 6- and 8-rings
levynite	4-, 6- and 8-rings
harmotome-Phillipsite zeolites	4- and 8-rings.

Intracrystalline molecule diffusion is primarily governed by rings of 8- and 12-(Al, Si)O_4 tetrahedra in all these structures.   However, these large rings may take various configurations, according to the structures in which they occur. Fig. 3 shows the stereochemistry of the 8-membered rings in the above structures in plan and elevation, and also the free diameters of these rings.   For all free diameters we are assuming regular (Al, Si)O_4 tetrahedra with edges of 2.7_2 Å, and therefore (Al, Si)—O bond distances of 1.66 Å.   The most symmetrically disposed of all the 8-rings is that in Linde Sieve A, with a free diameter of 4.3_6 Å. If the (Al, Si)—O bond distance is taken as 1.69 Å,[12] the free diameter is smaller, at 4.0_3 Å.[2]   All rings have free dimensions sufficient to allow certain molecules to pass through them ; but the free dimensions differ considerably, and accordingly so should the molecular sieve character.

* By " free " is meant not occupied even by the peripheries of the framework oxygen atoms.

Two additional conditions must be satisfied for these dimensions rigorously to determine molecular sieve action. First, when water is removed by heat and evacuation, the rings must keep their stereochemical configuration; and secondly, cations must not be so located as to block these rings. The first condition is approximately fulfilled; the latter is not in all cases. Thus, in aluminous, and therefore cation-rich, synthetic near-chabazites, sodium ions are so numerous, and so placed, as to prevent molecule diffusion, except of small polar molecules like water.[13]

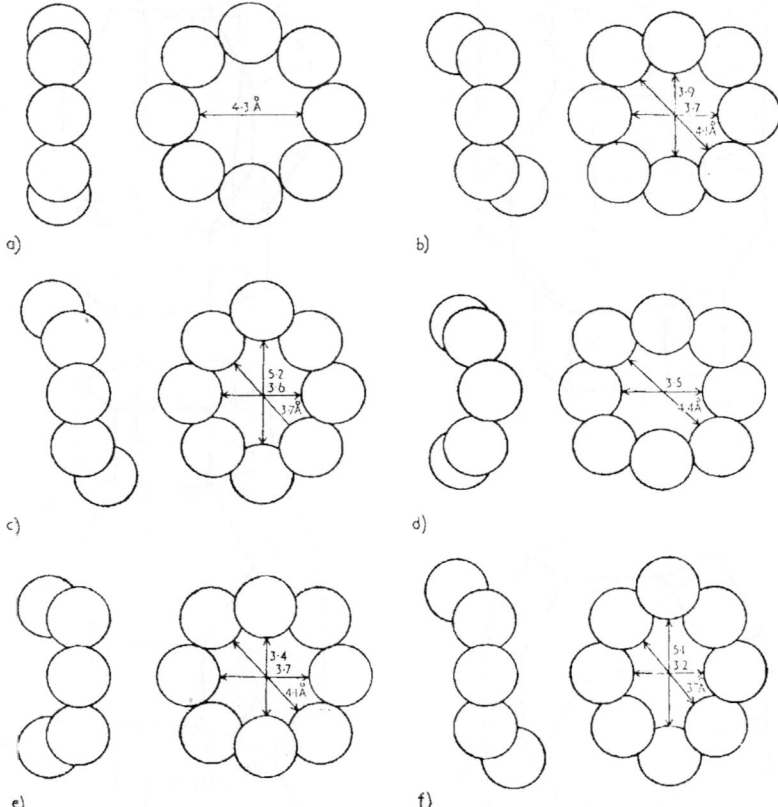

FIG. 3.—The stereochemistry and free dimensions of 8-membered rings in (*a*) Linde sieve A, (*b*) chabazite, (*c*) erionite, (*d*) cubic harmotome (Na—Pl), (*e*) gmelinite, and (*f*) levynite.

CHANNEL PATTERNS IN THE CHABAZITE GROUP OF ZEOLITES

Intracrystalline channel patterns have been described for analcite, sodalite-nosean minerals, cancrinite, faujasite and Linde Sieve A.[9, 10] It is these patterns which indicate the allowed diffusion paths within these structures. It is important that channels should interconnect to give a three-dimensional network. If they are all parallel as in basic cancrinite, or in sepiolite and attapulgite [14] it is possible through crystal defects to isolate a channel or group of channels and so inhibit or limit diffusion.[9, 10, 15] If, on the other hand, these channels form networks, the diffusing molecules can readily find paths around dislocations and stacking faults. We will now consider the channel patterns found in chabazite, gmelinite, erionite and levynite, in all of which large cavities of varying dimensions exist having the symmetries shown in fig. 4.

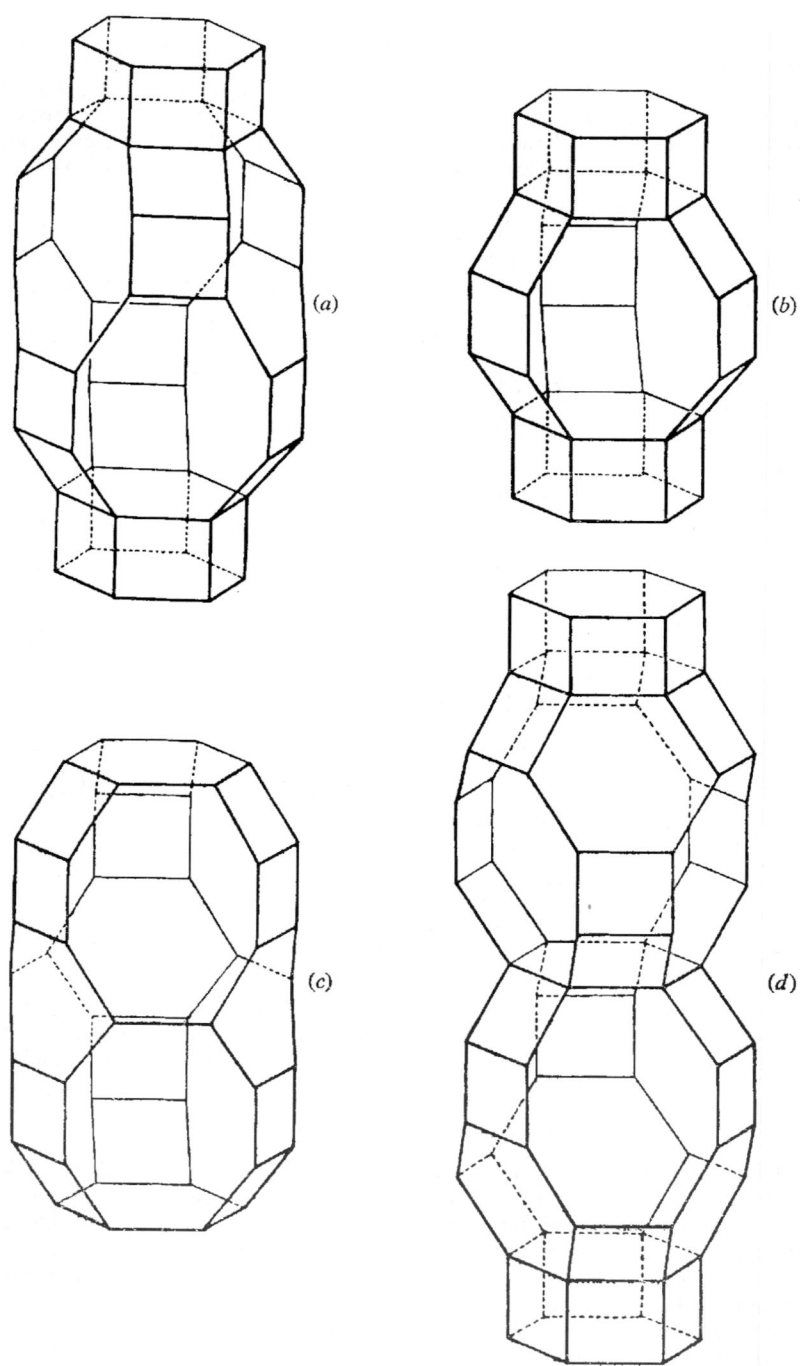

FIG. 4.—The cavities in (*a*) chabazite, (*b*) gmelinite, (*c*) erionite, and (*d*) levynite. In fig. 4*d*, two superposed cavities are shown. Al and Si atoms occupy each corner, but oxygen atoms are not indicated. All diagrams are drawn to the same scale.

CHABAZITE

In chabazite there are elongated cavities (fig. 4a), not wholly symmetrical and having major and minor free dimensions of ~ 11 Å and 6.6 Å. There is a slight waist in this cavity, and 6.6 Å refers to the free diameter of this waist. Forming part of the wall of each such cavity are six 8-membered rings, the minimum free dimension of which is 3.7 Å (fig. 3b). They are spatially distributed as are the six hydrogen atoms of ethane when the CH_3 groups are in positions of least energy. A window leads from any central cavity into one each of six other cavities, a window being common to two cavities. Thus the cavities are stacked with a co-ordination number of six. From this arrangement it can be seen that the actual pattern of channels referred to the hexagonal unit cell is that of fig. 5a. There are repeated

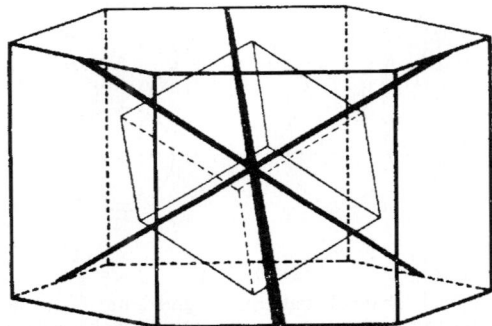

FIG. 5(a).—The orientation of diffusion channels in chabazite, in relation to hexagonal and rhombic unit cells.

crossing points, from each of which six channels diverge, and diffusion should be virtually isotropic. Each of the cavities is capped by a hexagonal prism at both ends. These hexagonal prisms, formed by six oxygen bridges which link two 6-membered rings, contain a roughly spheroidal free volume of diameter ~ 4.0 Å, which is approached through the 6-rings having free diameters of ~ 2.7 Å. However, there is no evidence that these play a significant part in molecule diffusion. As already known, chabazite occludes many species of critical dimensions up to those of *n*-paraffin chains.[16, 17]

GMELINITE

Gmelinite has a non-intersecting system of channels all running parallel to the hexagonal *c*-axis.[5] These channels are very wide, being circumscribed by 12-membered rings of free diameter about 6.4 Å. Diffusion in this channel system would be expected to occur very readily; in fact, however, the only sorption study [13] so far conducted with gmelinite indicated a behaviour like that of a rather sluggish chabazite. However, as noted previously, a non-intersecting system of parallel channels is always liable to blocking by stacking faults, while with intersecting channel systems, on the other hand, alternative diffusion paths around such blockages are available. Gmelinite does possess an intersecting channel system in planes normal to the *c*-axis, having the hexagonal pattern shown in fig. 5b. 8-Membered rings open from the wide channels into the cavities shown in fig. 4b which have free dimensions of ~ 7.8 Å normal to the *c*-direction and ~ 6.5 Å in this direction. Each cavity has then three such windows, and shares these as junctions with each of three of the wide channels parallel to the *c*-axis. Thus a hexagonal network of channels normal to this axis is produced, and this network is repeated in planes spaced at intervals of 5 Å along the *c*-direction and perpendicular to it, each third network being exactly superposable on the first. The free dimensions of the 8-ring windows are shown in fig. 3*e*. If there are stacking faults interrupting the wide channels the tempo of diffusion would depend upon the auxiliary hexagonal channel networks of fig. 5*b*, and as the smallest dimension of the 8-rings is about 3·4 Å the behaviour could well be that of a sluggish chabazite, as observed.[18] On the other hand, if structurally perfect crystallites were available, diffusion of quite large molecules parallel to the *c*-axis could occur.

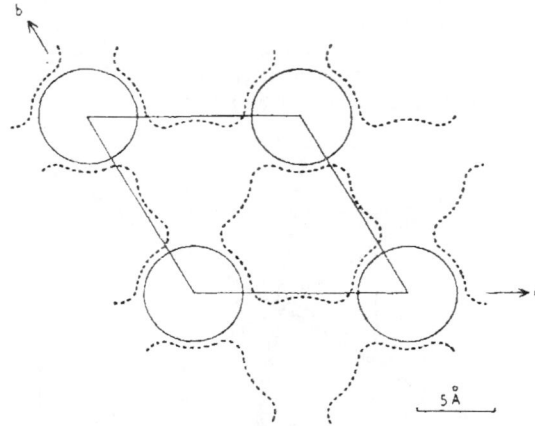

FIG. 5(*b*)—The hexagonal channel pattern in gmelinite in planes normal to the *c*-direction. Dashed lines give maximum free dimensions of channels and cavities, and full lines give minimum free diameter of channels in *c*-direction.

ERIONITE

In the erionite structure proposed by Staples and Gard,[8] columns of elongated cavities (free length ∼ 15·1 Å, free diameter ∼ 6·3 Å (fig. 4*c*)) lie along the *c*-direction of the hexagonal unit cell, one on top of another. As with chabazite there is a slight waist in the cavities, and the figure 6·3 Å gives the free dimension of this waist. Hexagonal networks are produced by their intersections with planes normal to the *c*-axis. There are also columns of a more compact type, likewise running in the *c*-direction. While this type of column is insufficiently open to permit molecule diffusion, the columns of open cavities allow such diffusion to occur throughout the structure.

Each of the elongated cavities is capped top and bottom by a 6-membered ring, which it shares respectively with the cavities just above and just below it. These common windows have free diameters of only ∼ 2·5 Å, and so as already noted molecule diffusion from one cavity to that immediately above or below it is unlikely. Six 8-membered rings form part of the wall of any such cavity. These rings have free dimensions of 5·2 Å and 3·6 Å (fig. 3*c*), the longer dimension being in the *c*-direction. They are arranged three in the upper half of the cavity, three in the lower, as required by the symmetry 3/*m*. Thus their spatial disposition is the same as for the six hydrogen atoms in ethane when the two methyl groups are rotated to the position of maximum energy.

Each window is common to two cavities, producing the packing of columns referred to above. However, the cavities in the three columns surrounding any given central column are displaced along the *c*-direction, relative to the cavities in the central column, by half the length of a cavity (7·5 Å). Thus, a molecule can diffuse along a cavity, through an 8-ring window into and along a cavity in one of the three adjacent columns, through another 8-ring window, and back into the next cavity in the original column. Accordingly, diffusion can proceed freely along the *c*-direction. However, a molecule could equally well diffuse along the hexagonal network of channels normal to the *c*-direction (fig. 5*c*), making use of the 8-ring windows to pass from one cavity to another. These hexagonal networks of channels lie in planes 7·5 Å apart but, in the manner noted above, molecules may freely pass from one plane to another. Diffusion has been observed to occur readily for atmospheric gases [19] in an outgassed synthetic near-erionite.*

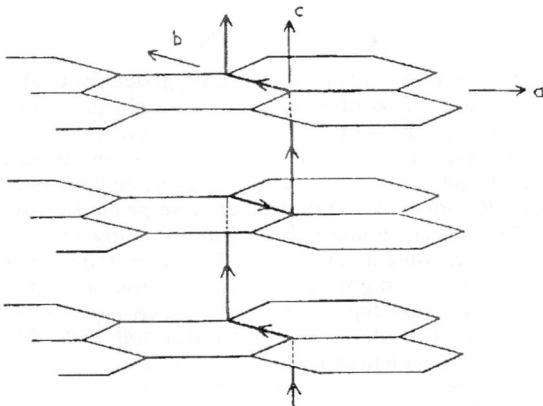

Fig. 5(*c*)—The arrangement of diffusion paths in erionite, showing the way in which molecules may move along the *c*-direction and normal to it.

LEVYNITE

The most important cavity (fig. 4*d*) in the suggested structure for levynite shown in fig. 1 and 2 is roughly spheroidal in shape with a free diameter of $\sim 7 \cdot 8$ Å for the inscribed sphere. A cavity is connected by a shared elliptical 8-ring window having major and minor axes of $\sim 5 \cdot 1$ Å and $3 \cdot 2$ Å (fig. 3*f*) with each of three

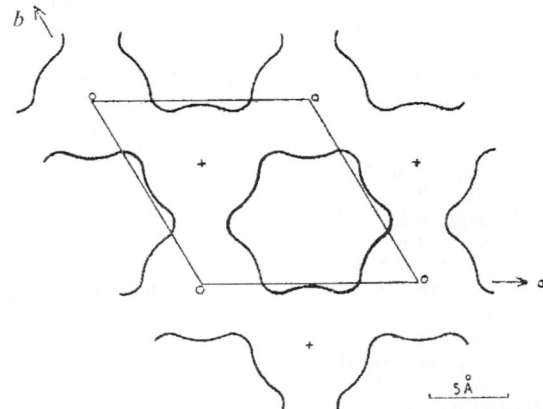

Fig. 5(*d*)—The hexagonal pattern of channels normal to the *c*-direction in levynite. ○ denotes a cavity centre in a reference plane, and ╪ a cavity centre above this plane. Thus, the network is puckered. The free dimensions are given by the line contour.

like cavities, so as to form the puckered hexagonal network of channels shown in fig. 5*d*. This network lies along the planes normal to the hexagonal *c*-axes, spaced at intervals of 7·5 Å along it. However, there is no diffusion path allowing passage of molecules from one such network to the ones just above or just below, except

* kindly supplied by Linde Air Products Co.

through 6-membered rings of free diameter $\sim 2 \cdot 7$ Å. Only one study of sorption by levynite has been made [20] the results of which accord with the foregoing description of the channel system and of the 8-ring windows. Thus at $- 186°C$ oxygen (critical dimension $\sim 2 \cdot 8$ Å) and nitrogen ($\sim 3 \cdot 0$ Å) are sorbed, but argon ($\sim 3 \cdot 8$ Å) is practically non-sorbed. This zeolite can effectively separate argon-oxygen mixtures.

DISCUSSION

It is seen that the structures which have been suggested for chabazite, gmelinite, erionite, and levynite, are not contradicted by the molecular sieve and diffusion behaviour so far observed, provided, in gmelinite, stacking faults interrupt the widest channels. If these structures can be refined and modified to describe the very slightly altered frameworks after the removal of zeolitic water, there seems no reason why delicate differences in molecular sieve properties should not be accurately described by the crystallographic method. Only in chabazite will diffusion be nearly isotropic, while in levynite it will be two-dimensional, in a series of planes normal to the *c*-axis. In gmelinite and in erionite, diffusion in such planes will differ from diffusion along the *c*-direction. Taken together with the channel patterns and structures of analcite, sodalite-nosean minerals, cancrinite, Linde Sieve A, and faujasite, it is evident that a diversity of very remarkable open cavities and of diffusion paths exists. The stereochemistry of zeolitic porous crystals is, however, still only partially explored.

The authors are indebted to Dr. J. A. Gard for an opportunity to examine the erionite structure proposed in ref. (8) prior to publication, and for helpful discussion. The levynite was kindly provided by Dr. G. P. L. Walker of this College.

[1] Reed and Breck, *J. Amer. Chem. Soc.*, 1956, **78**, 5972.
[2] Barrer and Meier, *Trans. Faraday Soc.*, 1958, **54**, 1074.
[3] Nowacki, *Experientia*, 1956, **12**, 418.
[4] Barrer, Bultitude and Sutherland, *Trans. Faraday Soc.*, 1957, **53**, 1111.
[5] Dent and Smith, *Nature*, 1958, **181**, 1795. Nowacki, Koyama and Mladeck, *Experientia*, 1958, **14**, 396.
[6] Barrer, Bultitude and Kerr, *J. Chem. Soc.*, 1959, 1521.
[7] Nowacki, Aellen and Koyama, *Schweiz. Mineralog. Petrograph. Mitt.*, 1958, **38**, 53.
[8] Staples and Gard, in press.
[9] Barrer, *Proc. Chem. Soc.*, 1958, 99.
[10] Barrer, Colston Papers, vol. X, *The Structure and Properties of Porous Materials* (Butterworths, 1958), p. 26.
[11] Strunz, *Neues Jb. Min., Mh.*, 1956, **11**, 250.
[12] Barrer and Baynham, *J. Chem. Soc.*, 1956, 2892.
[13] Smith, *Acta Cryst.*, 1954, **7**, 479.
[14] Bradley, *Amer. Min.*, 1940, **25**, 405. Nagy and Bradley, *Amer. Min.*, 1955, **40**, 885.
[15] Barrer and McKenzie, *J. Physic. Chem.*, 1954, **58**, 560.
[16] Barrer, *Quart. Rev.*, 1949, **3**, 293.
[17] Barrer, *Ann. Reports*, 1944, **41**, 31.
[18] Barrer, *Trans. Faraday Soc.*, 1944, **40**, 555.
[19] Barrer, unpublished measurements.
[20] Barrer, *Nature*, 1947, **159**, 508. *Trans. Faraday Soc.*, 1949, **45**, 358.

Catalysis

J. M. Thomas

*Department of Materials Science, University of Cambridge, UK CB2 3QZ and
The Royal Institution, 21 Albemarle Street, London, UK W1S 4BS*

Commentary on: **Studies of Cations in Zeolites: Adsorption of Carbon Monoxide; Formation of Ni Ions and Na$_4^{3+}$ Centres,** J. A. Rabo, C. L. Angell, P. H. Kasai and V. Schomaker, *Discuss. Faraday Soc*, 1966, **41**, 328-349.

The Discussion of the Faraday Society which I attended in Liverpool, April 1966, dealing with heterogeneous catalysis featured on its programme several of the funding fathers of the subject, notably E. K. Rideal, H. S. Taylor, G. K. Boreskov, J. Turkevish, D. D. Eley, S. Roginskii, G. M. Schwab, R. Burwell and G. C. Schuit; and it ranged freely between theory and experiment. At the Meeting, I was quite aroused by the presentation given by Jules Rabo from the Union Carbide Laboratory, not because I was at all interested prior to listening to him and reading his preprint in the topic of catalysis by zeolites—I had given hardly any thought to that subject beforehand—but because it dawned on me that the seemingly fiendishly complicated formulae and structures of zeolites were, perhaps, within my grasp. In truth, it was nearly a decade and a half later, following the successes that my electron microscopist colleagues and I achieved[1,2] in Cambridge in the early 1980s in atomically resolving the structures of zeolites, that I decided to enter deeply into physico-chemical (especially catalytic and structural) study of zeolites[3].

Repeatedly from the 1980s onwards, and especially when I was grappling with many of the then unknown aspects of zeolite science, I often consulted Rabo *et al.* The paper gave authoritative accounts of certain features pertaining to structure, adsorbability and reactivity of various kinds of cation-exchanged zeolites. It also highlighted some of the properties of these solids that were enigmatic or poorly understood.

For those whose interests lie far away from the physical chemistry of zeolites, here are some of the hard facts (in addition to those highlighted in the Abstract) that Rabo *et al.* revealed in their 1966 paper:

(1) Crystallographic proof of the existence of three well-defined sites S$_I$, S$_{II}$ and S$_{III}$ for the exchangeable cations (of the synthetic zeolite X and Y). Those two synthetic variants X and Y of the mineral faujasite (idealised formula {Na$_{48}$(Al$_{48}$Si$_{144}$O$_{384}$)}) in its dehydrated form for Si/Al ratio of 3) differ from one another only in their Si/Al ratios[4]. S$_I$ sites are at the centre of the so-called 'double six' units (i.e. hexagonal prisms). The so-called 'sodalite cages' are (Si,Al)$_{24}$O$_{36}$ truncated octahedra; and the S$_{II}$ sites are next to 6-oxygen rings, and the S$_{III}$ sites next to 4-oxygen rings (on the interior surfaces) of the sodalite cages.

(2) There are extremely large electrostatic fields inside X and Y zeolites, ranging from a low of 1.5 to a high of 6.3 V Å^{-1}, and this (as the paper outlines) plays a considerable role in favouring the formation and stability of carbonium ions (carbocations) inside the zeolites, thereby facilitating catalytic activity (in, for example, the thermal cracking of cumene to benzene and propene).

(3) Molecules of carbon monoxide are readily adsorbed on the exchangeable (extra zeolitic) cations, with IR frequencies specific to the cation.

(4) Alkali metal ion exchanged variants of both zeolites X and Y react with Na vapour to form cluster cations such as Na$_4^{3+}$ and Na$_6^{5+}$, with the production of intensely coloured products (reminiscent of the colour (F-) centres in alkali halides).

(5) Unusual ions (apart from Na_x^{x-1} clusters) such as the O_2^- radical are readily stabilised inside the zeolitic framework. And nickel in the +I oxidation state may be generated by reduction of $Ni^{(II)}$ species with H_2. In turn, strong adsorption of oxygen may follow owing to the reaction:

$$Ni^{(I)}(S_{II}) + O_2 \rightarrow [O_2^-Ni^{(II)}(S_{II})]$$

the complex on the right being more stable than $Ni^{(I)}(S_{II})$ itself.

In the body of their paper, Rabo *et al.* raise many relevant speculations, some of which I shall now briefly discuss because subsequent work has converted their speculation into hard fact.

(a) Thus, their discussion of the structural framework of zeolites X and Y prompts them to ponder on *the ordering of Al relative to Si.* For a long time after their work was reported there was little prospect of addressing the problem of Si,Al ordering in the connected (open-framework) SiO_4/AlO_4 tetrahedra. Fortunately for me, as I was a member of staff at the University College of North Wales, Bangor, I had become aware of the enormous potential of magic angle spinning NMR (MASNMR), which was invented by E. R. Andrew at Bangor[5]. In due course, following the advent of Fourier transform NMR methods, it became possible to determine directly the intricacies of Si,Al ordering (non-invasively) from high-resolution solid state ^{29}Si and ^{27}Al NMR in zeolites X and Y. Moreover, the complementary insights of computational studies shed further light on the phenomenon[7].

(b) In a prescient remark, dealing indirectly with *the intrinsic Brønsted acidity of* so-called tervalent-cation *zeolites* (e.g. $Ce^{(III)}Y$ or $La^{(III)}$-Y), Rabo *et al* state *"we expect that the effective electrostatic field of tervalent ions may always become strong enough on dehydration to cause hydrolysis, $M^{(III)}Y + H_2O \rightarrow (OH-M^{(III)})H^{(I)}Y"$.* Such a statement led others to speculate that the intrinsic Brønsted acidity of even divalent cation-exchanged zeolite Y could generate protons that were loosely attached to the oxidic framework (designated $H^+(z)$) of the zeolite, e.g.

$$Ca^{2+} + H_2O \leftrightarrow (CaOH)^+ + H^+(Z)$$

Convincing proof that such a process does indeed occur came from neutron diffraction measurements[8] (on deuteriated specimens of La–Y that Cheetham and I made in the 1980s).

(c) Rabo *et al.* also said: *"It is also conceivable that tervalent (and divalent) ions when fully dehydrated may shift to new positions".* Using time-resolved *in situ* X-ray powder diffractometry it has been possible to follow the precise course and destination of various ions (but especially of $Ni^{(II)}$ ions) in the cavities of zeolite Y both during dehydration[9] and during exposure to certain gases (such as C_2H_2 which, when the Ni ion progresses finally to the S_{II} site from its original S_I site, catalyses[10] the trimerisation of the acetylene to benzene).

(d) *Sodium vapour reduces NaY and NaX and forms the complexes Na_4^{3+} and Na_6^{5+}.* The multiplicity of the ESR spectrum obtained when NaY was exposed to Na vapour at 580°C leaves no doubt that Na_4^{3+} ions are formed. This remarkable fact first led Barrer and Cole[11] to confirm the veracity of Rabo *et al*'s report; and, in 1982, my then colleague (in Cambridge) Dr Peter Edwards and I began[12] a systematic study of the structural chemistry of this (and other analogue alkali metal cluster ions). Numerous other techniques (used largely by Edwards and his colleagues) were applied to the structural chemistry of these novel nanoparticle cations (see Fig. 1); and the results have been the subject of a recent summarizing review.[13]

In conclusion, I feel it important to state that, even though Rabo *et al.*'s paper was not the one in the Liverpool Discussion of the Faraday Society that had attracted me most to the Meeting, it profoundly (but much later) influenced a good proportion of my subsequent scientific career.

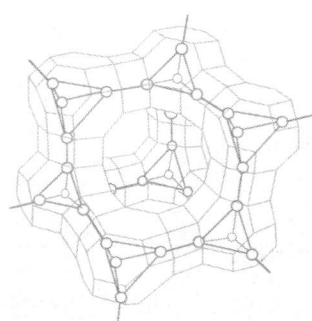

Fig. 1 Representation of an array of interacting Na_4^{3+} clusters, located in the sodalite cages of zeolite Na^+Y. Na_4^{3+} ions are formed within the zeolite according to the reaction:

$$M^0 + 4Na^+(zeolite) \rightarrow M^+ + Na_4^{3+}$$

where M^0 is the incoming alkali atom (either Na or K) and $Na^+_{(zeolite)}$ represent alkali cations already present in the host material. (Reproduced with permission from ref. 13).

References[†]

1 L. A. Bursill, E. A. Lodge, J. M. Thomas, Zeolitic structures as revealed by high-resolution electron microscopy, *Nature,* 1980, **286**, 111.
2 J. M. Thomas, G. R. Millward and L. A. Bursill, The ultrastructure of carbons, catalytically active graphite compounds and zeolitic catalysts, *Philos. Trans. R. Soc. London, Ser. A,* 1981, **300**, 43.
3 J. M. Thomas, G. R. Millward, S. Ramdas, L. A. Bursill and M. Audier, New methods for the structural characterisation of shape-selective zeolites, *Faraday Discuss.,* 1981, **72**, 345.
4 The faujasite framework is an open, negatively charged, complex ring system of composition $(Si,Al)O_2$ in which SiO_4 and AlO_4 tetrahedra are linked together by shared oxygens. Fig. 1 of the Rabo *et al.* article is more illuminating when drawn in colour.
5 E. R. Andrew, A. Bradbury and R. G. Eades. *Nature,* 1958, **182**, 1659.
6 C. P. Slichter, *Principles of Magnetic Resonance*, Springer Verlag, Berlin and New York, 1980.
7 J. Klinowski, J. M. Thomas, S. Ramdas, C. A. Fyfe and J. S. Hartman, A re-examination of Si,Al ordering in zeolite NaX and NaY, *J. Chem. Soc. Faraday Trans. 2,* 1982, **78**, 1025.
8 A. K. Cheetham, M. M. Eddy and J. M. Thomas, The direct observation of cation hydrolysis in lanthanium-zeolite Y by neutron diffraction, *J. Chem. Soc. Chem. Commun.,* 1984, 1337.
9 J. M. Thomas, C. Williams and T. Rayment, Monitoring cation-site occupancy of nickel-exchanged zeolite Y catalysts by high-temperature *in situ* X-ray powder diffractometry, *J. Chem. Soc. Faraday Trans. 2,* 1988, **84**, 2915.
10 P. J. Maddox, J. Stachurski and J. M. Thomas, Probing structural changes during the onset of catalytic activity by *in situ* diffractometry, *Catal. Lett.,* 1988, **1**, 191.
11 R. M. Barrer and J. M. Cole, *J. Phys. Chem. Solids,* **29**, 1755 (1968)
12 P. P. Edwards, M. R. Harrison, J. Klinowski, S. Ramdas, J. M. Thomas, D. C. Johnson and C. J. Page, Ionic and metallic clusters in zeolites, *J. Chem. Soc., Chem. Commun.,* 1984, 982.
13 P. P. Edwards, P. A. Anderson and J. M. Thomas, Dissolved alkali metals in zeolites. *Acc. Chem. Res.,* 1996, **29**, 23.

[†]To highlight the influence of Rabo *et al.*'s paper, those papers published by the author are given their titles.

Studies of Cations in Zeolites : Adsorption of Carbon Monoxide ; Formation of Ni ions and Na_4^{3+} centres

By J. A. Rabo, C. L. Angell, P. H. Kasai and Verner Schomaker *

Union Carbide Research Institute, Tarrytown, N.Y., U.S.A.

Received 31st December, 1965

Zeolite cations lying on the intracrystalline pore surface of the Linde X and Y molecular sieves are linked to only three oxide ions and consequently are not well shielded electrically. They therefore create very large electrostatic fields, extending into the main zeolitic cavities, causing carbonio-genic catalytic activity. This activity follows in magnitude the changes in the field, and is independent of the presence of OH groups ; no change in activity is observed even after 99 % of OH protons present after ordinary activation in vacuum at 500°C are removed. Bivalent cations exposed on the surface of the main intracrystalline cavities adsorb carbon monoxide with infra-red frequencies specific to cation ; the adsorption follows Langmuir isotherms, suggesting that a single carbon monoxide molecule can be independently attached to every surface cation. Certain transition-metal zeolites adsorb carbon monoxide much more tenaciously than alkaline earth cations, indicating a highly significant difference in their ability to form co-ordination bonds.

Univalent nickel ions can be prepared both on the intracrystalline surface and at fully co-ordinated positions by heating $Ni^{II}Y$ with alkali metal vapour. The surface Ni^I ions are chemically very reactive but thermally unstable ; those at fully co-ordinated positions are thermally stable even at 400°C, and are inert to H_2, NH_3, and CO although they react with oxygen to form O_2^-.

The alkali-metal X and Y zeolites react with alkali metal vapour to form coloured products of non-stoichiometric compositions containing the paramagnetic centres Na_6^{5+} and Na_4^{3+}. The Na_4^{3+} centres are stable up to 500°C and they react reversibly with gases ; with oxygen they form O_2^- radicals.

Many anhydrous zeolites are adsorbents, having an enormous intracrystalline surface associated with a homogeneous system of pores and cavities. Synthetic zeolites Linde Molecular Sieve Type X and Linde Molecular Sieve Type Y can be made to show [1, 3] a wide range of carboniogenic catalytic activities—i.e., activities apparently due to the formation of carbonium-ion-like intermediates—depending on appropriate selection of Si/Al ratio and charge and size of zeolitic cation introduced by ion exchange into the sodium-containing, as-synthesized materials. The carboniogenic activity increases with increase in cation charge and with decrease in cation size and has therefore been attributed to the unusually strong accessible electrostatic fields of the fraction of the cations that lie on the intracrystalline surface and, in essence, to polarization of the substrate molecules in these fields.

We briefly review in this paper the existing knowledge of the structures of the X and Y sieves, recapitulate the arguments on their catalytic activity, and indicate how we were led to predict or recognize a number of remarkable new properties, all intimately involving the zeolitic cations, including strong characteristic gas adsorption and special chemical activities, e.g., special susceptibilities toward reduction. Each bivalent surface cation seems to be able to bind a single carbon monoxide molecule. The resulting shift in the C—O stretching frequency depends on the cation and can be used to follow the distribution of cations between "surface" and "hidden" sites. The unusual reducibility of surface cations is

---

* present address : Department of Chemistry, University of Washington, Seattle, Washington, U.S.A.

illustrated by the reduction of Ni^{II} ions in Ni^{II}-exchanged Y-zeolite to Ni^{I} by added alkali metal, but hidden Ni^{II} can also be reduced to Ni^{I}. Another surface effect is shown when alkali-metal zeolites react with alkali metal to form products that contain complex ionic centres of unusual composition, e.g., Na_4^{3+}, that in the zeolite nevertheless have a high degree of thermal stability. These complex surface centres also co-ordinate various gases strongly.

Throughout the course of our work on the adsorption effects, catalytic properties, and general chemistry of these zeolitic surfaces we have profited by depending on the ionic model and simple physical principles for guidance and inspiration. We therefore believe it is appropriate to follow this line of thought in describing the results in this paper, even though it is insufficient to deal accurately and comprehensively with our systems and the phenomena.

STRUCTURAL FRAMEWORK

Despite differences in origin, composition, and chemical and physical properties, the mineral faujasite and the synthetic zeolites of interest in this paper—Linde Molecular Sieve Type X and Linde Molecular Sieve Type Y—are related in crystal structure to the extent that the basic framework is the same for all three. Their structures have been described in their general aspects [1-4] but are all too little known in their detail. Because the basic framework of these materials is primarily involved in the variations in the properties that are here our principal concern, namely, the net electric fields, valences, and electron affinities of the zeolitic (exchangeable) cations, its relevant aspects will now be reviewed. We must emphasize, however, that the only anhydrous material that has been studied by X-ray diffraction is single-crystal faujasite that had been repeatedly exchanged with calcium ion and then dehydrated. We of course cannot be sure of the extent to which we are justified in applying the resulting information to the synthetic materials. Any speculation on fine details of the structures, for example on the ordering of Al relative to Si, immediately leads one to recognize many probable differences among faujasite, X and Y. As a probable result of the chemical as well as structural differences there are differences in certain properties, both micro and macroscopic, between the Molecular Sieves X, Y and the faujasite. The absorption capacity of X and Y is greater than the faujasite,[5a] the Y adsorbs a " plug gauging " compound (triethylamine) which is excluded from X, and the Y has a far superior steam stability than the X.[5b]

In any event, the faujasite framework is a highly rigid, notably open, negatively charged, complex ring system of composition $(Si,Al)O_2$ in which SiO_4 and AlO_4 tetrahedra are linked together by shared oxygen ions. The stainless-steel and plastic-spaghetti model pictured in fig. 1 shows that the framework can also be regarded in terms of larger units linked together by oxygens. Such larger units, e.g., are the $(Si,Al)_{12}O_{18}$ hexagonal prisms and the $(Si,Al)_{24}O_{36}$ truncated octahedra (" sodalite baskets "). Also shown in fig. 1 are the three possible cation positions of greatest interest here: S_I, S_{II} and S_{III}. Site S_I is hidden from the zeolite surface, being intimately surrounded by the ions of the framework, whereas S_{II} and S_{III} are on (or in) the zeolitic surface, the S_{II} next to 6-oxygen rings on the three-fold cubic axes and the S_{III} next to 4-oxygen rings on four-fold inversion axes. An X-ray study [5] of a dehydrated Ca^{II}-exchanged single crystal of faujasite (Si/Al $\sim 2\cdot5$) revealed that sites S_I (16 per unit cell) and S_{II} (32 per unit cell) are occupied by the calcium ions (and any remaining original zeolitic cations). The average population deduced for S_I was about 1 and for S_{II} about $\frac{1}{2}$: the calcium ions seem to strongly prefer S_I over S_{II}. A cation at S_{II}, lying on an almost planar, scalloped O_6 ring, has only three-fold oxygen-ion co-ordination (Ca—O $= 2\cdot34$ Å), whereas in S_I, positioned between two puckered O_6 rings, it has six-fold octahedral co-ordination to oxygen ions (Ca—O $= 2\cdot42$ Å). At S_{II}, calcium also has three significant next-nearest oxygen neighbours (Ca—O $= 2\cdot97$ Å).

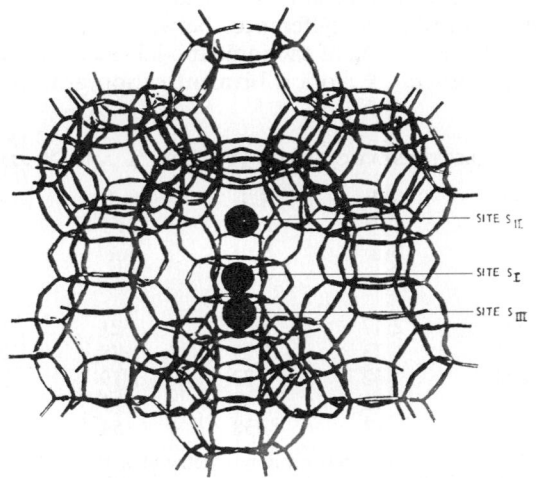

SITE S_{II}
SITE S_I
SITE S_{III}

FIG. 1.—Cation positions in faujasite.

The essential difference between X and Y and between the charge distributions around S_I and S_{II} then follow. In our discussion Type Y is understood to be NaY. Exchanged forms in which sodium is exchanged with other cations are designated here as CaY, ZnY, etc. Consider a CaY with Si/Al = 2. A calcium ion at S_I is surrounded by 12 tetrahedra, four of which, on the average, contain aluminium and bear a net negative charge. This calcium ion can be said to balance two of the four negative charges, leaving two of the charges—half a charge per AlO_4—to be balanced by the zeolite cations in S_{II}. In S_{II} the zeolitic cation is surrounded on one side by six tetrahedra, on the average two AlO_4 and four SiO_4, having an average net total of one negative charge. All the tetrahedra are equivalently related to S_I and S_{II} sites. Consequently, half of the S_{II} have a full positive net charge centred on the cation, while the other half have a full negative net charge distributed on the oxygen ions. By the same argument, the S_{II} sites in the CaX zeolite with Si/Al = 1 would be effectively neutral : the bivalent cation at S_I would face six AlO_4 groups, each of which would have a net negative charge of 2/3, whereupon each S_{II} with its three AlO_4 would have an effective negative charge of $3 \times 2/3 = 2$. The same conclusions may be reached in another way : CaY has 32 calcium ions per unit cell, 16 at S_I and 16 at S_{II}; if the 32 units of balancing charge for the S_{II} cations are considered to be localized on the 32 S_{II} sites, here regarded as equivalent, each must have one unit of charge, half enough to neutralize a calcium ion. On the other hand, a CaX with Si/Al = 1 has 48 calcium ions per unit cell, of which 32 match the charges of the 32 S_{II} sites they occupy.

Quantitative indications on the strength of S_{II} cations were obtained by Pickert *et al.*,[1] who calculated the electrostatic field along the three-fold cubic axis in the zeolite cavity near a surface cation at S_{II} for a particular fully ionic model at a Y zeolite with Si/Al = 2. Further calculations were later made by E. Dempsey *

* present address : Socony Mobile Research Laboratory, Princeton, N.J., U.S.A.

J. A. RABO, C. L. ANGELL, P. H. KASAI AND V. SCHOMAKER   331

on the same assumptions for a X zeolite of Si/Al ratio 1·0 and for positions in a cavity along a four-fold axis through a cation at S_{III}. The fields (table 1) are larger (i) in the Y than in the corresponding X, (ii) with bivalent than with univalent cations, and (iii) near S_{III} than near S_{II}, provided that both are occupied. The electrostatic shielding by the framework is unusually small at S_{II} and S_{III}, although still great enough to account for substantial differences between S_{II} and S_{III} and between X and Y. Yet, despite imperfections of the ionic model the actual fields in the cavities at distances as great as several Å from a bivalent cation are probably still as great as 1 V/Å.

TABLE 1.—ELECTROSTATIC FIELDS (V/Å) IN X AND Y ZEOLITES [a]

cation charge	+1	+1	+1	+1	+2	+2
Si/Al	1 : 1	1 : 1	2 : 1	2 : 1	1 : 1	2 : 1
site	S_{II}	S_{III}	S_{II}	S_{III}	S_{II}	S_{II}
weight [b]	4	6	4	2	4	2
r, Å						
2·0	1·52	2·47	2·30	3·21	5·65	6·30
2·5	·61	1·23	1·27	1·88	3·22	3·85
3·0	·23	·62	·77	1·19	1·97	2·51
3·5	·08	·31	·52	·79	1·14	1·73
4·0	·02	·12	·38	·54	·69	1·26

[a] calculated by E. Dempsey for positions on a symmetry axis at the distances *r* from the centre of a cation S_{II} (see text) for ideally ionic, fully exchanged models of the X and Y zeolites (unpublished research at Union Carbide Research Institute).
[b] number of equivalent occupied sites per $(Si/Al)_{24}O_{36}$ basket.

Around every cavity there are four tetrahedrally arranged S_{II} and six octahedrally arranged S_{III}. Since the S_I and S_{II} are more than sufficient to accommodate the bivalent cations the less attractive S_{III} sites are probably only populated in the univalent-cation form of the zeolite.

It is thus evident from the structure and the electrostatic field calculations that the order of preference for strong cations, such as the alkali and alkaline-earth cations, is $S_I > S_{II} > S_{III}$. This also suggests that bivalent ions will replace univalent ions at the most preferred sites and that the higher the valence of a cation at a surface site the higher will be both the electron affinity of the cation and the field near the cation. And the stronger the field, the greater will be the polarization of adsorbed molecules and the tendency for reduction of the cation. Some of these questions relate more naturally to the electrostatic potential at a cation site than to the electrostatic field at points near the site. Nevertheless, we have so far dealt with the field, because it is the more important by far for our concept of the carboniogenic activity of the surface cations in catalysis.

CATION-SPECIFIC ADSORPTION OF CARBON MONOXIDE

Carbon monoxide adsorbed on bivalent-cation X and Y zeolites shows a cation-specific infra-red stretching band as well as two bands that are cation non-specific to a remarkable degree, occurring unchanged or almost unchanged also with univalent-cation and decationized preparations.[6] The specific bands do not appear if even a small amount of water is present, presumably because the cations bind H_2O more strongly than CO.

The non-specific bands were shown to correspond to adsorption at two different kinds of ubiquitous sites, but these sites were not further characterized. The specific bands, however, were shown to have possibilities for the study of the cation–CO

complexes to which they are due and of the cation distribution and cation environment in the zeolite. For example, the shift of the band from the gas frequency
was shown to be directly proportional, for a wide selection of closed-shell bivalent
cations, to the calculated field at the expected position of the carbon atom in the
M—C—O complex. For the bivalent transition-metal cations Fe^{II}, Co^{II} and Ni^{II},
on the other hand, the shifts scattered rather widely from the shift against field curve,
as if not only a simple charge-polarizability bonding but also special covalent or
ligand-field bonding were involved. The specific band appeared even when the degree
of exchange of added ion for original sodium ion was less than enough to fill all the S_I,
showing that these cations did not realize the full expected preference for the hidden
site. Finally, the specific band is harder to pump away than the non-specific and
harder to pump away at low temperature than at room temperature.

We have now determined the pressure dependence of the peak optical densities
of the specific bands for vacuum-activated samples of CaY, ZnY, BaY, CoY, NiY
and CaX. We have also measured the total gas adsorptions for CaY, ZnY, BaY
and NaY and have made use of earlier data on CO adsorption by CaX,[7] NaY,[8]
CoY [8] and NiY.[8]

The experimental extinctions (mg^{-1} cm^2) and gas adsorptions (s.t.p. ml g^{-1}) are
shown in fig. 2, 3 and 4 for some of the samples, together with curves calculated
for Langmuir specific adsorption plus linear non-specific adsorption. Table 2 lists
the compositions, Langmuir and non-specific adsorption constants, and some further
data for all the samples. The notation is

$$D = D_0 n_0 Kp/(1 + Kp) \qquad (1)$$

for the extinction curves, and

$$n = (n_0 Kp/(1 + Kp)) + Cp \qquad (2)$$

for the gas-adsorption curves, where K is the Langmuir combination constant, D
the peak intensity of the specific band, n the amount of CO adsorbed, and C the
slope of the non-specific adsorption.

TABLE 2.—ADSORPTION OF CARBON MONOXIDE ON SEVERAL BIVALENT-CATION ZEOLITES,
SAMPLE COMPOSITIONS, LANGMUIR COMBINING CONSTANTS K AND NUMBERS n_0 OF
EFFECTIVE SITES, NON-SPECIFIC ADSORPTION COEFFICIENTS C, AND
RELATIVE MOLECULAR OPTICAL DENSITIES D_0

zeolite	exchange %	K mm^{-1}	C [a] cm^3/g mm	D_0 [b]	n_0 cm^3/g	n_0 sites/unit cell	M^{II} in S_{II} [c]
CaY	84	0·35	0·0144	0·148	13·7	7·6	8
ZnY	66	·50	·0142	·152	4·81	2·8	3
BaY	80	·00308	·0130	·113	24·6	16·4	6·8
CoY	78	103	·0135	·612	2·36	1·6	6·2
NiY	76	18·7	·0178	·380	3·86	2·7	5·6
CaX	98	0·108	·0558	·370	12·7	8·0	26·6

[a] for the NaY sample, C was found to be 0·0138 ml/g mm.
[b] with D the optical density per mg of a disc sample about 1 cm^2 in cross section.
[c] assuming 16 M^{II} in S_I and the remainder in S_{II}.

All the extinction data fit the Langmuir curves (eqn. (1)) well or fairly well, implying adsorption on approximately equivalent, non-interacting or only weakly interacting sites. The combining constants are generally consistent with the frequency
shifts, for the closed-shell ions, but for Ni^{II} and Co^{II} they are relatively very much
stronger and their order is inverted. When allowance is made for a non-specific

J. A. RABO, C. L. ANGELL, P. H. KASAI AND V. SCHOMAKER   333

adsorption proportional to the CO pressure and nearly the same for all the Y samples including the NaY, the gas adsorption data also follow Langmuir curves (eqn. (2)) with the same respective combining constants as were determined from the specific extinctions. The CaX adsorption does not fit very well and the CoY and NiY adsorptions do not extend to low enough pressures to afford more than a purely

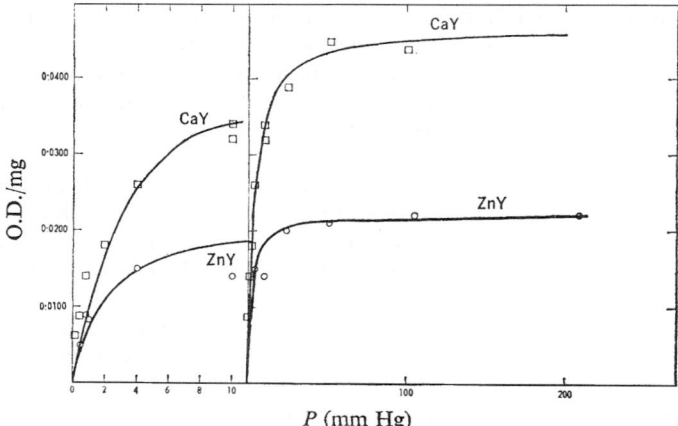

FIG. 2.—Peak intensity against pressure for the specific infra-red stretching band of CO adsorbed on CaY and ZnY. The solid lines are calculated from eqn. (1). □ and ○ are experimental values.

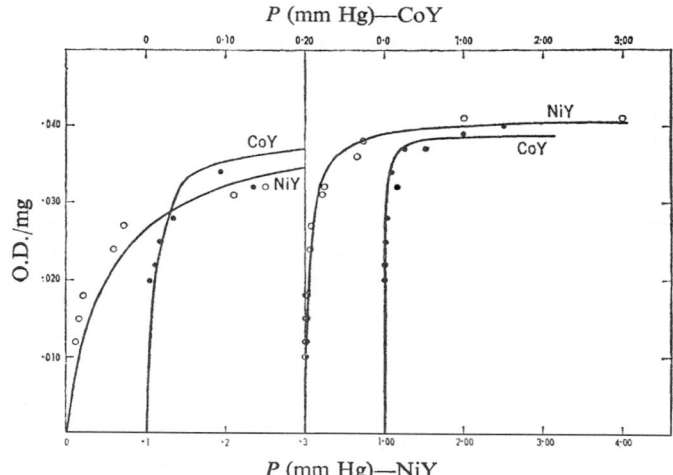

FIG. 3.—Peak intensity against pressure for the specific stretching band of CO adsorbed on CoY and NiY. The solid lines are calculated from eqn. (1). ○ and ● are experimental values.

qualitative check on the combining constants. Finally, the two kinds of data together afford relative determinations of the molecular extinction coefficients of the $M^{II}CO$ complexes and of the effective numbers of adsorption sites. Except for CaX the extinction values take the same order as the combining constants.

The numbers of effective sites all correspond at least roughly with the degrees of exchange and the postulated structures, i.e., with the assumption that all M^{II} ions in excess of 16 per unit cell go into S_{II}. The correspondence is good for ZnY

334   STUDIES OF ZEOLITE CATIONS

and for CaY, which is the only case for which there is independent comparable
structural information, viz., Dodge's study of Ca^{II}-exchanged faujasite. For BaY
the number of effective sites appears to be much greater than the number of Ba^{II}
ions in excess of 16 per unit cell, but this was hardly unexpected, since the large Ba^{II}
radius might well keep barium ion from penetrating the sodalite basket and, thence,
the hexagonal prism that encloses S$_I$.   For CaX, NiY and CoY the reverse is true:
either the numbers of bivalent surface ions in these zeolites disagree with our present
expectations or the old gas-adsorption results are for some reason in error.

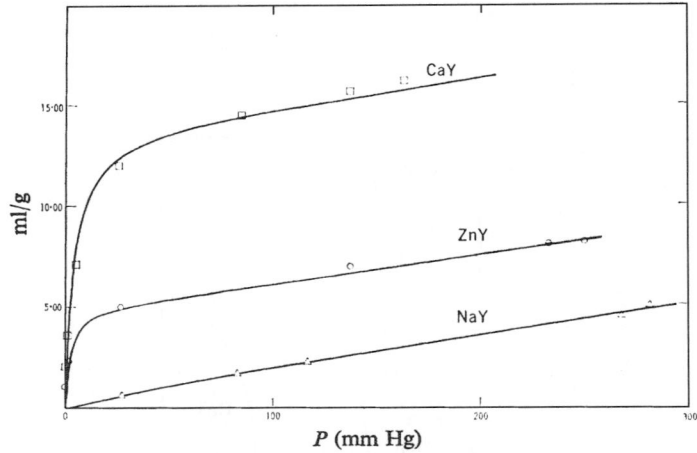

Fig. 4.—Pressure dependence of the adsorption of carbon monoxide on NaY, ZnY and CaY.
The solid lines are calculated from eqn. (2). □, ○, and △ are experimental values.

REMOVAL OF OH GROUPS IN ZEOLITE CATALYSTS

It has been suggested [3, 1] that the large electrostatic field near S$_{II}$ cations in the
zeolite cavities is responsible for the carboniogenic catalytic activity of bi- and
ter-valent-cation X and Y zeolites. As outlined above, this field is greater for
multivalent cations than for univalent cations and greater for a Y zeolite than for
an X zeolite, all in agreement with the observed differences in catalytic activity.
Unfortunately, all the X and Y preparations cited showed weight-loss on ignition
corresponding to 0·1-1·0 wt. % of OH groups,[1] which raises the question of whether
the hydroxyl protons were significant for the carboniogenic activity.   The OH
groups in zeolites have been studied extensively,[9-11] and it has been established that
Y zeolites activated by heating in vacuum to 550°C usually contain three distinct
structural OH groups with infra-red stretching frequencies of 3745, 3640 and 3540
cm^{-1}.   The small 3745 cm^{-1} band is chemically inert: it is not affected by adding
water or benzene, and its intensity and frequency are independent of the cation.
The other two, much stronger bands *are* shifted both by water and benzene, but
their frequencies are only slightly affected by change in cation.[9]

To check on the possible significance of the protons we have attempted by heat
treatment to remove the OH groups from samples of CaY and Ce^{III}Y using the
3640 and 3540 cm^{-1} bands to measure OH removal.   In one series of experiments,
the processed pellets of the zeolite samples were heated to the desired activation
temperature in vacuum in a specially designed quartz cell and subsequently transferred within the same cell into a 1 mm cavity between two NaCl windows.   This
process was repeated after every heat treatment as long as any of the OH bands,

with the exception of the chemically inactive 3745 cm⁻¹ band, remained. Fig. 5 shows that the OH groups persist up to about 650°C but are removed to less than about 1 % of their former number between 650 and 710°C. Samples of each of the zeolites so heat-treated were also exposed to water vapour at room temperature and then activated in vacuum at 500°C; spectra were taken after every step. The added water did not restore the 3640 and the 3540 cm⁻¹ bands, and it can be readily pumped off at 500°C. All these samples were also X-rayed before and after heat treatment, but no changes could be seen in the diffractometer traces.

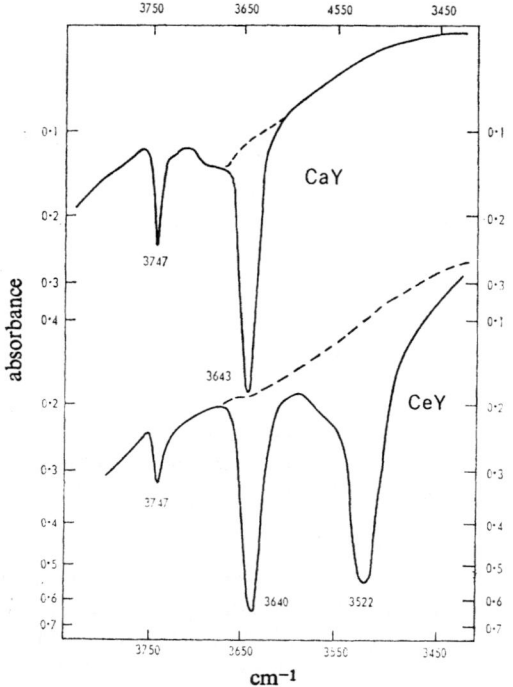

Fɪɢ. 5.—Infra-red spectra, OH region, of CaY and Ce^{III}Y.
———— activated at 500°C;   — — — — activated at 700°C

Two samples of each of three zeolites (NaY, CaY and Ce^{III}Y), one activated at the standard 550°C and the other (with the exception of NaY) at 700°C, were tested for cumene cracking in a 5-cm³ reactor attached to a gas chromatograph. One-hundred-microgram samples of cumene were injected into the stream of helium carrier gas. In this system [12] at low temperatures the catalyst bed also acts as an interfering chromatographic column; it therefore had to be run at temperatures above 400°C in order to permit good resolution of the product peaks. With NaY less than 10 % of the cumene was converted to benzene and propylene at 450°C while with CaY and Ce^{III}Y the conversion was greater than 99 % for both sets of catalyst samples. Because of the high conversion, it is not possible to say whether the CaY and Ce^{III}Y activities changed at all on OH removal; but they did not fall to the NaY level, which we take here as the standard for a zeolite catalyst lacking the special activity characteristic of the addition of multivalent cations. It is there-fore clear that the enormous rise in activity on replacing univalent cations with bivalent cations is due directly to the cations themselves and is not mediated by the $0 \cdot 1$-$1 \cdot 0$ wt. % of structural OH groups that are present after normal activation: this experiment has lowered the range for equivocation on this point to less than $0 \cdot 01$ wt. % of OH.

Tervalent-cation zeolites, e.g. $Ce^{III}Y$, probably require special attention, however. While we think that most bivalent cations will be stable in the anhydrous X and Y zeolites in either S_I or S_{II}, we expect that the effective electrostatic field of tervalent ions may always become strong enough on dehydration to cause hydrolysis, $M^{III}Y + H_2O \rightarrow (OH^- M^{III}) H^I Y$. The resulting OH groups on the tervalent cations might be more significant proton donors than the OH groups normally present in a bivalent-cation Y. The latter OH groups, according to the infra-red studies,[9] are *not* linked to the surface cations. In the $Ce^{III}Y$, on the other hand, there is indeed strong evidence for OH groups linked to the cation (band at $3522\ cm^{-1}$). It is also conceivable that tervalent ions when fully dehydrated may shift to new positions; and since in Y zeolite the number of S_I positions is larger than the number of tervalent ions, possibly only a small fraction will remain at surface sites. Consequently, the catalytic activity may diminish in certain cases to an extent depending on the activation conditions.

REDUCTION OF ZEOLITES BY HYDROGEN

Our discussion of the structure of the X and Y zeolites suggests that the surface cations should have uniquely high electron affinities and related tendencies toward reduction. Indeed, zeolites have been treated with various reducing agents,[13, 14] and a lack of thermal stability of certain cations has been demonstrated by Yates,[15] who showed that Cd and Hg metal can be vaporized from the corresponding $Cd^{II}X$ and $Hg^{II}X$ zeolites by heating at 140-$500°C$ in hydrogen.

Several years ago we observed another interesting reaction between noble-metal-loaded bivalent-cation zeolites and hydrogen gas. Unlike the non-loaded forms, these zeolites consumed large quantities of hydrogen (fig. 6). This reaction first attains appreciable rates at temperatures that show a strong inverse dependence on cation electronegativity; e.g., $250°C$ for $Ni^{II}Y$ and $Co^{II}Y$, $400°$ for $Mg^{II}Y$, and $500°$ for $Ba^{II}Y$. The highest hydrogen uptakes reached values of about a molecule for every two multivalent S_{II} cations. Since the Y zeolites without noble metal in general did not consume any hydrogen at these temperatures ($Ni^{II}Y$ and one or two other transition metals were exceptional), we attributed this effect to activation of the hydrogen on the very finely dispersed noble metal, followed by reaction with the zeolite, probably indeed with the surface cations. Furthermore, since in this reaction the H/Pt atomic ratios climbed to values as high as 20-30, we assumed that either the noble metal forms a surface metal hydride [16] complex (PtH_2) and migrates in the zeolite cavities, functioning as a hydrogen transfer catalyst, or the protons formed migrate by jumping from oxide ion to oxide ion. We hoped to find that the multivalent zeolite cations in S_{II} were reduced to the univalent state, while OH groups were formed by addition of protons to framework oxide ions. However, we found ferromagnetic $Ni°$ and $Co°$ in the hydrogen reduced Pd- or Pt-loaded samples of $Ni^{II}Y$ and $Co^{II}Y$ and were unable to obtain convincing e.s.r. evidence of the presence of univalent cations (e.g., Mg^I, Ca^I) in the hydrogen-treated alkaline earth forms. In any case, since noble-metal-loaded alkali metal zeolites do not react at least up to $500°C$, the reaction with hydrogen does depend on the bivalent cations; they therefore surely play a role in the reaction, even though the ultimate site of reduction has not as yet been defined.

J. A. RABO, C. L. ANGELL, P. H. KASAI AND V. SCHOMAKER   337

REDUCTION OF Ni^{II} IONS IN Ni^{II}Y BY ALKALI-METAL VAPOUR TO
FORM Ni^{I}

After the ideas discussed above had led to the hope that novel materials could be prepared by reducing multivalent surface (S$_{II}$) cations in Y zeolites to unusual lower valence states, and after the discovery of large hydrogen uptake by noble-metal-loaded multivalent-cation Y zeolites had given promise that this hope might be realized, we began experimentation with various reducing agents and many different

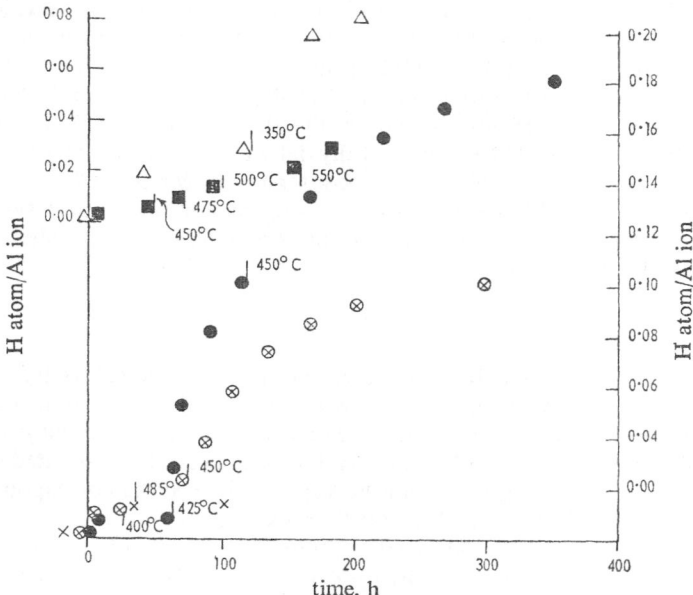

FIG. 6.—Reaction of noble-metal-loaded zeolites with hydrogen. The initial temperature on exposure to hydrogen was 300°C in all cases.

		final atom ratio H : Pt or Pd
●,	CaY×1·25 wt. % Pt	13·1
⊕,	MgY×0·5 wt. % Pd	9·6
△,	CoY×0·5 wt. % Pt	14·0
■,	BaY×0·1 Pd	15·0
×,	MgY	

zeolite preparations. Many attempts were made in collaboration with Dr. R. K. Müller, but none of these systems led to such good results as the ones to be described in the rest of this paper, although Ni^{I} was obtained and definitely characterized in some of the early experiments with Ni^{II}Y. The most remarkable results have been obtained with sodium vapour or, in some cases other alkali-metal vapours. The present section describes results on Ni^{II}Y, covering first the reduction technique and some of the key experiments, then the significant details of the resulting e.s.r. and optical spectra, which we ascribe to Ni^{I}, and finally some of the other properties of the reduced materials.

REDUCTION TECHNIQUE

In a typical reduction of Ni^{II}Y, sodium was placed at the bottom of a quartz tube, the zeolite sample, preactivated in vacuum at 575°C, was supported 2 in.

338   STUDIES OF ZEOLITE CATIONS

above it on a small wad of glass wool held by indentations in the tube, the top of the tube was connected to a pumping system and continuously evacuated, and a full-length heating mantle was brought to 585°C to vaporize the sodium and heat the zeolite bed. At this temperature, the vapour pressure of sodium is about 10 mm. The usual reaction time was 1 h, after which the heating mantle was removed and the tube allowed to cool.

REDUCTION OF Ni^{II} (72 %)Y

In the first experiment Ni^{II}Y was exposed to a large excess of sodium (Na/Ni > 4) for 1 h. In addition to a strong, broad background due to Ni°, the product showed a small e.s.r. signal at $g = 2.065$ which we attribute to Ni^{I}. In the next experiment Na/Ni was only 0.1, and the tube was heated at 585°C for 16 h. In this case only the signal due to ferromagnetic Ni° was observed, which suggested that on excessive heating the Ni^{I} ions disproportionate: $[2 Ni^{I}]Y \rightarrow Ni^{II}Y + Ni°$.

In order to optimize the amount of the reducing metal it was considered that there should be only one sodium atom for every S_{II} nickel ion, to avoid any excess of sodium that would further reduce the Ni^{I} to the metal. For the first experiments we assumed that only the S_{II} nickel ions would be reduced. In the Ni^{II}Y used in this study, 72 % of the original sodium ions were replaced by nickel ions, and we assumed that about 30 % of the nickel ions would take surface positions while the rest would occupy S_{I}. Allowing for some loss of sodium through reaction with the glass wool and the quartz tube, we chose $Na°/Ni^{II} = 0.4$ as standard in reducing the Ni (72 %)Y. The reduction was carried out at 575°C in 1 h. With this ratio the zeolite turned from pink to green, and although the e.s.r. spectrum was still mainly due to Ni°, the Ni^{I} signal (fig. 7) was much larger than from the previous samples.

Two Ni (72 %)Y samples were reduced at 575°C for 1 h in vacuum, one by sodium and one by caesium. Both samples turned green and gave an e.s.r. signal at $g = 2.065$, but the sample treated with sodium had the more intense colour. The amounts of alkali metal consumed were calculated from the amounts of hydrogen that were evolved on leaching the reduced samples in acid: $Na/Ni_{total} = 0.12$ and $Cs/Ni_{total} = 0.14$. On the basis of the e.s.r. signal, the amount of Ni^{I} in these samples was about 3 % of the total Ni. Three-quarters of the alkali-metal content therefore remained unreacted or was consumed in reducing Ni^{II} to Ni°. If all the Ni^{I} occupied S_{II}, the fraction of the Ni^{I} ions at the surface positions was about 10 %. The e.s.r. signal and colour of these samples remained unchanged even after several months at room temperature.

Samples of Ni (72 %)Y were also reduced at 250°C for 108 h, a long reaction period chosen to allow the alkali metals to vaporize even at this low temperature. With Na°, K° and Cs°, these samples turned respectively green, brown and yellow-gray, and e.s.r. indicated 1 % Ni^{I}, no Ni^{I} and 1.5 % Ni^{I}, while the amounts of Ni° were much smaller than in the samples treated at high temperature. After 4 days at room temperature, the $Ni^{I}(Cs°)$ signal (the signal obtained by using caesium as reducing metal) vanished and the Ni^{I} (Na°) signal decreased greatly. By acid leaching, Na°/Ni was 0.59 and Cs°/Ni was 0.54, much larger than was obtained at 585°C.

REDUCTION OF Ni^{II} (5 %)Y

In the experiments with Ni (72 %)Y we assumed that sodium reduced only the Ni^{II} ions at S_{II} and that the S_{I} ions were unaffected. To check on the reducibility of Ni^{II} ions at S_{I}, samples of a Ni (5 %)Y were also treated with alkali-metal vapour.

We recall that bivalent cations in general prefer S_{I} to S_{II}. With only 8/3 nickel ions per unit cell, Ni (5 %)Y therefore should have almost all its Ni at S_{I}. Since

J. A. RABO, C. L. ANGELL, P. H. KASAI AND V. SCHOMAKER   339

Ni^{II} should be more difficult to reduce at S_I than at S_{II}, a large excess of sodium was applied at 575°C for 1 h in vacuum. The sample turned from pale pink to pale green and developed a sharp e.s.r. signal (fig. 8) at $g = 2.094$, apparently shifted from the $g = 2.065$ of surface Ni^I because of the difference in crystal field between S_I and S_{II}. The relative sharpness of the signal indicates decreased interaction between nickel ions, as expected. By acid leaching, $Na°/Ni$ was 4·0.

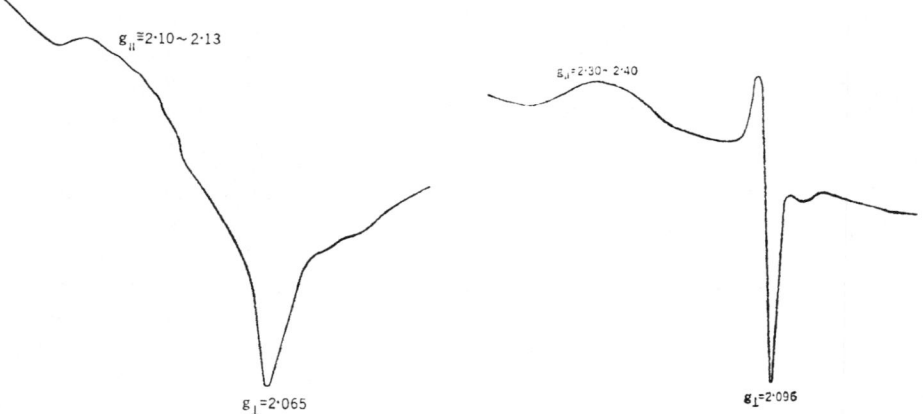

FIG. 7.—E.s.r. spectrum of surface Ni^I ions in Ni^{II} (72 %) Y reduced by sodium vapour.

FIG. 8.—E.s.r. spectrum of the fully co-ordinated (hidden) Ni^I ions at S_I in Ni^{II} (5 %) Y reduced by sodium vapour.

In order to confirm that the novel stability of Ni^I is peculiar to the zeolite substrate, a γ-alumina loaded with 5 wt. % of $NiCl_2$ was treated with sodium vapour at 575°C for 1 h in vacuum. The brown product gave a $Ni°$ e.s.r. signal but no Ni^I signal.

E.S.R. SPECTRA OF Ni^I IN REDUCED $Ni^{II}Y$

As mentioned earlier, the symmetry at S_I is slightly distorted octahedral, the site being surrounded by 6 oxide ions. Therefore, Ni^I at S_I and Cu^{II} (also $3d^9$) in a similar environment are expected to have similar e.s.r. spectra. The spectrum of Cu^{II} ion would be the more complex, however, because of hyperfine interaction with the magnetic nuclei Cu^{63} and Cu^{65}. In contrast, all but 1·2 % (Ni^{61}) of the Ni nuclei are non-magnetic. The e.s.r. spectrum of a sodium-reduced Ni (5 %)Y sample in which most of the Ni ions are believed to be at S_I is shown in fig. 8. This powder-pattern spectrum is characteristic of a system with an axially symmetric g-tensor:

$$g_{\parallel} = 2.30 \sim 2.40 \; ; \; g_{\perp} = 2.096 \pm 0.003.$$

These values are closely similar to those found for Cu^{II} in various tetragonally distorted octahedral ligand fields.* Their deviations from the free-electron value

---

* The comparison should best be made against the spectra of $Cu^{II}Y$ zeolites. Such spectra have been reported by Nicula, Stamires and Turkevich (*J. Chem. Physics*, 1965, **42**, 3684), who, however, made no attempt to interpret the e.s.r. spectrum of dehydrated $Cu^{II}Y$ in terms of specific cation positions. Our independent study with a thoroughly dehydrated $Cu^{II}Y$ sample is still in progress. Although a complete assignment of the spectrum has not yet been achieved, because of the complexity of the hyperfine structures, it is clear that the Cu^{II} ions are distributed between at least two different kinds of sites.

should then be given by

$$\Delta g_{\parallel} \cong 8\lambda/\Delta E_{\parallel}\ ; \quad \Delta g_{\perp} \cong 2\lambda/\Delta E_{\perp} \tag{1}$$

where λ is the spin-orbit coupling constant of the Cu^{II} ion, and $\Delta E_{\parallel}$ and $\Delta E_{\perp}$ are the respective energy separations of the d_{xy} and d_{zz} (or d_{yz}) orbitals from the lowest orbital, $d_{x^2-y^2}$.

Site S_{II} lies outside a puckered six-oxygen ring and has C_{3v} symmetry. In de-hydrated Ca^{II}-exchanged faujasite,[5] it lies $2.34\,\text{Å}$ from the three nearest oxide ions but only $0.57\,\text{Å}$ above the plane defined by them ($<O-Ca-O \sim 114°$). For Ni^{I} ions at this site there are then three d-orbital energy levels of which the d_{z^2} level (now calling the z-axis the C_{3v} axis and resorting to the limiting planar case for labels) and d_{xz}, d_{yz} level are clearly expected to be above the d_{xy}, $d_{x^2-y^2}$ level. Let us suppose that the degeneracy between d_{xy} and $d_{x^2-y^2}$ is removed by some further mechanism. The expressions given above are then applicable to the g tensor of Ni^{I} at S_{II} also (fig. 7). We believe that this spectrum is dominated by the Ni^{I} (S_{II}) signal super-posed on a broad background due to nickel metal. Analysis of the Ni^{I} signal gives

$$g_{\parallel} = 2.10 \sim 2.13 \quad \text{and} \quad g_{\perp} = 2.065 \pm 0.003.$$

It is easy to understand that $\Delta E_{\perp}$ should be larger for S_{II} than for S_{I}, and hence that $\Delta g_{\perp}$ should be the smaller at the surface sites. A subtle point is the mechanism by which the degeneracy between the $d_{x^2-y^2}$ and d_{z^2} at S_{I} and the degeneracy between $d_{x^2-y^2}$ and d_{xy} at S_{II} are removed. Jahn-Teller distortion might be responsible. However, these degeneracies are most prob-ably removed by the extra negative charge associated with each AlO_4 tetrahedron. The existence of a rather well-defined $g_{\perp}$ component for Ni^{I} ions at S_{II} might then be taken as evidence for a well-ordered arrangement of Al and Si atoms within the hexagonal rings. That there are two unique sites for Ni^{I} ions is further supported by the e.s.r. spectrum of another Ni (5 %)Y re-duced by Na vapour which shows Ni^{I} ions at both sites (fig. 9) and, as will be discussed later, by the differences of their temperature stabilities and of their reactions with ad-sorbed gases.

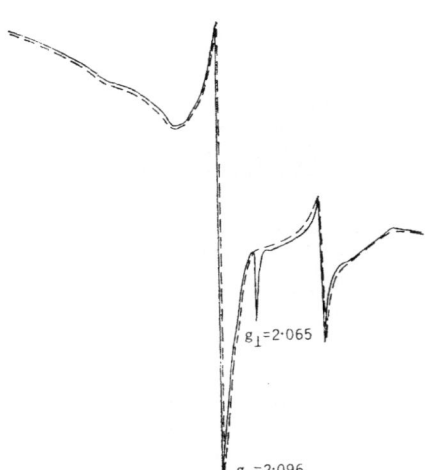

$g_{\perp}=2.065$

$g_{\perp}=2.096$

FIG. 9.—E.s.r. spectra of Ni^{I} ions at S_{I} and S_{II} in Ni^{II} (5 %) Y reduced by sodium vapour, ——; after exposure to carbon monoxide, — — —.

OPTICAL SPECTRA OF Ni^{I} IN REDUCED Ni^{II}Y

Anhydrous Ni^{II} (75 %)Y is pink, but on reduction turns bright green. The reflection spectra of Ni (5 %)Y and Ni (75 %)Y were examined in the visible region before and after reduction by Na vapour (fig. 10 and 11). The Ni(5 %)Y spectrum shows no prominent peak either before or after the reduction and therefore may be used as a standard for background. The broad absorption band centred at 450 mμ that is observed with Ni (75 %)Y before the reduction must then be attributed to Ni^{II} at S_{II}. Upon reduction, this band diminishes while a new band appears at around 650 mμ. The new band we attribute to an $^2E_g \rightarrow ^2T_g$ transition of Ni^{I} at S_{II}. As would be expected for iso-electronic d^9 ions, Cu^{II}Y and Ni^{I}Y show similar optical spectra (fig. 12).

J. A. RABO, C. L. ANGELL, P. H. KASAI AND V. SCHOMAKER   341

The spin-orbit coupling constant for a free Ni^I ion is 600 cm^{-1}. If one uses the reduced value $\lambda = 500$ cm^{-1} for Ni^I in the zeolite, taking into account the effect of the neighbouring oxide ions, the expression given above for $\Delta g_\perp$ and the value of $\Delta g_\perp$ observed for Ni^I at S$_{II}$ yield $\Delta E = 15,000$ cm^{-1}, in excellent agreement with the observed optical transition.

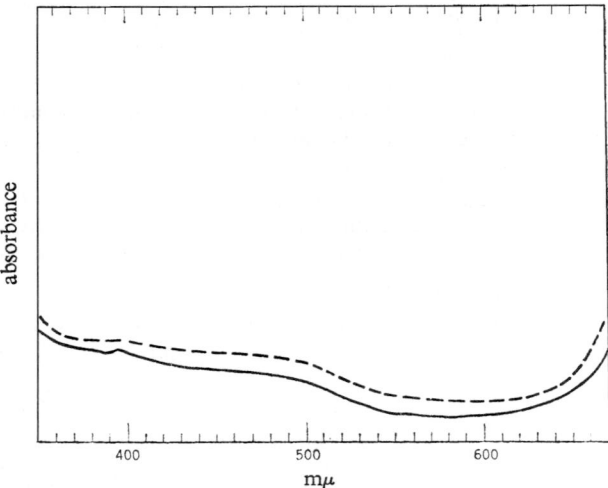

FIG. 10.—Optical spectra of activated Ni^{II} (5 %) Y, — — — —; and of Ni^{II} (5 %) Y reduced by sodium vapour, ———.

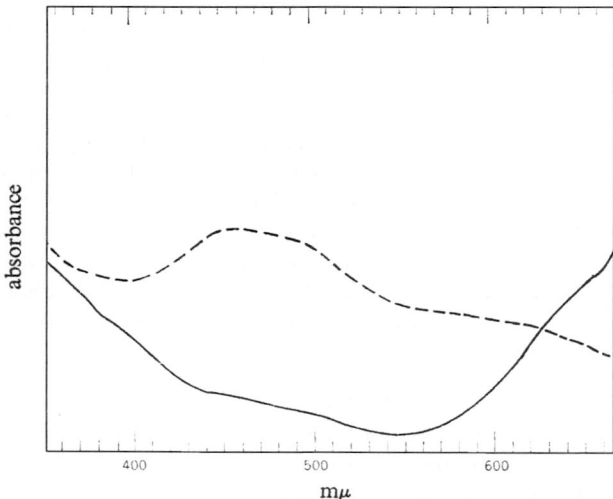

FIG. 11.—Optical spectra of activated Ni^{II} (75 %) Y, — — — —; and of Ni^{II} (75 %) Y reduced by sodium vapour, ———.

EFFECT OF HIGH TEMPERATURE

Since the Ni^I formed in Ni^{II}Y showed excellent stability at room temperature, we also studied it at high temperatures. It was observed occasionally that the e.s.r. signal of reduced Ni (5 %)Y increased on heat treatment. To investigate this phenomenon a Ni (5 %)Y sample with about 6 % of the Ni^{II} reduced to Ni^I was evacuated and sealed in a tube. After each heat treatment the intensity of the e.s.r. signal was examined :

temp. °C	time, h	Ni^I signal
100	18	slightly larger than original
200	24	twice the original
300	24	6·5 times the original
350	18	7 times the original *

* $Ni^I/Ni_{total} \sim 0.4$.

The last signal remained unchanged for months at room temperature. The sample was then heated to 500°C for 24 h, whereupon it showed a large Ni° peak and a considerably diminished Ni^I peak.

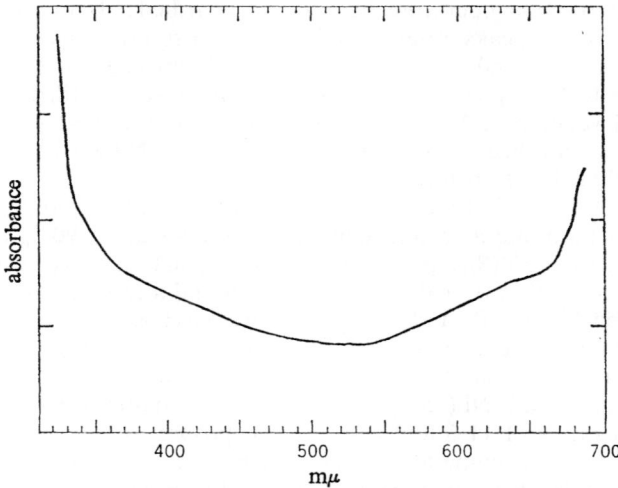

FIG. 12.—Optical spectrum of $Cu^{II}Y$ activated at 300°C. The remaining water content to be expected was therefore in the range 1-5 %.

A Ni (5 %)Y reduced by a large excess of sodium showed two e.s.r. signals, one large with $g_\perp = 2.094$ and one small with $g_\perp = 2.063$ (fig. 9), indicating the presence of Ni^I both at S_I and S_{II}. On heating at 100°C, the $Ni^I(S_{II})$ signal slowly vanished while the other slightly increased. A sample of Ni (5 %)Y reduced by caesium was heated in vacuum gradually to 560°C; up to 400°C the $Ni^I(S_I)$ increased gradually by about ten-fold, after which it remained unchanged at 460°C for 24 h. At 540°C the $Ni^I(S_I)$ decreased and the Ni° signal appeared.

Several times $Ni^I(S_{II})$ was found in small quantities when Ni (75 %)Y was flash activated at 585°C. However, acid leaching released no H_2, indicating that this Ni^I was produced by electron dislocation rather than by any change in overall chemical composition. The signal is thermally unstable, like the one produced by sodium vapour. The reduced Ni (72 %)Y showed a similar temperature dependence : the stability at room temperature was excellent, but at and above 100°C the $Ni^I(S_{II})$ signal and the strong green colour vanished and Ni° appeared.

EFFECTS OF ADSORBED GASES ON Ni^I

The profound difference in co-ordination between $Ni^I(S_{II})$ and $Ni^I(S_I)$ lends itself to study of their abilities to co-ordinate with various ligands. The expectation is that $Ni^{II}(S_I)$ will have virtually none. Samples of reduced Ni (5 %)Y, Ni (75 %)Y and Ni (25 %)Y were used with O_2, CO, H_2 and NH_3. Before and after admitting each gas the e.s.r. spectra were recorded.

The reduced Ni (5 %)Y used for CO adsorption showed a large e.s.r. peak at $g_\perp = 2.093$ ascribed to $Ni^{II}(S_I)$, a small peak at $g_\perp = 2.063$ ascribed to $Ni^{II}(S_{II})$, and a third peak not due to nickel. On admitting CO at room temperature the $Ni^{II}(S_{II})$ peak vanished while the $Ni^{II}(S_I)$ peak was unaffected. The reduced Ni (75 %)Y had a strong $Ni^{II}(S_{II})$ peak at $g_\perp = 2.063$. On admitting CO this peak vanished, while a small peak, previously masked by the much larger one, appeared at $g_\perp = 2.094$. Since this g value is identical with the one observed with reduced Ni (5 %)Y, we ascribe it to $Ni^{II}(S_I)$.

Oxygen was adsorbed on a Ni (5 %)Y reduced by caesium. Before admitting oxygen, the e.s.r. spectra indicated $Ni^{II}(S_I)$ as well as a trace of $Ni^{II}(S_{II})$. On admitting oxygen at room temperature both peaks vanished; however, on pumping for 30 min the original peaks re-appeared with almost unchanged intensity. On pumping further at 100 and 250°C, the $Ni^{II}(S_I)$ diminished to 1/40 in intensity while a peak due to O_2^- appeared. At the same time, the $Ni^{II}(S_{II})$ vanished. On further heating in vacuum at 425°C, the $Ni^{II}(S_I)$ grew to about 80 % of the original intensity, while O_2^- vanished. On heating above 500°C, $Ni^{II}(S_I)$ began to diminish and the signal assigned to ferromagnetic Ni° appeared.

With reduced Ni (25 %)Y (Na°/Ni = 0.68) on exposure to O_2 at 25°C, all signals vanished; but on pumping at room temperature for an hour, 90 % of the $Ni^{II}(S_I)$ was recovered while the $Ni^{II}(S_{II})$ grew to twice the original intensity; and on heating at 100°C in vacuum, both Ni^{II} peaks vanished again and a large O_2^- signal appeared. Samples of reduced Ni (5 %)Y and Ni (75 %)Y were exposed to NH_3 at room temperature; according to the e.s.r. spectra the $Ni^{II}(S_{II})$ was destroyed and metal was formed. The $Ni^{II}(S_I)$ was not affected at all.

Similar Ni(5 %)Y and Ni (75 %)Y samples were reduced with sodium with respective Na°/Ni ratios of 11.6 and 0.96. On exposure to H_2, the $Ni^{II}(S_{II})$ slowly diminished even at room temperature and vanished at 100°C with concurrent Ni° formation, while the $Ni^{II}(S_I)$ was unchanged even after 5 days at room temperature.

REDUCTION OF NaY AND NaX WITH SODIUM VAPOUR: THE COMPLEXES Na_4^{3+} AND Na_6^{5+}

In the course of studying the reduction of various cation-exchanged zeolites by sodium vapour, it was discovered that NaY upon exposure to sodium vapour at 580°C turns bright red and shows a unique e.s.r. spectrum consisting of 13 peaks with intensities gradually increasing toward the centre (fig. 13).

Development of pink colour and a similar e.s.r. spectrum were observed earlier [17] when NaY was exposed in vacuum to γ-rays or X-rays. Both the colour and the e.s.r. spectrum were attributed to electrons trapped in the large zeolite cavities, each electron being shared among four sodium ions tetrahedrally arranged at the four surrounding S_{II}. We believe that the new, red, non-stoichiometric NaY prepared with sodium vapour contains similar Na_4^{3+} paramagnetic centres.

The centres produced by radiation are stable at room temperature, but are quickly destroyed at about 200°C. The chemically prepared centres, on the other hand, show remarkable stability: the red material evolves no sodium on heating to 500°C under high vacuum, and afterwards the colour and e.s.r. spectrum are not significantly changed. Introducing oxygen to the chemically prepared material at room temperature results in instant disappearance of both the colour and the e.s.r. signal ascribed to Na_4^{3+} and in the appearance of a strong new e.s.r. signal attributable to the O_2^- radical. The bonding with oxygen is reversible: heating at 500°C in vacuum removes the O_2^- and restores the Na_4^{3+}. The g value ($1\cdot999\pm0\cdot001$) and the hyperfine interaction with the sodium nuclei ($A = 32\cdot3\pm0\cdot2$ gauss) are identical for the chemically prepared Na_4^{3+} centre and the irradiated material.

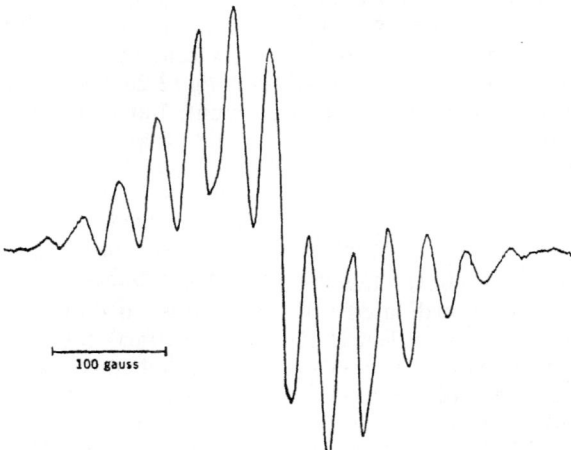

100 gauss

FIG. 13.—E.s.r. spectrum of the Na_4^{3+} centre in Y zeolite.

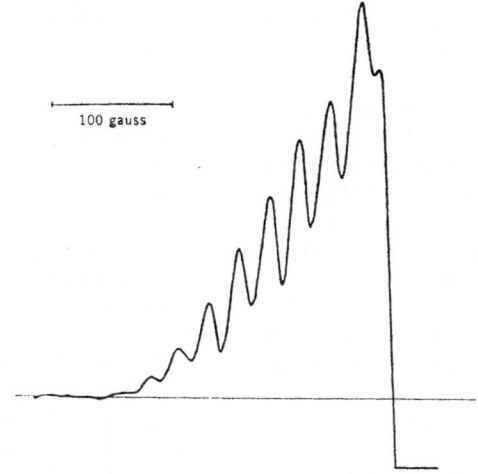

100 gauss

FIG. 14.—E.s.r. spectrum of the Na_6^{5+} centres in X zeolite (first half of spectrum).

This success prompted us to treat NaX similarly. The NaX used has a Si/Al ratio of about 1·19 and a correspondingly larger content of Na^I than the NaY. On exposure to sodium vapour, this NaX turns dark blue and shows a simple e.s.r. spectrum consisting of 19 lines (fig. 14). This spectrum corresponds to a centre similar to that obtained with NaY but involving six equivalent sodium ions instead of four. The g value is the same for the blue centre as for the red, but the hyperfine interaction per Na nucleus is reduced to about 2/3. Each large cavity in the structure is surrounded by six S_{III} arranged octahedrally in the directions of the fourfold cubic axes as well as by the four S_{II} arranged tetrahedrally in the directions of the three-fold cubic axes. The blue centre, we believe, is an electron shared among six sodium ions all situated at S_{III}: the S_{II} of the cavity are then necessarily vacant. This assignment suggests that in NaX, in contrast to NaY, S_{III} is occupied in preference to S_{II} in order to accommodate the increased number of sodium ions.

Magnetic susceptibility measurements on NaY reduced by sodium vapour showed paramagnetism corresponding to 50-80 % occupation of the cavities by Na_4^{3+} centres. On the other hand, estimates of spin concentration from the e.s.r. spectra of similar preparations gave only 10^{19} to 10^{20} spins/g (2-20 % occupation), depending upon the sample. This is hardly good agreement, but is reasonable, since e.s.r. detects only well-defined centres whereas all of the paramagnetic centres contribute to the susceptibility.

DISCUSSION

As has been reviewed above, the " surface " cations in anhydrous X and Y zeolites are incompletely co-ordinated with oxide ions, so that for molecules in the zeolite cavities the (enormous) electric field of a surface cation is not effectively shielded out by the fields of all the other ions. Essential consequences of this special accessibility of the cations are the carboniogenic activities and the specific complexing of, e.g., carbon monoxide. The shielding is still expected to be appreciable, however, and different for X and Y, and the differences between X and Y in catalysis and gas adsorption seem to correspond.

The low co-ordination of a surface cation by oxide ions should also lead to higher (less negative) electrostatic potential at the surface cation site than at the site of a fully co-ordinated cation (which we also call hidden cation) and so should cause a surface cation to have greater effective electron affinity than a fully co-ordinated cation, at least to the extent that the structures in question can be considered to be fully ionic. This concept follows immediately from the Born-Haber considerations for building up an ionic crystal from its component atoms. We used it, loosely, to explain our discovery that multivalent surface (S_{II}, S_{III}) cations could be reduced to subvalent states not ordinarily found in molecules or crystals, that this reduction is easier in Y zeolite than in X, and that it should depend on the sequence of electron affinities of the cation. Such thoughts led us to $Ni^I(S_{II})$ but left us surprised to get $Ni^I(S_I)$ and Na_4^{3+}: they are useful but incomplete. Without doubt, polarization is much more important for a cation in a surface site than for a fully co-ordinated, highly symmetrical site. The bonds are probably covalent to various degrees. Evaluation of the Madelung sums, especially with respect to the full process of introducing both electron and sodium ion, is necessary. Our knowledge of the structures is insufficient: for a small cation at S_{II} the primary co-ordination probably becomes planar rather than pyramidal, and major unexpected rearrangements may possibly occur.

We now comment on the reduction process $Ni^{II}Y + Na^\circ \rightleftharpoons Ni^I Na^I Y$ in which a bivalent cation is replaced by two univalent cations, of which the Ni^I remains at the original Ni^{II} site. In the Madelung energy, the sum of terms involving a Ni^I, Na^I pair and distant parts of the crystal will usually equal the corresponding sum for the original Ni^{II}, while a repulsion term between each Ni^I and its Na^I neighbour is added that has no explicit counterpart. The repulsion terms between neighbouring Ni^I, Na^I pairs, moreover, will be greater than the counterpart terms for the original Ni^{II} ions. In addition, each new Na^I may or may not go into a site equivalent to the Ni^{II} site with respect to interaction with the nearby part of the framework. Equivalence will prevail for reduction of S_{II} cations if the new Na^I also occupy S_{II}. The net change in crystal energy will then strongly oppose the reduction, but the process finds its support in the enormous difference between the ionization potentials of Ni^I and Na°: $Ni^{II}_{gas} + Na^\circ_{gas} \rightleftharpoons Ni^I_{gas} + Na^I_{gas} + 13 \cdot 1$ eV. In the reduction of Ni^{II} at S_I the new sodium ions again probably occupy S_{II}, which is less favourable for the crystal energy than S_I; the net change in crystal energy is therefore also less favourable than for the reduction of Ni^{II} in S_{II}.

Reduction of surface ions may or may not require activation energy, but reduction at S_I requires that electrons penetrate the surrounding negatively charged oxide ions, which will require significant activation energy.

The chemical stability of univalent nickel will depend also on the rate of disproportionation, $[2 Ni^I]Y \rightarrow Ni^{II}Y + Ni^\circ$. Since nickel atoms probably migrate freely on the zeolite surface even at moderate temperatures, they will agglomerate and form the metal, which will make the overall reaction substantially irreversible. In any case, disproportionation requires electron jumps between Ni^I ions either direct or indirectly, and increasing the distance between Ni^I ions, i.e., lowering the concentration, will reduce the rate. At S_I, Ni^I is much less mobile than at S_{II}, and the electron jumps presumably also require higher activation energy. Furthermore, metal formation will be prevented or strongly hindered because the mobility of Ni° atoms will also be less at S_I than at S_{II}: to move from S_I to S_{II}, they have to penetrate two six-oxygen rings, each too small to give them free passage. Hidden Ni^I is therefore expected to be much more stable than surface Ni^I toward disproportionation.

DISCUSSION OF THE EXPERIMENTAL RESULTS

SURFACE Ni^I IONS

The step $Ni^{II}_{surface} + Na^\circ \rightleftharpoons Ni^I_{surface} Na^I$ is probably followed in the presence of excess alkali metal by the step $Ni^I_{surface} + Na^\circ \rightarrow Na^I + Ni^\circ$. However, the first reaction probably can be carried out stoichiometrically. This reflects qualitatively the great difference in energy between the following two gas reactions:

$$Ni^{II}_{gas} + Na^\circ_{gas} \rightarrow Ni^I_{gas} + Na^I_{gas} + 13 \cdot 1 \text{ eV}$$

$$Ni^I_{gas} + Na^\circ_{gas} \rightarrow Ni^\circ_{gas} + Na^I_{gas} + 2 \cdot 5 \text{ eV}.$$

Not only large surplus of alkali metal leads to Ni° formation, but also long exposure at high temperatures (100°C and up) through the disproportionation $[2Ni^I]Y \rightarrow Ni^{II}Y + Ni^\circ$. Since our study of the surface ions was done with a rather highly exchanged (72 %) $Ni^{II}Y$, the rate of electron exchange between Ni^I and Ni^{II} is fairly fast and leads to disproportionation of the surface Ni^I. Much less formation of Ni° by disproportionation would occur in a NiY containing less nickel. Nevertheless, the hidden sites will be preferentially filled by Ni^{II} ions. The low stability of the surface Ni^I prepared at low temperature (250°C) is due to excess sodium stored in the zeolite cavity system on reduction. This sodium then reacts with the surface Ni^I even at room temperature.

The experiment with reduced Ni (25 %)Y, which first showed an increase in surface Ni^I at room temperature on exposure to oxygen, may support the suggestion that Ni^I is more stable relative to Ni^{II} in S_{II} than it is in S_I: the increase may be the result of the reaction $Ni^I_{hidden} + Ni^{II}_{surface} \rightarrow Ni^{II}_{hidden} + Ni^I_{surface}$. This reaction is expected only between adjacent ions and is therefore unlikely in the Ni (5 %)Y. Its non-appearance with Ni (75 %)Y, on the other hand, may result from disproportionation of the highly concentrated surface ions.

Surface Ni^I has a remarkable chemical activity. It co-ordinates with H_2, NH_3, CO, olefins, O_2, and probably with many other ligands. With NH_3 the metal is formed even at 25°C. This is not surprising because NH_3 is a very strong ligand and likely to weaken the binding of the Ni^I cation to the zeolitic surface and thereby lower the activation energy of migration. In addition, ionization of Ni^I by electron jump will be further strongly assisted by the strong-electron-donor ligand.

The reaction of Ni^IY with H_2 to form metal is of special interest. It occurs already at 25°C, while the H_2 reduction of $Ni^{II}Y$ itself occurs only at and beyond 200°C: the hydrogen molecule is more readily activated on Ni^I than on Ni^{II}.

CARBON MONOXIDE co-ordinates much more strongly with surface Ni^I than with Ni^{II}: adsorbed CO can be pumped off $Ni^{II}Y$ at 25°C but from Ni^I requires at least 200°C. Since this desorption requires temperatures at which surface Ni^I normally disproportionates it appears that Ni^I is stabilized by the carbonyl ligand. This strong bonding with CO was expected since CO is a very weak electron donor that favours co-ordination with atoms in the zero-valent state or in negatively charged complexes.

The C—O stretching frequency on Ni^{II} in $Ni^{II}Y$ is 2217 cm^{-1}, while on Ni^IY, presumably on the Ni^I ions, it is 2188 cm^{-1}. This suggests that the C—O bond is weaker in Ni^ICO than in $Ni^{II}CO$. This is reasonable because Ni^I is expected to be a stronger electron donor than Ni^{II}, any electrons donated to the carbonyl group going into antibonding orbitals.

OXYGEN is very weakly adsorbed on or near surface Ni^I at 25°C—the reversible loss of signal is presumably only a magnetic effect. At the same time, there is an unexplained stronger adsorption of oxygen somewhere else in the structure. At 100°C, O_2^- ions are formed:

$$Ni^I(S_{II}) + O_2 \rightarrow [O_2^- Ni^{II}(S_{II})].$$

To remove the O_2^- requires pumping at 200-300°C, and at this temperature the bare Ni^I ions disproportionate to Ni^{II} and $Ni°$. The $[O_2^- Ni^{II}$—$(S_{II})]$ complex is evidently more stable than $Ni^I(S_{II})$ itself.

HIDDEN Ni^I IONS

The sharp increase in the hidden-Ni^I concentration on heat treatment was unexpected. We now understand that two processes concur during the reduction of Ni^{II} (5 %)Y: reduction of hidden Ni^{II} ions, and reaction between surface sodium ions and sodium atoms to form the Na_4^{3+} complexes. This is reasonable because with as low as 5 % cation exchange most of the zeolite cavities are still filled with sodium ions capable of forming the Na_4^{3+} complex. This assumption is confirmed by recent experiments in which Ni (5 %)Y on reduction first turned pale red and showed the characteristic spectra of the Na_4^{3+} centres. The paramagnetic electron of the Na_4^{3+} centre so formed evidently then transfers at high temperatures to hidden Ni^{II} by the reaction $Na_4^{3+} + Ni^{II}(S_I) \rightarrow 4Na^+ + Ni^I(S_I)$, which results in a 5- to 10-fold increase in the concentration of hidden Ni^I.

The excellent yields of hidden Ni^I ions (~ 50 %) are due to the structural protection against disproportionation, which will also explain the enormous difference in heat stability between the hidden and the surface ions (~ 460°C against 100°C).

Since hidden Ni^I at S_I is octahedrally co-ordinated with six oxide ions and rather tightly surrounded by other ions besides, it cannot co-ordinate in classical fashion with any other ligand. It does react with oxygen at and above 100°C, however, giving up an electron to form superoxide ion: the O_2^- signal appears and the Ni^I signal vanishes. On heating at 460°C in vacuum the reverse occurs, the Ni^I signal reappearing and the O_2^- signal vanishing. Hidden Ni^I should not react in this way with NH_3, CO, or H_2, at least to the extent that these molecules are not known to form chemically significant negative ions. It is also reasonable that this process requires high temperatures: an electron has to penetrate a structure filled with oxide ions. The reversibility of the reaction is evidently due to a fairly even balance between the effective ionization potentials of $Ni^I(S_I)$ and $O_2^-(S_{II})$ and the absence of any competing processes. This is an interesting reaction, possibly the first of its kind to be observed on a macroscopic scale.

CONCLUSION

The experimental results and arguments put forward in this paper indicate that the carboniogenic catalytic activity, cation-specific carbon monoxide adsorption, reduction of Ni^{II} ions, and formation of the Na_4^{3+} centres are all related to the large electrostatic field in the cavities or to the corresponding large electron affinity of the surface cations. Changing the field by changing the size of the bivalent cations results in the expected change in catalytic activity. Similarly, the combination constant for carbon monoxide follows the changes in field for the alkaline earth and zinc cations in both X and Y zeolites. However, the simple electrostatic argument does not explain the enormous difference in the combination constant between alkaline-earth and transition-metal cations. Furthermore, there is a suggestion that the simple electrostatic concept upon which we attempted to generalize the structural distribution of bivalent cations, which has so far been established by X-ray diffraction only for calcium ions, may break down with transition-metal cations: in any case, fewer Co^{II} and Ni^{II} surface ions are implied by the CO-adsorption results than were expected. The large differences in combining constant between Ni^{II} and Co^{II} and the closed shell ions are probably specific effects of their open-shell structure. With the help of the significantly higher affinity of bivalent surface cations as compared to oxide ions, carbon monoxide adsorption emerges here as a quantitative method of measuring the fraction of cations at surface positions and to some degree determining their co-ordination.

The reduction of Ni^{II} ions and the stability of the corresponding Ni^I ions can be explained by arguments based on structure and electrostatic theory. But the chemical activity of Ni^I, particularly the highly increased bond energy with carbon monoxide, shows that the properties of the open-shell electronic structure are of great importance also.

The Na_4^{3+} centres bear similarities to the colour centres of solid-state physics. Their remarkable stability probably results, in first approximation, from the larger magnitude of the concerted electron affinities of sets of surface cations in a cavity, as compared to the electron affinity of a single cation. But full discussion of the combination of cation orbitals occupied by the delocalized electron and of every significant contribution to the ionic energy sum seems to be required for a better understanding of why and with just what effect these Molecular Sieves take up sodium vapour. The formation of simple, symmetrical Na_6^{5+} centres in X zeolite was surprising but has been rationalized satisfactorily except for a difficulty of structural conception that remains for both the X and Y materials, namely, the lack of a plausible assignment of sites for all the sodium ions.

The authors express their appreciation to Dr. E. Dempsey for his permission to publish the data in table 1 and to Mr. Gary Skeels, Mrs. Maria Howell, and Mr. Paul Schaffer for their skilful and creative assistance in carrying out the experiments.

J. A. RABO, C. L. ANGELL, P. H. KASAI AND V. SCHOMAKER   349

[1] Pickert, Rabo, Dempsey and Schomaker, *Proc. 3rd Int. Congr. Catalysis*, 1965, p. 728.
[2] Breck, *Abstr.*, 134th Meeting of the Amer. Chem. Soc., Chicago, Sept., 1958.
[3] Rabo, Pickert, Stamires and Boyle, *Actes Duexieme Congr. Int. Catalyse*, 1960, p. 2055.
[4] Broussard and Schoemaker, *J. Amer. Chem. Soc.*, 1960, **82**, 1091.
[5] Dodge, R. P., unpublished research at Union Carbide Res. Inst.
[5a] Breck and Flanigen, unpublished research, Union Carbide Corp., Linde Div., Tonawanda Lab.
  [b] Breck, U.S. Patent 3,130,007, (1964).
[6] Angell and Schaffer, *J. Physic. Chem.*, in press.
[7] Neddenriep, R. J., unpublished research, Union Carbide Corp., Linde Div., Tonawanda Lab.
[8] Kruerke, U. and Rabo, J. A., unpublished research at Union Carbide Res. Inst.
[9] Angell and Schaffer, *J. Physic. Chem.*, 1965, **69**, 3463.
[10] Bertsch and Habgood, *J. Physic. Chem.*, 1963, **67**, 1621.
[11] Carter, Lucchesi and Yates, *J. Physic. Chem.*, 1964, **68**, 1385.
[12] Emmett, *Adv. Catalysis* (Academic Press), 1957, **9**, 645.
[13] Castor, U.S. Patent no. 3,013,986.
[14] Breck, U.S. Patent no. 3,013,992.
[15] Yates, *J. Physic. Chem.*, 1965, **69**, 1676.
[16] Rabo, Pickert and Schomaker, *Proc. 3rd Int. Congr. Catalysis*, 1965, p. 1264.
[17] Kasai, *J. Chem. Physics*, 1965, **43**, 3322.